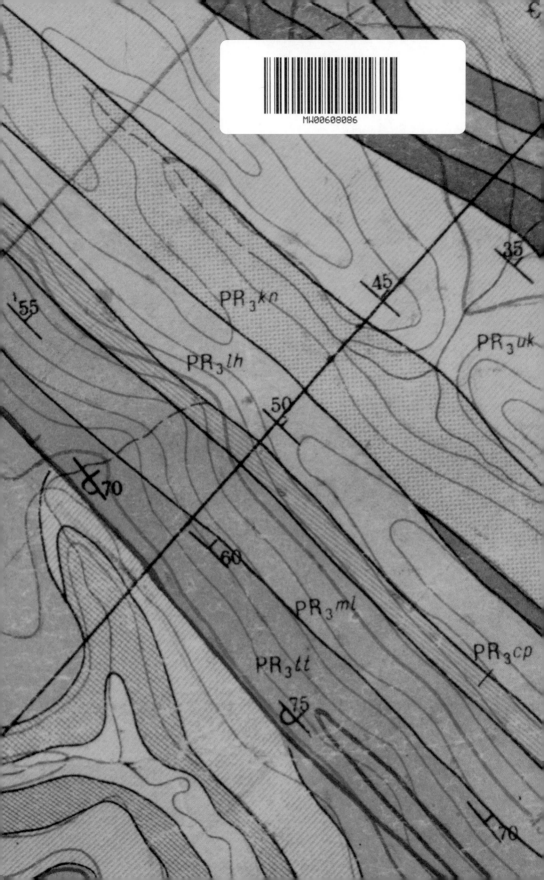

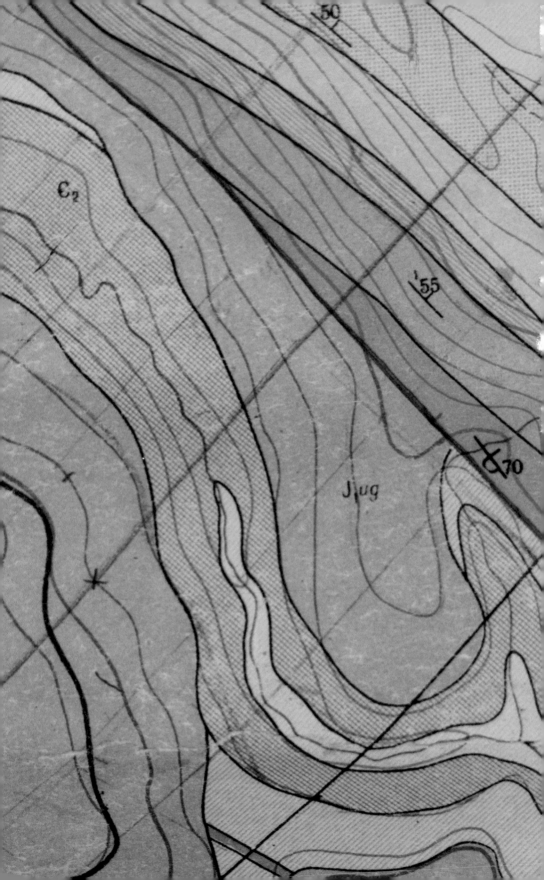

RADICAL BY NATURE

RADICAL BY NATURE

The Revolutionary Life of
Alfred Russel Wallace

JAMES T. COSTA

PRINCETON UNIVERSITY PRESS

PRINCETON & OXFORD

Published by Princeton University Press
41 William Street, Princeton, New Jersey 08540
99 Banbury Road, Oxford OX2 6JX

press.princeton.edu

All Rights Reserved

Library of Congress Cataloging-in-Publication Data

Names: Costa, James T., 1963– author.
Title: Radical by nature : the revolutionary life of Alfred Russel Wallace / James T. Costa.
Description: Princeton, New Jersey : Princeton University Press, [2023] | Includes bibliographical references and index.
Identifiers: LCCN 2022040771 (print) | LCCN 2022040772 (ebook) | ISBN 9780691233796 (hardback) | ISBN 9780691233789 (ebook)
Subjects: LCSH: Wallace, Alfred Russel, 1823–1913. | BISAC: BIOGRAPHY & AUTOBIOGRAPHY / Science & Technology | NATURE / General
Classification: LCC QH31.W2 C647 2023 (print) | LCC QH31.W2 (ebook) | DDC 508.092 [B]—dc23/eng/20220831
LC record available at https://lccn.loc.gov/2022040771
LC ebook record available at https://lccn.loc.gov/2022040772

British Library Cataloging-in-Publication Data is available

Editorial: Eric Crahan, Barbara Shi
Jacket Design: Heather Hansen
Production: Danielle Amatucci
Publicity: Alyssa Sanford, Kate Farquhar-Thomson
Copy Editor: Wendy Lawrence

Jacket image: Wallace's longhorn beetle, *Batocera wallacei*. Illustration from *Arcana naturae, ou Recueil d'histoire naturelle* by James Livingston Thomson, 1859. Paris: au Bureau du Trésorier de la Société entomologique de France, vol. 1, plate VI. Illustrated by Hercule Nicolet. Biodiversity Heritage Library. Courtesy of the Ernst Mayr Library of the Museum of Comparative Zoology, Harvard.

This book has been composed in Arno Pro and Alegreya.

Printed on acid-free paper. ∞

Printed in the United States of America

10 9 8 7 6 5 4 3 2 1

For Leslie, my partner in all things

CONTENTS

PREFACE

Multitudinous

Do I contradict myself?
  Very well then I contradict myself,
  (I am large, I contain multitudes)

—WALT WHITMAN, SONG OF MYSELF (1855)

IN *SONG OF MYSELF*, nineteenth-century American poet Walt Whitman declared that he "contained multitudes," singing of a largeness of spirit, perspectives, beliefs, and interests so expansive that there was room within for contradictions, which he acknowledged with equanimity. This is an apt description, too, of the accomplished naturalist and humanitarian Alfred Russel Wallace—which is surely why, in 1904, writer and critic G. K. Chesterton was hard-pressed to decide between Wallace or Whitman as "the most important and significant figure of the nineteenth century." This book, an homage to Wallace to mark the bicentennial of his birth, aims to inspire an appreciation for those "multitudes": for Wallace the preeminent field naturalist, evolutionist, traveler, biogeographer, explorer, and best-selling author as well as Wallace the sometime surveyor, builder, essayist, reformer, and social critic. For Wallace the spiritualist and devotee of séances as well as Wallace the husband, father, and friend. For Wallace the feted and famous as well as Wallace the ostracized radical, pushing back against the establishment, scientific and social.

If we had to choose one word to sum up Wallace, "radical" might be the most appropriate. Not a radical of the bomb-throwing persuasion, certainly—he was not one to tear down received truths or institutions gratuitously. No, this radical was more of the envelope-pushing persuasion, an explorer, philosopher, observer, and activist holding up a mirror to society—a humanitarian naturalist with a penchant for out-of-the-box thinking who sought truths about the natural world *and* the human condition. The two were of a piece for Wallace, after all, the boundary between the human and nonhuman worlds permeable depending on the angle the question was viewed from. That was very Wallace: his was a life marked by borders, boundaries, and lines of delineation literal and figurative—lines he drew and lines he erased, lines he respected and lines he transgressed, lines he discovered, and lines he thought he discovered. Consider this distinctly Wallacean tangle of lines . . .

Wallace was born something of a stranger in a strange land, the "little Saxon" in a Celtic Welsh borderland—disputed territory, even—but then, as a traveler living and working (and more than once nearly dying) among the locals in the distant reaches of distant lands, he came to appreciate the common humanity of all peoples. He came of age on one side of a social boundary, a working man alternately apprentice surveyor and carpenter, teacher, and builder, yet the brilliant autodidact crossed that line in his rise to the highest levels of scientific achievement and social standing, with international acclaim, medals and awards bestowed by the most learned of the learned societies and even the crown, and honorary degrees conferred by august institutions. He drew lines for a living as a sometime surveyor but later, sensible that these were meant to dispossess, disavowed them as a land nationalist and socialist, eloquently calling for their abolition. He was a committed materialist who came to see the physical world as incomplete, sensing a divide separating the material from a kind of spiritualistic promised land beyond. He came to see that just as political and cultural boundaries shift in time and space with the rise and fall of kingdoms and empires, so, too, does this dynamic play out in the natural world, a world where he detected remarkable lines, the ghosts of geographies past. He discovered, famously, an astonishing line of demarcation between two great faunal realms that speak of the history of Earth and life but also intuited that they shape-shift in deep time, their boundaries ebbing, flowing, dissolving, and forming anew as the planet cycles and species change. And in the context of that evolutionary vision was his even more famous discovery of the *mechanism* of species change—a discovery that saw him conquer the fiercely defended species barrier, only to erect another cordoning off the human mind. Yes, Wallace was multitudinous, all right: capacious enough to contain contradictions and radical enough that every one of them was startlingly original.[1]

———

But this is all well known about Alfred Russel Wallace, yes? So why this book? Why now? Fair enough. It is a reasonable question, but I would say in reply that much of this is *not* well known—not well *enough* known, to be sure. Yes, in recent decades a dozen or more books on Wallace have appeared—most notably, such fine works as Peter Raby's *Alfred Russel Wallace: A Life* (2001), Michael Shermer's *In Darwin's Shadow: The Life and Science of Alfred Russel Wallace* (2002), Martin Fichman's *An Elusive Victorian: The Evolution of Alfred Russel Wallace* (2004), Ted Benton's *Alfred Russel Wallace: Explorer, Evolutionist, Public Intellectual* (2013), and certainly Ross Slotten's indispensable *The Heretic in Darwin's Court: The Life of Alfred Russel Wallace* (2004). Not to mention helpful and informative anthologies by Jane Camerini (*The Alfred Russel Wallace Reader: A Selection of Writings from the Field* [2002]) and especially Andrew Berry (*Infinite Tropics: An Alfred Russel Wallace Anthology* [2002]) and enriching edited volumes that offer deeper dives into diverse facets of Wallace's broad interests: Charles Smith and George Beccaloni's *Natural Selection and Beyond: The Intellectual Legacy of Alfred Russel Wallace* (2008) and *An Alfred Russel Wallace Companion* (2019), which I had the privilege of co-editing with Charles Smith and David Collard.

Yes, these are worthy works that teach us a great deal, but they are not the story of Wallace and his life and times I really wanted to tell, or the style I wanted to tell it in. In light of the treasure trove of new insights into Wallace's life and thought unearthed by new scholarship since the 2013 centenary of his death—newly available notebooks and manuscripts, the Wallace Correspondence Project, newly discovered Wallace writings, and more—in honor of Wallace's two-hundredth birthday, I wanted to tell an *updated* story of his life, as he lived it, in a narrative that traces the arc of the remarkable adventures, poignant personal life, and breathtaking sweep of thought of this singular human being. And I wanted to cast this dynamic life against the backdrop of the dynamic planet and evocative landscapes Wallace exulted in, as well as within his cultural context. The book you are holding is thus neither detailed contextual analysis nor critical biography per se; such works by science historians have their place, but that is not this book. Rather, as a professional biologist thoroughly conversant in the science and Wallace and Darwin scholarship for nearly three decades I aim for high standards of scholarly rigor both scientific and historical while aiming to tell a good story, to do right by this remarkable individual's life, and to inspire—Alfred Russel Wallace's life story surely is nothing if not inspiring! My style is conversational, intimate, as told over a pint or two (or three). Or is that a whiskey? Or both? ... It's a long story, after all—an epic tale of an epic and fascinating life well lived. Better just leave the bottle.

———

There are other motivations for this book: the question of Wallace's place in the sun and the lessons he offers us. Those dozen or more works on Wallace of the past couple of decades? A respectable enough number—any of us would be lucky to be so remembered a century after our death—but probably three orders of magnitude fewer than the works on Darwin. This is not to throw darts at Darwin. As even Wallace recognized, the Sage of Down had it all figured out long before, and his laurels are well deserved. But as I have argued elsewhere, Wallace and Darwin were *together* our first guides to evolution. Wallace's discovery was fully independent of Darwin's and his journey getting there perhaps all the more remarkable given his disadvantages: the self-taught Wallace overcame astonishing odds to become one of the most—if not *the* most—respected scientific voices of his time. But unlike Darwin, who was laser focused on his science, to good effect, Wallace was multitudinous, a diffraction grating for ideas. Far from laser focused, he pursued myriad scientific interests (often to very good effect too) and social campaigns (some applauded today, others not so much). But worse, he was a devotee of the seeming nonsense of spiritualism and its slippery slope to theistic evolutionism. And thus was Wallace's eclipse complete.

Or nearly so. Indeed, Alfred Russel Wallace may be the least known of scientific luminaries, the most obscure of the great naturalist-explorers, but his star is brightening. Does it matter that the much-lauded "First Darwinian" fell into relative obscurity? I would argue that it does matter. I show that Wallace not only *contained* multitudes but is *of* the multitudes—a man whose life of triumphs, tragedies, and personal qualities holds lessons for us today. Far more than simply a model of up-by-the-bootstraps pluck, wit, and determination, Wallace's generous spirit, sense of justice, and embrace of non-Western peoples of diverse faiths, cultures, mores, and customs set him apart from most of his contemporaries. He has been aptly described as a "working man's naturalist," a Victorian version of an intrepid backpacker, a homespun philosopher-collector who traveled on the cheap, lived among the locals, and honored their customs and beliefs (even as he learned infinite patience in the process) while making some of the greatest discoveries in the history of the biological sciences. Traversing thousands of miles first in Amazonia and then in the Malay Archipelago, Wallace brought a plethora of rare and precious species to scientific light, in the process financing his bold pursuit of grand philosophical questions: no less than the nature and origin of species.

But we honor him today not only for his perseverance, incisive scientific observations, and watershed contributions—notable among them co-discovering the principle of natural selection and founding the field of evolutionary biogeography—but also for his enduring humanity and lifelong activism for social justice. True, Wallace's earnest and trusting nature (at times bordering on gullible . . .) backfired at times, with results ranging from near-tragic to comedic to merely eye-roll-inducing. And true, Wallace lived in the era of colonial empire and very much benefited from the colonial enterprise that facilitated his travels and collections. Yes, it is important to understand Wallace in the context of his time and places and equally important to understand that he is much more than "of his time": that his life is a study in stick-to-it-ness against all odds, a man whose genius, perseverance, equanimity, humility, and generosity offer invaluable lessons for today's aspiring naturalists—and all of us, really.

—*Cullowhee, North Carolina, and Princeton, New Jersey*
May 2022

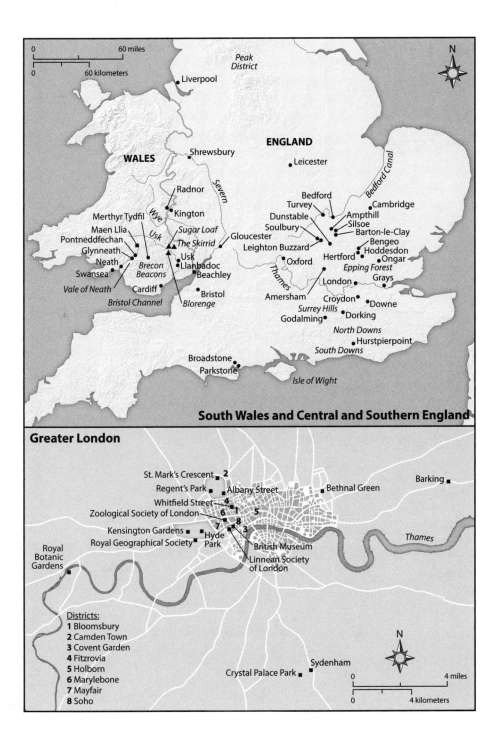

South Wales and Central and Southern England

0   60 miles
0   60 kilometers

N

Peak District

Liverpool

WALES

Shrewsbury

ENGLAND

Leicester

Bedford Canal

Radnor

Severn

Merthyr Tydfil   Kington

Maen Llia   Sugar Loaf
Pontneddfechan   The Skirrid   Gloucester
Glynneath   Usk
Neath   Brecon   Llanbadoc
Swansea   Beacons   Beachley
Vale of Neath   Cardiff   Bristol
Bristol Channel   Blorenge

Wye
Usk

Bedford
Turvey   Cambridge
Dunstable   Ampthill
Soulbury   Silsoe
Leighton Buzzard   Barton-le-Clay
Bengeo
Oxford   Hertford   Hoddesdon
Epping Forest   Ongar
London   Grays
Amersham   Croydon   Downe
Surrey Hills   Dorking
Godalming   North Downs
Hurstpierpoint
South Downs

Thames

Broadstone
Parkstone   Isle of Wight

Greater London

St. Mark's Crescent   2
Regent's Park   Albany Street   Bethnal Green   Barking
Whitfield Street
Zoological Society of London   4
6   1   5
7   8   3
Kensington Gardens   Hyde   British Museum
Royal Geographical Society   Park
Royal Botanic Gardens   Linnean Society of London

Thames

Districts:
1 Bloomsbury
2 Camden Town
3 Covent Garden
4 Fitzrovia
5 Holborn
6 Marylebone
7 Mayfair
8 Soho

Crystal Palace Park   Sydenham

N

0   4 miles
0   4 kilometers

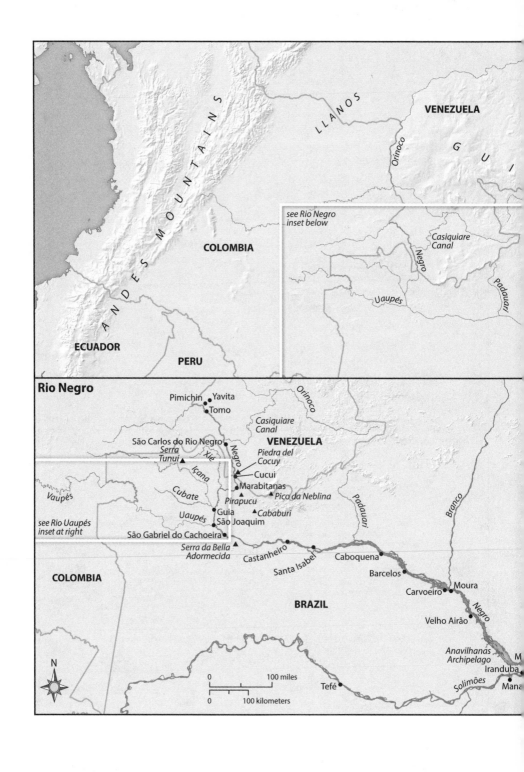

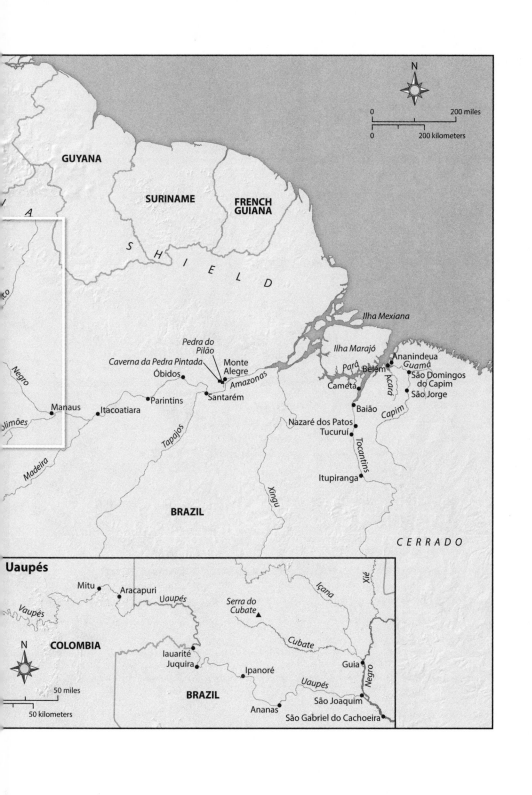

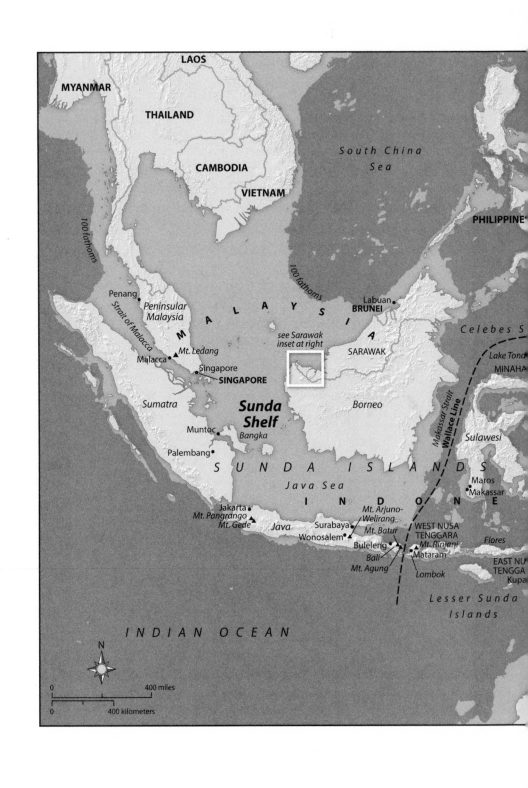

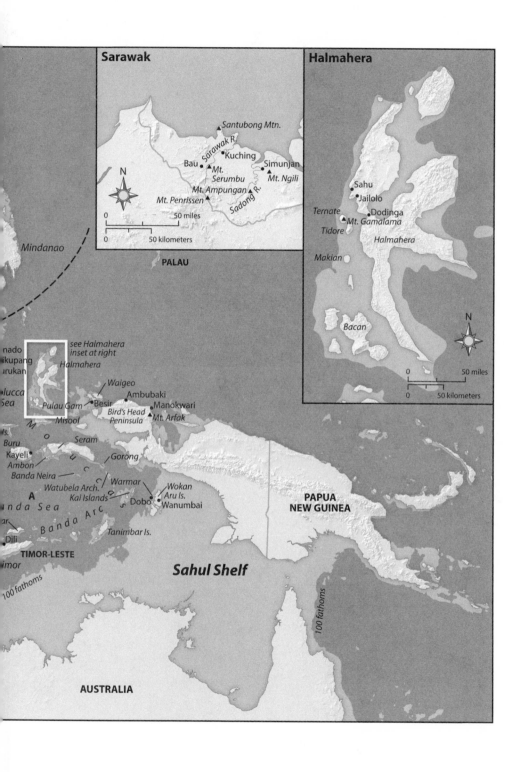

Sarawak

Santubong Mtn.

Sarawak R.

Bau • Kuching

Mt. Serumbu

Simunjan

Mt. Ngili

Mt. Ampungan

Mt. Penrissen

Sadong R.

N

0   50 miles

0   50 kilometers

PALAU

Halmahera

Sahu

Jailolo

Ternate • Dodinga

Mt. Gamalama

Tidore

Halmahera

Makian

Bacan

N

0   50 miles

0   50 kilometers

Mindanao

nado

ikupang

urukan

see Halmahera inset at right

Halmahera

Waigeo

lucca Sea

Ambubaki

Pulau Gam • Besir

Manokwari

Bird's Head

Misool

Peninsula

Mt. Arfak

M

o

Seram

l

Buru

u

Kayeli

c

Ambon

c

Banda Neira

a

Gorong

Watubela Arch.

s

Warmar

Wokan

A

Kai Islands

Aru Is.

nda Sea

Dobo

Wanumbai

Banda Arc

ar

Tanimbar Is.

PAPUA NEW GUINEA

Dili

TIMOR-LESTE

imor

100 fathoms

Sahul Shelf

100 fathoms

AUSTRALIA

CHAPTER 1

A HAPPY BUT DOWNWARDLY MOBILE FAMILY

KNOWING WHAT we know of Alfred Russel Wallace's later achievements in biogeography and evolution, it seems most appropriate that he was born atop a boundary in space and time. He came into the world on 8 January 1823, in a modest cottage in the hamlet of Llanbadoc, near Usk, South Wales, on the banks of the River Usk. The eighth of Thomas Vere Wallace and Mary Ann Greenell Wallace's eventual nine children, he was one of just six who survived to adulthood. Three sisters died in infancy or childhood, a tragedy not uncommon in those days and the likely reason why his parents had the infant Alfred Russel quickly "half baptized" in the nearby Llanbadoc church, as a precaution until a proper "full baptism" could be done. A fourth sister, Eliza, died in early adulthood, and his surviving siblings included William, the eldest (some fourteen years older), Frances ("Fanny," ten and a half years), John (four and a half years), and his younger brother Herbert ("Edward"), born in 1829 when Alfred was six.

Their home, now called Kensington Cottage, was a modest but handsome house on the west side of the river, backed up against a long, steep, north–south-running ridge just a quarter mile from the fine five-arched brick bridge leading to the Usk town center. Picturesque and bucolic, there was nothing outwardly remarkable about the site—but in fact this lovely place of Wallace's birth is a borderland of deep time, a place marking continental collisions, ebbing and flowing ancient seas, uplift, deformation, and untold eons of erosion, all yielding the curious geography of Wallace's early childhood. There, between ridge and river, Wallace was born atop the Llanbadoc Fault, a deep fracture in

1

the earth's crust that the River Usk found on its meandering journey from the uplands of the Brecon Beacons to the Bristol Channel.

This fault lies at the eastern edge of the great Usk Inlier, a more or less oval-shaped formation truncated to the northwest and measuring about four miles at its widest east to west and eight miles at its longest north to south, the whole dating to the Silurian period some 420 million years ago.[1] In geological terms an *inlier* is basically a formation of older rock surrounded by younger rock, typically formed by the erosion of overlying younger rocks to reveal the older ones beneath. One way they can form—true of the Usk Inlier—is from the horizontal layers of rock being squeezed from the sides and pushed up into an arched dome, in this case beginning about 350 million years ago. As erosion slowly but surely does its work, the bowed strata are exposed as a series of more or less concentric bands of rock in a definite age sequence: oldest in the middle, successively younger bands of rock to the outside. The different kinds of rock making up the layers differ in hardness and so erode at different rates. The tougher rocks are worn away a bit more slowly than the softer ones, over time becoming higher ground—just like the long ridge behind the cottage of Wallace's early childhood, a bit of ancient Silurian seabed tilted sharply and teeming with fossil bryozoans, corals, and brachiopods. This wall-like bank of Wallace's earliest memories comprises the youngest, outermost rocks of the Usk Inlier, 420-million-year-old limestone projecting above terrain just across the river, lower but younger still and with a wholly different geology: Devonian-period old red sandstone stretching for miles around, a piece of old Avalonia, as paleogeographers now know that ancient continent, named for King Arthur's island paradise.

Wallace could not have known any of this history, of course, not just because he was so young but because the science of geology itself was still in its early childhood. That doesn't mean we cannot appreciate the resonance: the man whose greatest contributions to science were insights into the interplay of the geological and biological forces giving rise to the ever-ramifying evolutionary tree through time while shaping species distributions as we see them today—the man of the eponymous Wallace Line, demarcating two of the planet's great biogeographic realms—was himself born atop a great divide, a boundary marking the meeting of continents and other slow-motion cataclysms of the distant Paleozoic, creating the singular geography of his childhood.

———

Kensington Cottage, Usk, ca. 1900.

At the time it was the geography that left the longest-lasting impression on his mind. His recollections of early childhood in Llanbadoc and Usk were strongly visual, and he commented in his autobiography how all the main features of *place*—the cottage bounded by river and steep bank, the old bridge, a quarry further up the ridge, the distant mountains—were more vivid in his mind than the people in his life. He well remembered scrambling up the steep bank many a time with his siblings—including one occasion when, inspired by Thomas Day's children's book *The History of Sandford and Merton*, a favorite, his older brother John led them up and over the ridge on an adventure: "John provided himself with the matchbox, salt, and potatoes, and having climbed up the steep bank behind our house, as we often did, and passed over a field or two to the woods beyond, to my great delight a fire was made, and we also feasted on potatoes with salt, as Sandford and Merton had done."[2]

It was one of many happy memories of his childhood home in remote Wales despite the financial duress that had brought the family there to begin with. His father, Thomas Wallace, qualified as a solicitor but never practiced, preferring literary and artistic pursuits as a young man. He was a man of taste, fond of theater and wordplay, but also something of an idle socialite, living off an inheritance and frequenting fashionable spa towns like Bath in season. In 1807 he married Mary Ann Greenell, of a prosperous Hertford family, and by 1810 the couple had two children. When the realities of a growing family motivated Thomas to seek additional means of income, they moved to Marylebone, the dynamic central London district where several notables, fictional and real, have taken up residence over the years. The artist J.M.W. Turner and polymathic mathematician and engineer Charles Babbage lived there at the time of the Wallace's residence, and Charles Dickens, Frederic Chopin, Elizabeth Browning, and even Sherlock Holmes were residents at various times later in the century (Baker Street was a short walk from Wimpole Street in Marylebone, where Arthur Conan Doyle had his ophthalmology practice); Paul McCartney and John Lennon were among Marylebone's twentieth-century luminaries. Rather than resort to law practice, Thomas Wallace embarked upon the first of what was to become a succession of disastrous business ventures, starting up a new large-format illustrated magazine of art, antiquities, and literature that was in his son's words "one of the most risky of literary speculations." All too predictably, it soon came to grief, owing to the cost of the lavish engravings and stubbornly low subscription rates. In the meantime the family continued to grow, with two more children born in Marylebone. The family soon moved to Southwark, South London, which was a bit more affordable. But additional mouths to feed and further deteriorating finances soon induced the family to move once more, this time to a place "where living was as cheap as possible." Rural South Wales it was, to the picturesque town of Usk, Monmouthshire, where Alfred and then his brother Herbert Edward came along. In his autobiography, Wallace commented on just how cheap the living was: rents and provisions of all kinds were half the going rates in London, and his father further provisioned the family from their own garden and taught the children himself. It was surely, he later thought, the happiest time of his father's life.

And most likely his mother's too. As a child Wallace knew little to nothing of his parents' financial travails, probably because they themselves were undaunted. By all accounts their marriage was quite a happy one, marked by great mutual affection and respect. No, what young Alfred was sensible of was security and joy at that juncture of his life. Their father read aloud in the evenings— Shakespeare, the poetry of William Cowper, Sandford and Merton, and of course the staple fairy-tales and legends: "Jack the Giant Killer," "Little Red Riding Hood," "Jack and the Beanstalk," "Aesop's Fables," and more. Wallace recounted being struck by Aesop's fable of the fox and the pitcher. More commonly known as the crow and the pitcher, in this tale a thirsty crow puzzles over how to reach water at a frustratingly low level in a narrow pitcher. The clever bird's solution is to drop pebbles into the pitcher one by one, displacing the water until its level is high enough to drink. Whether fox or crow, this trick "seemed quite like magic" to the three- or four-year-old Wallace. He decided to try the experiment himself. He poured an inch or two of water into a bucket and with a little spade scooped in stones and gravel (and probably some soil). It proved to be a failed experiment: "Instead of the water rising, it merely turned to mud; and the more I put in the muddier it became, while there seemed to be even less water than before." The moral of this story for Wallace was never to believe experiments out of storybooks, but it does show an in-quisitive turn of mind.[3]

Again, place was indelible in this memory—the scene of the experiment, the small backyard between the kitchen and the steep, rocky bank, "has always been clearly pictured before me," Wallace later wrote. The river, too, remained vivid in his mind. He recalled fishermen bobbing in the River Usk in their coracles, traditional single-person vessels resembling a large floating walnut shell. Or maybe "turtle shell" is more apt: a coracle is typically carried on one's back, and men transporting them resemble some bipedal version of a giant tortoise. Constructed of split willow sticks tied with bark and covered with waterproofed animal hide, coracles are designed for shallow rivers and were traditionally used for river fishing in Wales, the English West Country, Ireland, and Scotland. The name is derived from the Welsh *cwrwgl*, which has cognates in the Scottish and Irish Gaelic "currach," still used today.

Wallace and his siblings did some fishing too, but not from coracles. Great slabs of limestone from a quarry near their home, where the steep ridge of the Usk Inlier comes close to the river, provided convenient fishing platforms for the kids. He remembered the fearsome thrill of blasts at the quarry, where larger charges used some time in the past had flung huge slabs into the river. Equipped with old saucepans and washbasins, Alfred and the others excitedly scooped up young eellike lampreys making their way in shoals back to the sea. Lampreys are anadromous fish, spawning in the gravelly shallows of freshwater streams and rivers but living most of their lives in a marine environment. They make for good eating, and the Wallace kids' catches were typically fried up for supper, to Alfred's delight.

Another vivid memory was beautiful and romantic Usk Castle, where family friends lived in the gatehouse attached to the ancient castle ruins. Stra-tegically situated on a hill overlooking the town's north side, the Norman castle (still there today) dates to the early 1100s, though the commanding posi-tion of the hill was recognized by the Romans, who earlier had a fortress on the same site. Picturesque and evocative, the ancient castle inevitably conjured

A Welsh fisherman carrying his coracle turtle-style.

up visions of knights, giants, and prisoners in dismal dungeons to young Alfred. While most kids must be content with pretend castles in their playacting, the Wallace kids and their friends staged their pretend battles on the parapets of a real one.

Alfred's companions in his daily exploits at this age were typically his brother John and one or two of his sisters. John was the one constant playmate, as two sisters, Mary Anne (yes, spelled differently from his mother's name) and Emma, died in childhood at the ages of five and eight, and Frances (nicknamed Fanny) and Eliza, being some eleven and thirteen years older than Alfred, were more nannies than playmates. Their oldest brother, William, already fourteen when Alfred was born, had left home to become an apprentice surveyor in Kington, Hertfordshire. His visits home were cause for celebration in the close-knit family, and Wallace recalled the esteem his brother was held in. Besides his talents as a budding surveyor and businessman, William was a young man of some literary and scientific taste, even becoming involved, like their father, in a publishing venture, a monthly magazine of literature, science, and local events. The magazine may not have been the financial debacle that his father's was, but it was evidently not successful insofar as it does not appear to have lasted long. Alfred recalled his brother

Romantic Usk Castle, ca. 1838.

showing the family copies of the magazine, pointing out one article in particular that he may have authored and using diagrams to convey how the reflections of distant hills were sometimes visible in the river depending on small differences in water level. It may say something that Wallace recalled this despite his lack of understanding of the principles involved—it was a curious natural phenomenon of place.

Those distant hills were very much fixtures of place too, and Wallace well remembered the beautiful view up the river valley where the distinctive peaks of Sugar Loaf, Blorenge, and the Skirrid, in what is now spectacular Brecon Beacons National Park, marked "the beginning of the unknown land of Wales, which I also heard mentioned occasionally."[4] For in some ways, the Wallaces were strangers in an unusual but beautiful and welcoming land: the family was not of Welsh extraction, and as a child the flaxen-haired Alfred was nicknamed "the little Saxon" by the locals. Indeed, their very home was uncertain territory. The status of the county of Monmouthshire had long been disputed, at times considered part of Wales, at times part of England, a dual identity reflected in the county motto: *Utrique fidelis*, "Faithful to both." It seems appropriate, then, that the landscape of Wallace's birth was a borderland twice over, a geological one of deep time underlying a politico-cultural one on a human timescale. Monmouthshire's split personality persisted for centuries, until the county was firmly situated in Wales by virtue of the Local Government Act of 1972.

Such borders may be more political than natural, yet they can leave their imprint in the form of dual if not divided cultures, languages, and psyches. The question of Wallace's "nationality" as Welsh or English is a point of contention among some today, but though Wales was the land of Wallace's birth, he is most fairly considered an Englishman—as he regarded himself to be—though one with affection for Wales and the Welsh people.[5] Given Wallace's affinity for languages, it is a pity that he never learned to speak Welsh, though he became competent enough at reading it. He would surely have become adept had he been able to remain in Wales longer, but his childhood idyll ended in 1828 at the age of five when his mother came into an inheritance from her stepmother, Rebecca Greenell. The family soon relocated to her hometown of Hertford, in England.

————

Getting there was memorable in itself—a journey that today takes about three hours by car and under five by train was a multiday undertaking, though following much the same route beginning with the passage from Wales to England across the broad estuary of the River Severn. The Severn is Britain's longest river and also happened to be the river, far upstream, of Charles Darwin's youth as it courses through the border market town of Shrewsbury, the highest navigable point. An especially high tidal fluctuation—possibly the world's largest after the Bay of Fundy in Canada—and fast, changeable currents combined with high and unpredictable winds made the mile-long Severn crossing a dangerous proposition even under steam. Wallace recalled the passage as "a little awful," and he had good reason to be apprehensive. Their route was known as the Old Passage, crossing at the narrowest point from Beachley on the Welsh side, near where the River Wye joins the Severn, to Aust on the English, essentially the same passage point used since the days of Roman Britain.

Although a steam ferry service had opened in 1827, the Wallaces went by sail, young Alfred recalling the small boat heeling sharply and the party having to stoop to stay clear of the boom as it swung back and forth. That was the most dangerous way to attempt what are surely the region's most treacherous waters, with many a boat lost in the attempt over the years. In the eighteenth century,

Daniel Defoe, who knew something about shipwrecks, was alarmed at the "sorry boats" on offer in Aust. "The sea was so broad, the fame of the Bore of the tide so formidable, the wind also made the water so rough, and which was worse, the boats . . . appeared so very mean" that he and his party refused to take the "ugly, dangerous, and very inconvenient ferry," electing to use a safer passage far upriver at Gloucester.[6] The steam ferries were safer than sail but still dangerous—a decade after the Wallaces' safe crossing, the Beachley-Aust ferry sank with all aboard on 1 September 1839, and another was lost five years later.

We might consider this the first of Alfred Russel Wallace's many dangerous sea voyages. Fortunately, it proved uneventful, if scary, and the family made their way to London, where they first stopped to visit relatives in Dulwich, south-central London near their previous Southwark home. As Thomas Wallace made arrangements for their Hertford home, Alfred stayed temporarily at a boarding school in Ongar, Essex, where he recalled both misadventures (accidentally sending a stone lawn roller careening downhill into a pond) and an intriguing bit of natural history: belemnites, the fossilized internal guard, or rostrum, of extinct squid relatives. Located at the tail end of the living animal, where they likely played a role in balance, the hard bullet-shaped rostrum is all that remains of these creatures that swarmed the Jurassic and Cretaceous seas that covered much of Britain. Wallace and his friends picked the "thunderbolts" out of the gravel—ancient lore held that belemnites fell in thunderstorms—no doubt seeking choice specimens to fill a box or jar in perhaps his first collection. He would have known nothing of their true origins, but they excited curiosity enough even as worn and broken tubular fragments. Sometimes smooth-sided and sometimes rough, in cross section a central hollow was visible around which radiated glittering lines like so many crystalline wheel spokes.

It was not long before the family moved into number 1 Saint Andrew's Street, Hertford, the bustling market town of Hertfordshire just north of London. His mother's family had lived in the area for generations as solidly middle-class professionals and tradesmen, with a host of lawyers, architects, mill owners, and the occasional alderman and mayor. Situated in the heart of town, the house (now no. 11, a doctor's office) was a sturdy three-story brick structure, half of a kind of duplex with a covered passage between mirror-image houses. It did not take young Alfred long to meet the neighbors: a little boy about his age peered over the garden wall and greeted him with a "Hallo! What's your name?" It was George Silk, who was to become a lifelong friend.

About a year later, the family moved to a more spacious house just up the street on Old Cross (now no. 23, a barbershop). This one was heaven, with a side yard, a flower-filled garden in the back, and, most excitingly, a stable with a loft that soon became Alfred and John's headquarters. "Almost like a robber's cave," Alfred later recalled, "our greatest delight." It was their lair, hideaway, lab, and shop, where they spent untold hours playing, reading, and inventing.

But the great outdoors was their main theater of fun. Again his sense of place was strong, his memories filled with scenes of streams and rivers with great working mills coursing through a varied landscape of farms, woodlands, and flower-filled meadows. "One of the most pleasantly situated county towns in England," Wallace declared, a rolling and verdant landscape emblematic of Blake's "green & pleasant Land." The Hertford geography of Wallace's memory was a map of favorite play spots and wonders crisscrossed by rivers, lanes, and footpaths. Located on the western side of the East of England, Hertford lies at the confluence of four river valleys, where the River Lea, the main river through town, is joined by the Rivers Beane and Rib from the north and the Mimram from the west. The east-flowing Lea turns to the south as the canalized Lea Navigation, coursing toward London and the Thames. A favorite swimming hole in the Beane was the site of Alfred's first brush with death not long after the family's arrival when a cavorting friend pushed him into the water. Struggling, he may well have drowned had it not been for his brother John quickly jumping in to save him. Though scary at the time, the incident did not much alter his affinity for the rivers, or water generally. The four and a half years separating Alfred and John held less significance as John became his closest companion in explorations and exploits.

Favorite haunts in and about town were vividly recalled. There was Hartham Common, today the same broad park on high ground between the Lea and the Beane that Wallace knew as a "first-rate" cricket field and playground—he would surely be impressed with the range of sports on offer there now, from football, rugby, and tennis to kayaking and canoeing on the rivers. It also boasts a gym and swimming pool. Immediately beyond Hartham and the Beane to the north was a steep, wooded slope that Wallace, John, and friends knew as the Warren, atop which the lovely village of Bengeo sits. Just to the west of town along the Mimram were Hertingfordbury and Panshanger Park, once the estate of the earls of Cowper. Wallace does not mention the grand Panshanger House, still standing at the time. Rather, a sight grander still to him was the awe-inspiring oak tree dating to the time of Queen Elizabeth I. Already some 19 feet (5.8 m) in circumference in Wallace's youth, "one of the sights of the district," the venerable tree had grown to about 25 feet (7.6 m) before its deterioration led to its removal in 1978.[7] In Hertford, too, they had a castle, although not nearly as evocative as the one in Usk. The town has medieval origins, with records of tenth-century earthwork fortifications guarding the ford over the Lea against Vikings and later a castle built by the Normans and reconstructed by Henry II in the twelfth century. By the nineteenth century, only parts of the walls and the beautiful gatehouse, itself rather imposing, remained of the old castle. The kids scrambled up the parapet and could imagine marauders kept at bay by the moat that had once girdled the castle, flowers marking where water diverted from the Lea had frustrated would-be invaders—and, who knows, perhaps marked some of their graves. Then there was their "racing field" near Bayfordbury, a favorite play spot perhaps close to today's modern observatory and greenhouses of the University of Hertford, and the "chalk cave" near elm-lined Morgan's Walk, a deep hollow in a chalk bank well hidden by overhanging shrubs and well stocked with candles, a tinderbox, potatoes, and sundry other provisions, where Alfred, John, and their coconspirators fancied they were brigands hiding out in a secret lair. To slake their thirst, they could steal out to the bubbling brick-lined spring in the field to Dunkirk's Farm, just at the end of Morgan's Walk: "We seldom went this way without running down to it to take a drink of water and admire its purity and upward bubbling out of the earth."

The area was renowned for the purity of its springs—notably, Chadwell Spring, a large and circular bubbling spring that gives rise to the New River, which is not a natural river but a remarkable eighteenth-century aqueduct that follows the one-hundred-foot contour some forty miles to Islington, in London. The spring was famous for its turquoise blue-green waters, a tint that says something about the area's geology: chalk and limestone bedrock topped with chalky soil and gravels, reflecting at least two epochs of geological history. The chalk and limestone were laid down in Cretaceous seas (the very name derived from "creta," or "chalk" in Latin), while the much younger gravels are the product of the slow grinding and conveyor-belt transport of rock by Pleistocene-era glaciers. Dissolved minerals and suspended calcium carbonate from the bedrock scatter light at the blue end of the spectrum, conferring a vivid blue-green color to our eye. While Wallace recalled the "exquisite shades of blue and green in ever-varying gradations" of this spring, he also lamented in his autobiography that it had since been ruined by ill-considered well drilling in the area, altering the hydrology: "Thus does our morbid civilization destroy the most beautiful works of nature." Indeed, for some time in the early twentieth century the spring was dried up altogether, its subterranean waters diverted. It wells up again today but is no longer the "exceedingly beautiful" color Wallace remembered. The chalk was a universal feature of the landscape of Wallace's youth, never far belowground and surfacing in stark white outcroppings here and there. "In the total absence of any instruction in nature-knowledge at that period, my impression, and that of most other boys, no doubt, was, that in some way chalk was the natural and universal substance of which the earth consisted, the only question being how deep you must go to reach it."[8]

———

The prodigious "nature-knowledge" that Wallace later became famous for had its origins here, but not in the way one might suppose. It was a slow osmosis, the product of the odd seed of incidental remarks and observations chancing upon the fertile soil of his mind. That fertile soil was enriched mainly by play, books, and a loving homelife and very little by formal instruction. School was to be endured. About a year after the family moved to Old Cross, Wallace started attending Hertford Grammar School, run by headmaster Clement Henry Crutwell, "a rather irascible little man." John was already attending, smoothing the transition. The school, founded in 1617, had a single long classroom for about eighty boys, an open fireplace on either end, desks for four teachers on the sides, and rows of desks for the boys down the center. Instruction consisted of the usual staples of Latin, history, geography, and a bit of French, all with a heavy (and tedious) emphasis on memorization. The school day started at 7 AM and on three days of the week continued until 5 PM—beginning and ending in twilight, if not darkness, in the depths of winter, when the boys were expected to provide their own candles by which to work. "Buzz Wallace," as he was known to his schoolmates, enjoyed hearing "Old Cruttle" the headmaster declaim Homer or Cicero far more than "blundering through" the forty or fifty lines he and his schoolmates were often assigned.[9] "When we were called up, it was all a matter of chance whether we got through well or otherwise." The word "painful" appears seven times in Wallace's recollection of his school days, but he evidently performed well enough considering that, a few years later, he assisted by tutoring the younger students in reading, writing, and arithmetic, though that was not a role he relished. After 313 years, in 1930 the growing school moved to more spacious grounds and was later renamed to honor founder Richard Hale, a prosperous seventeenth-century merchant. Yes, school was to be endured, but for all that this most famous "Old

Boy" of the Richard Hale School would be touched that he, too, is honored there now, lending his name to one of the school's six houses and, more poignantly if utterly unimaginable to the young Wallace, an annual scholarship to support student travel and study abroad. What better tribute to one of the greatest scientific travelers of modern times?

As Wallace himself later acknowledged, his real education occurred outside school, as is so often true in families that encourage eclectic reading and give kids free rein to pursue creative interests. Both boxes were checked with the Wallace family. For all his lack of ambition, Thomas Wallace kept the house well stocked with books, further aided by taking a position at the town library at one point. The town boasted several societies or book clubs supported by annual subscriptions, circulating books among members and in some cases extending borrowing privileges to nonsubscribing local families. Not one but two reading rooms well supplied with newspapers, reviews, and magazines were available to boot, one frequented by the "gentlemen of the county" and the other for the general populace.[10] A steady stream of books and magazines flowed through the house as a result, including classics, histories, plays, and travelogues: Milton, Pope, Defoe, Fenimore Cooper, Byron, Scott, Swift, Goldsmith, Bunyan, Dante, Cervantes, Shakespeare, Mungo Park, and more. Serials like Dickens's *Pickwick Papers* were much anticipated, and the family devoured issues of the *Rambler*, the then new *Spectator*, and the great favorite *Hood's Comic Annual*. Thomas Wallace would read aloud at home, and when he worked in the library, Alfred would often join him—especially once John left for London—helping fetch or shelve books but usually off reading in a corner.

Unsurprisingly, perhaps, the family's eclectic literary tastes seemed to go hand in hand with toleration, at least to a point. While they were fairly orthodox members of the Church of England, attending church twice on Sundays, their circle of friends—close enough friends that the Wallaces would sometimes attend their services—extended to Dissenters and Quakers. Bored with the prevailing silence of the Quaker meetings, Alfred found the Dissenters' chapel far more exciting. The spontaneous prayers and attestations, passionate singing, and vigorous preaching were a welcome departure from the sedate proceedings of the Anglicans, let alone the Quakers. The experience even piqued some religious feeling within him, but lacking "sufficient basis of intelligible fact or connected reasoning to satisfy my intellect," the feeling did not last long and never returned—though some thirty-five years later he would become another kind of dissenter as a spiritualist, which had quasi-religious overtones.

Imbibing all he experienced, as kids do, at the time Alfred's exposure to the non-conforming religious communities of the town surely left its mark as part of a growing social awareness. In those years, too, he had occasion to witness sessions of the court of assizes, recalling sheep rustlers on trial, aware that the penalty could well be transportation—exile to some far-flung penal colony for life, a form of punishment that ended in the 1850s. The nine-year-old Alfred surely felt the palpable excitement coursing through the town at the passage of the Great Reform Act in 1832, celebrated by a great outdoor feast for the working-class families of Hertford. The bill changed the electoral system rather dramatically, eliminating centuries-old traditions like the forty-shilling franchise (wherein the right to vote was based on the ownership and value of property) and the many pocket or "rotten" boroughs, which amounted to reserved seats (and thus influence) in Parliament even if the borough had few or no inhabitants.[11] His father's disapproval of the act may have given Alfred his first inkling of political division and the winds of social change, very much on display when the radical member of Parliament Thomas Slingsby Duncombe was ceremoniously "chaired" through the streets after his electoral victory.[12]

Alfred was much later to propose his own radical social and political reforms (of which his father surely would have not approved), but here and now, as a kid, what Alfred really lived for day-to-day and recalled most vividly later in life were the endless diversions with his brother in the beloved loft over the stable, their private lab and lair in the few years the family lived on Old Cross. John was something of a natural engineer, with a talent for mechanical contraptions and carpentry. He would surely have explained the workings of the great linseed mill in town that so fascinated his little brother, who vividly recalled its great rotating vertical millstones and curved scoop continually sweeping back and forth, grinding the seed into an ever-finer meal. The adjacent stamping mill for compressing the linseed meal into oil cakes was more awesome still—some two dozen or more great vertical rammers cycling up and down, striking and rebounding from the molds at different rates in a mechanical clockwork din as deafening as it was oddly musical. Alfred remembered that time tinkering and experimenting with his brother as "certainly the most interesting and perhaps the most permanently useful" of his whole childhood.

William Clarke's *The Boy's Own Book*, an encyclopedia for the "amusement and instruction" of Britain's "men in miniature," was their go-to manual for all manner of inventions and games.[13] First published in 1828, the popular how-to gave detailed instructions for making things that would give a modern publishing attorney nightmares. Stocking up on gunpowder, sulfur, charcoal, iron filings, and saltpeter, for example, John and Alfred were all set for homemade fireworks: squibs, firecracker strings, Roman candles, and revolving Catherine wheels (spectacular when they didn't just burst into flames) were all favorites, especially on holidays like Guy Fawkes Day. He did not recall anyone getting injured, even when "now and then" some hapless friend had crackers or squibs explode in a pocket. Nor did they get hurt, fortunately, firing off the six-inch brass cannon they got in a trade, especially considering that they liked to pack the barrel "to the very muzzle" before carefully snaking a trail of powder a few feet away, giving them a bit of time to dash off to safety after lighting it. The ear-ringing explosion would send the cannon jumping into the air. The miniature "key cannons" they constructed were fairly harmless by comparison. Using the hollow shank or stem of old brass skeleton keys as a barrel, the little guns made a satisfyingly loud report: "By filing a touch-hole, filing off the handle, and mounting them on block carriages, we were able to fire off salutes or startle our sister or the servant to our great satisfaction." More innocuous were the popguns they made with hollowed-out elder branches and the elaborate miniature spring pistols that fired peas—so skillfully made that John sold them for a shilling or more at school. They had more constructive, even educational, toys, too, of course: John and Alfred made their own cricket balls, and cherry-stone chains and ornately carved bread seals were favorites. Their father purchased a model wooden building-block bridge illustrating the principle of the arch and keystone, and the family pored over large dissected maps of Europe and England, challenges that had the added benefit of instructing the kids in geography. Alfred attributed his lifelong love of maps to those puzzles.

He thought their father was at his most content those few years on Old Cross, gardening, making beer and wine from their own large and productive grapevines, working at the library, reading to the family. This is not to say it was idyllic: Alfred had a dangerous bout of scarlet fever, and he remembered his family's acute grief when his older sister Eliza succumbed to tuberculosis in 1832, at age twenty-two. Around this time, too, their remaining sister Fanny left home to become a governess with a family in the nearby village of Hoddesdon. All was not well with family finances either, but that was not something he was even dimly aware of—though that awareness soon grew.

———

If he had not known before, Alfred knew that something was amiss when the family moved again. The trouble started in late 1833 or 1834, a perfect storm of financial disaster brewing. Mary Ann Wallace's brother-in-law Thomas Wilson, a solicitor and one of the executors of her father's estate, imprudently invested what remained of the family's already modest assets in a speculative building project in London, only to go bankrupt. Somehow Mary Ann's inheritance—and that of the children—also became a casualty of the bankruptcy, drastically reducing the family's income. Things went from bad to worse as Thomas Wallace's own savings were lost in ill-considered investments, and the family was forced to exchange their comfortable house on Old Cross for part of an old house near All Saint's Church, the former vicarage now part post office and part residence. Other moves soon followed—about this time a dizzying series of changes rapidly unfolded for Alfred in a relatively short period of time. Precise dates are unclear, but in the space of the few years from about 1834 through 1836, his sister Fanny left home to perfect her French in Lille, John was packed off to London as an apprentice carpenter, and the family moved to a smaller house on Saint Andrews Street, then into a portion of an old house near Saint Andrews Church. This last at least had the double virtue of having Alfred's friend George Silk once again living next door and a large fruit-laden mulberry tree in the garden that he and George loved to climb, where they would "feast luxuriously."

Mary Ann Wallace was beside herself with worry over the family's sinking fortunes, especially the question of the children's modest share of their grandfather's bequest. She wrote increasingly urgent letters to her brother-in-law: "The object of this is not to harrass [sic] you—but to request of you to inform me How I ought to act with respect to the claims my children have on you as their Grandfather's Executor." She trusted to his honor that he would "do the best for my dear children and will acknowledge the debt due to them." Fanny needed funds to remain in France, as she was never paid her inheritance, and what do to about John—he owed half a year of board in his position as apprentice carpenter and would be discharged if it was not paid. And poor William was afraid of showing his face in London where "that Elkin the apothecary has threatened to arrest him for his debt of £20. . . . It would be William's utter ruin if anything of that nature was to occur."[14] She turned to Louisa Draper, the daughter of Richard Draper, family friend and the other executor of the estate, for advice, imploring that she "not feel offended at this application in behalf of my poor children, they have but little, and it is hard that little (their all) should be lost! It is a delicate matter to know how to act between friends, but in such a matter as this I must act the best for my children by doing everything in my power to recover that which seems lost through the failure of one of the

Trustees. . . . My situation is a most painful one we are harrassed [*sic*] in every way."[15]

Funds were eventually forthcoming, but it took awhile, and even then it was too little, too late to keep the family together. Fanny returned from France, and Alfred was sent to board with twenty or thirty other boys at Old Cruttle's house on Fore Street for about six months until Fanny returned to her position as governess in Hoddesdon. As home finances got tighter and tighter, Alfred had to help cover his school fees by tutoring the younger boys—to his great embarrassment. By early 1837 the family was forced to move yet again, leaving Hertford for a small abode called Rawdon Cottage in Hoddesdon, close to Fanny. It was too small for both Alfred and Edward to live there, and they could no longer afford Alfred's school or boarding fees. They reluctantly removed Alfred from school and packed him off to join John in London, a stopgap measure until William could take him on as an apprentice surveyor back in Wales. It was the best thing that could have happened to fourteen-year-old Alfred Wallace.

At the very time Alfred arrived on Robert Street, off Hampstead Road, sharing both room and bed with John in the home of Mr. Webster, the master builder to whom John was apprenticed, a young man twice his age had just moved into rather nice accommodations exactly one mile due south at 36 Great Marlborough Street. Charles Darwin, just five months back from his voyage around the world, was delighted to move in around the corner from his beloved brother Erasmus. It is an uncanny parallel, the impecunious teenage surveyor's-apprentice-to-be and the well-to-do young gentleman naturalist living one mile apart, both arriving in March 1837.[16] That was the very month that Darwin had his transmutational epiphany, the dots suddenly and clearly connecting and pointing to the truth that species must change. It was a time when Alfred Russel Wallace's mind was about to be profoundly opened, too, setting him on his own path to eventual epiphany, one that would inevitably intersect with Darwin's. But that was not for another twenty-one years, and much was to happen to both of them before then.

CHAPTER 2

TAKING MEASURE
IN THE BORDERLANDS

ARRIVING IN CENTRAL London could only have inspired awe in fourteen-year-old Alfred Russel Wallace, having moved from a market town of perhaps ten thousand to the heart of a metropolis of over two million. His brother John had been living under the roof of Mr. Webster, to whom he was apprenticed—and who was also his father-in-law-to-be—for about two and a half years by then. Webster's small firm of sawyers and carpenters on Albany Street did all manner of construction and joinery, from milling the lumber to crafting windows, doors, cupboards, and staircases by hand, 6:00 AM until 5:30 PM, with an hour and a half for meals, six days a week. Alfred was not expected to do much; he was there temporarily, an unobtrusive and low-cost boarder sharing a room with John until he could start training with their older brother William in the surveying trade. His days were spent hanging around Mr. Webster's carpentry shop, helping with odd jobs and taking in the workmen's banter, gaining some insight into their lives. On evenings and free days, he was surely shown the sights of the district by John, especially nearby Regent's Park, the east side open for only two years by then, with a wide green prospect encircled by long rows of elegant townhouses, the terraces. They would have viewed Jenkin's Nursery in the Inner Circle with the lake beyond, and their explorations would likely have taken them along Regent's canal at the north end of the park too, where the menagerie of the Zoological Society of London was kept. Only fellows were permitted entry at the time, however, so at best they would have had only tantalizing distant views of the exotic creatures there. John also showed him the posh shops, where they admired the window displays, explorations that likely took them down Tottenham Court Road to Leicester Square and beyond, maybe detouring just to the west at times to take

The Zoological Gardens at Regent's Park, ca. 1828.

in the grand neoclassical edifice in progress at the British Museum, then back up via Piccadilly and Regent Street in a loop around Soho. This route likely crossed paths with a distracted Charles Darwin, an as-yet-unknown young man in a hurry who at that time could often be found frequenting the Royal College of Surgeons at Lincoln's Inn Fields, the Zoological Society at Leicester Square, and the Geological Society off Piccadilly.

But more often than not, their evenings were spent at the Hall of Science just a short walk away on John Street, off Tottenham Court Road. This was the heyday of the halls of science and mechanics' institutes sweeping Britain— new kinds of institutions promoting self-improvement through education that started in Edinburgh and Glasgow in the early 1820s.[1] Initially, they were a kind of working men's free technical college, with open lectures on scientific principles, phenomena, and the latest discoveries. But they soon evolved into equal parts library, technical college, and community center where science, politics, and social reform went hand in hand. Science was synonymous with reform, education, and progressive ideas, which to many came to mean anti-religion and pro-rationality, anti-establishment and pro–working class. Higher education was the purview of the elite in those days, yet the less privileged classes demonstrated a tremendous appetite for knowledge by signing up in the thousands for lectures and classes, most of them free. By 1837, when John introduced his impressionable little brother to the Hall of Science on John Street, there were hundreds of mechanics' institutes across Britain, where coffee and radical ideas flowed freely.[2]

By far the most memorable experience of teenaged Alfred Wallace's eye-opening engagement with the John Street Hall of Science was imbibing the teachings of reformer and utopian socialist Robert Owen and his acolytes. He even once had an opportunity to hear the venerable Owen himself, then a "tall spare figure" in his mid-sixties with a "very lofty head, and highly benevolent countenance and mode of speaking."[3] Owen, who like Wallace was born in Wales, was a textile manufacturer, a philanthropist, and, perhaps most famously, a social reformer renowned (reviled by some) as the founder of the socialist movement in Britain. Far ahead of his time, Owen campaigned for educational and labor reform (e.g., the Cotton Mills and Factories Act of 1819), pioneering such "outrageous" policies as an eight-hour workday, childcare for workers (rather than sending children to labor in the factories), improved workplace health and safety, and universal education. The parallels between Owen and Wallace, besides their Welsh birthplace, are curious: both were second to last of a brood of siblings, received little formal schooling before being sent off to learn a trade, read voraciously and were largely self-taught, and came to embrace secularism, socialism, and ultimately spiritualism. (Both tended to make money only to lose it too, but the scale and circumstances differed greatly.)

Owenism was just entering its most important phase when Wallace arrived in London, its principles first laid out in a series of essays published as "A New View of Society: Essays on the Formation of the Human Character" two decades earlier, in 1813–1814. First and foremost was Owen's conviction that "man is a compound being, whose character is formed of his constitution, or organisation at birth, and of the effects of external circumstances upon it, from birth to death" and that "such original organisation and external influences [are] continually acting and reacting each upon the other." In modern terms the idealist Owen resolutely believed that *nurture* trumped nature when it came to developing human potential, and his goal was to improve quality of life in childhood and beyond to foster a crime- and strife-free society. He railed against the landowners and clergy who ensured that British society remained rigidly stratified, and insofar as the crime, squalor, and degeneracy afflicting the lower classes were a direct product of their conditions of life, he lay the responsibility for these ills at the feet of the elite.

Walking the walk as well as talking the talk, Owen launched social experiments on a grand scale. The first was at New Lanark, his cotton-spinning mills on the River Clyde, south of Glasgow, Scotland (now a United Nations Educational, Scientific, and Cultural Organization [UNESCO] World Heritage site), where in 1800 he first instituted such reforms as the eight-hour workday, an elementary school, and a school for workers (the Institute for the Formation of Character). It was a great success, showing that productivity and profitability need not be at odds with the humane treatment of workers; indeed, the mills at New Lanark were in operation until 1968. But that was just the beginning. New Lanark was Owen's proof of concept, and he followed in the mid-1820s by investing most of his fortune in a scaled-up version in the United States: a "Village of Unity and Mutual Cooperation" called New Harmony, on the banks of the Wabash River in Indiana.

It was a breathtaking utopian experiment, a new model for society that would, he declared to an overflow crowd of worthies gathered in the U.S. Capitol, "commence a new empire of peace and good will to man, founded on other principles, and leading to other practices than those of the past or present, and which principles, in due season, and in the allotted time, will lead to that state of virtue, intelligence, enjoyment, and happiness, in practice, which has been foretold by the sages of past times."[4] Unfortunately for Owen, the devil is always in the details, and there was a disconnect between his principles and planning and between his vision for reform and that of the people he aimed to reform. The secular, fiercely anti-ecclesiastic element of the Owenite's vision was perhaps the most problematic for the movement in 1820s America, then in the midst of the second Great Awakening, a time of extreme religious fervor marked by its own brand of social utopianism.[5] The New Harmony experiment failed after just two years, but many believers in the project stayed on, including several of Owen's adult children. The town became, if not a socialist utopia, a scientific center thanks in part to their efforts—two of his sons became noted geologists (one was also the founding president of Purdue University), and another went into politics and helped found the Smithsonian Institution in his capacity as a U.S. congressman. Owen's American friend and supporter William Maclure, president of the Academy of Natural Sciences of Philadelphia, also deserves credit: Maclure persuaded a distinguished group of artists, educators, and naturalists to move to New Harmony with him, and the idealists were soon plying the Ohio in the keelboat *Philanthropist*, dubbed the "Boatload of Knowledge."[6]

Most of them stayed on after the collapse of the social experiment, but Owen was soon back in London. Undeterred by the failure of New Harmony, by the mid-1830s Owen started a weekly called the *New Moral World*, a socialist newspaper that advocated for unions, peaceful revolution, and utopian communities, and launched ambitious trade unions with names befitting the high hopes and ideals they represented, like the Grand National Consolidated Trades Union (1834), the Association of All Classes of All Nations (1835), and the Universal Community Society of Rational Religionists (1839)—you get the idea. This last was founded in the same year Owen launched his final grand utopian social experiment, Harmony Hall, later Queenwood College, in Hampshire.[7] It, too, was to fail within a few years, but the movement lived on (and still does today, in the form of the many successful co-ops and laws regulating working conditions, public health, and education).

In 1837, when Alfred Russel Wallace heard Owen, the future was bright with the prospect of not merely reform but socialist utopia—1837 was, in fact, the start of a new era of British Owenism. Consider that no fewer than twenty-three provincial Owenite branches opened around the country in 1837 and another twenty-two in 1838. The Metropolitan Institute opened on John Street in 1837 (Owenite branch no. 32, Wallace's Hall of Science, which became known as the John Street Institute), as did the Society of Materialists (branch A1) just a couple of blocks away on Cleveland Street. Dozens of Owenite free-thought societies flourished across London *alone* in the years between 1837 and 1866. The John Street Institute was just one of many "infidel halls," as they were called by detractors like the *London City Mission Magazine*. It was electrifying to the fourteen-year-old Wallace, who was already skeptical of organized religion: at the John Street Institute, he later wrote, "We sometimes heard lectures on Owen's doctrines, or on the principles of secularism or agnosticism, as it is now called. . . . It was here that I first made acquaintance with Owen's writings, and especially with the wonderful and beneficent work he had carried on for many years at New Lanark. I also received my first knowledge of the arguments of sceptics, and read among other books Paine's 'Age of Reason.'"[8]

The force of Paine's *Age of Reason* resonated with Wallace, with its lucid and lyrical smackdown of organized religion, decrying church corruption and manipulative priestcraft: "All national institutions of churches, whether Jewish, Christian or Turkish, appear to me no other than human inventions, set up to terrify and enslave mankind, and monopolize power and profit."[9] Which is not to say the tract was atheistic; as indicated by the subtitle (*Being an Investigation of True and Fabulous Theology*), Paine advanced deistic arguments, advocating reason over supposed revelation, "natural religion" over organized. This was very much part of Owenite philosophy, perhaps best expressed by the eldest Owen son Robert Dale Owen. Wallace remembered reading the younger

Owen's *Lecture on Consistency*, a damning indictment of eternal damnation, one might say. A key take-home—"The only true and wholly beneficial religion was that which inculcated the service of humanity, and whose only dogma was the brotherhood of man"—echoed Paine's own personal creed: "My country is the world, and my religion is to do good."[10] "Thus," Wallace wrote, "was laid the foundation of my religious scepticism."[11]

That skepticism went hand in hand with Wallace's fervent belief in both the inherent dignity of the individual and the inherent injustice of the prevailing social and political system. Owen's influence on Wallace cannot be overstated. He was Wallace's "first teacher in the philosophy of human nature" and "first guide through the labyrinth of social science." Asked late in life by a New York journalist if he would name his top ten "chief humanitarians" of the nineteenth century, it comes as no surprise that Owen topped the list as "a true lover of mankind, the wisest and most practical of workers."[12] The seeds of the socialist policies that Wallace later fiercely and eloquently advocated for, especially their more utopian elements, were certainly planted in those formative months in London and to some extent even colored Wallace's view of the natural world and his later reading of clergyman and early political economist Thomas Robert Malthus, as we shall see.[13] In dedicating fully two-thirds of the twenty-three-plus-page chapter 6 in his autobiography to Owen, the "greatest of social reformers and the real founder of modern Socialism," Wallace trusted that his readers would not mind—ditto for the space I have given here to Owen's outsized influence on Wallace.[14]

———

Alfred's stint in London ended that summer when he left John behind to become an assistant surveyor with William in Bedfordshire. His education at the Hall of Science and Mr. Webster's carpentry shop provided a fine foundation for what came next: effectively, a six-year-long apprenticeship not only in surveying and mapmaking but in observation, reflection, and discovering science. Ironically, some of the very social forces and policies that the Owenites railed against were also responsible for William's livelihood and, by extension, Alfred's as his older brother's trainee-assistant. There was plenty of survey work to be had at the time, some of it owing to the expanding railways and canals, but far more important were the seismic changes in land use stemming from enclosure and, most immediately, the commutation of tithes. He had no idea that he was an accessory to a crime, as he later saw it.

Enclosure, or "inclosure" in the archaic spelling of the time, was the act of enclosing what had been common land from time immemorial. Enclosure by fencing or hedging off land had been going on for centuries on a local level, but it accelerated greatly in the seventeenth and eighteenth centuries. By this time enclosure by parliamentary act had become common, with nearly four thousand such acts passed between 1750 and 1819 alone. But the Inclosure Act of 1773, "for the better Cultivation, Improvement, and Regulation of the Common Arable Fields, Wastes, and Commons of Pasture in this Kingdom," elevated enclosure practically to the level of national policy: "In every parish or place in this kingdom where there are open or common field lands, all the tillage or arable lands lying in the said open or common fields shall be ordered, fenced, cultivated and improved."[15] As this language suggests, the impetus was ostensibly agricultural improvement, and enclosure did encourage the introduction of agricultural innovations incompatible with the common use of land, such as strict systems of crop rotation (the new four-course rotation system of wheat → turnips → barley → clover and rye was a novel technique that eliminated the need for a fallow year), as well as experimentation with fertilizers and plowing techniques. But the monetary gain from boosted productivity was surely an underlying motivation for landowners (who could charge more rent from tenants) and well-established tenant farmers with access rights (whose farms thus increased in value), both of whom pushed for enclosure. The losers were the peasants and less well-off tenants used to eking out their existence on common land by grazing a few cows or planting a meager plot.[16]

The commutation of tithes was seen as another blow to the laboring classes. For centuries the people working the land paid an annual rent in kind, traditionally one-tenth of the produce of the land in the form of crops, livestock, wool, fish, timber, honey, and so on. The church was originally the major landlord, but over time (especially after the dissolution of the monasteries) this expanded to include many lay landowners, often aristocrats, who could even buy and sell tithe rights. But the tithe system was seen as inefficient and led to endless disputes over what was tithable and how and when payment should be made. Tithes were always a burden to the already struggling poor laborers, but commutation to cash payment was even more so. Advocates of enclosure had long argued for monetary rent payment rather than tithes on the premise that it led to improved agricultural productivity, so many enclosure acts also stipulated tithe commutation on newly enclosed lands—a double blow to the agricultural laboring classes. The Tithe Act of 1836 aimed to sweep away remaining tithes for good, changing "all uncommuted tithes, portions and parcels of tithes . . . and prescriptive and customary payments" to an annual adjustable monetary rent pegged to the seven-year average price of wheat, barley, and oats.[17] Rents varied quite a lot from parish to parish, and some parts of parishes were exempt from rent and some were not, so the precise location of boundary lines became all-important—and surveyors were in high demand.

The impetus for all this surveying work did not register with Alfred at the time, of course: "Such ideas never entered my head. I certainly thought it a pity to enclose a wild, picturesque, boggy, and barren moor, but I took it for granted that there was *some* right and reason in it, instead of being, as it certainly was, both unjust, unwise, and cruel." He was to come to regret the role he unwittingly played in, as he saw it, the dispossession of people who had the least to begin with: "And to carry out this cruel robbery, how many of the poor have suffered? How many families have been reduced from comfort to penury, or have been forced to emigrate to the overcrowded towns and cities, while the old have been driven to the workhouse, have become law-created paupers?"[18] But at the time, he loved the work, from the long days of roaming beautiful countryside to the satisfying precision of the applied mathematics and mapmaking of the craft.

It was a peripatetic existence and the perfect job for a budding naturalist beginning to take more serious notice of the world around him, with a bit of anthropology to boot. The next year or so was spent in Bedfordshire, the county just north of London, where William landed a series of jobs mainly surveying parishes for the commutation of tithes. Barton, Turvey, Silsoe, Soulbury, Leighton Buzzard—they stayed at small country inns or boarded with local families, Alfred periodically walking the twenty or thirty miles back to Hoddesdon for short holidays with their parents. There was the Coach and Horses public house in Barton-le-Clay, Bedfordshire (where Alfred would often "sit in the tap-room with the tradesmen and labourers for a little conversation or to hear their songs or ballads, which I have never had such an opportunity of hearing elsewhere"[19]), the Tinker of Turvey in the nearby Bedfordshire village of Turvey, Mr. Carter's inn in Silsoe, and others.[20] They worked long days, rising early for breakfast and heading out with surveying equipment in tow as well as "a good supply of bread-and-cheese and half a gallon of beer" for a hearty lunch under a hedge. Part of this demanding but satisfying routine for William was to enjoy a smoke after lunch, pulling his pipe from his pocket. Alfred figured he ought to take up smoking, too, joining his brother in a few puffs until one day he overdid it: "I had such a violent attack of headache and vomiting that I was cured once and for ever from any desire to smoke"—in retrospect probably a good thing for him and for science![21] Besides learning the surveying trade (and not to smoke), Alfred remembered that time as an awakening of sorts, as he began to take note of the local geology and devoured introductory books on mechanics and optics, part of the Library of Useful Knowledge series published by the Society for the Diffusion of Useful Knowledge. The brainchild, so to speak, of the whip member of Parliament Henry Brougham, the Society for the Diffusion of Useful Knowledge was founded in London in 1826 with the aim of producing inexpensive educational volumes on a range of scientific subjects for self-improvement—literature for the mechanics' institutes.[22]

Geology loomed large: he was in a borderland of deep time again, where some one hundred million years of the Mesozoic era unfold across the eroded, terraced landscape of central Bedfordshire—not that he could have known this, yet the landscape spoke of eons. The Upper Cretaceous chalky hills above Barton, pushing 600 feet (182 m)—where William showed Alfred fossil *Gryphaea*, a genus of extinct oysters, and belemnites, the oblong stony thunderbolts of his childhood—give way to lower and earlier Cretaceous formations just to the north around Ampthill, near Silsoe. Ampthill in turn sits atop a four-hundred-foot sandstone ridge and looks out over the broad Marston Vale to its north, a gently rolling plain of still older fossil-rich clays of the Upper Jurassic. The compact town of Bedford lies just at the north end of the vale, where the River Ouse has sliced even deeper into the past, to mid-Jurassic limestone less than one hundred feet above sea level composed of distinctive microspheroids of calcium carbonate.

He remembered the roadworks further south near Dunstable where workmen were excavating the chalk to improve the slope for the cross-country Holyhead Road that came in from the northwest to ascend the Dunstable Downs (known as Chalk Hill today) and on into town via Watling Street. William had procured a nice specimen of chalk at the excavation, and he and a fellow surveyor they met devised a way to measure its specific gravity, which fascinated Alfred: "This little experiment interested me greatly, and made me wish to know something about mechanics and physics." Later, in Soulbury, where they boarded in a red-brick schoolhouse with the schoolmaster and his sister, he could not help but notice the large limestone boulder right out front, a local landmark still there today at the same intersection.[23] In his autobiography he surmised (correctly) that because this type of rock was not found in that region, it had been carried scores of miles by glacial ice and deposited

School house and Boulder Stone, Soulbury.

Soulbury's glacial erratic, an evocative geological traveler.

there. There were many so-called glacial erratic boulders like this dotting the countryside. He later mused that when he and William had resided in Soulbury, the locals would have attributed the rock to some legend of old, as the existence of ice ages and continental glaciers had not yet been discovered.[24] Which explanation would the villagers have found more incredible? To Wallace the answer was clear: he was exhilarated by scientific reasoning and method, and he was living at a time of breathtaking scientific advances and insights, not least those of the geologists who were at that very moment pushing back the limits of time as they mapped and named the alien topography of the earth's past. Much of that work was unfolding right there in the borderlands and in nearby Wales, where geologists Roderick Murchison and Adam Sedgwick had delineated what they had dubbed the Silurian and Cambrian periods in 1835 and would announce the Devonian (named for Devon, England) in 1840.

Yes, geology was absolutely fascinating but so were other fields he longed to learn about—botany perhaps foremost. In his rambles he noticed nature more and increasingly recognized how little he knew—not even the common names, let alone the Latin ones, of the wildflowers, trees, and shrubs he encountered daily. He recalled marveling as a boy back in Hertford when a local lady announced finding a rare plant in the neighborhood, *Monotropa*. It was no doubt *M. hypopitys*, variously called Dutchman's pipe, yellow bird's-nest, or false beech-drops, a pale, waxy zombie of a plant, lacking chlorophyll and making its living by parasitizing certain soil fungi. "How nice it must be to know the names of rare plants when you found them," he thought. He was to do more botanizing a bit later, after a nine-month detour as an apprentice watchmaker. The surveyor he and William had met in connection with measuring the specific gravity of the Dunstable chalk, Mr. Matthews, turned out to do surveying on the side, his main profession being watch- and clockmaking. With survey work slacking off, it was decided that Alfred might try his hand at that craft, and he boarded with the Matthews family at Leighton to learn. It was a good experience but far from satisfying. His prospect went from the green and rolling Bedfordshire countryside to the portal of a loupe, taking the measure of cogs and springs rather than fields, meadows, and woodlands. He left before the obligation of a formal apprenticeship would have kicked in to return to surveying with his brother—what he later regarded as "the first of several turning-points of my life." It was autumn of 1839, and Alfred was just a few months shy of seventeen years old.

———

The peripatetic lifestyle of a surveyor continued, moving back and forth across the borderlands, sextant, compass, rod, and chains in tow. With Kington as their base, over the next two years they surveyed parishes and estates in Shropshire, England, and in Radnorshire (now Powyshire), Brecknockshire, and Glamorganshire in Wales. A small, pleasant market town between the River Arrow and its tributary, the Gilwern Brook, Kington was a between place, lying in western England for the past millennium or so but even longer in eastern Wales—found, as it is, on the west side of Offa's Dike (Clawdd Offa in Welsh), the great eighty-plus-mile-long earthworks marking the England–Wales border that was begun perhaps as early as the 500s. It is also in the shadow of the prominent Hergest Ridge astride the England–Wales border now immediately to the west. Alfred makes no mention of visiting, but the proximity of the nearly 1,400 foot (426 m) elongated ridge of fossil-rich Silurian rocks, sprinkled with glacial erratic boulders, was surely irresistible to such geology buffs.

Between explorations he wrote family and friends, celebrating the advent of the penny post in verse in one letter to John—"Hurrah, Hurrah for the Penny Post | For now we may write like fun | And not feel a shock | when the Postmans knock | Proclaims that a letter is come." His letters to John and his childhood buddy George Silk were exuberant, with evocative accounts of his countryside rambles, gossipy commentary, and rapid-fire questions pumping them for news. To George, back in Hertford, he wondered how Mr. "Crut'll" gets on ("Has he any more young 'uns yet?") and if there are any pretty girls in town ("There are a pretty fair lot here—I suppose you will be looking out for a *wife* soon"). He gave comical accounts of life with his landlord and landlady, "Alderman" Wright the gunmaker and his wife: he rotund and slow talking and she superstitious and talkative, with "a pretty considerable tarnation long tongue."[25]

The surveying all this time was enlivened by his growing interest in understanding his geological surroundings, from curiosities like the igneous and steep-sloped Stanner Rocks (now named a National Nature Reserve, home to many rare plants) to the striking iron oxide–stained Old Red Sandstone formations so common in South Shropshire, now understood as Devonian deposits with cross-bedded layers so clear they evoke great sand dunes frozen in time (which they are). This sandstone is also the rock of the lofty Brecon Beacons, the highest peaks in South Wales. Ascending the massif, Wallace was surprised by its striking sheared-off twin summits. He could only make sense of it much later. Reading history in the Book of Landscape was to take time, but learning Lyellian ABCs, the basics of reading the history of landscape so compellingly articulated by the distinguished geologist Charles Lyell beginning in the 1830s, starts just this way, with learning to see: we need to perceive pattern to try to infer process since we cannot ask questions about what we do not even notice. He would discover Lyell soon enough. To William's likely frustration, surveying around places like Ludlow, Shropshire, went at a snail's pace as Alfred could not resist inspecting each and every rock outcrop they came across. He recognized that the Ludlow area was special geologically, a boundary zone between two great formations: "At every bit of rock that appeared during our work I used to stop a few moments to examine closely, and see which of the formations it belonged to."[26] Indeed, had he been able to spend much time in Ludlow he would probably have realized that he could learn the local geology by "reading" the very buildings, as virtually all the construction materials of the town—from bricks, blocks, and ornamental stone facades to roof tiles, flagstones, and curbs—consisted of rock quarried from the surrounding formations.[27]

He found Welsh culture, history, and especially Gaelic fascinating, too, even attending chapel not to receive spiritual edification but to hear the "rich and expressive language" intoned by the minister. These interests sometimes intersected with geology, in a sense, as when he made a pilgrimage to the headwaters of the River Llia far beyond the Vale of Neath where the large sandstone slab Maen Llia stood sentinel over the surrounding valleys. No glacial erratic, this was an imposing monument erected by some ancient people. Just who, and when, was lost to the mists of time: maybe the pre-Roman Silures or their neighbors the Ordovices, ancient Celtic tribes who lent their names to the Silurian and Ordovician periods.[28] Maen Llia stood in an ancient borderland. "These strange relics of antiquity have always greatly interested me," Wallace later wrote, "and this being the first I had ever seen, produced an impression which is still clear and vivid."[29]

The brothers eventually landed in South Wales at Neath, Glamorganshire, only some thirty-nine miles from Alfred's birthplace of Usk as the rook flies (or the surveyor's chain runs), crossing the River Usk en route and celebrating his (relative) homecoming in verse:

> From Kington to this place we came
> By many a spot of ancient fame,
> But now of small renown,
> O'er many a mountain dark and drear,
> And vales whose groves the parting year
> Had tinged with mellow brown;
> And as the morning sun arose
> New beauties round us to disclose,
> We reached fair Brecon town;
> Then crossed the Usk, my native stream,
> A river clear and bright,
> Which showed a fair and much-lov'd scene
> Unto my lingering sight.[30]

The next two years in Neath, from late 1841 through late 1843, saw something of a transformation in Alfred Russel Wallace. He boarded for much of that time at the Bryn-coch (Red Hill) farmhouse just north of town, his landlord one David Rees, a hale and hearty Welsh farmer who was also bailiff of the Duffryn estate. At nineteen and twenty years old, it was a time of growing confidence in himself, not just in terms of his interests and knowledge (with a snowballing appetite for more) but in his desire to share that knowledge. He later pointed to this crucial period himself: "I have some reason to believe that this was the turning-point of my life, the tide that carried me on, not to fortune but to whatever reputation I have acquired."[31]

In the sometimes lengthy periods between surveying jobs, with his brother off trying to drum up work among the parishes of the district, Alfred was left to his own devices. His interests were broad: besides geology he became interested in astronomy, constructing his own telescope to admire the lunar landscape and Jupiter's satellites, became expert in the use of a pocket sextant and trigonometrical survey techniques, and increasingly turned to botany, "more and more the solace and delight" of his countryside rambles and more and more raising his awareness of, as he put it, "the variety, the beauty, and the mystery of nature as manifested in the vegetable kingdom."[32] He acquired a cheap introductory botany book published by the Society for the Diffusion of Useful Knowledge and started pressing plants for a home herbarium, identifying them as best he could. Itching to know more, an ad for John Lindley's *Elements of Botany* caught his eye as he thumbed through one of Mr. Rees's copies of the *Gardener's Chronicle*, and he decided to splurge and order a copy, only to find it was more academic treatise than handy field guide. The friendly bookseller, Charles Hayward, took pity and lent him a copy of Loudon's popular *Encyclopaedia of Plants* so he could copy out the key characters of the British species into the wide margins of his Lindley volume, making his own identification guide.

Around this time he also acquired a copy of William Swainson's *Treatise on the Geography and Classification of Animals*. His annotations of both Swainson's *Treatise* and Lindley's *Elements* are telling. Swainson, a noted ornithologist, was a fan of William Macleay's short-lived "quinarian" system of classification, which took a quasi-mystical approach to ordering (read: shoehorning) species into nested sets of five groupings, each arranged in a circle. Macleay and his followers, such as Swainson, thought his scheme was "natural," in the natural theology sense of getting us closer to the imagined plan of the Creator. Wallace strongly disagreed. "A most absurd and unphilosophical hypothesis!!" he declared in one of his many marginal annotations, continuing: "To what ridiculous theories will men of science be led by attempting to reconcile science to scripture!" Despite this failing, the Swainson volume contained a wealth of valuable information, as Wallace's many other annotations show. As for Lindley, it became the go-to manual for Wallace's newfound passion of botany, and in a show of youthful enthusiasm, he copied lengthy quotes from Darwin's 1839 *Journal of Researches* onto the title page: just as a person who understands details of musical notes, melody, and harmony can better appreciate a composition as a whole, "I am strongly induced to believe that . . . so he who examines each part of a fine view may also thoroughly comprehend the full and combined effect. Hence a traveller should be a *Botanist*, for in all views plants form the chief embellishment"—with Wallace's emphasis on "botanist."[33]

He learned from his mother that William was not pleased with his growing plant mania and in fact dismissed it as a waste of time—probably after William found out how much his brother spent on that Lindley book: ten shillings and six pence, nearly thirty-two pounds in today's money and about two days' wages then! He was irritated that his sister Fanny had commented on his slacker botanizing, too, provoking a defensive response: "However much you may think I may waste my time on Botany & such like studies," he huffed, "it would give me much pleasure to hear of your taking to them—for it is impossible to express the pleasure you would derive from them in going to a tropical country." He went on to wax rhapsodic about the flora and fauna of the faraway tropics, Guyana in particular, and extolled the "permanent pleasure," "interesting occupation," and even "delightful interest" botany afforded. And besides, he lectured his sister, "Some knowledge of it is beginning to be considered universally as a necessary portion of a good education."[34] Their father, Thomas, thought this humorous and copied it out in a notebook, noting it was from Alfred "to his only Sister in defence of the science of Botany" and also coming to Fanny's defense: "NB It should be observed that the Sister did not despise the science of Botany as an unworthy pursuit, but wished her brother to set apart a portion of his leisure hours to the study of the modern Languages which she had found so useful in travelling on the Continent."[35]

Alfred's letter shows that, beyond merely expressing his passion for plants, he was beginning to think big, wistfully imagining the botanical beauties of the tropics. His mention of Guyana of all places is curious. Fresh in his mind was probably Charles Waterton's popular *Wanderings in South America* (1825), which recounted the naturalist's explorations of Guyana (then British Guiana), where he managed his uncle's estates near Georgetown—a work found on the shelves of the Neath Library at the time, according to a catalog compiled in 1842.[36] He could have easily named Brazil instead, having by then read Darwin's account of the *Beagle* voyage—also in Neath Library—in which Darwin rhapsodizes about tropical South America.

Although Alfred does not say much about it, Neath proved to have remarkable cultural, including scientific, assets for a town of its size. Besides a library of some thirty-seven hundred volumes that included an impressive selection of scientific texts, reports, and periodicals, there was a Philosophical and Literary Society established in 1834 (featuring a library and a lineup of notable lecturers), and about the time Alfred and William arrived in town, plans were already underway for a mechanics' institute—plans that came to fruition in October 1843, when the institute opened in a room in the Town Hall.[37] The neighboring town of Swansea, just a few miles away, boasted even greater intellectual capital as the home of the Royal Institution of South Wales, as well as having a Literary and Scientific Society and mechanics' institute of its own, each with a museum and library. Members of the Neath societies were granted access to those in Swansea, and Alfred took advantage of the perk.

It is no coincidence that this same period marks Alfred's earliest known writing and lecturing efforts. One of these was on his increasingly favorite subject of botany, prompted by being bored to tears "by a local botanist of some repute," who gave a lecture that "seemed to me so meagre, so uninteresting, and so utterly unlike what such a lecture ought to be, that I wanted to try if I could not do something better."[38] The dull speaker was surely James Ebenezer Bicheno, botanist and former secretary to the Linnean Society of London, who had relocated to the area as co-founder of the nearby Maesteg Ironworks. Bicheno was more successful as a practitioner than a communicator of science. Wallace himself as yet had little experience as a speaker, but this draft lecture shows him in fine-spirited (and opinionated) form, passionate in his commitment to science and his educational ideals.[39]

Another indication of his growing confidence was a letter he wrote around this time to the polymathic British inventor William Henry Fox Talbot, pioneer of early photography. Having read a paper by Talbot titled "On the Improvement of the Telescope" in the *Reports of the British Association for the Advancement of Science* for 1843, Wallace sent the inventor a paper proposing original technical solutions to the manufacture of large flat and curved telescope mirrors using mercury—ideas that were ahead of the technology of the time but anticipated modern electroplating to coat mirrors as well as spinning liquid-mirror telescopes.[40] It is not clear if Talbot ever responded, and Wallace's paper does not appear to have been publicly presented or published, but the initiative Wallace took here is noteworthy in showing how even at that age—and given his relative poverty and lack of formal education—he did not shy away from reaching out to well-established figures if he felt he had something of intellectual value to bring to their attention, much like his later efforts to communicate to Charles Lyell, then the leading geologist of Britain. This reflects a distinctive personality trait in Wallace: he was always willing to engage others in discussion or debate whenever he felt that he had a firm grasp of the matter, regardless of reputation or social standing. This quality reflects something of the Owenite in Wallace—an emphasis on the power of reason with a concomitant deemphasis on class or social standing.

This does not mean, however, that he was immune from the arrogance of youth, and a certain condescending smugness, otherwise so uncharacteristic, is certainly detectable in some of his earliest writing. Around this time he also penned—and panned—"The South-Wales Farmer," an article sent on spec to an editor for a London magazine.[41] Based on his years living in rural Wales with and among Welsh farmers, he felt he was in a good position "to compare the high-class farming of the home counties [England] with that of the ignorant Welshmen." You see where this is going. The piece was unflattering, to say the least, a patronizing condemnation of the supposed ignorance, poverty, superstition, loose morals, and incompetence of the hill farmers. He may have been attempting humor of the black sort, wryly noting that these farmers viewed weeding their fields as "an unnecessary refinement" or how the chinks in the rough plank flooring of their flea-infested houses "conveniently" allowed everything going on in the rooms above to be heard "and much seen," but he comes off as rather obnoxious to modern sensibilities.

The premise for the piece seems rather insensitive, too, for a proto-Owenite. He figured that since, in light of the recent "Rebecca disturbances" in Wales, this was a class of people that "excited much interest," he would provide their portrait for curious readers. The Rebecca riots, or uprisings, were so called because the rioters disguised themselves as women, representing the biblical Rebecca of Genesis 24:60: "And they blessed Rebekah and said unto her, Thou art our sister, be thou the mother of thousands of millions, and let thy seed possess the gate of those which hate them." The uprisings thus started in 1839 as attacks on tollgates but soon escalated with protests (some deadly) raging through 1842 and 1843 against the dire poverty of rural Wales and the perceived ruinous government and church policies responsible for it.[42] One might have thought Wallace would be more sympathetic to the downtrodden farmers. He did acknowledge their virtues in his conclusion, a mere two paragraphs to cap fifteen pages of uncharitable criticism. The editor rejected the article, and Wallace did not pursue it any further—which was just as well.

Wallace's other early writing efforts seem much more in the generous and optimistic Owenite spirit, bearing on his commitment to science and the pursuit of knowledge for self-improvement. His first, "On the Best Method of Conducting the Kington Mechanic's Institution," was submitted for an essay contest held by the institution and won a prize.[43] The essay, which was included in a history of Kington published in 1845, is noteworthy for his focus on building a first-rate scientific library, starting by limiting general literature so that as many publications in "Natural History and Natural Philosophy" as possible could be obtained. Subscribe to actual scientific reports, Wallace states, like those of the British Association meetings, and procure reference works such as the *Cabinet Cyclopaedia* and Loudon's *Encyclopaedia of Agriculture*. As for books, the collection

THE WELSH RIOTERS.

"Rebecca and her daughters" attacking a tollgate, 1843.

should include a broad selection of standard works: Lyell's *Principles of Geology*, Murchison's *Silurian System*, Lindley's *Natural System of Botany*, Kirby and Spence's *Introduction to Entomology*, Brande's *Chemistry*, Humboldt's *Personal Narrative*, George Combe's *Constitution of Man*, and more. Once these are in place, the institute should actively "promote an interest in scientific and literary pursuits." Lectures are a must, in addition to which members might take turns reading original papers or initiating a scientific discussion, with recommended reading. In this way intellectual seeds are sown that may bear the fruit of great discovery and invention: "All who have become really great have had the desire, and in some degree at least, the means of obtaining knowledge." Wallace the idealist saw a great social good in this. Promoting individual happiness as well as benefiting the community, nurturing the next Herschel or Watt, would repay society a hundredfold. With this conclusion Wallace came full circle with the Owenite message of the essay's epigraph: "Knowledge is power."[44]

That message certainly resonates with the subject of another of his essays, "The Advantages of Varied Knowledge," of which only fragments survive. Thought to be from a lecture dating to about 1843, this essay is more reflective, offering a philosophical perspective on the pursuit of knowledge. We have an almost moral imperative, Wallace argues, to strive to understand the world around us.

In a curious anticipation of threads of Wallace's later spiritualist arguments yet also using language with almost religious overtones, he suggests we are not "fulfilling the purpose of our existence while so many of the wonders and beauties of the creation remain unnoticed around us" and while so many laws of nature are "unknown and uncared for" by us.[45] We ought to know, and do, better, he urges: we do not lack the power to seek knowledge but the will. Indeed, we owe it to past generations; how *could* we "allow this great store of mental wealth to lie unused, producing no return to us, while our highest powers and capacities rust for want of use?" And what will be gained if we apply ourselves? Well, can we doubt for an instant that by improving the "nobler faculties of our nature in this world," we will be all the better equipped "to enter upon and enjoy whatever new state of being the future may have in store for us?"[46] Heady stuff for a twenty-year-old surveyor's apprentice.

———

Thomas Vere Wallace died in Hoddesdon in April 1843, aged seventy-two, and the already precarious financial straits of the family only got worse. They hung on for a time but ultimately had to disperse: In the following year their mother, Mary Ann, moved to Isleworth to take up housekeeping for a well-off family, and Fanny emigrated to the United States to teach at an Episcopal college at Montpelier Springs, near Macon, Georgia. Young Edward had to leave Mr. Perry's Academy and was apprenticed to a trunk maker in London. John was still working in London, and William remained in Neath, struggling to keep his business afloat—most of the tithe commutation surveys were complete by then and work was scarce. Change finally came for Alfred too: his stay in Neath came to an end as the stubbornly slow surveying business forced William to lay him off in late 1843. He spent Christmas of that year visiting Hoddesdon in preparation for a move to London, where he stayed with John for awhile as he looked for work, thinking he would apply to teach at a school somewhere. He had just turned twenty-one, and while he came into a small inheritance of one hundred pounds, it was an uncertain time. Though he could not know it then, he would soon return to Neath but not before yet another chance occurrence altered the course of his life dramatically. He had been traversing literal borderlands for some years by now; what awaited him next was an introduction to borderlands of a more philosophical kind as he tested the boundaries of orthodox and unorthodox science.

CHAPTER 3

BEETLING AND
BIG QUESTIONS

ALFRED RUSSEL WALLACE'S prospects were not so bright at the start of 1844, his twenty-first year. Throughout the ups and downs of the past seven years crisscrossing the Welsh–English borderlands, he had regularly returned home—not to the place of his birth nor the house of his fondest childhood memories in Hertford but to small red-brick Rawdon Cottage in the tiny village of Hoddesdon: to where his parents were, near his sister Fanny. Where his parents were was home. It was where he went for respites and festive holidays, to get treated to new clothes once a year, to be nursed back to health when he became deathly ill with a lung infection after falling into a bog that time, taking two months to recover with generous doses of parental affection (surely more beneficial than the leeches prescribed by his old-school doctor). It was where he could again hear the cadence of his father's voice reading aloud Scott or Shakespeare or some other mainstay or that of his mother as she read her favorite stories from the Bible to the family on Sundays. But his father was now dead, Rawdon Cottage theirs no more, the family's dispersal complete. His mother and brothers were scattered from London to Wales, and by late summer his sister was plying the waters of the Atlantic aboard the *Quebec*, heading to the United States.

William was unable to find enough surveying work to keep his brother on, so after a final bittersweet Christmas together in Hoddesdon, the newly unemployed Alfred headed to London to stay with John and seek work, the first time he truly had to fend for himself. He was no doubt anxious; what were his prospects? Wallace says nothing about the dire economic and social situation then affecting people like him, working people of little means. The mid-1840s were a time of inflated food prices, thanks to the so-called Corn Laws that

THE TORY GOVERNMENT ON A SEA OF TROUBLES.

Satire of the social movements creating a "Sea of Troubles," 1843.

shielded the privileged landowners and large farmers from foreign competition in cereal grains, sparking widespread protests by the Anti–Corn Law League and other groups. The Chartist movement, campaigning for economic relief and the political reform that the Great Reform Act had failed to deliver on, had reached a crescendo in 1842 with strikes and violent protests across the country, leading to swift and at times violent response by the government, which imprisoned the movement's leaders. As Wallace came of age in January 1844, the fiery Chartist leader Thomas Cooper, convicted of seditious conspiracy, was just then completing his epic poem *The Purgatory of Suicides* from jail, memorializing the radical movement and its suppression. Meanwhile, the ever-tightening squeeze of the enclosure laws, as inexorable as a constricting snake, drove more and more people to flee the rural districts for the cities, a slow suffocating depopulation of countryside. What awaited them

were sweatshops and squalid housing lacking in sanitation, misery memorialized by Charles Dickens.

How did the twenty-one-year-old Wallace see himself getting on? He was
nothing if not resilient, though this was not among the positive personality
traits he saw in himself some sixty years later when he offered "Remarks on
My Character at Twenty-One" in his autobiography. There his perceived flaws
outnumber the assets nearly three to one! While on the one hand he felt he
had good reasoning skills, a talent for clear writing, and well-developed senses
of both aesthetics and justice, on the other hand, he dinged himself for lacking
wit and any facility for music, mathematics, languages, or public speaking. Oh,
and he also suffered from lack of confidence, and the list continues with his
"want of assertiveness and physical courage," "delicacy of the nervous system
and of constitution," and, topping it all off, "general disinclination to much
physical exertion, physical or mental." He makes himself out to be quite the
dullard in this portrait of mediocrity, but his rather severe self-assessment
reflects another trait, one that can be both virtue and liability: modesty to a
fault. Also, being a glass-half-full kind of guy—should he not have included
"optimism" in the positive column? He did see virtue in adversity: the very
traits causing his shyness and reticence may have caused some "unpleasant-
ness," but on balance they were beneficial, allowing him to become immersed
in "wild nature" among "uncultured man," where the opportunities to read,
study, ponder, and reflect led to flashes of insight.

This self-assessment was a rather unflinching take on himself in 1844 as he
prepared to find gainful employment. With surveying and mapmaking, the
profession he had the most experience with, then in a slump, he thought he would
try his hand at teaching. His first interview did not go well when he was asked
to translate a passage from Virgil. He probably winced at memories of "blundering through" his forty or fifty lines of *The Aeneid* under Old Crut'll's baleful
eye, where "it was all a matter of chance whether we got through well or otherwise." In this case he did not get through. But his second interview went much
better, and he was soon offered a job teaching reading, writing, and arithmetic
plus a bit of surveying and drawing at the Collegiate School in the town of
Leicester, where the kindly Reverend Abraham Hill presided.

He arrived around Easter 1844. It was a comparatively happy time: boarding
with the Hills, he had a comfortable room where a fire was lit every afternoon
in winter, and he had time enough to prepare for his classes and pursue his
own interests. He probably was not thrilled when the reverend discovered that
he knew a little Latin and asked him to take the bottom class, but he soon

brushed up on the basics to stay a step or two ahead of his students. Ever one for self-improvement, Wallace took up the offer of the headmaster, who had studied mathematics at Cambridge, to give him lessons in algebra, trigonometry, and calculus. He only got so far in the latter before the "almost trackless wilderness" of the equations tied him in knots. He badly wanted to understand higher mathematics but just could not make sense of it. He later reflected that no matter how much he tried, under the tutelage of Reverend Hill or anyone else, he probably would never have gained much facility with mathematics— comparing this with musical talent, he took a "you've got it or you don't" attitude. There is an element of determinism in Wallace's thinking here, but it was a time when phrenology—the idea that the bumps and contours of one's cranium reflected differing degrees of development in the various lobes or regions of the brain beneath—was all the rage. In the further belief that tastes, talents, and aspects of character mapped to the brain, a bump (an imagined enlarged lobe) here or indentation (a wimpy one) there could only mean these traits were also well developed or underdeveloped. Like many of his time, Wallace accepted these notions, and they likely fueled his interest in *mesmerism*, as he was not only fascinated with books like George Combe's influential *Constitution of Man* but discovered that he had a talent for "mesmerizing" people himself. Mesmerism—hypnotism in modern terms—is named for eighteenth-century German physician Franz Mesmer, who studied a phenomenon he called "animal magnetism," posited to be an invisible natural force (*Lebensmagnetismus*) inherent to all living organisms. After attending a lecture and demonstration at the mechanics' institute in Leicester, Wallace tried his hand at the art. To his surprise he had some success, and with further encouragement by the Reverend Hill, he soon found he could mesmerize some of his pupils, inducing focused attention and getting them to respond to suggestion. It was uncanny; it seemed to Wallace that there was indeed some psychic force at play. At the time the magnetism analogy was irresistible: Who could understand the invisible yet palpable force seemingly emanating from certain (magnetized) metals, affecting others? Or the mysterious relationship between magnetism and electricity (then called galvanism)? Here *something* seemingly emanated from one mind to affect another. In fact, the parallel between mesmerism and metallic magnetism and galvanism was more than analogy. Many believed there was a direct correspondence and would even use different kinds of metals in "mesmeric" experiments to channel energy or animal electricity. This is the very subject of Wallace's earliest known publication, a letter to the editor in the 10 May 1845 issue of the London magazine the *Critic*, which had a regular column titled "Journal of Mesmerism." Wallace writes: "Seeing in your journal an account from a correspondent, of some mesmeric experiments with metals, I beg to send you a brief account of a few experiments I have made on the same subject, the results of which, however, do not agree with those of your correspondent."[1]

His fascination with mesmerism and phrenology no doubt contributed to the spiritualist convictions Wallace later famously espoused, but more importantly, it surely contributed to his iconoclastic streak. If this trait may have sometimes led him astray in our view today—on spiritualism, say—it also led him to great scientific insights, as he put more faith in his own power of reasoning and observation than mere weight of authority for a mainstream view. As he expressed in his autobiography, the real significance of his little adventure in mesmerism was that it convinced him "once for all, that the antecedently incredible may nevertheless be true."[2] Think about that wonderful turn of phrase: what once seemed incredible may come to be recognized as true after all—not so much owing to more and better data, I might add, but more often by seeing the question with new eyes. The episode further convinced him that the pronouncements of "scientific men should have no weight whatever against the detailed observations and statements of other men . . . who had witnessed and tested the phenomena." Valuing direct experience and a willingness to challenge others, authority or not, if he thought he was right, were key elements of Wallace's personality. We see an early expression of these traits in that letter to the editor of the *Critic*.

It is also interesting to consider how the determinism suggested by phrenology squares with Wallace's strongly held beliefs in the human capacity for improvement and the beneficial effects of environment. Wallace's belief is actually not far from the modern understanding that we are born with certain capacities, most average but some with rather greater or lesser acumen for different sorts of skills—music, for example, or drawing, or quickness of learning, or facility with languages. In some cases hard work might compensate for average ability, and one can become quite expert; in other cases not so much. But regardless, poor environmental conditions can keep us from reaching our full potential to begin with, whatever that is, while healthy living conditions and opportunities for mental and physical development can only help.

The Leicester Collegiate School certainly seemed to provide just that. It was a well-run school and a harmonious working and living environment. Later he recalled long excursions with the schoolboys and teachers, once traveling overland by coach to see the evocative nine-hundred-year-old

Kenilworth Castle in Warwickshire and another time going by coach and canal, emerging from a long dark canal tunnel to the green, picturesque countryside around Wirksworth, Derbyshire—Lutudarum to the Romans, the center of their lead-mining works. There the explorers would have climbed one thousand-foot Crich Hill, where, from the site of the present-day war memorial tower, they would have had a commanding view of the Derwent Valley and eight counties all around, as well as the adjacent limestone quarry, started by the Romans.

More typically, Wallace and a few of the students would embark upon local rambles out in the beautiful Leicestershire countryside. A favorite was Bradgate Park, just northwest of town on the edge of Charnwood Forest, "a wild, neglected park with the ruins of a mansion, and many fine trees and woods and ferny or bushy slopes." Once a medieval deer park, the 850-acre Bradgate Park is indeed wild looking: a rolling landscape of windswept hills studded with jagged rock outcrops. The ruin Wallace recalled was not the Old John tower, which is a ruin of a fake ruin—a folly built in the 1780s by the earl of Stamford to look like a romantic old ruin high on a hill in the park. Rather, it was the genuine ruin of the once stately Bradgate House, completed around 1520. One of the earliest of the great English country estates to be built entirely of red brick, Bradgate was the birthplace of Lady Jane Grey, whose grim fate poignantly adds more than a touch of melancholy to the romantic beauty of the tumbled-down mansion.

Frequenting Bradgate, Wallace surely noticed that the local geology was quite different from that of Kington or Neath. It was another kind of borderland of deep time, with billion-year-old volcanics and crystalline diorites—hard whitish quartz and feldspar heavily peppered with jet-black minerals—thrust up and tilted, exposed on the hills of the park amid younger rocks of the early Paleozoic. No fewer than four rock formations of differing age and composition lay close to the Bradgate House ruins, along a great fracture in the earth's crust that modern geologists have dubbed the Groby Reverse Fault. Much of this geological landscape is well hidden by a coating of glacier-deposited gravelly till, but the outcrops and quarries dotting the area reveal distinct rock types that speak of cataclysmic events of the distant past. If Wallace puzzled over the odd sausage-shaped markings of tightly packed concave layers in the sandstones and mudstones of Bradgate, he did not mention it in his writings. But he must have noticed them and would have been fascinated to learn that these mysterious trace fossils, dubbed *Teichichnus* in the 1950s and interpreted as the remains of shallow marine worm burrows, came to be recognized as worldwide markers of the Cambrian period, the very dawn of the Paleozoic era.

He may not have noticed them, come to think of it, because he was just then gripped by a new obsession: beetles. A new world had recently been opened to Wallace by a kindred spirit he met in town named Henry Walter Bates. Two years Wallace's junior, Bates, like Wallace, was born to a well-read middle-class family, in Leicester. Like Wallace, too, his formal schooling ended about age thirteen when he became apprenticed to a hosiery manufacturer (the family business and the dominant industry in Leicester at the time)—but unlike Wallace, he had rather less interest in the family trade. The two shared a passion for the natural world. They met in the library of the mechanics' institute or the public library—where else?—and were soon swapping collecting exploits, prized finds, and recommended reading. Bates was an amateur entomologist with a keen interest in butterflies and beetles. Ogling Bates's sizable beetle collection, Wallace was doubly amazed—not merely at their diversity and beauty but in learning that Bates had collected all those beetles right there, in and around Leicester: "If I had been asked before how many different kinds of beetles were to be found in any small district near a town," he later wrote, "I should probably have guessed fifty or at the outside a hundred, and thought that a very liberal allowance." Yet amazingly: "I now learnt that many hundreds could easily be collected, and that there were probably a thousand different kind within ten miles of the town."[3] Wallace did not realize it at the time, but he had just been introduced to one of the largest group of animals on the planet: imagine, beetle species rival in number all other animal species *combined*. That means that even in a relatively species-poor region like the British Isles, beetles are a (relatively) abundant and diverse group. Today nearly forty-one hundred beetle species in over one hundred families have been recorded for the United Kingdom. Compare that to the fifty-odd species of tree and maybe a few hundred herbaceous plants.

Wallace was practically salivating, eager to stow his plant-collecting kit and gear up for beetling—though of course he never entirely gave up botanizing. But entomology was new and exciting, and more than a guide, Bates was a role model, having actually published short notes on insects he had collected in a new and respected scientific journal, the *Zoologist*, plus a fascinating letter to the editor asking whether he would admit "communications of a general critical nature" bearing on systems of classification. (Yes, the editor "would be obliged" to hear the opinions of his readers on the subject.)[4] Bates made equipment recommendations and turned Wallace on to James Francis Stephens's *Manual of British Coleoptera*, hefty with accounts of all the British species—Wallace got a good deal, too, paying the wholesale price through Reverend Hill's bookseller. Leicester is at nearly the geographic center of England, situated between two hilly districts in the broad valley of the sinuous River Soar with its network of myriad tributaries and branches, and Alfred and Henry roamed through town and country collecting on their free days. Wonderfully varied beetling habitat was to be found within a short ramble inside and beyond the town limits, with innumerable meadows, fens, parks, and pools in the valley, surrounded by an undulating jigsaw puzzle of woodlands and farmland, each neatly separated by bushy hedgerows—havens and highways for wildflowers and critters, including beetles.[5]

When they were not on the beetle prowl, nets, bottles, and beat sheets in tow, the two could perhaps be found in the library of the mechanics' institute or one of the other local societies. One of their friends was an officer in just about all of them, facilitating access. By the time Wallace landed in Leicester in 1844, the town also boasted a Literary and Philosophical Society and general news room, and an Athenaeum Society was established that following year. Only the Literary and Philosophical Society survives today (the venerable and vibrant Lit. & Phil.), but all played a collective role in the education and edification of townspeople from all walks of life in the mid-nineteenth century. If so many subscription societies seems a bit excessive, or redundant, in a town the size of Leicester (population just shy of thirty-nine thousand in 1841), that is because it was. But for good reason: their origin reflects the political and social divides of the time.[6] The Leicester Mechanics' Institute, founded in 1833, served not only as an educational outlet for artisans aspiring to the middle class, offering varied lectures and classes in the (old-school) three R's plus French, Latin, music, and drawing, but also as an increasingly fractious forum for advocates of political and social reform. In the words of one historian, by the mid-1830s "the poisonous divide in Leicester was not just about where you worshipped or who you voted for: even the place you met had a political significance."[7] Repeatedly attacked as "a mere political union" in the local press, the membership of the mechanics' institute declined as its more liberal middle-class supporters (including its founders) sought a fresh start elsewhere. Two years later, in 1835, they helped establish the Leicester Literary and Philosophical Society, for "the reading and discussion of papers on literary and scientific works, and the formation of a museum and library of scientific works."[8] Theological and political subjects were pointedly banned, and the president's chair was to be alternately filled from the Conservative and Liberal Parties each year.

The difference in philosophy between the two societies did not stop them from cooperating: the fledgling museum of the Lit. & Phil. was initially housed with the mechanics' institute in New Hall, a beautiful 1831 neoclassical building designed by local architect William Flint. (Today this building and the elegant adjoining semicircular edifice, originally a Dissenter's chapel, function as a lending library and adult education center, appropriately.) It soon outgrew this space, and when Parliament passed the Museums Act in 1845, making funds available for the establishment of public museums in larger municipalities, the Mechanics' Institute was one of three petitioners to the Leicester Town Council to support the Lit. & Phil.'s museum application, which was granted. The museum was established in 1849 (and is thriving today). Wallace and Bates had left town by then but were surely aware of the plan afoot—the Museums Act may have been derided by some as the "Beetle Act," indulging the interests of a bunch of nerdy insect collectors—but that is precisely what would have delighted the keen young entomologists!

Little did they know then that they would each return, years later, to give lectures themselves at the Literary and Philosophical Society, Bates in 1862 and Wallace in 1878. But in 1844–1845 Bates and Wallace were simply young men keen on beetles, and all the Leicester societies had subscription fees that were mostly beyond their meager means. At the time the impecunious Wallace and Bates could perhaps most easily afford the mechanics' institute, costing two shillings per quarter for "artisan" subscribers (compared with a pound or more for full membership). Although a bargain—at that time more library and less pulpit or political stage, with an impressive collection of well over two thousand volumes, plus some three dozen newspapers and periodicals—most of the books they were interested in could be found in the well-stocked public library, accessible for a smaller subscription fee. It was there that Wallace read several works that, he later wrote, "influenced my future."[9]

The first one he mentioned was Alexander von Humboldt's *Personal Narrative* documenting the great naturalist's travels in South America between 1799 and 1804—"the first book that gave me a desire to visit the tropics." This was a bit of an understatement: Humboldt's book had an electrifying effect on readers like Wallace (and Darwin). It was *the* travelogue of the time, chronicling an epic journey from the verdant eternal summer of the lowland tropics to the glacier-capped high Andes, observing, measuring, seeking data points enough that the dots would reveal pattern and pattern process: nothing less than new laws of nature were out there awaiting discovery. Nothing was outside Humboldt's purview: botany, zoology, geology, astronomy, meteorology, ethnology—all subjects, not incidentally, that excited the twenty-one-year-old Alfred Russel Wallace—pursued with as much an aesthetic as scientific eye: "Nature herself is sublimely eloquent," Humboldt declared in his famous *Personal Narrative*. "The stars as they sparkle in the firmament fill us with delight and ecstasy, and yet they all move in orbit marked out with mathematical precision." He inspired holistic, big-picture thinking, seeking interconnections, relationships, and the need to travel, explore, and personally experience the wonder and mystery. Wallace got a "desire to visit the tropics" all right.

Wallace also recalled reading "Prescott's 'History of the Conquests of Mexico and Peru,' Robertson's 'History of Charles V.' and his 'History of America,' and a number of other standard works." "But," he concluded, "perhaps the most important book I read was Malthus's 'Principles of Population,' which I greatly admired for its masterly summary of facts and logical induction to conclusions."[10] These volumes are telling—he may or may not have realized the resonance between the historical accounts of Prescott and Robertson and the Reverend Malthus's landmark essay on population: all dealt with conflict, struggle, and the displacement of some peoples by others. It is hard to imagine that such readings would not have further resonated with the climate of conflict he had been witnessing for the past half dozen-plus formative years in the English-Welsh borderlands: Rebeccaite uprisings and Chartist riots, strikes, enclosure, agitation of Dissenters and Nonconformists, and the acrimony of conservative Tories and liberal Whigs, not to mention the plight of the Irish then in the grip of the Great Famine, with millions forced to emigrate. The Malthusian doctrine would surely have lodged deeply in his mind, to resurface some dozen or so years later.[11]

Besides his voracious reading and collecting, Wallace delighted in his time teaching at the Collegiate School, evidently becoming very much part of the school community. There were regular rambles with the students and celebratory midsummer prizes (to which he contributed lines of verse), and he even wrote a comic play called *Guy Faux* for the end-of-year festivities, his humor shining through with funny anachronisms. It was all good: the fine school with its congenial and supportive headmaster, lectures and books and the mechanics' institute, reveling in "beetle-mania" with his newfound buddy Bates. He might have stayed indefinitely had not tragedy struck again.

Early in the new year, 1845, his brother William suddenly died—a complete shock to the family. He had been in London giving expert testimony as a surveyor for a parliamentary committee considering proposals for the route of the South Wales Railway. It was a time of rapid railway expansion fueled by ballooning financial speculation, more driven by opportunities to move goods than people: create more expedient trade and mail routes with Ireland and beyond and move coal and iron from South Wales to waiting markets.

A prospectus was first issued in 1844, proposing a line more or less following the southern Welsh coast, joining the Great Western Railway line from Bristol and Gloucester. One issue debated at that time was where to cross the River Severn above the treacherous Bristol Channel, and how: Should it be by bridge or tunnel, and should the crossing be closer to the channel at Fretherne where the river is wider but the route shorter or upriver a ways at Gloucester? Knowledge of the terrain and likely grades of the routes was important, all in William's bailiwick. But returning to Wales from London that January evening, he opted for—or could only afford—a third-class open car and caught a bad chill that quickly progressed to what was probably pneumonia. Alfred headed to Neath to sort out his brother's affairs.[12] He soon decided to take on running his own surveying business and even expand it to include basic architecture, construction and engineering, having convinced John to give up London and join him. By Easter he sadly but resolutely left Leicester behind, but his year-plus there was, he later reflected, "perhaps the most important in my early life."

––––––

The years 1845 and 1846 were busy for the Wallace brothers thanks to the railroad boom, and their business success happily allowed most of the remaining family to reunite. After first boarding with the family of photographer Thomas Sims on New Street for some months (there befriending Thomas Sims Jr., who was later to marry their sister Fanny), they rented a cottage near the old Norman-era Church of Saint Illtyd, where William was buried, at the north end of Neath overlooking the river and canal with a view to the valley beyond. First their mother and then, in autumn 1847, their sister Fanny joined them; Edward was now in Neath too, William having secured him a job in the pattern shop of the Neath Ironworks, much more to his liking than the London luggage maker. Fanny had been in the United States almost exactly three years, first as a teacher at the Montpelier Institute in Montpelier Springs, northwest of Macon, Georgia, and then as headmistress at a start-up school in Robinson Springs, near Montgomery, Alabama. Her voyage to the States, a twenty-eight-day passage, did not get off to a good start: the *Quebec* nearly wrecked in the English Channel twice, first when a much larger ship bore down upon it in heavy fog and then when it was driven perilously close to the French coast by fierce winds. If that were not enough, the captain fortuitously saved the ship from burning—following his nose to a closet, smoldering cotton burst into flames as soon as he cracked the door. But they made it to New York, then sailed aboard the *Exact* down to Savannah, Georgia, where she was unnerved by all the shops displaying coffins in their windows, the going commodity in that "unhealthy season." They hastened to Macon by rail and then bumpy stage-coach to the heart of Monroe County, deep in the piney woodlands of Georgia. She made friends; comfortable and happy, she gave lessons in music, arithme-tic, and English grammar twice a week to twenty-four young ladies and taught French in the evenings. But just half a year later, she traded Montpelier Springs for Robinson Springs further west, in Alabama, where she was recruited to help found a new school. Young and talented teachers fresh from the mother country were in demand.[13]

Fanny's letters home are full of optimism, hope, and some naivete: optimism for her prospects and hope to succeed in her new venture—and to persuade Alfred, John, and eventually the rest of the family to join her in this young country, where a real living could be made with talent and perseverance. But despite urging John and Alfred in letter after letter to move to Alabama, inten-sifying her pleas after William's death, she despaired of ever convincing them. John, as we will see, did eventually emigrate, heading to California, a genuine forty-niner. Alfred eventually made it to the United States too, taking a victory lap some forty years later and happily reuniting with his brother. We will get there, in chapter 13, but in the mid-1840s neither had any intention of leaving the homeland despite Fanny's pleading. Thinking she could coax her brothers across the Atlantic may have been a bit naive on Fanny's part, with their recently widowed mother to think about, but her real naivete lay elsewhere. The schools she was teaching at were built and maintained by slave labor, and while Fanny was no out-and-out apologist for the institution, she was also determined not to offend her hosts and employers and so seemed to view the lives of the enslaved through decidedly rose-colored glasses: They were happy, were they not, and grateful to be so well cared for? But as she watched a group of enslaved men, women, and children being taken to market one day, the tears that welled up in her eyes suggest she knew better.[14] She wished she could wave a magic wand and emancipate the slaves, she said. But then the imagined vengeance of the enslaved against "even the kind hearted masters" stayed her hand. "Vain are we mortals," she chastised herself, "always aiming at a height we seldom reach."[15]

In September of 1847, Fanny returned home, probably owing to homesick-ness with no prospect of the family joining her but perhaps also to growing heartsickness at the malaise of slavery. The letters from her family do not survive, so we do not know exactly why Alfred and John steadfastly refused to emigrate. But they were making a go of it in Neath; their business venture was doing pretty well, and their mother had come to live with them—hardly a time to uproot. By the time she arrived, Alfred and John had several business successes to their credit, and Alfred had not only immersed himself in the intellectual life of Neath and Swansea but steadfastly continued his self-education and beetling.

Alfred's first surveying success in his new locale stemmed from the railroad boom. He landed a well-paid job with a Swansea-based firm—two guineas a day plus expenses!—leading a crew to survey a route that largely followed today's A465 highway all the way up the Vale of Neath and east to the town of Merthyr Tydfil, a major center of ironworks on the edge of the Brecon Beacons. He was in his element out in the elements—long days of surveying, rain or shine, made longer by frequent entomological distractions (since he surveyed with collecting gear in tow and could not resist chasing down a fine beetle or butterfly). Though the job certainly benefited him at the time, he later came to realize what was behind the growing railroad mania: a stock market bubble, peaking in 1846. Only a small fraction of the 9,500 miles (15,300 km) of the new railroad proposed in a dizzying 260 separate parliamentary acts were completed (the Vale of Neath Railway among them, opening in 1851), with many speculators going bust when the bubble burst.[16] He had only a vague sense at the time that something is wrong with a system that permits such waste. Even worse was his firsthand lesson in the human costs of the tithe commutation.

He landed a commutation survey of the rolling eighteenth-century Gnoll Estate east of Neath, now a lovely one-hundred-acre public park. But he had no sooner finished and submitted a meticulously drafted map of the estate when he was told to go collect the taxes owed. Alfred despised this work, like a hated excise man of old, struggling in broken Welsh to extract cash from dirt-poor and confused subsistence farmers who spoke no English. He had to defend his work, too, as legal notice had to be given of the new apportionment maps, and public appeals sessions were held at which Alfred and his co-apportioners had to be present to answer objections.[17]

Far more satisfying were his and John's building contracts, including a small cottage for one client and a new building for the Neath Mechanics Institute on quiet Church Place. Co-founded in 1843 by the kindly industrialist William Jevons, owner of the Cwmgwarch Venallt Iron Works, the Neath Mechanics Institute had been meeting in the Town Hall, a building suggesting a Greek

---

TITHE COMMUTATION.

PARISH OF NEATH.

NOTICE is hereby given, that the MAP and DRAFT APPORTIONMENT of the PARISH of NEATH, in the County of Glamorgan, is deposited at the Office of Mr. Wallace, Surveyor, New-street, Neath, and can be seen by all parties interested therein.

The APPEAL MEETING will be held at the CASTLE INN, NEATH, on the 30th day of MARCH instant, at ten o'clock in the forenoon.

ALFRED R. WALLACE, } Apportioners.
DAVID REES.

A. R. Wallace "Tithe Commutation" legal notice, the *Cambrian*, 13 March 1846.

temple with a four-columned portico flanked by pediments. The ideals of the institute, one of at least eight established in South Wales in those days, was shared by the community, and many local shops closed at 8 PM on Tuesdays and Fridays (this was well before the standard eight-hour workday) to allow their employees to attend evening lectures and classes.[18] In August 1846, £600 was allocated for a plan for a new building, just a few doors up from Town Hall, and the Wallace brothers were awarded the contract to design it—Jevons had a high opinion of their late brother William and held Alfred in high esteem.[19] Jevons in fact regularly lent Alfred books from his extensive library and coaxed him into lecturing on elementary physics and other subjects at the Neath Mechanics Institute. At the time Alfred was a bit unsure of himself as a lecturer and wondered how effective he was (remember that list of his perceived deficiencies?). But imagine his delight when, some fifty years later, he received a letter from one Matthew Jones, "anxious to know if you were the same Mr Alfred Wallace that taught in Science Classes—Evenings—to the working engineers and Artificers of 'Neath' Abbey and 'Neath' in South Wales." Jones had high praise: "I benefited more while in your Class—if you are the same—Mr A Wallace—in practical Mechanics Thermo Dynamics Statics &c &c than I ever was taught at School and I have often wished I knew how to thank you or the teachers of the Neath Mechanics Institute for the Good I and others received from them"—the kind of note that warms the heart of any teacher.[20]

As good a job as he did in his lectures, Alfred was still more student than teacher—or was he both, as autodidact who taught himself as he read and

Neath Mechanics Institute, designed and built by John and Alfred Wallace in 1847.

studied and observed? Busy as he was with surveying and periodic construction projects, he was fully engrossed with books, beetling, and attending lectures in both Neath and nearby and larger Swansea, where there was a mechanics' institute, the Literary and Improvement Society of Working Men, and, most importantly, the Royal Institution of South Wales, where prominent and accomplished naturalists regularly lectured.[21] All the while he kept up his correspondence with Bates, back in Leicester. "Dear Sir," most of his letters begin, followed by all manner of entomological business from collections to book recommendations to cabinet design: "I send you with this a box of duplicates which I hope will contain some acceptable specimens. . . . What Books do you get from the 'Ray Society'? . . . I discovered a fine larva of the Elephant Hawk moth. . . . I have considerable thoughts of Setting up a Cabinet myself this winter. . . . I quite despair of ever getting a good collection of moths. . . . I shall be much obliged to you for any of the following of which you can send me good specimens. . . . I have recommenced a plan which I began 2–3 years ago but discontinued. That of keeping a Natural History Journal. . . . I got some [insect] Pins some time since & use them for all of my coleoptera. . . . Hoping you will write me a letter full of Entomological news."[22] He was a bit annoyed when Bates's responses were not timely: "I trust," he remarked wryly after one two-month lapse, "nothing very calamitous has befallen you to retard your Entomological correspondence."[23]

Alfred went on long rambles (sometimes with John, who loved herps and birds), taking it all in: beetles, botany, and his first love, geology. "I cannot call to mind," he later recounted, "a single valley that . . . comprises so much beautiful and picturesque scenery, and so many interesting special features, as the Vale of Neath."[24] It was on one memorable overnight excursion with John that he had one of his first entomological triumphs. Following Alfon Pyrddin, a tributary of the River Neath far up the Vale, they sought Sgwd Gwladus waterfall and the rocking stone, a glacial erratic of Carboniferous sandstone so finely balanced that it could be pushed slightly back and forth (but unfortunately "rocks" no more). There he came upon the beautiful bee beetle *Trichius fasciatus*. Then, as now, these fuzzy yellow-and-black-striped scarabs were rare in Britain, and Wallace described to Bates how he "had the good fortune to meet with one of the most beautiful & local of the British Coleoptera."[25] He later dashed off a note reporting this and other captures to the *Zoologist*, where it was duly published in the April 1847 issue—though the editor did not think much of most of his finds:

> *Capture of Trichius fasciatus near Neath.*—I took a single specimen of this beautiful insect on a blossom of Carduus heterophyllus near the falls at the top of Neath Vale.—*Alfred R. Wallace, Neath.* [The other insects in my correspondent's list are scarcely worth publishing.—E. Newman].[26]

He was aware of geology's role in creating the "special features" of this remarkable landscape. The lower valley lies within the great South Wales Carboniferous coalfields, with coal deposits alternating with sand-, mud-, and siltstones in a pattern geologists now call cyclothems. Traveling up the vale past Glynneath to the village of Pontneddfechan and beyond is to go back in time, as the coal basin gives way to older and harder coarse-grained sandstones, great upthrust folds of limestone, and, finally, massive formations of Old Red Sandstone, passing from the Carboniferous to the Devonian periods. The area is famous today as the South Wales waterfall country, where four rivers course down deep wooded gorges to join the River Neath, forming a stairstep series of waterfalls great and small, from "little Niagaras" (in Alfred's words) to tumbling cascades to ninety-foot free falls. It is a landscape that speaks of cataclysm, most evident in the strongly tilted limestone formations like Dinas (Fortress) Rock and nearby Bwa Maen, the Stone Bow, its thick strata bent double in an impressive geological fold. There are great limestone arches and the awe-inspiring Porth-yr-Ogof (Gateway to the Cave), where the River Mellte disappears underground. No wonder the remarkable western Brecon Beacons National Park, with Pontneddfechan at its doorstep, is now recognized by UNESCO—Fforest Fawr Geopark, the first geopark of Wales.[27]

Yes, Alfred was alive to these geological marvels, but he may not have understood that he was in a borderland of time and space again. It is a mash-up on the northern end, a transition zone of ancient accordioned landscapes. The Vale of Neath leading to it lies along a great fracture in the earth's crust, a fault called by today's geologists the Neath Disturbance. Stretching from Swansea northeast along the Vale and beyond as far as Hereford in western England, the disturbance itself is, true to its name, a record of tumult, where colliding continents squeezed and buckled strata laid down long before in shallow seas and river deltas. The rocks can bend only so far, eventually cracking with great blocks of landscape grinding past one another to create a topography of undulating frame-shifted formations. Later sculpted by cycles of advancing and retreating glacial ice, the modern landscape of the long Vale of Neath slowly emerged, the landscape a product of a line of weakness in the earth's crust: an ancient fault line now traced by the River Neath. The fault is still geologically active too: the earthquake of 17 June 1906, one of Britain's strongest of the past century, was centered on the Vale of Neath.[28]

Although some of Alfred's letters to his friend Bates discuss the fascinating local geology, most are dominated by lengthy accounts of coleopterous captures—successes and failures, dups and new rarities, beetles to admire and gloat over, beetles to long for, and of course beetles to trade, as the two agreed to exchange monthly lists. But it becomes clear in the letters that Alfred is not thinking merely as a covetous collector but as a *student* of his collections. Their "arrangement" becomes important to him. The group is so diverse that getting a handle on all of them is impossible. "I think it much better," he suggests, "to attach yourself to some particular families first to investigate thoroughly." A "perfect series" of certain well-represented families would be instructive—but of what?[29] Pattern, process. In letters discussing the design of drawers and cabinets for arranging specimens, Wallace almost casually asks Bates, "Have you read 'Vestiges of the Natural History of Creation'? or is it out of your

line?"[30] The juxtaposition is not coincidental: the infamous *Vestiges*, a sensational book anonymously published the year before, argued for transmutational change of species one into another, part of a grand cosmic vision of
material change in all things, stars and planets and species and human society.
Transmutation, as evolution was termed back then, was not a new concept—it
had been kicked around for a few centuries, most recently and (in)famously by
the French naturalist Jean-Baptiste Lamarck in the late eighteenth and early nineteenth centuries. But the idea was never taken seriously for want of any plausible mechanism—and even more problematically, its materialism. Largely
owing to this materialistic aspect, the concept of transmutation advanced in post-
revolutionary France came to be strongly rejected in Britain, not merely
because it undermined the veracity of scripture and therefore religion generally but because, in doing so, it undermined the very fabric of society.[31] The
two went hand in hand in church-and-state Britain, after all, where the reigning
monarch was (and is) the head of the Church of England. As transmutation
became politicized, to many establishment figures it was scandalous and seditious in equal measure. Thus was the sociopolitical context in which *Vestiges*
appeared in 1844, like a literary comet flashing suddenly onto the intellectual
scene—and as with comets in the days of old, it was dazzling and grand to
some yet repulsive and terrifying to others.[32]

But what was so scary about *Vestiges*? The book's greatest sin lay in the fact
that the "natural history of creation" in its title is all about relationships, transitions, and *transmutations* by natural law, explicitly rejecting the actively engaged God of scripture. This is not to say the book was atheistic, as its most
vehement critics falsely charged. No, the Divine Author is very much in there:
it was more deistic, with a role for a hands-off Creator who does not personally
supervise the origin, actions, and fate of each and every species and individual.
Deism's creator works through natural law (of "His" own device), a concept
born of the Enlightenment, a time of astonishing scientific insights in all fields,
an age of wonder where the very limits of space and time, of the possible, were
being pushed back and back and back.[33] The world beneath our feet and the
heavens overhead and everything in between was suddenly and mind-
blowingly expanding in that remarkable time, from the microscopic to the
astronomical to the paleontological. It was a time of progress, of elucidating
"natural laws" that regulated with mathematical precision the clockwork universe and the unearthing, literally, of the history of our world. It is understandable that when European astronomers charted the Southern Hemisphere stars
so long unknown to them, they would immortalize their instruments of

discovery and invention in the very constellations they drew: unsurprisingly, Microscopium and Telescopium are among thirteen constellations introduced in the 1750s by French astronomer Nicolas-Louis de Lacaille in celebration of Enlightenment science and art.[34] The Frenchman might well have added Geologorum Malleus, the geologists' hammer.

This was all well and good, as even the most pious High-Church Anglican appreciated the beauty of natural laws and was impressed by progress in the arts, sciences, and industry. It was what was made of those laws, what one saw as their philosophical implications, that made all the difference. Just how hands-on is God, after all? According to Christian orthodoxy, *very* hands-on, up close and personal. To the unorthodox and deistically minded, how inefficient and unnecessary for a creator to have to actively intervene and orchestrate literally everything when tidy natural laws of nature could take care of the nuts and bolts of running the universe. A mechanical clock provides a good analogy. Wind it up and it automatically ticks away the minutes and hours, with no need for us to stand there turning those clock hands ourselves. We let the laws of physics do the work for us. That was more the creator of Mr. Vestiges: we can logically infer, he argued, that just as the formation of the earth and other planets of the solar system did not require "any immediate or personal exertion on the part of the Deity" but rather resulted from "natural laws which are ex-tensions of his will," so, too, must the grand panoply of organic beings have arisen. "How," he asks, "can we suppose that the august Being who brought all these countless worlds into form by the simple establishment of a natural principle flowing from his mind, was to interfere personally and specially on every occasion when a new shell-fish or reptile was to be ushered into exis-tence on *one* of these worlds? Surely this idea is too ridiculous to be for a mo-ment entertained." Impious? Hardly: "To a reasonable mind," continued Mr. Vestiges, "the Divine attributes must appear, not diminished or reduced in some way, by supposing a creation by law, but infinitely exalted."[35]

Insofar as *Vestiges* argued for a natural, material origin of all things, species included, according to a law of development, what did this mean for humanity? That was clear: *Vestiges* even discussed human origins in the context of devel-opment from other, "lower" life-forms—for example, in the chapter on the "Early History of Mankind," where the author suggestively points out that according to one theory "we should expect man to have originated where the highest species of quadrumana are to be found." And who are we to criticize the creator? "For it may be asked, if He, as appears, has chosen to employ in-ferior organisms as a generative medium for the production of higher ones,

For good reason, *Vestiges of the Natural History of Creation* was published anonymously.

even including ourselves, what right have we, his humble creatures, to find fault?" (emphasis mine).[36] To Britain's divines this was beyond the pale; it was this rejection of the mainstream Christian belief in an immanent God who regularly and actively intervenes in the affairs of all things and who has special regard for the one species created in "His" image that so outraged the establishment. Of course it had to be published anonymously—the author of so vile a book would certainly be ruined socially and financially and probably jailed. Yet, even as preachers denounced the book as shameful, disgusting, satanic, and worse, the brouhaha and sensational press it garnered had the effect, as these things often do, of provoking even wider readership than if it had been simply ignored. Prince Albert even read it aloud to Queen Victoria.[37] The book's anonymity added to its allure, and speculation ran rampant—was it authored by a fallen cleric? Some anarchic atheist? A woman?[38] Paradoxically, *Vestiges* popularized and thus normalized an ostensibly reprehensible notion. It was transmutation for the people.

As the book reverberated through society, it sure had a seismic effect on Wallace. The seed of this heterodox hypothesis fell on fertile ground, an iconoclastic mind already primed by anti-establishment Owenite reform and the anti-ecclesiastical arguments of Paineites. The *development hypothesis* seemed to him more than plausible, an idea to test by collecting evidence. Bates's reply has not been found, but he was evidently lukewarm on *Vestiges*. Dismissing it as a "hasty generalisation" (read: half-baked idea), he provoked a bit of an upbraiding from Wallace, who called it far from hasty and an "ingenious hypothesis, strongly supported by some striking facts and analogies." Sure, further evidence is needed—more research, additional facts. "It at all events furnishes a subject for every observer of nature to turn his attention to; every fact he observes must make either for or against it, and it thus furnishes both an incitement to the collection of facts & an object to which to apply them when collected."[39] Wallace followed this with a fascinating torrent of thoughts on the subject. Did Bates not know that "many eminent writers give great support to the theory of the progressive development of species in Animals & plants"? Wallace cited "interesting & philosophical" works such as Sir William Lawrence's *Lectures on Physiology, Zoology and the Natural History of Man* (1819) and *Researches into the Physical History of Man* (1813) by James Cowles Pritchard, both treatments of human variation and its origin. He discussed the nature of species and varieties in his letter—touching on issues that lie at the very heart of the systematic arrangement of species for study—and concluded that even "the venerable Humboldt supports in almost every particular its theories."[40]

In another letter written the following April, in 1846, he was pleased to learn that Bates appreciated Charles Lyell's *Principles of Geology* and Darwin's *Journal of Researches* too: "I first read 'Darwin's Journal' 3 or 4 years back & have lately reread it—as the Journal of a scientific traveller it is second only to 'Humbolts [*sic*] personal narrative' as a work of general interest perhaps superior to it—He is an ardent admirer & most able supporter of Mr Lyell's views—His style of writing I very much admire, so free from all labour, affectation, or egotism & yet so full of interest & original thought."[41] He was going to comment on "representation & analogy"—again, relevant to the arrangement of specimens—but ran out of paper. These letters reveal the depth and breadth of Alfred's personal program of study at Neath between 1845 and 1847: exotic scientific travel, grand natural processes and laws, profound philosophical questions.

———

When Fanny returned home in September 1847, she took her brothers on a celebratory holiday to London and Paris, a trip that included visits to two of the world's great museums: the British Museum (which then still housed the natural history collections) and the Muséum National d'Histoire Naturelle. Writing to Bates afterward, Wallace described the grand sights and customs of

Alfred Russel Wallace, age twenty-four.

Paris, as one would (even giving a lecture on the subject at the mechanics' institute), but he knew that Bates would want to hear all about the Jardin des Plantes and the great Muséum, offering detailed descriptions of the arrangement and presentation of specimens. But it was his five hours in the insect collection of the British Museum, where he was able to get up close and personal with a global collection for the first time, that left the greatest impression. He was blown away by the astounding diversity among scarabs alone, declaring the endless variety of *Phanaeus* rainbow scarabs, jet-black *Copris* dung beetles, enormous Goliath beetles, and the brightly patterned flower beetles (subfamily Cetoniinae) "inexpressibly magnificent." Fabulous beetles collected from around the world, all neatly pinned and labeled in orderly row after row, drawer after drawer, cabinet after cabinet—precisely what is needed to do a thorough investigation! But of what? The nature of species and varieties, yes, but now Wallace's thinking was trending to even bigger and more philosophical—if fraught—questions. His letter to Bates concludes with a remarkable declaration: "I begin to feel rather dissatisfied with a mere local collection. Little is to be learnt by it. I should like to take some one family, to study thoroughly—principally with a view to the theory of the origin of species. By that means I am strongly of [the] opinion that some definite results might be arrived at."[42]

"A view to the theory of the origin of species." A year earlier, the two amateur naturalists had briefly discussed the idea of traveling abroad to pursue their collecting passion. Now there was an added motivation for such travel. *Vestiges* had opened their eyes and inspired a transmutation of another kind, transforming them from keen (if bumbling, in the case of Wallace) enthusiasts into naturalists determined to contribute to the grand scientific issues of the day. Wallace was twenty-four, Bates twenty-two, neither with much by way of funds. But, inspired by vestigian "development" and smitten with Humboldtian visions of teeming tropical paradises, regions of boundless diversity and exquisite beauty where all manner of species could be easily collected and studied, they began to dream. And then to scheme.

CHAPTER 4

PARADISE GAINED . . .

"A WILD SCHEME!" Those were the words that came to the mind of Charles Darwin's no-nonsense father, Robert, when his wayward son proposed journeying around the world for years aboard a Royal Navy vessel—just when he seemed on the verge of a respectable career as a clergyman. Those words might aptly describe the ambitious plan that young Wallace and Bates had just hatched too: a journey to tropical South America, where they would collect insects for both sale and study. Indeed, theirs might be the wilder of the schemes, in view of the fact that they had precious little by way of money, formal education, professional standing, or connections. But they made up for each of those deficiencies in their fashion: they had saved a bit of cash, enough to get them to South America, and once there hoped they could pay their way with the sale of specimens. They may have lacked much *formal* education, but as we have seen, they were certainly not uneducated, having devoured the books and lectures available at the mechanics' institutes and the like for years by then. They were talented amateurs, budding naturalists with something of a philosophical bent, as up on the taxonomic minutiae of their favorite groups as on the current ideas and theories that naturalists like Humboldt, Lyell, Lawrence, Darwin, Pritchard, Swainson, and others debated. As for professional standing and connections to open doors and to help them prepare, indeed they had nothing but their youth, earnestness, and naivete—and a bit of luck.

The desire to travel, collect, and study had been building in the young naturalists for quite some time. As for where to go, they were surely predisposed to favor South America after reading Humboldt and Darwin, and the book that pushed them over the edge in that direction was William Edwards's *A Voyage up the River Amazon: Including a Residence at Pará* (1847), no doubt read at the mechanics' institute library. It was a book of purple-tinged prose, a bit longer on rhapsody than accuracy when it came to the flora and fauna but sincere for all that. An American whose livelihood was coal mining but whose passion was butterflies (he later authored the lauded *Butterflies of North America*), Edwards traveled up the Amazon as far as modern-day Manaus in 1846 and was smitten. In his memoir he expressed surprise that so few travelers had journeyed to the "Southern continent," so promising to "lovers of the marvellous"—a land of the "highest of mountains," "brightest skies," and "mightiest of rivers" rolling "majestically through primeval forests of boundless extent."[1] Edwards's "land of sunshine, of birds, and flowers" was filled with "gambolling" coatis, "frolicking" monkeys, squirrels that "scamper in ecstasy . . . unable to contain themselves for joyousness." It was a place where "birds of the gaudiest plumage flit through the trees," and the "Myriads of gaily coated insects" found by day swap places at night with "myriads of fire-flies" and large luminous lantern flies streaking by like meteors. He marveled at the profusion of butterflies, "almost all gaudy beyond anything we have in the North," especially the iridescent-blue morphos renowned for their striking metallic brilliance.

If his accounts of the profusion of exotic and beautiful species were not enough of an inducement to eager young travelers, Edwards also waxed lyrical about the agreeable and helpful people, ease of travel, healthy climate (plus a bounty of natural medicines), and ease of mastering the language, not to mention "beautiful Indian girls" flitting by "like visions." For any naturalists or sports lovers among his readers who were inspired to travel there, he helpfully offered pointers on gear and other useful items. The book sealed the deal: they immediately resolved to travel to South America and wrote to Edward Doubleday, curator of butterflies and moths at the British Museum, for advice. Wallace had likely met Doubleday the previous year, when he had visited the museum with Fanny and John. Doubleday was then hard at work on what was to become his magnum opus, *The Genera of Diurnal Lepidoptera*, an exquisitely illustrated folio series co-authored with entomologist John O. Westwood. Congenial and encouraging, Doubleday informed Wallace and Bates that northern Brazil was little known, and there was demand for as many different insects, birds, mammals, and snails as they could collect, easily paying their expenses.

They each had a modest savings, plus Bates's dad offered to lend a bit of money. In March 1848 they met in London to study the collections in the British Museum, noting gaps that needed to be filled, and Doubleday even introduced them to Edwards himself, fortuitously in London at the time, who quickly offered further encouragement, advice, and letters of introduction.

They were also referred to Edwards's fellow American Thomas Horsfield, author of *Zoological Researches in Java, and the Neighbouring Islands* (1824). A fellow of the Royal Society and active in both the Zoological and Entomological Societies of London, Horsfield was now keeper of the East India Company museum in London and only too glad to lend his expertise, advising the novices on the design of specimen cases for shipping. These establishment contacts also helped set Wallace and Bates up with an excellent agent, Samuel Stevens—kindly, diligent, and himself an amateur entomologist and regular contributor of notes to the *Zoologist*. They could not have been in better hands. Aptly described as a natural history impresario who represented a number of traveling naturalists, Stevens ably looked out for his clients, advertising and finding good-paying homes for their collections, updating the scientific community on their travels, insuring their shipments, and keeping them well supplied with cash and the latest scientific news from home, all for a 20 percent commission on sales plus 5 percent to cover insurance and shipping.[2] He also offered sound practical advice as an expert collector himself.

Undoubtedly, their contacts also played a role in gaining them an audience with Sir William Jackson Hooker, founding director of the Royal Botanic Gardens at Kew. Though not hired on as one of Hooker's many field collectors, they promised to send consignments of plants back to Kew and asked Hooker for an all-important official letter: "It would serve to shew that we were the persons we should represent ourselves to be, & might much facilitate our progress into the interior."[3] He obliged them with not one letter but two, one to the Foreign Office, enabling them to obtain passports, and one for the Brazilian authorities, to facilitate their travels in the country.

Such letters were essential for opening otherwise locked bureaucratic doors and securing local assistance. Despite being virtual unknowns with little travel experience, Wallace and Bates certainly benefited from the global network of "imperial science" made possible by centuries of trade and colonialism, backed up by the military might of Britain at that time. The scientific institutions at home—museums, universities, menageries, botanical gardens, herbaria, learned societies, and the like, together with the human network of curators, faculty, patrons, editors, publishers, sales agents, insurers, and the banks that connected them—were linked by steamship and sail to far-flung colonial settlements and outposts. These outposts were themselves linked by barge and canoe, mule and horse to even farther-flung camps, trading posts, and villages where deep local knowledge resided—in the form of Indigenous guides and collectors, the ultimate guarantors of safe passage in the storied realms where coveted rarities were found. We think of collectors like Wallace and Bates as the laborers on the front lines of field collecting. But although they were, of course, very much boots on the ground, given their reliance on locals they were actually middlemen between those with deep local knowledge, often Indigenous, and the collectors and institutions back home.[4]

Their scheme was taking shape: passage was booked on a ship out of Liverpool, the barque *Mischief*, bound for Pará at the mouth of the Amazon. Bates's parents invited the two to spend a final week with them in Leicester, where they continued their crash course in specimen collection and preservation, practicing their shooting, skinning, taxidermy, labeling. . . . They had plenty of experience with insect pinning and such, but neither had done much with birds or mammals. Books like William Swainson's 1840 *Treatise on Taxidermy* were a good how-to on preservation techniques, but their time was about up, their ship was scheduled to sail soon, and their education was about to shift to on-the-job training. After attending the wedding of their friend John Plant, the two headed to London, with a brief detour to Derbyshire to see the orchid and palm houses at Chatsworth House, the magnificent estate of the Duke of Devonshire. The remarkable gardener, architect, horticultural magazine publisher, and engineer Joseph Paxton had pioneered new greenhouse designs made possible by advances in cast iron and glass manufacturing. At Chatsworth, Paxton oversaw construction of the Great Conservatory, then the largest glass structure in the world and a model for his design of the Crystal Palace of the Great Exhibition of 1851 (for which he was knighted). In 1848, when Wallace and Bates visited, Paxton had begun construction on the Conservative Wall, a magnificent series of eleven greenhouses 331 feet long and 7 feet wide, stair-stepped up a hillside.

They arrived in Liverpool the following day after a "cold and rather miserable journey outside a stage-coach," and got their luggage and equipment aboard the ship. One more night in "the scantiest" accommodations and they were off, setting sail on 26 April 1848.

It is unclear how Wallace's family felt about all this. It had not been that long since they had finally been reunited in Neath, and now Mary Ann was anxiously seeing the beginnings of another and perhaps permanent dispersal, centrifugal social and economic forces slowly but surely forcing them apart again. With Alfred gone and their business closed, John decided to make a go of dairy farming. He rented a cottage with some pastureland closer to town, his sister and mother joining him for a time, while Edward continued at the Neath Iron Works. But sure enough, one by one the rest of the family followed

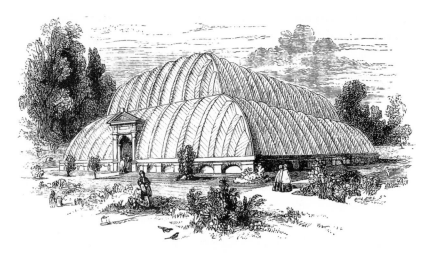

The Great Conservatory at Chatsworth.

Alfred's lead and dispersed: just a year after Alfred set sail, John was on his way to the California goldfields, unable to make a living dairy farming, and Edward resolved to quit Neath for the Amazon to assist his brother, arriving by July. In February, Fanny had married budding photographer Thomas Sims, the eldest son of John and Alfred's landlord in Neath, and moved to Somerset, where Mary Ann joined them after John's departure.[5] There would be a partial reunion yet, but it would have been difficult indeed for Mary Ann Wallace to envision, with her surviving children now scattered among three continents.

———

The intrepid young travelers Alfred Russel Wallace and Henry Walter Bates, twenty-five and twenty-three years old, respectively, were quickly inducted into the ranks of the seafarers thanks to a gale that blew up in the Bay of Biscay, giving them the widely shared and miserable experience of acute seasickness. They were alternately prostrated in their berths and retching over the railing for about the first ten days. But the seas smoothed out, and the fast three-masted ship flew with the trade winds. They entered another kind of border-land, exquisitely beautiful, where the infinitude of sky meets that of ocean. By day they marveled at the vast Sargasso Sea—an in-between place in some ways more land than sea, a vast slowly rotating "island" of reddish-brown sargassum weed teeming with arthropods, mollusks, worms, fish, and myriad other

denizens—while at night the ethereal glow of the Milky Way above was answered by a sea suffused with the collective light of untold billions of phos-
phorescent organisms. The first sign of approaching land, still far over the
horizon, was a change in the color of the sea, deep blue becoming greenish,
then olive, then giving way to an olive-yellow color as they approached the
coast—seawater burdened with sediment, the soil of the lofty Andes some
two thousand miles away.

Twenty-nine days out from Liverpool, they landed at Pará (now Belém), a
mere degree and a half south of the equator, at the confluence of the Marajó
and Guamá Rivers to the southeast of the forty-mile-wide mouth of the Amazon itself. It was the morning of 26 May 1848. What were Wallace's first impressions from the harbor? The morning sun illuminated a compact, low-lying
town, now the Cidade Velha (old town), with whitewashed buildings and
church steeples catching the light and a blocky eighteenth-century fortification
in the foreground, the Forte do Presépio, built (and rebuilt) by Portuguese
colonists hoping to stave off Dutch and French incursions. The "luxuriant
tropical productions" of tall palms and plantains made the sight "doubly beautiful," but the town must have also seemed rather insignificant, backed up
against a looming green wall of forest. Now a busy port city of over 2.5 million,
Belém was then a town of but fifteen thousand or so people. When they could
finally go ashore, Wallace found it "outlandish," with rough streets and handsome though "ruinous" buildings, Italianate in design with yellow-and-blue
washes about the doors and windows, surrounding weedy squares. Palms and
fruit trees abounded as well as people of "every shade of colour."[6] Exotic, to be
sure, but where were all those gamboling coatis and frolicking monkeys? There
was not much by way of gaudy birds flitting through the trees or myriads of
brightly colored insects.

On the whole it was a bit deflating after the great expectations inspired by
Humboldt, Darwin, Edwards, and other travelers of the tropics. "The weather
was not so hot, the people were not so peculiar, the vegetation was not so
striking, as the glowing picture I had conjured up in my imagination."[7] Many
a traveler has had this experience on first encountering the tropics, especially
the rainforest, where immensity and density conspire to give a false impression
of homogeneity. It takes time to train the eye, to see, and then the green ocean
of forest resolves into breathtaking diversity. It is a world of extremely high
diversity but low density, so it also takes time to encounter the wonders of the
flora and fauna and, in the case of the latter, to capture it. Wallace later realized
that this is why the travel narratives he had read were misleadingly chock full

of marvel after marvel and chastised "travelers who crowd into one description all the wonders and novelties which it took them weeks and months to observe," inevitably producing "an erroneous impression on the reader, and [causing] him, when he visits the spot, to experience much disappointment."[8]

The prospect improved when they moved about a half mile outside of town to a small village called Nazaré—now a mere bus stop on a square in sprawling modern-day Belém, then a hamlet close to the green wall of the forest, within which was a vast riverine labyrinth with myriad braided waterways great and small coursing through a seemingly infinite forest. They lucked out in hiring an ex-slave named Isidoro as a cook, who proved to be quite knowledgeable about plants and their medicinal and other uses, and they were further helped by Vicente, perhaps enslaved, who had a real talent for catching *bichos*—insects, reptiles, and other small animals. Their education continued: Portuguese, flora, fauna, the conventions and rhythms governing life in a small colonial outpost in the tropics, and the rhythms of the forest. Their first exotic captures whetted their appetite: eight species of swallowtails, three of fabulous blue morphos, bird-catching spiders, mantids, and three species of beautiful *Epicalia* (now *Catonephele*) butterflies, strikingly ornamented with deep-blue and orange bands on rich velvety black. They made two excursions to Manguari (now part of Ananindeua, a municipality on the east side of Belém), where their letters of introduction from Edwards got them an invite to visit the rice mills. Walking the twelve miles to the site, they passed through a deep forest of towering, buttressed trees festooned with epiphytes, looking up and seeing through Humboldt's eyes the fine foliage of the distant tropical canopy silhouetted against blue sky and recalling the Prussian naturalist's vivid descriptions of sights like this. Here they collected plants adorned with plants ("parasites on parasites, and on these parasites again") and encountered their first monkeys, vipers, and a host of beautiful birds—toucans, kiskadees, wading jacanas. And they bagged prized butterflies: the exquisite clear-winged Andromeda satyr (*Citherias andromeda*), more metallic-blue morphos, stunning metalmarks. They also soon learned the hard way how quickly the ubiquitous marauding ants would make trophies of their catches. Putting his collecting box aside for a short conversation one day, Wallace was horrified to find it swarming with voracious ants that had already carved up over a dozen of his choice specimens. But the glass-half-full Wallace saw a benefit nonetheless: "I had great difficulty making them quit their prey, and gained some useful experience at the expense of half a successful day's captures."[9] Indeed, he and Bates had to quickly come up to speed on how to secure hard-won specimens from the constant depredations of the ants, which, teeming in their untold billions, represented a collective stomach as vast as it was insatiable. (One trick of the trade was to stand table legs in cups of water, creating little moats that helped prevent the ants from scaling the parapets.) Besides their bounty of insects and birds, the two had quite the plant collection too. Three months into their Amazonia adventure, they shipped off their first consignment to England: a large chest packed with 3,635 insect specimens representing over fourteen hundred species for Samuel Stevens and a whopping twelve chests bearing one hundred dried plants, mainly palms, for William J. Hooker's new Museum of Economic Botany at Kew.[10] They may have felt like they got off to a slow start, but this first consignment was prodigious.

Having dipped a toe in the ocean of the Amazonian forest at Manguari, Wallace and Bates went ankle-deep soon after, with a monthlong trip down the River Tocantins in the company of Canadian Charles Leavens, one of the rice mill managers. Leavens was also a timber dealer in search of valuable Spanish cedar trees (*Cedrela odorata*), then, as now, a hardwood prized for its aromatic and decay-resistant wood. With a letter of introduction from Senhor João Augusto Correio, comandante of Belém, to smooth their way, the three set off up the Tocantins at the end of August 1848 in a hired *vigilinga*, a sturdy twenty-four-foot two-masted type of canoe with a toldo, a palm-thatched shelter, at one end.[11] One of the largest rivers in South America, the Tocantins courses south to north over 1,500 miles (2,500 km), from the Cerrado grasslands to the Amazon Basin through a broad valley bound by steep rocky bluffs, all that remain of ancient, once lofty sandstone plateaus. It was another borderland in deep time, a double borderland, even: a region where the South American and African continents were once joined as part of the vast southern supercontinent of Gondwanaland and a geologically varied district where three ancient provinces of the continent meet—and have corresponding formations in Africa. Wallace took note whenever the geology, usually well hidden by a thick veneer of tropical verdure, revealed itself.

The explorers traveled some 150 miles upriver, as far as the rapids of Aroyas, present-day Itupiranga. It was no mean feat to fight increasingly stiff currents and rocky rapids as they rose in elevation. It was an object lesson in the difficulties of back-country collecting, with mishaps like Wallace losing his spectacles while frantically fleeing the wrath of small, vindictive wasps; unhelpfully dropping his gun into the water; or, thinking they had successfully killed an alligator, seizing a leg of the still very much alive and enraged reptile and nearly swamping their canoe. Then there was their indifferent field help. The people

Forest scene on the Amazon.

they hired (or tried to), Indigenous Miriti-Tapuyo and mixed-race *pardos* and caboclos, were on their own time and had their own priorities, and the enormous exertion required to alternately paddle against stiff currents and drag a large canoe over furious rapids for oddly clad foreigners chasing bichos—as laughable as it was pointless—was not high on the list. They could be extremely helpful but seemed reliable mainly in their unreliability, disappearing without a word at any time, if they could be persuaded to join the party to begin with. They can hardly be blamed—dealings with the *brancos* (whites) rarely turned out well for the locals.

It was modestly successful for all that, this short foray that Bates later called a "tourist's gallop."[12] They bagged quite a few new butterflies, including striking longwings, sulphurs, and swallowtails, as well as beetles and cicadas, river snails and mussels, and a mixed flock of bird specimens: parrots, goatsuckers, jacamars, and the decidedly prehistoric-looking hoatzin, one of the world's few leaf-eating birds. Others, like the brilliant blue macaw, *araruna* to the Indigenous people, remained tantalizingly out of reach. But it may be telling that Wallace spent as much time inspecting the local geology as chasing down birds and insects. He noted the appearance of the metamorphosed volcanic and sedimentary rocks as they made their way upriver: coarse conglomerates and dark igneous rocks near the site of the modern Tucuruí dam, then crystalline sandstones near the site of the abandoned settlement of Alcobaça. "Vast

masses of volcanic rock" at the falls and the thundering rapids of Tapai-
unaquára and Guaribas caught his attention, where "much twisted and confused" strata with "volcanic masses rising up among them" gave "evident
proofs of violent volcanic action at some former period." The evidence of cataclysms past excited wonder in Wallace, but more than this the geological observations inspired bigger-picture thinking—in particular, he wondered about
the link between the local geology and the prized blue macaw that had so far
eluded them. These birds become rare downriver, disappearing altogether
below Baião, while upriver they are abundant. "What can be the causes which
so exactly limit the range of such a strong-flying bird?" he wondered. "It appears with the rock, and with this there is no doubt a corresponding change
in the fruits on which the birds feed." These are ecological musings revealing
a sense of both the interconnectedness of species and that of the earth and the
life upon it. Relationships and interdependencies exist in relation to geography: it is all about lines, borders, boundaries.

The explorers reluctantly headed back downriver, serenaded by howler
monkeys, cicadas, frogs that uttered decidedly unfrog-like calls—one evoking
the approach of a distant train, another a hammering blacksmith—and the
constant whine of mosquitoes. Pickings were slim, but they bagged a frigate
bird, a hawk, and more fine swallowtail and longwing butterflies. Halfway back
they stopped at the cacao plantation of the local comandante, Senhor Seixus,
where Wallace's Amazonian dream almost came to an abrupt end when he
carelessly went to pick up his gun by the muzzle and it fired. The blood came
before the pain, and he was lucky to lose only a small piece of his hand near
the wrist, missing the artery (and the people standing behind him). When
Wallace made it back to Belém, he was in a world of hurt, with his hand seriously inflamed. He was treated by a doctor, who put his arm in a sling with
strict orders not to do a thing with his hand for two weeks. He couldn't even
pin an insect and was "consequently rather miserable." But the silver lining to
this potentially fatal cloud was time to observe, and think. As was so often the
case when Wallace had enforced downtime for one reason or another, his
wheels were always turning, leading to striking insights. In this case, while
recovering at the estate of the Swiss consul, Monsieur Borlaz, and trying to
collect small birds belonging to various groups that apparently procure food
in the same way, he reflected on adaptation, taxonomic relationships, distribution, and competition. By now, having collected a great many birds as well as
insects and having dissected many to inspect their gut contents, a conundrum
emerged. All works of natural history, it seemed to Wallace, dwelled on the

marvelous adaptations of animals to their environment—their foraging, habits, behavior. But there was something else going on—"Naturalists are now beginning to look beyond this, and to see that there must be some other principle regulating the infinitely varied forms of animal life." What about unrelated groups of birds and insects with scarcely any resemblance to each other yet living in the same area, feeding on the same food? "It must strike *every one*," he declared, that these groups "cannot have been so differently constructed and adorned for that purpose alone."[13] It flew in the face of the received wisdom of the day, which held that each species is specially adapted in every way, including to its particular food source. Yet all those unrelated species with strikingly different forms end up eating basically the same things in the same places, directly competing with each other. Food is actually limited, not superabundant. Hence when a given tropical tree is in fruit, "birds of the most varied structure and of every size will be found visiting the same tree." They take what they can get, vying for what's available. If each species is not specially designed for its niche in life, what principle regulates "the infinitely varied forms of animal life?"

These thoughts are given almost as an aside, yet they reveal an ongoing quest to understand pattern and process. What did Bates think about these musings? We do not know; something happened on the Tocantins trip, and the two decided to go their separate ways. A mutual friend, botanist Richard Spruce, who arrived the following summer to collect for William Jackson Hooker at Kew, commented in a letter that they quarreled.[14] Maybe so—Bates is barely mentioned in Wallace's later narrative of the voyage, and Wallace is only politely referenced a few times in Bates's. Whatever precipitated the separation, Wallace and Bates were rather tight-lipped about it, though much later in life they intimated that their decision to split up was based more on pragmatism than acrimony. Divide and conquer: collecting in different regions minimizes duplication of effort, which makes sense for biogeographically attuned collectors whose finances depended on salable rarities. In any case, although they certainly remained friends through life, at the time that friendship may have been a bit strained.

Before separating, the two did, however, work together to assemble their second consignment for Stevens, who duly advertised it and published an extract of their letter: "Messrs. Wallace and Bates, two enterprising and deserving young men, left this country last April on an expedition to South America to explore some of the vast and unexamined regions of the province of Pará, said to be so rich and varied in its productions of natural history."[15]

For the next five months, Wallace stuck to the mouth of the Amazon, mainly to the great island of Marajó and the much smaller Mexiana Island just to the north. The two straddle the equator, low landscapes of open, uneven campo—ranchlands dotted with horses and cattle, bordered by thick forest bound at the shore by natural levees. Overseen by enslaved African cowboys, or *vaqueiros*, the open ranches surely presented a strange sight, with emaciated horses and cattle harried and blood-streaked with vampire bat bites and often picked off by jaguars and alligators. But Wallace was focused on the birds: raucous flocks of parakeets, green with white and orange on their wings, swooped in and out of great fruiting trees, and sleek black anis and long-tailed cuckoos that sounded like rusty hinges sailed from tree to tree. Insects were scarce in that season, but bird collecting was good—not only eagles, hawks, herons, and storks galore but also elegant scarlet ibises and roseate spoonbills, jewel-like sabrewing hummingbirds, spectacular red-breasted meadowlarks and troupials, and handsome toucans dressed in neat black-and-white evening wear, brandishing enormous colorful bills. There he dined on alligator tail and pirarucu, the enormous freshwater fish. The astonishing number of alligators at one site inspired musings on the nature of the fossil record: some geologists interpret the large numbers of alligator fossils as evidence that these reptiles were superabundant, living in a mainly watery world at a time before there were many land animals. But that would be a hasty conclusion: "But, as it is evident that the remains of these alligators would be found accumulated together should any revolution of the earth cause their death, it would appear that such descriptions are founded upon insufficient data, and considerable portions of the earth might have been as much elevated as they are at present."[16] A most Lyellian analysis, and one that his friends at the mechanics' institute back home would appreciate. In fact, they were on his mind, and he wrote them not long after this excursion, as he had promised to do when he left, but it was more a sketch of the place without scientific musings—landscape, climate, flora, fauna, and the diversity of the people and their customs. Summoning his Humboldtean muse, he waxed lyrical about the immense forest, where "no one who has any feeling of the magnificent and the sublime can be disappointed." His description is worth quoting in full:

The sombre shade, scarce illumined by a single direct ray even of the tropical sun, the enormous size and height of the trees, most of which rise like huge columns a hundred feet or more without throwing out a single branch, the strange buttresses around the base of some, the spiny or furrowed stems of others, the curious and even extraordinary creepers and climbers which wind around them, hanging in long festoons from branch to branch, sometimes curling and twisting on the ground like great serpents, then mounting to the very tops of the trees, thence throwing down roots and fibres which hang waving in the air, or twisting round each other form ropes and cables of every variety of size: and often of the most perfect regularity. These, and many other novel features—the parasitic plants growing on the trunks and branches, the wonderful variety of the foliage, the strange fruits and seeds that lie rotting on the ground—taken altogether surpass description, and produce feelings in the beholder of admiration and awe. It is here, too, that the rarest birds, the most lovely insects, and the most interesting mammals and reptiles are to be found. Here lurk the jaguar and the boa-constrictor, and here amid the densest shade the bell-bird tolls his peal.[17]

The only philosophical aside in his letter was a long condemnation of slavery: "A wrong to the slave and an evil to the Slaveholder and a clog upon the prosperity of the Country in which it exists." The injustice of the institution was rarely far from his mind. Around this time, too, he wrote to the family. His little brother Edward was still unhappy working in the Neath Abbey Iron Works, and a bid to take on pupils and teach French like his sister Fanny was not going well. Alfred had encouraged his little brother to join him in Brazil, and he reluctantly decided to give it a go, writing to his sister: "I hope this will be a better speculation for me; we are doomed to be a scatter'd family, and if it must be so, if circumstance has so ordained it let us meet it bravely, and with honest hearts go forth, resigned and cheerful under the dispensations of Providence."[18]

As Edward prepared to sail for South America, Alfred moved to another house in Nazaré, his separation from Bates complete despite the two living practically next door in the little village. There Wallace met a Congolese man named Luiz with a wealth of collecting know-how. Luiz had served as an enslaved field assistant to Austrian naturalist Johann Natterer, who spent some eighteen years collecting in Brazil between 1817 and 1835. Freed by Natterer as he left Brazil with an enormous collection of preserved animals of all kinds, Luiz made a good living as a hunter, successful enough to now own a bit of land and enslave a couple of individuals himself. Alfred hired him at two shillings two pence per day to hunt birds with him along the Guamá River, snaking east from Belém, as far as São Domingos do Capim at the confluence of the Guamá and its larger "tributary," the Capim. He wanted to travel this river as much to see the famed tidal bore, the pororoca, as for the collecting—it was getting a bit monotonous, the incessant skinning, pinning, pickling, labeling, packing, and fretting to constantly safeguard his hard-won specimens from scavengers or that insatiable collective ant stomach. This curious natural phenomenon was just the diversion he needed. At its highest around the time of the equinoctial new and full moons, the pororoca is a tidal surge that travels hundreds of miles upstream in the Amazon and adjacent rivers in a succession of waves a good twelve to thirteen feet (four meters) high in an audible roar. It is a surfer's dream, the "endless wave" that now attracts competitors from around the world for the longest ride, greeted in São Domingos do Capim with prominent signs: *SORRIA, você está a terra da pororoca* (SMILE, you are in the land of the pororoca).[19] How could any naturalist resist? Wallace was thrilled: the great wave came "in a sudden rush," traveling rapidly upstream, and "lifted our canoe just as a great rolling ocean-wave would do."[20] True to form, he made careful observations, and concluding that the prevailing explanations were wrong, he offered his own, complete with diagram, in his Amazonian travel narrative a couple of years later.[21]

The collecting had its charms too, traveling by canoe through "wild, unbroken and uninhabited virgin forest" amid metallic-blue morphos and darting green kingfishers in the company of Indians and enslaved individuals working for Senhor Calisto, an accommodating landowner who helped provision the expedition. Wallace appreciated the assistance, though he had qualms with the forced servitude, even under the most benign treatment. His argument against slavery at that time was based on the supposed improving effects of struggle, responsibility, and self-dependence: enslavement keeps individuals "in a state of adult infancy—of unthinking childhood," he wrote.[22] It is the "struggle for existence" and the desire for self-improvement—personal gain, power, adulation—that hones the faculties and inspires genius. But the enslaved have no hope of improvement, nothing to aspire to. It is interesting to consider that this idea of the improving effects of struggle is not too far afield from his and Darwin's later ideas on natural selection.

Far up the Capim, the party dined on stewed jacu, a pheasant-like guan, and slung their hammocks between ancient rainforest trees where, by flickering campfire light, the Indians recounted their hunting exploits and narrow escapes from jaguars and snakes. Wallace, gazing up dreamily from his hammock at the distant canopy, could hardly distinguish the winking stars from the fireflies.

Reading by the light of the bioluminescent click beetle *Pyrophorus* (*Elater*) *noctilucus.*

These were not ordinary fireflies but click beetles of the genus *Pyrophorus*. Sporting bright, light organs on either side of their thorax, like running lights, they were so bright that he could read by the light of one held over a newspaper.

In little more than a year, Wallace had traveled very far indeed.

––––––

Edward arrived in Belém at the beginning of July 1849 aboard the brig *Britannia* in the chance company of the botanist Richard Spruce and his assistant. Born in Yorkshire and trained to become a teacher like his father, Spruce had the same combination of wanderlust and taste for natural history as Wallace and Bates and soon abandoned that intended profession. Unlike them, he was quite a seasoned traveler and field collector by the time he headed to Brazil in the employ of William J. Hooker. Within a month Spruce had several hundred pressed specimens for Hooker, two species of palm germinating and the fruits and flowers of several others drying, and was experimenting with getting epiphytic *Fernandezia* orchids to take to orange trees—which, if it worked, could be a nifty way to propagate them at Kew.[23] The affable Spruce was soon heading up the Amazon to Monte Alegre and Santarém. The Wallace brothers planned to head upriver too, but not with Spruce—Alfred was initially a bit standoffish, wanting to go it alone, maybe owing to his argument with Bates.

Timing was everything for traveling upriver, especially before the introduction of steamers in 1853. The rivers were strongly tidal even hundreds of miles upstream, and so they were dependent on the flood tides each day. Seasonally, the winds could be a help. In the dry season, roughly from August through December, a thousand miles could be covered in a mere forty days riding the brisk *vento geral*, the east trade wind. But the same trip took more than twice as long in the wet season, January through July, when the winds failed, and the rivers were swollen and raging. Then it was excruciatingly slow going upstream in their small *montarias* (another type of canoe) and vigilingas. The fierce current was bad enough, but the river was often freighted with huge tangles of uprooted trees and other vegetation swept down. Storms were frequent, with violent lightning-lit squalls with winds that generated furious waves. The tumultuous weather had to be waited out, and in between the boats would be slowly towed upriver, secured by cable from the mast to the nearest sturdy tree and hauled forward by the hired hands. The cables would be set and reset, inching the boat upstream in the face of the violent current. It was best to avoid travel in the rainy season if at all possible, but the Wallaces did not have that luxury.

After the usual delays, the brothers secured passage on a small empty boat heading upriver, but Edward was probably less than delighted with their accommodations—what you might call "substeerage"—the hold, still reeking of the salted fish just unloaded in Belém. Setting out in early August 1849, they at least had favorable winds in the Rio Pará, then the going got slow as they paddled, poled, and hauled their way against the current through the labyrinthine *igarapés* and *furos* around Marajó Island, channels hemmed in by what has often been described as a forest wall. It gave the impression of traveling through a deep green gorge, as Bates aptly described it.[24] Two weeks later, they finally entered the majestic Amazon. It may as well have been an ocean, as it was so wide that the banks were lost to sight once one was out in it. "It was with emotions of admiration and awe that we gazed upon the stream of this mighty and far-famed river. Our imagination wandered to its sources in the distant Andes, to the Peruvian Incas of old, to the silver mountains of Potosi, and the gold-seeking Spaniards and wild Indians who now inhabit the country about its thousand sources." It was breathtaking, beholding the "accumulated waters of a course of three thousand miles," and it blew his mind to realize that all the rivers and streams "for a length of twelve hundred miles drained from the snow-clad Andes" were gathered in the broad ocean of the river now rolling before him.[25]

They set sail for Santarém, a trading outpost about five hundred miles inland at the confluence of the turbid Amazon and clear-water Tapajos Rivers. It was a tidy little town with streets of grass (lacking wheeled vehicles or even many horses), the obligatory fort, a handsome church with two square towers, bright yellow-and-white houses with green doors, and lots of "quite amphibious" Black and Indian children splashing at the beach—"not quite a Modern Babalon [sic] as to size and appearance," Edward joked in a letter to Fanny, "although the grass growing in the streets might remind you of some deserted city of the ancients."[26] Santarém was the home of the colorful Scotsman Captain Hislop, who presided over a nightly Amazonian salon for expatriate Americans and Brits. Wallace had letters of introduction from his contacts in Belém. He would have liked the more official one he had asked Stevens to get for him the previous May from the Foreign Office, to facilitate his travels in the interior, but it had still not arrived.[27] The letters they had were fine, and congenial Hislop was happy to help. It is worth remembering that few travel truly alone, and Alfred and Edward were assisted by a network of Brazilians and expats like Captain Hislop—a local judge, comandantes, traders, a shopkeeper, cattle ranchers, a padre—and always the enslaved and the Indians, who did the hard work.

With their collective help, Alfred and Edward found accommodations here, borrowed a montaria there, and had a helping hand with the collecting, cooking, skinning, and paddling. And many were happy to take the collectors on excursions, too, or let them tag along on outings or trading jaunts. They took advantage of one such invitation soon after arriving in Santarém, when the local *juiz de direito* (magistrate) offered the Wallace brothers use of a sturdy canoe to travel downriver some fifty miles to Monte Alegre, on the north bank, perched high on a hill reached by slogging through shifting sands between thick, thirty-foot candelabra cacti, an outlandish contrast to the rainforest scenes to which they were accustomed. The village had not recovered from the depopulation resulting from the Cabanagem of the previous decade, a popular rebellion of the oppressed caboclos (those of mixed Indigenous Brazilian and European ancestry), pardos (those with triracial European, Amerindian, and African descent), and other *ribeirinhos*, denizens of the ubiquitous ramshackle riverfront stilt houses. As their misery at the hands of the whites came to a head, their fury overwhelmed Belém and triggered a veritable civil war between 1835 and 1840, with decimated towns and burned ranches and plantations across the province, leaving tens of thousands of people dead. Now Monte Alegre consisted of so many dilapidated houses arrayed around a half-finished sandstone church in a central square and "weeds and rubbish on every side, with sometimes a few rotten palings round a corral for cattle."[28] A few inhabitants were cattle ranchers, but the economy was mainly based on cacao. The friendly judge provided them with letters of introduction to his friend Senhor Nunez, a local shopkeeper who proved welcoming and very helpful. They stayed about a month in a mud house at a cattle ranch, where the food was as good as the collecting: turtle stew washed down with fresh milk. The rooms where they slung their hammocks were closed up and stifling, but for good reason: legions of bloodthirsty mosquitoes emerged at night, and they were forced to place pans of smoldering cow dung outside their door in a bid to keep them at bay. The trick had limited success.

Alfred was keen to visit the area for the beetles, which had been in frustratingly short supply thus far. He was disappointed on that front, but the trip was a great success in other ways. A bonus to the excellent bird and butterfly collecting was a bit of geology and archaeology in the nearby *serras* (mountains). Those low peaks just to the southwest and north of the river are Paleozoic outcrops, the southernmost margin of the great Guiana Shield, the nearly two-billion-year-old uplifted continental craton and one of three that make up the South American Plate. Wallace was traveling another borderland in deep time: a beautiful landscape of Carboniferous-era sandstone high above the river, eroded into honeycombed formations and sentinel hoodoos wearing broad caps of more resistant rock—notably, the famous Pedra do Pilão, a pinnacle visible in the distance from the village. It is a land of springs, waterfalls, and caves, including what is perhaps South America's most famous archaeological site, the Caverna da Pedra Pintada, or Painted Rock Cave, which the Wallace brothers visited with the help of Senhor Nunez. Now protected within Monte Alegre State Park, this sandstone cave and its associated formations give evidence of a Paleo-Indian presence dating to about 11,200 years ago, far earlier than humans were believed to have made it to the Amazon Basin.[29] Alfred and his brother admired the spectacular cave paintings and pictographs—hundreds of stick figures, hand stencils, and geometric designs, the earliest known in the Americas. Edward later described their "journey of discovery" in a letter to their mother: "We stopped there three days, climbed up the mountains, grazed our shins over the broken rocks, copied the curious figures that were drawn on them—in short performed all those wonderful feats with our legs & hands which wonder seeking travellers generally do."[30]

Edward hurt his leg not long after, laying him up for two weeks. But Alfred made the most of their stay there: the "shady groves" of the hills around Monte Alegre "formed our best collecting-ground for insects." Two rarities in particular hinted at those puzzling patterns of distribution: the closely related brush-footed butterflies *Callithea* (now *Asterope*) *leprieuri* and *C. sapphira*, a dazzling species with metallic bluish or greenish undersides and black with deep blue or orange on the upper surface. One occurred on the north side of the river, where Monte Alegre lies, while the other was found on the south side, including Santarém and its vicinity. "Here," Wallace wrote much later, "I first obtained evidence of the great river limiting the range of species."[31] Another line; why? That may not seem terribly significant, finding one species here and another there. But it would not be long before he came to see that the geographical distribution of closely related species just may say something about what astronomer Sir John Herschel called "that mystery of mysteries," the origin of species.[32] At the time he was only just beginning to cotton on to such patterns, and he would later regret not being more careful in noting from just which side of the great Amazonian rivers different species and varieties he collected had come. For now the butterflies of Monte Alegre and Santarém would be another observational dot waiting to be connected to others.

Soon after arriving back in Santarém, they met back up with Richard Spruce, who soon became a fast and lifelong friend. Little did they know that Bates had passed through, too, while the Wallace brothers were at Monte Alegre, but he soon moved on upriver. Wallace packed up a consignment to ship to Samuel Stevens, his first since separating from Bates. Beetles may have been elusive, but he had a bounty of birds and especially butterflies—nearly all the species differed from those he had found around Belém. He called Stevens's attention especially to the gorgeous and hard-won *C. sapphira*, "the *most beautiful thing* I have yet taken. . . . It is very difficult to capture, settling almost invariably high up in trees; two specimens I climbed up after and waited for; I then adopted a long pole which I left at a tree they frequented, and by means of persevering with it every day for near a month have got a good series." Stevens duly published the letter, adding an enticing footnote about "the rare Callithea Sapphira" to prospective buyers: "Hitherto only one example appears to have existed in the collections in this country."[33] Almost as an afterthought, Wallace also enclosed alligator vertebrae, thinking someone might find them useful for comparing with fossil reptiles, and a specimen of the great Victoria waterlily, *Victoria amazonica*—the very emblem of the botanical exuberance of the tropics that he and Bates had marveled at in the Chatsworth greenhouses just before their departure a year and a half earlier, and now beheld in nature, filling the marshes below Monte Alegre.

It was now November 1849, and the rainy season was starting a bit early. Keeping on the move in an effort to expand their collections, the naturalists hopscotched their separate ways upriver to Barra, now the city of Manaus. Óbidos, fifty miles upriver where the Amazon's channel is narrowest, was the first stop. Alfred and Edward arrived to find that Spruce had only just gotten there the previous day even though he had left Santarém ten days earlier than the Wallace brothers. It was a good example of the vagaries of travel on the Amazon. Spruce had passage in a large boat but was beset by a lack of wind and an owner who refused to sail at night, while Alfred and Edward's small leaky canoe made it in three days. (They did not know it, but Bates beat them all, taking just a day to reach the town from Santarém, and was long gone by the time his fellow countrymen made it.) Alfred and Edward pushed on to Vila Nova da Rainha (modern Parintins), where they were assisted by Padre Torquato de Souza, the sometime traveling companion of Prince Heinrich Wilhelm Adalbert of Prussia during his expedition up the Xingu, another large Amazon tributary. The small outpost of Serpa, or Itacoatiara, was next (where the Wallaces unknowingly passed Bates again), and they finally reached Manaus, at the great confluence of the Rio Negro and the Amazon, on the last day of the year. The largest blackwater river in the world, the Rio Negro flows nearly 1,500 miles (2,400 km) from its headwaters in Colombia and Venezuela. It announces itself even before arriving at the river itself. Owing to temperature and density differences, the inky black waters of the Rio Negro, the color of which comes from an abundance of tannins and other compounds, actually flow side by side with the yellowish sediment-laden waters of the Amazon for several miles after they meet, myriad café au lait eddies slowly melding them into one.

Today a city of well over two million, Manaus did not look like much in those days, having been all but destroyed in the Cabanagem. There were two churches, in disrepair, and the old fort that had given the town its initial name, Barra, had been reduced to a mound and fragment of wall. Low red-tile-roofed houses, white or yellow with green doors like in Santarém, lined the regularly laid out but unpaved streets, which were "undulating, and full of holes." Wallace took a dim view of the locals: even the "more civilized inhabitants," mainly traders, "have literally no amusements whatever"—unless you count drinking and gambling—and "most of them never open a book, or have any mental occupation." No wonder, Wallace concluded, that "morals in Barra are perhaps at the lowest ebb possible in any civilised community." By "civilized" he meant

Western, European, but his comment shows he appreciated that this was relative in more ways than one. He would soon come to deplore the corrupting influence of the morally dubious elements of Western society that too often had the most interaction with Indigenous people, and praise the merits of traditional Indigenous lifestyles. But it would be a century and a half before Western society would begin to come to terms with the corrosive effects of even what Wallace would have considered the most benign forms of colonialism. For a multitude of cultures actively or passively extirpated, including the former inhabitants of the Manaus region, it would be too late. Poignantly, the town's name derives from the Indigenous Manaós people, a culture that had lived there for perhaps millennia, their story now relegated to the city's Museu do Indio.

The Wallaces arrived on the very last day of the decade, quickly welcomed with open arms by Senhor Henrique Antonij, Italian expatriate, well-regarded merchant, and beloved friend to all travelers. Senhor Henrique, as he was known, received the brothers "with such hearty hospitality as at once to make us feel at home," as he had done for William Edwards before them. (At Spruce's suggestion Senhor Henrique was later honored with the genus *Henriquezia*, a genus that includes "the finest tree on the Rio Negro," bearing "a profusion of magnificent foxglove-like flowers."[34]) What is more, they were now in the very heart of the Amazon, a thousand miles from Belém, on the very edge of terra incognita for Europeans. Alfred wasted no time—despite the season he immediately organized an initial monthlong expedition up the Rio Negro in search of the celebrated umbrella bird, *Cephalopterus ornatus*, a large crow-like species with an oversized Elvis pompadour and an inflatable wattle that frequented the river islands above Manaus. Edward stayed behind while Senhor Henrique gave Alfred a letter introducing him to Senhor Balbino, who lived near modern-day Iranduba on the west side of the river, about three days' paddling from Manaus. Balbino in turn introduced Alfred to an Indian family, who gave him a "small room with a very steep hill for a floor" and three doorways, one of which doubled as a window.[35] It was all he needed, as he was out collecting most of the time: Wallace and his hired hunters worked from before sunrise to late at night, and their catches included not only umbrella birds (twenty-three, all told!) but several specimens of the rare white bellbird (*Procnias albus*), a curious species that sports a long, dangling wattle and produces an earsplitting bell-like call considered to be the loudest of any bird.[36]

Returning to Manaus in February, the brothers were reunited with Bates, and the three of them once again collected together, between torrential downpours, until Bates departed for the outpost of Ega, four hundred miles up the Rio Solimões, as the vast branch of the Amazon from Manaus to Peru was called. The friendship of Wallace and Bates had held, and their parting was more amicable this time.[37] The Wallace brothers, in the meantime, continued enduring the rainy season, awaiting the long-anticipated official letter from the Foreign Office. Day after day, week after week, no news. They made the best of it, the wait enlivened by the company of the other naturalists and travelers stuck there too, including one English bird collector whose Indian assistant taught them the use of a blowpipe, a deadly weapon made from hollowed-out palm stems fitted with a mouthpiece and equipped with needle-sharp darts dipped in poisonous curare (derived from the bark of the vine *Strychnos toxifera*, the source of the pesticide strychnine).

He packed up the latest consignment for Stevens, not a bad lot considering the season. Besides those bellbirds and umbrella birds, there were parrots, hawks, a rare toucan, trumpeters, partridges, manakins, ducks, and aracaris, as well as an assortment of fish, reptiles, and of course insects. Stevens excerpted his letter for publication in the *Annals and Magazine of Natural History*—good advertising for the intrepid naturalist far afield.[38] To the Zoological Society, he also communicated a short paper Wallace had penned—his first paper from the field—on the habits of the umbrella bird: "Having had the opportunity of observing this singular bird in its native country, a few remarks on its characters and habits may not perhaps be uninteresting, at a time when a consignment from me will have arrived in England."[39] In this letter, Wallace also revealed an ambitious plan. He would next ascend the Rio Negro practically to its source, penetrating into the Republic of Venezuela, and on his return he would make for the high Andes—ascents into paradise, the storied realms where the venerable Humboldt explored! He figured he probably could not start "for the frontiers" until June or July at the earliest. It would be a two-month voyage up the Rio Negro, especially as he planned to survey as he went, and he would stay in the upper reaches for about a year.

Having sent his fourth consignment of specimens downriver bound for England, Wallace was back to waiting out the rainy season. To shake off the monotony, he joined a small group for an excursion to Manaquiri, just up the Solimões, where he had an invitation to visit the estate of Senhor Antônio José Brandão, Senhor Henrique's father-in-law. En route he marveled at the tremendously swollen rivers, now in the prodigious seasonal flood stage that makes a waterworld of the lowland forest, the *gapó* (called *igapó* today), "one of the most singular features of the Amazon."[40] This vast flooded forest extends twenty to thirty miles on either side of a seventeen hundred-mile stretch of the Amazon, the waters reaching as deep as forty feet for six months of the year. Wallace remarked on how the Indigenous people recognized gapó forests as distinctive, with species not found in the terra firma forest—something borne out by modern floristic analysis. He noted the distinctive animals, too, attracted to the abundant fruits of gapó trees, and he would have been delighted (though perhaps not surprised) to learn that the seeds of these trees are mainly dispersed not by birds and mammals, nor even the water itself, but by fish! The blooming and fruiting of many gapó tree species is timed to coincide with the great inundation, and dozens of fish taxa are coadapted to feed upon and disperse the annual fruit bonanza, a remarkable set of ecological and evolutionary relationships not well characterized until the early twenty-first century.[41]

The party paddled all day, long stretches in the deep shade of the flooded forest, gliding past great cylindrical tree trunks rising from the water like columns, Wallace said, the gloom relieved here and there by elegant gleaming-gold *Oncidium* orchids seemingly suspended in the air. A large genus commonly known as the golden dancing or golden shower orchids, each bears a number of bright-yellow flowers on long, slender stems. Now and then they suddenly emerged into bright sunshine as they left the forest and crossed broad lakes dotted with yellow-flowered bladderwort (*Utricularia*) and blue-flowered pickerelweed (*Pontederia*). They stopped for the night, moored to a massive floating tree trunk held fast by vegetation, kindling a fire to roast fish and boil coffee. It would have been quite a nice aquatic encampment, but it turns out they were encroaching on the floating real estate of a large colony of fire ants—insects whose name is inspired by their eye-wateringly painful sting, not any affinity for fire. The agitated ants did not take kindly to the intruders or their smoky fire; swarming into the canoe, they made the unwitting humans pay dearly.[42]

The next day they reached Manaquiri and Senhor Brandão's estate, a somewhat ramshackle farm of cane and tobacco fields, orchards, and pasture. There Wallace spent the next two months observing and collecting what he could, with help. He fell into a routine: up early he would enjoy some coffee and a simple breakfast of porridge or chocolate and milk, then go collecting in the forest or sit down to work on material at hand—pinning insects, skinning birds and monkeys, and once dissecting a manatee that was brought to him. He took note of the behavior of the vultures perched nearby as he worked on his specimens. The ever-hungry birds kept a sharp eye out for any "snacks" he might produce while preparing his skins—and that it was sight and not scent that guided them he was sure, noting how if they saw him toss a morsel that got hidden somehow from view, they would hop about searching for it in vain. For his own meals, lunch and dinner featured similar fare: delicious *tambaqui* fish (*Colossoma macropomum*) was the staple, sometimes varied with fowl, deer, manatee, or other game and served with rice, beans, and corn bread. He often ate with Senhor Brandão and his daughter. A former magistrate in Manaus, Senhor Brandão and his family had narrowly escaped death at the hands of a raiding party during the Cabanagem two decades earlier, but his farm was nearly destroyed. Years later it had still not much recovered; now a widower, his heart was not in it. Now he lived with his servants and his last unmarried daughter, spending his days tending to his farm and reading whatever books he could get his hands on—including in French, which he had taught himself to read. Wallace's great admiration for Senhor Brandão was perhaps exceeded by that for his rather attractive daughter. He first made the acquaintance of the senhorita and her father in Manaus—she was one of several young women he met. Dressed in elegant French muslins, not a hair out of place, adorned with flowers, she was a vision. Now the vision was here, incongruously: "It seemed rather strange to see a nicely-dressed young lady sitting on a mat on a very mountainous mud floor, and with half-a-dozen Indian girls around her engaged in making lace."[43] Wallace does not say much more about Senhorita Brandão in his writings, but he may have been far more taken with her than he let on: in recent years a letter from his brother John to their mother came to light, with an awfully intriguing comment:

> The account of Alfred's intended marriage is certainly news & may perhaps in some measure account for his not writing to me, all his spare time being now occupied with other thoughts. However I am glad he has found some one in that distant land to be a helpmate & companion to him, I cannot say so much of myself as I have seen none yet in my wanderings who can compare with "the merry maids of England so beautiful and fair."[44]

What—intended marriage!? John was clearly responding to news his brother had relayed to their mother from distant Amazonia. Given the timing, just about the only occasion for Alfred to have been in the company of a young lady long enough to even begin to contemplate marriage was his two-month stay at Senhor Brandão's estate in Manaquiri. Maybe it's not too surprising—it is not hard to imagine the twenty-seven-year-old, living mainly among a bunch of less than refined guys, being entranced, to say the least, at the sight of such a lovely young woman as Senhorita Brandão. And as we shall see in chapter 5, there were times in his Amazonian travels when he did fantasize about settling down to raise a family in his tropical idyll.[45] If this interpretation of John's letter is correct, a letter bearing the big news from Alfred to his family back home has not yet been found to confirm it.

As the rainy season approached its end, Wallace became restless. Bidding Senhor Brandão and his daughter adieu, he returned to Manaus in May 1850, keen for the mail and for continuing his explorations upriver. His plans did not include Edward, however. It had become increasingly clear that his younger brother was not cut out for the rigors of the tropics, nor was his heart in the natural history. The plan was to leave him behind at Manaus, where he would try to earn enough money collecting to make his way back to Belém and book passage home. Alfred departed Manaus in August 1850, giving Edward ten pounds, all that he could spare. Edward wrote home: "I am a thousand miles from Pará, and my present plan is as follows; to hire a hunter immediately, and go for a couple of months into the country to make a collection of Birds and Insects which will be sufficient to pay my voyage to England, and I hope I have a few pounds in my pocket besides." There had evidently been some talk about him joining his brother John in California, but Edward was unimpressed: "When I arrive in England, I have my plans which I can better *tell* than *write*. I do not like the Californian Scheme for many reasons, much obliged to you for mentioning it the same." He wished he was "a little more unpoetical" but concluded that "I am what I am, I must try and do the best for myself I can." He signed off on a hopeful note, quoting Shakespeare: "'Trifles light as air' be gone!!—I have bussiness [*sic*] before me. . . . P.S. You may expect me home at Christmas."[46] It was not to be; in retrospect, Edward's own lines, penned back at Santarém, foreshadowed his tragic end far from home:

> With days of sunny pleasure,
> But, oh, with weary nights,
> For here upon the Amazon
> The dread mosquito bites—
> Inflames the blood with fever,
> And murders gentle sleep,
> Till, weary grown and peevish,
> We've half a mind to weep!

> But still, although they torture,
> We know they cannot kill,—
> All breathe to us in whispers
> That we are in Brazil.[47]

He could not know that the bite of the "dread mosquito" can inflame the blood with fever in more ways than one, ways that can indeed kill—the role of mosquitoes in vectoring disease would not be established for another fifty years. In his ascent to paradise, Alfred could not know that he would never see his brother again.

CHAPTER 5

. . . AND
PARADISE LOST

IT WAS LATE AUGUST 1850, and the sturdy thirty-five-foot canoe of trader João Antonio de Lima was loaded up and ready to head upriver. Alfred Russel Wallace's agonizing months enduring the rainy season and preparing for his yearlong journey to the upper reaches of the Rio Negro had turned into agonizing weeks awaiting the arrival of a canoe bearing mail and vital funds from home. He really could not leave without them; still the canoe did not arrive. Senhor Henrique helped convince Senhor de Lima to wait one more day—news traveled faster than canoe on the Amazon, and word came that the much-anticipated boat of Irish trader Neill Bradley was arriving imminently with mail. Such were the vagaries of the post at that time: from sailing ships and steamers to barges and canoes, bags of mail made their way slowly and uncertainly across oceans and up rivers, where traders and other chance couriers sailed, poled, and paddled their way upstream—in Amazonia, by montaria, vigilinga, and small dugout *igarité*.

The accommodating de Lima waited; finally, on the verge of nightfall, a relieved Wallace received a packet: some twenty letters in all, some well over a year old by then—news at last from family and friends in England and California (where his brother John the Forty-niner had gone to seek his fortune) and from his Australian cousins! It was the first mail he had received since leaving Belém. He read them into the wee hours, dozed a bit, then roused at 5 AM to answer as many as he could. The clock was ticking; the agonizing wait now gave way to frantic last-minute details as Senhor de Lima prepared for departure. Wallace dashed off final letters, picked up second-thought items, put together a box of specimens for Stevens, and went over plans with his brother one last time before parting ways, expecting to meet again one day back home in Wales, half a world away. Finally: "It was on the last day of August, 1850, at about two o'clock on a fine bright afternoon, that I bade adieu to [Manaus], looking forward with hope and expectation to the distant and little-known regions I was now going to visit."[1] At the order of Senhor de Lima and with a final push of long poles against ramshackle dock, the trading canoe pulled out into the river and pointed northwest. Wallace was twenty-seven years old.

Garrulous and grizzled, de Lima was an old river hand, regularly plying some seventeen hundred miles from Manaus to Guía and back, trading in all manner of items from his floating general store: axes and knives, fabric and thread, buttons and beads, that sort of thing—and well stocked, too, with cooking supplies and *caxaça* (cachaça), the local rum. His wife, a Portuguese Indian *mameluco*, and their children traveled with him. Wallace was shocked to learn she was de Lima's second wife, the first an Indian woman heartlessly turned out because she could not teach their children Portuguese. A servant named Old Jeronymo helped with the cooking, and an Indigenous crew poled and paddled the canoe. It was a comfortable boat—roomy enough amid the boxes and gear to sit or stretch out under the toldo, the thatched palm-leaf shelter, or sit outside in the cool of the morning and evening, taking in the rhythms of the wide, dark river. The Rio Negro may have been a tributary of the Amazon, but the word hardly does justice to the mighty river: gigantic in its own right, miles wide and gently flowing—at least in the lower reaches— and mercifully mosquito-free, as the larvae do not fare well in blackwater.

They had something of a routine. Starting out each day before dawn, some- times earlier, they landed after daylight for coffee, biscuits, and butter, then were off again until breakfast around ten or eleven o'clock, dining on whatever they caught by hook or gun that morning—chicken-like curassow or guan, or a twenty- or thirty-pound *pirahíba* catfish, or some other delicious river fish that Wallace characteristically sketched before it was cooked. Underway once more in the early afternoon until well after dark, they would find a spot on terra firma to hang their woven hammock *redés* if they could, sleeping until four or five o'clock in the morning, when they would embark again. Imagine the long canoe gliding silently through inky waters still and deep, Wallace propped on a box of gear, "enjoying the fresh air and the cool prospect of dark waters around us" as he eyed the swooping goatsuckers and drifting butter- flies.[2] As still and deep as the river, the forest was a green ocean of shaded depths, out of which flocks of garrulous parrots would unexpectedly erupt. At night they must have felt they were sailing in some ethereal realm, those glittering star fields above mirrored in the river below. Wallace dropped off to sleep to the chorus of tree frogs, the distant cries of howler monkeys, and the crickets and katydids gearing up their concerts. It was an odd and beautiful sort of borderland, between river and sky—the sky itself divided, with hemispheres familiar and unfamiliar to Wallace: the sometime astronomy buff would have recognized as old friends the constellations in the upriver direction, filling the northern sky, while those behind were still strangers, reminding him that he was far, far from home.

They passed through the vast watery maze of the Anavilhanas Archipelago, a seeming infinity of thickly vegetated river islands—narrow, snaking, whimsical doodles of green great and small, shaped by the river's flow. The archipelago is part of today's Parque Nacional de Anavilhanas, a national park that forms part of the Central Amazon Conservation Complex. They glided unknowingly past the lands of the fierce Waimiri-Atroari, or Kinja people, who did not make contact with Westerners until 1884, and made periodic stops trading or visiting de Lima's friends at scattered *sitios*, or country houses, and a succession of dilapidated half-abandoned towns and villages. The village known to Wallace as Ayrão, now Velho Airão, was one of the first he visited. There he observed sandstone outcroppings along the river, now recognized as Paleozoic sedimentary rocks atop the far older granitic rocks of the Guiana Shield. The sandstones become more crystalline and intersperse with granite just beyond, at Pedreiro (now Moura), more or less opposite the mouth of the whitewater Rio Branco, at the eastern edge of sprawling modern-day Jaú National Park. It is an area renowned for the elaborate petroglyphs, carved by ancient peoples across unknowable expanses of time, decorating the rocks along the river, populated with human and animal figures amid beautiful geometric mazelike spirals, some bordered like Greek *meandros*.[3] Ever fascinated by such antiquities, Wallace made extensive observations and drawings.

Carvoeiro, Barcellos, Caboquena, Santa Isabel, Castanheiro . . . one "ruinous" and "miserable" village after another, depressingly all depopulated by the Cabanagem and its bloody aftermath, with just a few colorful, if disreputable, characters remaining. But the natural history got more and more interesting. Barcellos, now spelled with one "l," exemplified the uneasy tension between the human desolation and the natural history: it was the would-be capital of the Rio Negro, but he found the town almost deserted, abandoned blocks of marble imported from Portugal for the construction of grand public buildings emblematic of the dashed hopes and plans for the place. But it was here, too, that he first noticed an elegant, slender palm tree that grew in small clusters at the river's

Petroglyph sketches by Wallace, Rio Uaupés.

edge. He made a sketch; it was a *Mauritia* palm, a new species, he thought (he later dubbed it *Mauritia gracilis*—it turned out to be one already named by Humboldt, *M. aculeata*, but he would describe several new palms before he was done). Soon after, the change in geology got his attention: stepping out of the canoe onto a "fine sloping table of granite" streaked with quartz veins, he was struck by the new character of the river. Was there a relationship between that and the appearance of the palm in the realm where granite commences?

Since their departure from Manaus, Wallace had also diligently employed his surveying skills. With compass, sextant, and watch in hand, he occupied himself with taking bearings at every opportunity, "not only of the course of the canoe, but also of every visible point, hill, house, or channel between the islands, so as to be able to map this little known river." Later, back home, he drafted a remarkably accurate map based on his measurements and presented it to the Royal Geographical Society.[4] More than a geography of the river, Wallace's map is also a geography of the people, geology, forests, and key species along it, showing the realms of the "Manao" and "Macu" Indians, the first appearance of those *Mauritia* palms and telltale granitic outcrops, the limit of the umbrella bird's range, and the extent of the alluvial plains along the river's south bank and the igapó along the north—the remarkable flooded forest stretching hundreds of miles, that inverted world where fish disperse the seeds of trees and vines as the terrestrial becomes aquatic six months of the year. He mapped the serras, the steep-sided granite mountains looming out of the rainforest: lone Cababuris (Cababuri), the long ridge of Pirapucó (now Pirapucu) north of the river, and the three dramatic peaks of Serra Curicuriarí (now Serra da Bela Adormecida).

As the river got narrower, rockier, and swifter, they swapped their large canoe for two smaller ones—piled high, they seemed nerve-rackingly unstable. Their Indigenous crew had to jump out periodically and push and pull the vessels around and between projecting rocks. But that was but a prelude to the real challenge: the first of the big rapids of the Rio Negro near modern-day Saõ Gabriel do Cachoeira, marking the fall line of the Guiana Shield, where a maze of rocks, ledges, and narrow channels funneling torrents of water had to be navigated. Crisscrossing the mile-wide river, the Indians did the hard work, diving into the rapids and swirling eddies with tow ropes in hand, hauling and pushing the heavy boats through sluices of thundering water, clambering up steep rocks to pass a rope to one another. Then came the most difficult stretch of the difficult passage: pushing, poling, and hauling the boats around towering rocks that channel the fierce current, trying to get to safety and rest in the calmer water in the lee of rocks and islands before attempting the next foaming torrent. It took hours to progress fifty yards. They stopped for the night, exhausted. It was more of the same the next day. But not for Wallace: while the Indians fought the maze of rock and furious water, he took in the sights along that "most picturesque" stretch of river: "The brilliant sun, the sparkling waters, the strange fantastic rocks, and broken woody islands, were a constant source of interest and enjoyment to me."[5] It is another reminder that even intrepid travelers like Wallace had considerable local assistance—as true today as it was then.

Incredibly, no life or boat was lost, at least on that ascent, although somewhere along the line his thermometers got broken. They finally reached the village of Saõ Gabriel, 710 miles from Manaus and inconveniently sited on yet more foaming rapids, such that the canoes had to be unloaded at some distance and portaged. There they presented themselves to the comandante for permission to continue. Then more rapids, for a stretch of thirty miles! But they eventually achieved a level and rapid-free section of river above the fall line, entering the region where the Rio Uaupés joins the Rio Negro, a sizable tributary originating in the uplands of Colombia (where it was called the Vaupés). The upper Uaupés/Vaupés was terra incognita for Europeans, countless waterfalls and rapids insulating it from all but the most adventurous of travelers. Wallace could not resist. But first, the destination of Senhor de Lima and his crew awaited: the tiny village of Nossa Senhora da Guia, just below the mouth of the Rio Içana, where Wallace was provided with a little two-room house. He was impatient to explore the surrounding forest, but it was a few days before he could get anyone to help—the crew's return was celebrated by a multiday *festa*, with drinking and dancing "from morning to night." He was reluctant to wander far into the trackless forest alone, but he eventually got some assistance, albeit half-hearted, and was fascinated at how the Indians hunted with ten-foot blowpipes and fished with timbo, a fish poison extracted from the root of the legume *Lonchocarpis utilis*. Chock full of toxic rotenoids, timbo yielded a bounty of fish that in true Wallace fashion were sketched before they became dinner, including electric eels, *Electrophorus electricus* ("eaten, though not much esteemed"). He collected a number of beautiful and interesting specimens at Guia, but it was not long before he set his sights on that spectacular upland rarity, the *Gallo do Serra*, or cock-of-the-rock, *Rupicola rupicola*. A crow-sized cotinga of the deep forest, males of this storied species are a brilliant orange red, sporting a dazzling semicircular fan of orange feathers atop the head, neatly outlined in a delicate tracing of black.

Three of Wallace's Amazonian fish drawings: *top, Hypostomus plecostomus; middle, Pterophyllum altum; bottom, Heros severus (modern names; not to scale).*

He set out in the company of two Baniwa Indians, Arawakan people also known as Walimanai. It was several days' journey to their village far up tributaries of tributaries, paddling the Rio Negro to the Rio Içana to the Rio Cubate, from black to blacker waters through endless meanders. They passed from common riverine caatinga scrub to the terra firme forest of lofty vine-festooned trees, a "luxuriant virgin forest, whose varied shades of green and glistening foliage were most grateful to the eye and the imagination."[6] Those "varied shades" reflected incredible plant species richness—just one given hectare might boast nearly three hundred tree species alone, consisting of some 125 genera in forty-five to fifty plant families—far greater than the tree diversity of all of Europe and Eurasia.[7] But it was mainly rare birds Wallace was hunting, and from the Baniwa village, they were another ten or twelve miles away, at Serra do Cubate, where he was told the *gallos* he sought were found in abundance.

Led by a party of Baniwa hunters, Wallace trekked through a forest of soaring buttressed trees, slender palms, and stately tree ferns, stopping at a remote abode where he was struck by the appearance of a young European Indian woman whom he realized must be the rumored daughter of Johann Natterer, who had left the area some twenty years prior. Nine days later, after precipitous ascents and descents, fitfully sleeping in caves, frustration over his incessantly snagged Western clothing (a liability in that environment), and a rewarding diversion hunting wild peccaries, they returned with twelve cocks-of-the-rock, plus an assortment of trogons, manakins, barbets, and ant thrushes, a respectable catch. The behavior of the manakins, caught by the Baniwa with snares carefully placed at "certain places, where the males assemble to play," fascinated him. They are lekking birds, with males congregating at the same sites year after year, vying for female attention with their displays. An intrigued Wallace was unfamiliar with anything quite like it at home: "Two or three males meet and perform a kind of dance, walking and jigging up and down. The females and young are never seen at these places, so that you are sure of catching only full-grown fine-plumaged males. I am not aware of any other bird that has this singular habit."[8] He did not realize that cock-of-the-rock males congregate in much the same way but high up in the canopy rather than on the ground like the manakins, which is why the Baniwa knew just where to find them. Wallace stayed among the hospitable Indians another two weeks and then made his way back to Guia to plan his next, bigger expedition: pushing even farther up the Rio Negro, into Venezuela. A party of Baniwa joined him—a padre was expected in Guia, an occasion for another festa and baptisms. Religion was a fluid amalgam of traditional beliefs and Christianity, the ritualistic aspects of both seemingly most appealing to the Indians.

It was now late December 1850, four months since leaving both Manaus and his brother behind. Edward had not made it far, at that very time writing Spruce from the little village of Serpa (now Itacoatiara), one hundred miles downriver from Manaus: "Whilst repairing my weary limbs amid the luxurious folds of a rede, drinking a fragrant cup of the sober beverage, and meditating (but cheerfully) upon the miseries of Human Nature, I received notice of your arrival in the Barra. . . . You are at last in that Promised Land, a land flowing with Caxáca [cachaça] and Fariniha . . . a land where a man may litterally [sic] and safely sleep without breeches, a luxury which must be enjoyed to be appreciated." He supposed Spruce would soon be heading up the Rio Negro, where his big brother Alfred was no doubt "glorying in Ornithological rarities, and revelling [sic] amid the sweets of Lepidopterous loveliness." For his part, he would soon be out of there, awaiting passage to Belém and then home, but he would think of Spruce in his snug redé when the "restless billow" of the rolling winter seas roared around his pillow.[9]

———

Back at Nossa Senhora da Guia, mail arrived with the padre, Frei Jozé dos Santos Innocentos, whom Wallace found to be something of a fraud—"He

told us he had great respect for the cloth, and never did anything disreputable—
during the day!"[10] He took a dim view of the "Do as I say, not as I do" attitude
of the padre, a veritable Don Juan who lined his pockets as he performed rites
for a shilling each and preached the gospel to the locals with de Lima and the
comandante smirking from the sidelines. Wallace's reading of Paine's eloquent
denunciation of exploitative priestcraft in *Age of Reason* surely came to mind.
But he was delighted to receive a packet of letters dating to the previous May
and July from his brother-in-law Thomas Sims and agent Stevens in England,
from his cousins in Australia, and from John in California. He revealed his next
steps in his response to Sims:

> My canoe is now getting ready for a further journey up to near the Sources
> of the Rio Negro in Venezuela where I have reason to believe I shall find
> Insects more plentiful and at least as many birds as here—On my return
> from there, I shall take a voyage up the great River Uaupes or another up
> the Isanna, not so much for my collections which I do not expect to be very
> profitable there, but because I am so much interested in the country & the
> people that I am determined to see & know more of it and them than any
> other European traveller.[11]

He had ambitious literary plans too, he confided: first, a travel memoir and
books on fish and palms. He had already drawn one hundred different kinds
of fish, all since leaving Belém, and some thirty of the forty-odd palm species
he had learned so far, taking copious notes on their natural history. Then, not
to be seen as a slacker, he planned a work on the "Physical History of the Great
Amazon Valley, Comprising Its Geography, Geology, Distribution of Animals
and Plants, Meteorology & the History & Languages of the Aboriginal
Tribes—to Be Illustrated by a Great Map Showing the Colour of the Waters, the
Extent of the Flooded Lands, the Boundaries of the Great Forest District &c,
&c." Regardless of the value of his collections, he hoped to at least get credit
as "an industrious & persevering traveller."

With some elementary Spanish under his belt, off he went at the end of
January, ascending the Rio Negro right into Venezuela, to the storied realms
explored by his role model Humboldt, the type specimen of the "industrious
& persevering traveller." Senhor de Lima helped arrange for four Baniwa Indians to accompany him, one of whom spoke a bit of Portuguese. They traveled
light: each had his blowpipe and poisoned arrows, along with a paddle, knife,
tinderbox, and hammock. Wallace's kit was anything but light: still intent on
surveying, he had his watch, sextant, and compass in tow in addition to his

gun, ammunition, and insect and bird boxes for collecting. And of course plenty of the local currency: fishhooks, salt, beads, and calico for trade and payment.

Still a mile wide, the river coursed north, then sharply east at the mouth of the blackwater Rio Xié, home of the Baré and Werekena people, "uncivilised and almost unknown" in Wallace's Eurocentric view. Lofty serras loomed in the distance, the great Neblina Massif straddling the border. He may have caught sight of the sharply tilted sandstone Pico de Neblina, Brazil's highest peak at nearly 10,000 feet (2,995 m), but usually shrouded in clouds. They swung north again at the outpost of Marabitanas with its ruined mud fort and soon beheld the sentinel serra of the triple borderland of Brazil, Venezuela, and Colombia: the monumental Piedra del Cocuy, a dramatic steep-sided granitic dome rising abruptly from the rainforest to over 1,300 feet (400 m). The little party camped on beaches of granite stones—comfortable enough except for the swarms of blackflies—where Wallace came upon more petroglyphs and one night was delighted to find his "old friend" Polaris, the North Star, glittering just over the horizon where the river channel pointed north. They caught more fish than birds on the journey, Wallace carefully sketching each one, a "glittering" acara here, a "bright-coloured" tucunaré there.

They now left Brazil behind for a stretch of river where the Colombian–Venezuelan border ran right up the middle of the channel, perhaps a quarter-mile wide. Some fifty miles past the Brazilian frontier, they arrived at São Carlos do Rio Negro, the furthest point inland that Humboldt and his companion Aimé Bonpland had reached some fifty years before, coming from the opposite direction, from the Caribbean via the Rio Orinoco. For Humboldt, too, this was "an unknown land . . . partly mountainous, and partly flat, receiving at once the confluents of the Amazon and the Oroonoko." It was the Rio Negro that joined those great river systems via what Wallace called "that singular stream," the fabled Casiquiare Canal, a winding river that forms a natural two-hundred-mile canal between the upper Orinoco and Negro rivers, linking the Caribbean and Atlantic drainages.[12] Rumored by missionaries and conquistadores since the seventeenth century, the canal's existence was confirmed by Humboldt and Bonpland in their explorations of 1800, taking seventy-five days to pass from the whitewater Orinoco to the blackwater Negro and back again. They first entered the Rio Negro not by heading down the Casiquiare but via blackwater tributaries of tributaries of tributaries of the Orinoco—the Atabapo, the Temi, and the Tuameni—making their way to the remote settlement of Javíta (now Yavita), followed by a four-day portage nearly ten miles through marshy forest to the east bank of the *caño* Pimichín, a large creek or small river ("as broad as the Seine opposite the gallery of the Tuileries," Humboldt helpfully noted for his European readers), *then* down to the Rio Negro. The Pimichín enters the Rio Negro some seventy miles above São Carlos. If his canoe "be not broken to pieces" in the process, Humboldt was told, "you will descend the Rio Negro without any obstacle" and from there could ascend the Casiquiare north back to the Orinoco. He and Bonpland managed the round trip in thirty-three days, "not without suffering, but without danger, and with facility."[13] Today a small airstrip can be found at Maroa, from where a twenty-five-mile road has been cut through the forest to Yavita.

Yavita, following in Humboldt's footsteps, was Wallace's destination—though the settlement had moved a few miles north since Humboldt's day, from the banks of the Tuameni to the south bank of the Temi. Wallace found it a tidy village of about two hundred Indians in the gently undulating landscape straddling river drainages—yet another borderland, a continental divide marked not by mountains but by endless forest that Humboldt described as an "immense quantity of gigantic trees." To get there, Wallace pushed on past the Casiquiare beyond ever-smaller outposts on the Rio Negro: San Miguel, Tomo, Maroa. Stopping briefly at Tomo, the little party was welcomed by the inhabitants, mainly different branches of the Baniwa/Walimanai people, and given use of a "stranger's house," a *casa de naçao*. Improbably, the denizens of Tomo and other villages beyond were engaged in shipbuilding, constructing large canoes and even two-hundred-ton schooners that were taken downriver in the wet season, laden with cargos of *piassába* palm, pitch, and *farhina*—the Amazonian staple made from grated and baked cassava. It was a one-way trip, as the boats were too large to bring back up through the rapids and falls, so the upland traders needed to commission new boats annually. This curious arrangement was a legacy of colonial Brazil and Venezuela, when shipwrights were sent by Portugal and Spain to establish shipyards in unlikely places—namely, remote forests, turning tall and fragrant *Ocotea cymbarum* trees—rich in essential oils—into trading vessels destined to ply the Amazon and Atlantic coast. The clever Baniwa, learning from the European shipwrights, soon became expert, starting an oral shipbuilding tradition without ever consulting a plan or drawing or sailing one themselves.

A "road" had been cut through the forest from Pimichín to Yavita, and Wallace arranged to pay local porters in salt for transporting his gear the next day. Wandering into the forest late that afternoon, he found himself face-to-face with the most feared and revered of South American animals: a black jaguar,

Panthera onca. He reflexively raised his gun but realized that, loaded with bird-shot, it would only enrage the powerful cat. They silently gazed at one another for a long moment, until the jaguar dissolved into the forest, a shadow melting into shadow. Wallace later wrote that he was "much too surprised, and occupied too much with admiration, to feel fear."[14] Exhilarated, he scampered out of the forest, not wanting to tempt fate with a second, and potentially fatal, encounter.

At Yavita he was ensconced in a convento, another kind of traveler's house, and eager to begin collecting and exploring. But, unfortunately for him, the wet season kicked in early that year, and on the very day of his arrival, a torrential rain set in that did not let up. The wet season was a challenge in more ways than one. Preserving his specimens became a race against mold and the ravenous maggots of swarming flies, while the proliferating sandflies preferred human blood, burrowing into any exposed skin and driving him half mad as he tried to work: "Often have I been obliged to start up from my seat, dash down my pencil, and wave my hands about in the cool air to get a little relief."[15] Down but not out, he persevered with the help of native hunters and paid kids in fishhooks for specimens. In short order he amassed a nice collection, with new species he had never seen before coming in daily. There were several species of great morpho butterflies and bizarre harlequin beetles (*Acrocinus longimanus*)—a hand-sized black-and-red-patterned long-horned beetle sporting front legs enormously elongated to match its equally long antennae. He drew a veritable school of new fishes stunned with timbo by the locals, despite sandfly-bitten hands "rough and as red as a boiled lobster, and violently inflamed."[16] He skinned a rare caiman for his collection and in the forest saw agoutis, coatis, monkeys, and a variety of birds and snakes. And the palms! He was ever more fascinated by their diversity, beauty, and—very importantly—geographical distribution. He drew and took notes on all he found—stately *inajá* palms (*Maximiliana regia*), prickly *caranaí* (*Mauritia aculeata*), and a curious one called piassába by the Baniwa (also known as *chiquichique* or *chiquichiqui*), which he recognized as a new species of *Leopoldina* that he dubbed *L. piassaba*. The heavy fibers that cover this short, stocky palm—making it look a bit like a botanical Cousin Itt—were harvested and made into quality brooms and brushes in Britain, and Wallace mused how those ordinary domestic implements back home depended upon a substance that grew in but a limited and remote district of Amazonia.[17]

Wallace's interest in palms and fish intersected with his interest in the Indigenous people and their culture. He admired their physique, acquired samples

Piassába palm (*Leopoldinia piassaba*), drawn by Wallace on the Rio Negro.

of their handiwork, recorded elements of their language, and observed their habits, modes of ornamentation and dress, and traditions—in large part a curious mix of Catholic and traditional rites and festas. One festa he described as an hours- to weekslong dance fueled by copious quantities of intoxicating beer-like *xirac*, adults and children whirling to monotonous chanting, drums, and reed flutes, "accompanied by strange figures and contortions."[18] But these experiences, more anthropological than convivial, could not make him feel less lonely—he was the sole white person in that remote outpost, barely able to make himself understood, an odd skinner, stuffer, and pinner of specimens who was perhaps the most exotic specimen of all. His native assistants, too, were veritable foreigners, belonging to a different Baniwa branch, and could not understand the locals who were known as the Baré.

He reluctantly decided to depart at the end of March—when the rains were *supposed* to have started—a decision partly motivated by his assistants absconding one night. They had gotten increasingly uneasy living among people they could not understand, even if they were cousins of a sort, and decided to head home. Now Wallace was even more alone. The incessant rain and loneliness put him in a reflective mood. Before departing Yavita he mused in blank verse on life, happiness, and contentment back home and here, among the Baniwa/Baré. Clothing and conventions of dress (or undress) were emblematic of the culture clash: the "graceful form" of the near-naked Indians was owed to the "free growth" permitted, and "no straps or bands impede." This was far preferable to the ill-fitting garments and shoes of English lads or the diabolical stays that confined the "waist, and chest, and bosom" of English lasses. All it takes is simple nutritious food, fresh air, regular bathing, and exercise, he declared, "to mould a beautiful and healthy frame," a view he was to stress much later in life as a social critic and activist back home. In his poem he again questioned the "civilizing" influence of European civilization. In principle it brings the "joys, the pleasures and delights" of the well-cultivated mind, an appreciation of beauty in nature and art. But in practice it seems that the meanest elements of European society end up having the greatest impact on Indigenous peoples—an impact almost entirely negative.

Here Wallace the Owenite was tapping into a deep Rousseauian well of discontent with the state of so-called civilized society and a longing for a back-to-basics (which often meant back-to-nature) lifestyle, a romanticized vision to be sure but no less sincere for that. Those ill-fitting garments and suffocating stays were the embodiment of society's shackles: "Man is born free, and everywhere he is in chains," wrote philosopher Jean-Jacques Rousseau in *The Social Contract*. To Rousseau, writing in Enlightenment France, the burdens of class structure, social strictures, even government itself originated in inequality, a devolution from the first usurpation of some people's rights by others to the systemic inequalities of today. That first usurpation, Rousseau argued, concerned land and resources; it was the invention of "property" and the envy and avarice it inflames that sent humanity down the slippery slope into social bondage.[19] Wallace saw that the meanest elements of his own society stemmed from "the thousand curses that gold brings upon us"—a society that encouraged legions of men to "live a lower life," knowing or wanting nothing but money and getting more of it. The "noble savage," an idealized primitivism predating Rousseau, had long represented life in that earlier unfettered state. So, yes, if it was a choice between living a life in thrall to wealth, low in "physical and moral health," or as an Indian of the trackless wilds, the romanticized life Wallace would choose was clear:

> I'd be an Indian here, and live content
> To fish, and hunt, and paddle my canoe.
> And see my children grow, like young wild fawns.
> In health of body and in peace of mind,
> Rich without wealth, and happy without gold![20]

The Commissario helpfully sent a half dozen Indians to replace his absconded crew; they would help Wallace make his way downriver at least as far as Tomo, where he hoped Senhor Dias would send others to help him as far as Nossa Senhora da Guia. After some last-minute trading for a few more specimens of fabulous harlequin beetles, he departed the rainforest village on 31 March 1851. The very next day, Richard Spruce, still in Manaus, reported in a letter to his mentor and agent George Bentham back in London that two Englishmen had just arrived gravely ill with malaria but that "Mr. Wallace" had sent word downriver from the "frontiers of Venezuela" that he was far beyond the fever-prone lower reaches, in a mosquito-free realm where he was

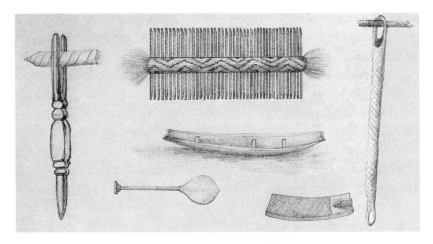

A selection of Wallace's artifact drawings from the Rio Uaupés: ceremonial cigar holder, comb, canoe, paddle, cassava grater, cassava strainer (*not to scale*).

"enjoying himself amazingly in a romantic and quite unexplored country."[21] Wallace would later encourage his friend to explore and botanize there, which Spruce did in 1854.

The return trip was going smoothly, though Senhor Dias was away from Tomo when Wallace arrived. He waited some days, staying with Senhor Domingos, who was overseeing the construction of a large canoe. Whiling away the time, he traded for a number of artifacts (finely crafted baskets, quivers, blowpipes and darts, and beautiful featherwork) and collected some nice specimens as he could—a handsome black-headed parrot and some very curious naked-back knife fishes (*Gymnotus*) that he promptly drew in his sketch pad. He also witnessed a remarkable festa, with painted and costumed locals whirling in a wild thirty-hour dance fueled by "shirac." At length Senhor Dias returned, and they procured a crew, paying a small band with calico, cotton, soap, beads, knives, and axes. Best of all, besides their able assistance with the canoes the Indians allowed him to record a vocabulary of their language, which he recognized as distinctive from any above or below Tomo. The return trip was far easier than the long slog upriver. Back in Nossa Senhora da Guia by the end of April, Wallace already had a plan for his next journey: far up the Rio Uaupés, terra incognita for European naturalists.

First, there was the usual waiting game, however. Senhor de Lima's canoe and promised assistants, who were expected from Manaus any day, did not arrive for a month. But he made the most of his time collecting: spiny rodent-like mammals, white-marked birds, palms, and more and more fish—he reckoned he had now drawn 160 species from the Rio Negro alone! Wallace ultimately made two voyages up the Uaupés, the great whitewater tributary of the Rio Negro that originates in the Colombian uplands. An uncharted realm, he dreamed of new rarities and of encountering the Indigenous people in a state of nature. His first excursion took some eight weeks, from early June through July 1851, right up to the Colombian border, followed by a second monthslong trip into Colombia, a journey made longer by deathly illness that could easily have ended it all right there. Between the two he restocked in Manaus.

––––––

His first journey started in early June 1851, and the current was fierce there at the mouth of the river at São Joaquim. It was the wet season, and the small party—Wallace, his friend Senhor de Lima, and two Baniwa Indians—could only inch upstream by pulling themselves along by shrubs and creepers through the flooded forest, trying (unsuccessfully) to dodge the wasps and ants they disturbed in the process. To all the other species collected or eaten—like the small anaconda sunning itself on a branch one moment and roasted for their dinner the next—it was the humans who were the scourge. The rich cultures of the many Indigenous groups along the thousand-mile-long river were still largely intact at that time, though they had suffered a century or more of "civilizing" missionaries and predatory slavers and human traffickers. It was to get far worse, culminating in a campaign, led by the Salesian missionaries, of violent destruction of their way of life in the early twentieth century. But at the time Wallace visited, the Turkanoan, Cueretu, and Arawak-speaking peoples along the Uaupés were more or less insulated, in particular those farther upriver, owing to the rigors of actually reaching them beyond the succession of falls and furious rapids. As we shall see, however, that protection had limits, as Wallace observed: de Lima proved to be one of the traffickers.

Each community lived in a great communal lodge called a *maloca* (spelled with a double "c" in Wallace's day), which could house a dozen or more families numbering over one hundred people. The first one Wallace encountered, of the Desana or Umukomasã people, was a long, tall structure around one hundred feet by thirty or forty feet, extending up to thirty feet or more high. It was "very strongly constructed of round, smooth, barked timbers, and thatched with the fan-shaped leaves of the Caraná palm. One end was square, with a gable, the other curved; and the eaves, hanging over the low walls, reached nearly to the ground."[22] Two long rows of supporting columns defined a broad aisle that ran up the center, between which ran rows of shorter columns that created side aisles where communal utensils were found: cassava graters, long woven tipitis to squeeze the grated cassava, ovens to make farhina, large pans and earthen vessels to brew the mildly alcoholic *caxirí*. A row of small apartments divided by thatched walls lined the secondary aisles on either side, each occupied by a family. The *Tushauá*, or hereditary chief, more ceremonial head than ruler, occupied privileged space with his family at the curved end. It must have been stunning, approaching that great maloca: peering into the dim light, hazy with smoke from the cooking fires within, Wallace would have slowly discerned people inside, looking in his direction with curiosity. The women hung back, naked and unornamented except for a kind of garter tight about the legs below the knee, but covering themselves up at the arrival of the foreigners. The men, too, were naked except for a *tururi* bark loincloth, their ears pierced with a piece of straw and their long black hair gathered into two yard-long sleek ponytails bound with monkey hair and topped with an elegant comb of palm wood and grass, a few toucan feathers adding a splash of color. Each wore a circular pendant of white quartz, personal ornaments of great symbolic value, painstakingly shaped and pierced with a hole laboriously made by slowly grinding with a plantain-leaf petiole and sand, mortar-and-pestle style. He was awed: "On entering this house, I was delighted to find myself at length in the presence of the true denizens of the forest," a people so different from all he had previously encountered that it felt like he "had been suddenly transported to another quarter of the globe."[23] He spent the night in the maloca. Drifting off to sleep to the sound of torrential rain outside and the quiet murmurs of the "naked Indians hanging round their fires, which sent a fitful light up into the dark smoke-filled roof," he admired how these families lived together in community and comfort, a veritable paradise.[24]

Wallace could not know, however, that at that very moment his brother Edward was gravely ill. That magical night in the Indian maloca would in fact be his brother's last night alive, as he succumbed by early the next afternoon to yellow fever. His brother had finally made it to Belém after all those months upriver and had booked passage to Liverpool on a ship soon departing, on 6 June. But his timing could not have been worse, as a yellow fever epidemic was then sweeping the town. Edward's symptoms first appeared on the 2nd of June, as Alfred, far upriver, prepared to leave Guia, and had greatly worsened by the time he landed at that great maloca on the Uaupés. The course of the disease was swift, and soon there was black vomit, the beginning of the end. Bates was with Edward almost to the last, getting him medical care, sleeping by his side, and nursing him, until becoming gravely ill himself. Bates stricken, the British vice-consul, Mr. Miller, took over caring for Edward, sending a letter to Alfred upriver to alert him to his brother's dire condition. Yellow fever soon killed Mr. Miller too. Bates was one of the lucky ones: he recovered and fulfilled the sad duty of writing his friends' mother, Mary Ann Wallace, with the terrible news that her youngest son had died; he was just twenty-two years old.[25] Bates did not write Alfred at the time, not knowing where he was, as he had had no word in many months, he later explained to Wallace's sister Fanny in a letter the following October. Prostrated by his own illness, Bates was not sure if the vice-consul had written Alfred either, but he had lately heard that Alfred was expected in Manaus: "I intend to write him by the first canoe that leaves for that place."[26]

But at the time, June 1851, Wallace was ignorant of all this as he pressed further into the paradise of uncharted realms. A little farther upriver, he and his small crew came upon another great maloca, of the Wanana people, where a ritual dance was being concluded. There were some two hundred people inside, largely naked but for paint—red, yellow, purple, black, and white applied in patterns of diamonds, spots, and stripes. The women wore narrow "aprons" of beads ("arranged in diagonal patterns with much taste"), while the men and boys were far more ornamented, with feathered armbands, coronets, and combs; beaded necklaces about their necks; rattles of dried gourd-like fruit about the ankles; and girdles of jaguar teeth about the waist. They brandished their weapons—spears, clubs, bows, and *curabís* (war arrows)—and made music with instruments of reed, bone, and turtle shell. Wallace was delighted and overwhelmed with the "wild and strange appearance of these handsome, naked, painted Indians," the sound of their incomprehensible tongue and exotic music, the taste of the caxirí constantly passed around in large calabashes. Aware that he himself was a spectacle, he could feel "a hundred bright pairs of eyes" continually directed at him from all about the maloca. Afterward he sketched and traded for articles—a decorated spear, carved holders for large ritual cigars, woven baskets, and more—and of course he was also ever on the lookout for birds, fish, and insects to acquire.

They pushed on, pausing briefly at one other maloca before stopping at São Jeronimo (modern-day Ipanoré), then as now a tiny outpost just below the first falls of the Rio Uaupés—raging rapids channeled through a narrow,

A maloca, or communal house, drawn by Wallace on the Rio Uaupés.

sloping gorge studded with great rocks, generating "immense whirlpools" and thunderous rolling waves, as Wallace described it. He had been ill with dysentery since leaving the first maloca; his symptoms worsening and lacking medication, he was beginning to get a bit alarmed—but not enough to pass up sketching new fish species. Or, the next day, to abandon his resolve to ascend the falls, which could only be accomplished by emptying his canoe of all its gear and portaging around the dangerous rapids, wading through adjacent flooded forest. He pressed for the frontier, traveling another week to reach the home of the Tariana people: Jauarité (now Iauarité), "Jaguar" falls, the second great rapids of the Uaupés. Beyond this point, along the border that follows the river and then heads into Colombia, the rapids and falls increase in frequency and ferocity. Wallace was feeling better, but it was unwise to press on in the wet season. They stayed a week in Iauarité, welcomed by the Tushauá and treated, at de Lima's request, to a spectacular "snake dance" in the large maloca there.

This time Wallace got to witness the preparations for the festa as well as the full dance itself, observing how the women decorated each other first, painting patterns of black and red all over their bodies, with circles outlining their breasts, curved lines about their hips, and bright vermillion spots on their faces. They then decorated the men, not only with paint but with bracelets, anklets, copper earrings, plumed combs, and cords of monkey hair adorned with egret or harpy eagle feathers hanging down their backs, each crowned with a spectacular *acangatára*, a headdress of red-and-yellow macaw feathers. Wallace noted that the Indians seemed to follow nature, so to speak, in regard to the ornamentation of the sexes, thinking of how male birds tend to be more flamboyantly ornamented than females—a subject that was to become of great interest to him and Darwin in the years to come.

For the snake dance, two thirty- or forty-foot "snakes" about a foot thick were fashioned from branches and bushes bound with cipó vine cords, with fearful "heads" consisting of bunches of *Cecropia* leaves painted red. Each was carried by a dozen men in an undulating back-and-forth procession just outside the great maloca, before an audience of three hundred or more assembled inside. Closer and closer they came, advancing and retreating repeatedly until the "snakes" entered the maloca, mock fighting and dashing past one another until escaping outside. There were other performances too. Circles of men and women stamped their feet in unison in one, and in another young men ritually drank a bitter liquid, hallucinogenic caapí, shaking their weapons and grimacing, some then running madly about.[27] Adding to the festive and otherworldly atmosphere was a continuous drone of music, and a great fire at the center of the maloca illuminated the costumed Indians, their shadows dancing on the walls and high-sloped ceiling. Three attendants ran about, keeping everyone's calabash full, while outside young boys showed off their prowess in jumping over and through fires. A thick ten-inch cigar twisted in leaves was handed around in a great carved two-foot-long forked holder. Wallace passed on that, he said, but drained his calabash with gusto and pleased the wife of the Tushauá by declaring it *purángareté* (excellent)—despite knowing this special caxirí was made with cassava "processed" by being chewed and spit out by "a parcel of old women." It was another astounding experience, and he "longed for a skilful [*sic*] painter to do justice to a scene so novel, picturesque, and interesting." Indeed, he was the first foreigner to witness and describe such a ceremonial dance.[28]

The party left Iauarité about 24 June and after a few days at Juquira were back at Ipanoré by July, where they stayed two weeks. Wallace carefully recorded his observations and obtained some first-rate collections by trade, barter, and hired hunters. Anthropologically, there was his array of artifacts and sketches, with extensive notes on the customs, languages, and lifestyles of the peoples he met, noting curiosities among their foods like the "occasional luxury" of termites and plump-winged leaf-cutter ants—grasped by the wings and eaten like strawberries held by their leafy green crown—but having to resort to earthworms in the wet season. (The worms made it easy for foraging humans, as they sought refuge from the floods by climbing trees, where they could be found mingled together in epiphytic tillandsias.) Among the outstanding birds he collected was a mixed flock of nearly one hundred *live* ones, including trumpeters, curassows, a host of parakeets, and nine handsome black-headed parrots (*Pionites melanocephalus*). He found several new-to-science butterfly species, including an exquisite clearwing, two swallowtails, satyrines, and more, and collected the odd mammal, including an anteater and a monkey. Always, there were fish and palms to sketch, and even though he was not collecting plants, he was dazzled by the "complete natural orchid-house" he came upon in one brushy area of cleared forest, where in short order he found thirty or more species ranging from elfin orchids no bigger than a moss to large epiphytic species: "I had never seen so many collected in one place."[29]

Amazingly, on top of the rigors of travel, note-taking, and collecting (itself quite a job procuring, pinning, skinning, processing, labeling, and safeguarding specimens . . .), he continued his survey of the river with his trusty prismatic compass, pocket sextant, and watch. He had long since broken his thermometers, so there was no means of estimating elevation above sea level, but he was able to estimate the heights of the few striking granitic peaks he could see rising above the forest of the lower Uaupés. These are plutons, ancient rocks pushed through even more ancient rocks, representing the core of the continent: modern technology reveals them to be Mesoproterozoic in age, 1.5-billion-year-old granites intruded through granites a half-billion years older still; so old the eroded landscape was now mostly flat to gently undulating, the geology mostly obscured by thick alluvial deposits and thicker forest. The local rocks are best seen in and around the rapids of the river, where they are exposed—not a good time or place for study, but he did his best to sketch dome-shaped masses of rock and rock samples, noting that they mostly contained very little mica but were streaked with veins of quartz. Larger quartz deposits were the source of the painstakingly shaped and pierced pendants of the Wanana men. He was told the prized stone came from far upriver, and he imagined maybe from the base of the Andes. That is possible, though masses of quartzite are found closer, at the 2,400-foot (733 m) Serra Tunuí, now one of thirteen named peaks in the vast thirty-thousand-square-mile Alto Rio Negro Indigenous Territory. It is easy to imagine trade from there to the malocas along the Uaupés via the smaller rivers Içana and Acari and then long winding igarapés.

His discoveries and collections made the trials and tribulations of travel worthwhile, the worst of which were the "countless myriads" of *piums* (blackflies), the agonizing *bichos do pé* (the chigoe flea, *Tunga penetrans*), and the horrific vampire bats that feasted on exposed legs and feet at night: one morning in Juquira they awakened to the "ghastly sight" of de Lima's legs "thickly smeared and blotched with blood," vampire bats being messy eaters. With a shudder, he tried to keep well wrapped at night no matter the tropical temperatures. But Wallace also rather matter-of-factly recorded another form of parasite, if not predator: reprehensible human parasites. His friend de Lima turned out to engage in human trafficking and even murder. It was a very dark side to the already murky business practices of the traders, technically illegal but countenanced and even supported by the authorities. Raiders, usually Indians, would be commissioned to travel upriver and attack the malocas of distant tribes, kidnapping young boys and girls and killing their families and others in the process or rounding up everyone not killed or escaped to press into service as slaves. On this trip Wallace learned that de Lima had hired and armed an Indian named Bernardo to kidnap two young girls for clients in Manaus. This is why they were waiting at Ipanoré, for Bernardo's return with his captives. Wallace credulously repeated the pathetic justification commonly given for this barbaric practice. "There is something to be said too in its favour," he writes, giving the victimizer's line that this can actually be for the victim's own good since the person would otherwise likely fall prey to the Indians themselves as they constantly war and wantonly kill one another. In so many words, he seems to be saying that the thirst for slaves by the whites has a silver lining: it induces the warring tribes to spare many captives they would otherwise kill so they can sell them, goes the logic. As for the kidnapped children, they are "brought up to some degree of civilisation," and although sometimes mistreated "they are free, and can leave their masters whenever they like."[30] As if. Wallace tended to have a blind spot when it came to certain friends—or fiends. Bernardo returned empty-handed that time; he had taken every precaution, he said, but suspicions were aroused. He was sure he would succeed next time.

———

Wallace's thoughts were more and more trending toward home, but he was not done exploring and observing. He wanted to undertake one last expedition before quitting Brazil and for some time had entertained the idea of a journey to the high Andes, further traveling in Humboldt's footsteps. Now he was having second thoughts: "From what I had seen on this river, there is no place equal to it for procuring a fine collection of live birds and animals; and this, together with the desire to see more of a country so interesting and so completely unknown, induced me, after mature deliberation, to give up for the present my intended journey to the Andes, and to substitute another voyage up the river Uaupés."[31] He would instead ascend at least to the *Yuruparí* (Devil) rapids, he decided, a monthlong voyage to the ultima Thule, a land beyond the known world where the White umbrella bird was said to make its home. Was there such a species, sister to the black species so common throughout upper Amazonia? And if so, how and why do such rarities occur where they do? Managing to collect so singular a bird would help make his name and provide more data on species, varieties, and their geographical distribution. He had to try! Not that this was an avian Moby Dick, a bird to obsess over. He was not altogether convinced that it existed, despite the assurances of traders and Indians, but if it did . . .

First, however, he had to journey fifteen-hundred miles to Manaus and back. At this point he had thousands of specimens to sort, pack, and ship to England: he had left many behind in Manaus when he had first ascended the upper Rio Negro nearly a year earlier, many others at Nossa Senhora da Guia following his expedition to Venezuela, and now he had his extensive Uaupés collections of just the past two months, which contained everything from hundreds of pinned insects, preserved skins, pickled fish, and Indigenous artifacts to a flock of incessantly squawking, cawing, screeching, twittering birds, some in cages and tamer ones wandering and flying about. He could not let his collections lie around months more, risking their ruin by damp and pests. Besides, he needed to restock, especially his supply of bartering items, and the timing seemed right—if he could complete this trip quickly, he reckoned, he could be back on the Uaupés by the start of the dry season, the best time to attempt to ascend the many *cachoeiras* (waterfalls and rapids) beyond Iauarité. If all went well, he might get to Manaus and back, continue to the upper Uaupés, and then return and descend all the way down to Belém, sailing for England all in a year, maybe by July or August of 1852. The plan cheered him tremendously, especially the thought of actually heading home—he had not realized how homesick he was becoming. *Home*—that distant land seemed a paradise to him now: "There was a pleasure in the mere thought, that made me leap over the long months, the weary hours, the troubles and annoyances of tedious journeys, that had first to be endured."[32] Never, he mused, did he long to be away in the tropics with half the earnestness he now felt about returning home.

Wallace headed down the Uaupés to the Rio Negro, then turned north to Guia to fetch his collections. That took a few weeks, at the end of which he still could not find enough men to help him descend the rapids safely. He relocated to São Joaquim, at the mouth of the Uaupés, hoping to hire more men, but that took another two weeks. *Patiencia* was the watchword for river travel in Amazonia. He was finally able to depart on 1 September 1851, hiring an expert pilot to safely navigate the Scylla and Charybdis of the São Gabriel rapids: "All depends on the pilot."[33] Successfully shooting the rapids, the rest of the trip proceeded smoothly, with two tame parrots in tow: a beautiful black-headed species, *Pionites melanocephalus*, sporting light orange on its neck and thighs, and a larger red-fan parrot, *Deroptyus accipitrinus*, with gleaming blue-and-red-striped feathers. There was good collecting as they went—a six-foot alligator, new fish to draw, small red-headed turtles. Friends and acquaintances hailed them from riverside sitios along the way, taking a "friend's privilege" to prevail upon him to procure a miscellany of items they needed in Manaus: a pot of turtle oil for one, a flagon of wine for another, a guitar for a third; the *delegarde* (chief of police), would appreciate a pair of cats, *muito obrigado*, and his clerk needed two fine-toothed ivory combs. They would pay later, in coffee or tobacco. If only he could have induced them to pay in rare birds and insects . . .

Arriving in Manaus on 15 September, he was reunited with Richard Spruce, who had been stuck there for months for want of hired men to crew his canoe. The botanist made the most of it, collecting all around while residing in Johann Natterer's old abode—hallowed ground for naturalists like him. Wallace joined him there and received a backlog of months-old mail—including an alarming letter from the British vice-consul, Mr. Miller, alerting him to Edward's grave illness. He was apprehensive and puzzled. The letter suggested that Edward was not likely to survive, yet there was no following letter reporting his death. But neither was there any letter from Edward himself, or anyone else with news of his brother. He was left in a terrible state of suspense, and Spruce was equally in the dark. Wallace stayed with Spruce just two weeks while he busily prepared his specimens for shipment and prepared for his much-anticipated "last hurrah" of a journey. These things took time. He first had to procure lumber to construct boxes for his insects and other specimens, then packing cases sturdy enough to withstand rough handling by a succession of stevedores and coachmen on the long journey from Amazonia to Stevens back in London. Leaving what was now his fifth great consignment of specimens for shipment with their old friend Senhor Henrique, Wallace was off again around 1 October 1851. Spruce accompanied him in his own small montaría for a day of collecting. To Wallace's delight he caught what he was sure was an undescribed species of *pacu*, a piranha relative, and Spruce was successful in finding some new shrubs and trees in flower. Stopping along the river, the naturalists worked up their specimens and notes, Spruce with his plant press and Wallace with sketch pad and pickling cask. Wallace pushed on, stopping at sitios for this and that—procure some large hooks, repair his gun, prepare turtle *mixira* (cooked meat preserved in oil)—and of course collecting: there were new fish, a fine matamata, the bizarre river tortoise *Chelus fimbriata*, and a full-grown male manatee.

These all took time, and in retrospect Wallace perhaps should have hastened his efforts to reach higher, healthier ground. He did not get far before he came down with a raging fever, but, self-medicating, he pressed on despite being in and out of a reverie, if not delirium, thinking about Edward's fate and that of his brother John in California. He got better after about a week, and the small party slowly paddled on. But then one of the Indians assisting him came down with fever, then the other. He was two days below São Gabriel and desperately needed new assistants and a pilot to help navigate the dangerous rapids. He downsized to a smaller canoe, and after several false starts—the difficulty of hiring a crew frustratingly compounded by paying handsomely in advance only to have his hired hands abscond—he was finally able to ascend the rapids. It was now mid-November; he made it back to São Gabriel and had just left on the next leg, heading to São Joaquim at the mouth of the Uaupés, when fever struck again. This time he was rapidly incapacitated, to the point where de Lima did not expect him to survive. His symptoms point to the deadliest form of malaria, caused by *Plasmodium falciparum*: daily shivering chills, raging delirium-induced fevers, and profuse perspiration in turn, each bout leaving him weaker and weaker until, emaciated, he was unable to speak intelligibly or move in his hammock. After some weeks he was able to send word to Spruce, who had finally made it just below São Gabriel and described Wallace's situation to John Smith at Kew: his friend was "almost at the point of death from a malignant fever," he wrote, now so weak that "he cannot rise from his hammock or even feed himself. The person who brought me the letter told me that he had taken no nourishment for some days except the juice of oranges and cashews."[34] Spruce went on to lament how the "fevers of the Rio Negro have proved fatal" to several people he knew, including Wallace's brother Edward—since Wallace had left Manaus, Spruce received the bad news from Belém. Now he hastened to see Wallace on his deathbed.

Even quinine did not seem to conquer the parasite, at least not quickly. But it surely helped, and slowly, slowly, he came back. First, he could speak and turn himself over in his hammock. Soon he could stand without aid, then walk across the room with a makeshift cane. With the return of his appetite and,

incrementally, his strength, Wallace was taken down to São Gabriel to see
Spruce. It was then that he was given the sorrowful news of Edward's untimely
death. He later learned his brother's final words: it was "sad to die so young."
Those words surely weighed heavily on his big brother.[35]

Two and a half months after first falling ill and still not fully recovered,
Wallace resolved to ascend the river. Despite the usual problems crewing a
canoe, he made it to Iauarité in under two weeks and soon entered a realm of
successive malocas and rapids. The geology puzzled him. Inspecting the rocks
at each rapid, he noted that they were granitic and yet looked stratified, like
sedimentary rocks, and so strongly tilted as to be nearly vertical. Further up,
he thought the granite looked like it had been remelted. He was essentially
correct: geologists would later recognize these rocks as shape-shifters, originally sedimentary sand and siltstones that were so greatly squeezed and heated
as to melt and recrystallize their constituent minerals, uplifting and tilting
them in the process. Such *metasedimentary* rocks, as they are called, speak of
erosion and deposition on a long-ago continent, then inexorable metamor-
phism deep within the earth, uplift, and eventual exposure by erosion, all playing out over timescales far more ancient than Wallace and his contemporaries
could have imagined at the time.

There was little time for geology, however, as the going was tough. Each
waterfall and rapid, the cachoeiras, had a name: *Uacú* (a fruit) cachoeira . . . *Uacará*
(egret) . . . *Mucurá* (opossum) . . . *Tyeassu* (pig) . . . Macáco (monkey) . . .
Baccába (a palm) . . . then macaw, gecko, armadillo, tortoise, toucan—a veritable menagerie of waterfalls and rapids great and small, representing all the
animals and many of the plants significant to the Indigenous people. His crew
strained against the raging current of each cascade day after day, heaving and
pulling the canoe with *espias* (tow ropes) of cipó vine. When they were lucky,
they didn't have to unload the canoe, but more often than not, that was too
dangerous, and they had no choice but to unload, hoist, reload, unload, hoist,
and reload, again and again up worse and worse rapids until reaching *Carurú*,
the innocent-sounding "water plant" rapids, named for the tasty *Podostema*
plants covering the rocks—prized even by the large black pacu fish, which the
locals readily caught with carurú bait. The rapid was a foaming torrent of
whitewater raging through mammoth rocks, quickly losing fifteen to twenty
feet of elevation. It was quite a sight. The river whipped white as it poured over,
around, and between a tumult of dark granite boulders festooned with bright-
green plants, thriving in their half-drowned state. Here they had to hoist the
canoe along the river's edge over dry rock, unloaded (again!) and pulled along

on a succession of cut branches and logs to avoid tearing up the bottom. It was "the greatest difficulty for my dozen Indians, their only resting-place being often breast-deep in water, where it was a matter of wonder that they could stand against the current, much less exert any force to pull the canoe."[36] They sent for help, and it took an eventual twenty-five men to raise the canoe above the rapids. The bottom needed repair despite their efforts. The canoe was just too large and cumbersome for the even more formidable cachoeiras to come, so he bargained with the Tushaúa to purchase his large *obá* (dugout canoe)— the more sensible watercraft of the upland rainforest rivers, put to the test over millennia by the many tribes who made their home along the river. It cost him an axe, shirt, trousers, two knives, and some beads.

When he paused in fretting about his canoe conundrum, Wallace realized that he was in a land most interesting: Carurú was home to the Turkano-speaking Kótirya people. He admired their large maloca, "very tastefully" adorned with a crosshatch pattern of diamonds and circles of red, yellow, black, and white, and sketched striking petroglyphs found on rocks at the river. Best of all, he was extremely fortunate to witness the playing of the so-called *Yurupari* music made with sacred instruments, great bassoon- and trumpet-like horns made of spirally twisted bark with a mouthpiece of leaves. He first heard them in the distance, coming closer and closer until a quartet of Indians appeared, playing the horns of different length, the longest of them about five feet or so and waving side to side and up and down as played. These were no wandering minstrels of the rainforest but medicine men announcing the start of a ceremony. At the sound of the first distant trumpet blast, the women ran for cover—according to their tradition, seeing one of these instruments, even accidentally, meant death, usually by poisoning. Tremendous religious significance was, and is, attached to these instruments—when, a few years later, Spruce managed to acquire one for Kew, he had to keep it well hidden deep under the cargo of his canoe, as he knew his crew would refuse to board a vessel bearing such a sacred object. Today the Yurupari instruments are still only seen and handled by men, brought out for an elaborate annual celebration of the new year and periodically for lesser ritual celebrations of the arrival of certain tree fruits. One of these is the fruiting of the lofty, buttressed *ucuquí* or ocoki tree, *Pouteria ucuqui* (Sapotaceae), also known to the Kótirya as *puch-pee-á*. Wallace learned firsthand just how esteemed ocoki fruit was when, paddling along one day, his crew "all suddenly sprang like otters into the water, swam to the shore, and disappeared in the forest. 'Ocoki,' was the answer to my inquiries as to the cause of their sudden disappearance; and I soon found they had discovered an ocoki-tree, and were loading themselves with the fruit to satisfy the cravings of hunger."[37]

On they pushed, farther and farther up the great river, Uaupés on the Brazilian side, Vaupés on the Colombian, until one great bend in the river where they left Brazil behind. It was terribly slow going, rapid after rapid: on average nearly six per day and no fewer than twenty-eight in under a week! By 11 March Wallace found himself among a different people: the Kubeo, or Kubéwa, also Turkoan speaking. He called their maloca Uarucapurí, most likely modern Aracapuri. These were "handsome . . . clean limbed and well painted" Indians, bearing white-beaded armlets and necklaces and large ear gages decorated with a white porcelain-like material. He learned that they were cannibals, too, consuming people of other tribes killed in battle and sometimes even raiding for that purpose. Further on he arrived at Goatsucker falls, Uacoroúa, a ten-foot drop that required once again completely unloading the canoe and hauling it up, during which an unluckily timed downpour sent them scrambling to shelter the cargo. Still at least a week from the fabled Yuruparí falls, he had about had it. The collecting was okay, but the white umbrella bird, the rarest of the rare that had lured him so terribly far, seemed more and more to be a fiction. Accounts of these birds grew more conflicting and less certain now that he had actually made it to the ethereal realm where, he had been assured, these ghostly birds of the deep forest are found. If it existed at all, it was an occasional color variant, he reluctantly concluded, a rarity of the curio type, not one to make London's scientific society swoon. On top of this, another rarity he was promised, a painted turtle he was sure was new, managed to escape, and no others could be found. His months lost to illness in São Joaquim weighed heavily; still weak and exhausted, he gave up on reaching his ultima Thule but was consoled by the fact that he had made it to lands no European traveler had reached before. The glass-half-full Wallace mused that the voyage had been favorable, all things considered, ascending without loss of life or limb (or canoe) a river "perhaps unsurpassed for the difficulties and dangers of its navigation" and surmounting some fifty cachoeiras, "some mere rapids, others furious cataracts, and some nearly perpendicular falls."[38] He went as far as Múcura, not far from modern-day Mitu, and stayed two weeks. Hiring hunters and trading with his stock of fishhooks and beads, his collection grew: a brown woolly monkey first described by Humboldt, several parrots, new fish to draw, assorted insects, and, as if to taunt him, plenty of jet-black umbrella birds. He also carefully recorded what he could learn of the customs and vocabulary of the Kubeo and bought several of their ornaments and implements.

On 25 March 1852, some six weeks since departing São Joaquim, Wallace left Múcura to begin the arduous journey downriver in his new obá and a borrowed canoe. Reaching Aracapuri the same day, however, where he hoped to hire a native pilot to navigate the succession of rapids and falls below, he found the maloca and surrounding village oddly devoid of men. It turned out that a pair of slavers—a trader, Chagas, whom he knew well, and an official called *Tenente* (Lieutenant) Jesuino—the newly named "director of Indians" for the rivers Uaupés and Içana and described by Wallace as "an ignorant halfbreed"—had led a large party of Kubeo men from Aracapuri to attack a maloca of the Carapanas people upriver. Wallace was able to persuade the son of the local Tushaúa to help pilot and with his help managed to make it to the great Carurú rapids with the canoes intact. There a fleet of canoes soon arrived— Chagas and Jesuino, with a group of captives destined to be sold or given as gifts in Manaus. The raiders had killed seven Carapanas men and one woman and taken twenty captives, all women and children but for a lone male, keeping them tied up. Shortly before departing Múcura, Chagas asked Wallace to lend or sell him and Jesuino the large canoe he had recently purchased from the Tushaúa. Wallace declined, his departure imminent, but now it was clear what the canoe was for, and he found himself targeted by the vindictive Jesuino for refusing to comply. Threatened by Jesuino, his young pilot absconded. He then instructed a party of other Indians to take Wallace partway downriver and abandon him, hoping he would be a goner. Which they did; he was left with one man and one boy for each canoe, which normally required six or eight strong paddlers to shoot the rapids. Incredibly, they made it to Iauarité, "much to the surprise of Senhor Jesuino" but not before almost losing a canoe—with Wallace aboard—as it was being lowered down the last of the falls by espias.

By the end of April 1852, Wallace made it to São Gabriel, briefly reuniting with Spruce, then pushing on. He had exited the Uaupés with fifty-two live animals in addition to his prodigious preserved specimens, artifacts, notes, and sketches. Now his menagerie had thinned—one of his marmoset monkeys had devoured two of his parrots, another had been lost going down one of the rapids, and several other birds had expired, probably for lack of proper food. He was struck by malarial fever again, thankfully not nearly as severe, developed a seriously inflamed foot owing to chigoe fleas, and coped with constant rain yet was determined to complete all he had set out to do, including his river-mapping project. There were a few stretches on one side or the other of the mile-wide river that he had been unable to survey on the way up, and he figured he would fill them in on his way back down. He was foiled in that effort for one important section, however, as he could not find a pilot to help navigate the east bank of the Rio Negro. He dared not go it alone, for fear of being lost for weeks in the labyrinthine infinity of islands and channels on that side of the river. It remained terra incognita to him, so near and yet so far. The Rio Negro map Wallace eventually published states, "This side of the river not known" in neat lettering along the east bank between the Rio Padauari and Barcelos.[39]

He reached Manaus on 17 May. Before he could leave with his collections in tow, there were a great many details to attend, so he rented a house and tried to tend to his animals and sore foot while trying to shift from navigating the river to the even more labyrinthine government bureaucracy. He soon discovered, to his horror, that the six large cases of specimens he had left with Senhor Henrique the year before had never been shipped to London! A full year's worth of precious collections had just been languishing there, evidently because, he sardonically wrote, "The great men of [Manaus] were afraid they might contain contraband articles, and would not let them pass."[40] He took a dim view of the pompous and semicompetent clerks and officials he had to deal with, but polite compliance was the only option. He duly unpacked his cases and paid duty on the contents, then jumped through the hoops necessary to secure the passport required to leave the country. For this he had to first advertise his intention to leave in the newspaper, then complete an array of forms and get them doubly signed and stamped—one office for some of the stamps, another across town for others. Finally, papers in order and having paid the requisite fees, "I was at liberty to leave [Manaus] whenever I could; for as to leaving it whenever I pleased, that was out of the question."[41] He acquired some additional live animals and lost others, so by the time he left Manaus on 10 June 1852, on a canoe arranged by the kindly Henrique, just thirty-four remained: five monkeys, two macaws, twenty parrots and parakeets of a dozen species, five small birds, a beautiful white-crested Brazilian pheasant, and a toucan. Well, thirty-three: he had no sooner boarded when the toucan was lost.

Home loomed large for Wallace: he was headed down the mighty Amazon for Belém at last, then across the ocean where his family and fortune awaited—fortune both financial and scientific, his astounding and hard-won natural history and ethnological collections guaranteed to fetch considerable sums, his intellectual riches guaranteed to earn him grand entrance into scientific society. His four years' worth of packed notebooks and sketchbooks, chock full of drawings of palms, fishes, landscapes, the Indians, and their crafts and artifacts, contained material enough to mine for years to come. He would write a travel memoir in the spirit of those of his heroes, Humboldt's *Personal Narrative* and Darwin's *Voyage of the Beagle*, as well as authoritative books on the novel fish of the Amazon basin and the remarkable palms and their many uses. There were papers to read at the learned societies on subjects geological, entomological, ornithological, and ethnological. Surely bigger philosophical questions too: geology, species, distribution—these were all of a piece for Wallace. One hint of his engagement with big-picture questions at the time comes from a newsy and congratulatory letter from Spruce, written from Peru almost a decade after their last parting, in Saõ Gabriel. By then Wallace's stature had grown, his remarkable role as co-discoverer with Darwin of the principle behind species change well known among naturalists. "If you recollect our conversations at Saõ Gabriel," his friend reminded him, "you will understand that I have never believed in the existence of any permanent limits—generic or specific—in the groups of organic beings. From what I can gather of the scope of Darwin's work I think I should push the doctrine even farther than he has done." Spruce went on to wonder, "By the bye, have you, acting on the principle of *Natural Selection*, yet taken unto yourself a Signorina . . . as you once hinted to me you proposed doing?"[42] Imagine them there in Spruce's little house, Natterer's old abode so "classic to the Naturalist," Ricardo and Alfredo as they called one another, musing on the nature of species and varieties over tumblers of cachaça, gently rocking in their redés in the cool of the tropical night.

But now, Ricardo was preparing to ascend the Uaupés as Alfredo descended to Belém, his final port of call before reaching England's shores.[43] En route he stopped at Santarem hoping to see Bates but was disappointed to have missed his friend by just a week—Bates was off to collect up the Rio Tapajos. He landed in Belém on 2 July, and he lost no time preparing for his passage home. He was warmly greeted by his old friend "Mr. C.," the Englishman who had assisted Wallace and Bates soon after they had first arrived in Belém, and found the city improved some, with a few handsome new buildings and avenues lined with almond trees. But the balmy weather belied the dangers of the disease still rampant in the coastal town: if the overcrowded cemetery where his brother was interred in a mass grave was not reminder enough of that, it was underscored by his own returned illness, bouts of malarial shivering and fever that alarmed him. He longed for England, his sister, brother-in-law, and mother; his friends in Neath; and of course scientific society: *imagine* his reception, he dreamed, as he strolled into the scientific salons of London bearing lively rainforest rarities, a jewel-like bird on one arm, perhaps a monkey on his shoulder! He had done it—four long years, surviving a host of dangers and disease, bearing with equanimity endless frustrations, complications, and machinations. He had followed in Humboldt's footsteps and beyond to realms no European had ever seen, where he had dwelt among storied tribes. And despite difficulty after difficulty, seemingly insurmountable at times, he had collected thousands of exquisite birds, insects, fish, reptiles, mammals, and more. *He did this*, the sometime surveyor and self-taught naturalist, albeit at a terrible price: it was inevitable that the tropical paradise of his imagination would be lost, driven before the wisdom and knowledge of hard experience. The exhilarating and often transcendent experiences were real enough but tempered with the harsh realities that included, most acutely, personal loss and the guilt that would always be his to bear.

He booked passage on the *Helen*, a 235-ton brig bound for London under the command of Captain John Turner. Ill and with mixed emotions—hopeful, wistful—Wallace "bade adieu to the white houses and waving palm-trees" of Belém on 12 July 1852, four years and forty-seven days after his arrival there with Bates. He was twenty-nine years old.

CHAPTER 6

DOWN BUT NOT OUT

HE SLOWLY CIRCLED the towering funeral pyre that was his ship, an unspeakable calamity the enormity of which had not yet registered. Nearly four weeks out from Belém, the weather fine and making good time, the captain caught a whiff of smoke one day after breakfast and asked Wallace to go investigate with him. The merchant vessel was carrying a large quantity of India rubber and "balsam of capivi," also known as copaiba balsam, an essential oil obtained from leguminous rainforest trees of the genus *Copaifera*, then as now a prized ingredient in varnishes and ointments. Prone to combustion, kegs of copaiba balsam oil are most safely packed in damp sand. Captain Turner's first mistake was packing them in rice chaff. The heat of friction from the ship's incessant motion soon ignited the oil in the already sweltering hold, but it may have slowly smoldered on in that airless space had the captain not made his second mistake: he and the mates foolishly fed the fire life-giving oxygen by opening hatches and cutting holes in the deck as they investigated the source of the smoke. In minutes the fire roared to life and by noon had gone from smolder to inferno, sending the crew scrambling to deploy the two lifeboats—a longboat with sails and a smaller gig, neither much used and so riddled with gaps between planks desiccated from years of baking in the tropical sun.

The call came to abandon ship, and the crew was in a frenzy of activity—get the boats lowered; find caulking, cork, oars and pins, spars for a mast, and sails that fit; roll out kegs for food and water; grab what you can save. The captain pulled together his charts and books, sextant, chronometer, and compass, the crew their clothes and other personal effects. Wallace descended to his cabin, "suffocatingly hot and filled with smoke," where a strange feeling of apathy descended over him. Time slowed, and the frantic cries of the crew grew distant. As if in a trance, he looked about, picked up his watch, a small change purse, and a sheaf of palm and fish drawings fortuitously laying out, and put them into a "small tin box" of shirts. He inexplicably left a large port-folio containing a great many other drawings as well as notebooks, books, and papers; retrieving any of his collections in the hold was out of the question. Snapping out of it, eyes watering and gasping for air, without a second thought he scrambled up on deck, where some of the mates were using a bucket to lower provisions and other items into the boats. The deck was getting hotter and hotter; his box was lowered down as he clambered over the rail, fighting to keep his balance with the constant rise and fall of the swell. Down he plunged by the rope, badly skinning his hands in the process, tumbling in among the confusion of items sloshing around in the leaking boat. He joined the mates in baling for their lives, his flayed and raw hands stinging painfully in the salt water as they battled to keep the boats afloat.

Now Wallace and the crew watched—stunned—from the comparative safety of the bobbing boats as the insatiable fire steadily consumed the ship. The shrouds and sails browned, then blackened, then burst into flame. The burning mainmast of the rolling ship snapped about twenty feet up, while the foremast, a pillar of flame, lasted an hour more. Flames leaped from the hatches, the decks were a mass of fire, and the bulwarks broke off and fell flam-ing into the sea with a hiss. Most of his menagerie was dead by now, but a few surviving monkeys and parrots (their wings likely clipped, preventing them from flying) made it to the bowsprit. Rowing as close as they dared, they tried in vain to coax the hapless animals into the boats. They had to give up, between the intense heat and the growing danger of being stove in by one of the many fallen half-burnt timbers tossed around by the waves. Fire crept up the bow-sprit. Most of the uncomprehending animals trapped there dashed into the flames to their doom; they managed to save a single parrot that had fallen into the sea. As night approached they moved a safe distance from the ship, but not too far, hopeful that the fatal flames would now be their salvation, a great beacon on the high seas visible for scores of miles around. Drenched and exhausted, they baled all through the night in the angry red glare of the blaze, embers and ash carried aloft by the heat raining down all around, the ship's ironwork glow-ing red-hot. If the glow and heat did not render the scene hellish enough, when by chance the heaving and rolling ship gave the awed men a view into the in-ferno of its molten interior where the hull had burned away, the sight must have been the very picture of the hell of their nightmares: "A huge caldron of fire," with the cargo of rubber and balsam "forming a liquid burning mass at the bottom," a "fiery furnace tossing restlessly upon the ocean" that surely re-called vivid images from Dante's *Inferno*, complete with rains of fire, boiling

Catastrophe at sea: the burning of the brig *Helen*.

pitch, and the "boiling crimson flood" of the rivers of hell, from within which "the parboiled sinners cried in pain."[1]

Their position was 30.5° N, 52° W. Reckoning they were some seven hundred miles from Bermuda, about a week's sail in a fair wind, they left the now mostly sunken, charred, and smoldering wreck of the *Helen* the next day. The enormity of Wallace's loss was beginning to sink in: the past two years of collections, thousands upon thousands of hard-won specimens burnt—think of it!—on top of the irreplaceable sketches, notes, and observations—the three most interesting years of his journal! So much for having one of the finest collections in Europe, striding victoriously into London's scientific salons, writing his many treatises and papers. As painful as this surely was, Wallace was as philosophical about these losses as he was the precarious position they found themselves in. Despite being constantly wet from spray and rain, blistered by the sun, and threatened with foundering in heavy squalls, as well as suffering strict rationing of their dwindling food and water, the ever-optimistic Wallace cheerfully noted that they did not have to bale as much after awhile (the saltwater-swelled wood now narrowing the gaps between the planks), not to mention the interesting birds and flying fish, curious jellies, and schools of "superb" dolphins of gorgeous green, blue, and golden hues: "I never tired of admiring them." And the meteors! Thanks to an early waxing moon, he saw several, very likely the earliest arrivals of the swift, bright Perseid meteors, streaking silently across the black-velvet vault of night amid myriad sparkling stars. "In fact," he realized, he "could not be in a better position for observing them, than lying on my back in a small boat in the middle of the Atlantic."[2]

The avid geologist in Wallace would have been equally interested in what lay beneath him. Knowledge of the seafloor was fragmentary in those days, especially that of the abyssal seas, but he would surely have marveled at the epic tale of earth history that ocean floor geomorphology is now understood to reveal— and that the region then below him played a major role in that discovery. They were shipwrecked to the west of the mid-Atlantic ridge, due south of the Corner Rise Seamounts and nearly atop the west end of the Atlantis Fracture Zone, a great east–west trending scar running perpendicular to the mid-Atlantic Ridge and one of many such gashes slashing across the deep ocean basin at near-regular intervals as the North American and African tectonic plates continue to slowly pull apart. Trying to make for Bermuda but buffeted by winds from the southwest, the shipwrecked band trended northwest, almost in a line with the New England Seamounts, a chain of lofty—though undersea—volcanoes in an age gradient spanning some 83 million to 103 million years old east to west. Wallace, who was to write much later on the nature of continents and ocean basins, was unknowingly traveling another fascinating borderland of time and space.[3]

After ten harrowing days at sea, they were rescued at 32.8° N, 60.45° W, having traveled some 450 nautical miles from the site of the wreck and still about 215 miles from Bermuda. They were fortunate that the ship *Jordeson*, bound for London from Cuba with a cargo of timber, happened upon them. Sun blistered, suffering "exceedingly from thirst," and almost in despair of being rescued, they were perhaps not quite as bad off as the becalmed crew of Coleridge's *Ancient Mariner*—"As idle as a painted ship, Upon a painted ocean"—but they could still relate to "Water, water, every where, And all the boards did shrink | Water, water, every where, Nor any drop to drink." But their travails were not yet past. The rescue doubled the number aboard the *Jordeson*, which Wallace reckoned was "one of the slowest old ships going" and not provisioned for that many mouths to feed. The refugees feared it was not especially seaworthy either. But he was delighted that he finally got to witness waterspouts, a "curious phenomenon" he had long wanted to see. It was surely an awesome sight, as not one but three sinister funnels snaked down from a line of black clouds heavy with moisture, whipping the sea into an explosive fury at their deadly touch: "I had much wished once to witness a storm at sea, and I was soon gratified."[4] But he may have regretted getting his wish: the tornado-like waterspouts are spawned at the leading edge of a storm front, in this case portending a fierce storm that blew for two days and two nights and nearly did them in, splitting the mainsail and rolling them with waves so high the sea poured over the bulwarks, smashed in the cabin skylight, and repeatedly plunged the bowsprit under water. The *Jordeson*'s Captain Venables rested with an axe at his side, answering Wallace's quizzical look by saying it was to cut away the masts in the event that the ship capsized. The incessant "click-clack, click-clack" of the pumps running day and night in a desperate effort to clear the hold of water and prevent them from sinking was more "disagreeable and nervous" than reassuring. More alarmingly, Captain Turner confided to Wallace that if they got "pooped by one of those waves"—flooding the cockpit and cabins— "we shall go to the bottom," adding that although they were not very safe in the *Helen*'s lifeboats, "I had rather be back in them where we were picked up than in this rotten old tub."[5] And it was literally rotten: Wallace's crewmates discovered they could pull out chunks of wood in the forecastle, which explained why they were getting soaked in their beds. Wallace tagged along as the captains went to investigate, finding "sprays and squirts of water coming in at the joints in numerous places, soaking almost all the men's berths."[6]

Latitude 49.5° N, longitude 20° W, 19 September 1852. Approaching the British Isles at last with "some prospect of being home in a week or ten days," Wallace breathed easy and wrote Spruce a full account of his travails.[7] He was understandably optimistic at this point, despite being on food rations, but as the ship limped into the English Channel heading for the port of Deal on the southeastern coast, then one of the busiest ports in England, yet another storm nearly sank them. "We had a narrow escape in the channel," he added in a P.S. to Spruce, "many vessels were lost in a storm on the night of the 29th." The morning light revealed four feet of water in the *Jordeson*'s hold. But then— "Oh! glorious day!"—on 1 October, eighty days out from Belém, they landed, the journey three times longer than expected but "thankful for having escaped so many dangers, and glad to tread once more on English ground." That night he was treated to beefsteaks and damson tart in the company of the two captains— "a paradise for hungry sinners."[8]

His kindly agent Samuel Stevens and Stevens's mother took care of Wallace that first week back, taking him round to a tailor for new clothes and feeding him a steady diet of nourishing food—Wallace felt he was in too frightful a state in his weak and emaciated condition to see his own mother and sister immediately. Fortunately, too, Stevens had insured Wallace's collections, for which the collector duly received £200, substantially less than he reckoned his rarities were worth but enough for a single man to live on, frugally, for a year or more in those days. But his losses were of course incalculable. They were more intellectual than material riches: he had fully expected that his copious notes, records, observations, and drawings, combined with his personal collection of insects and birds—"hundreds of new and beautiful species"—would make his name in scientific society and bring him a comfortable and edifying living. Their loss was a blow too heavy to really comprehend, made all the heavier by the thought of what those four long years of blood, sweat, and tears had cost him: desperately hard work ascending and descending a multitude of rapids and falls; endless haggling, repairing, procuring of boats, supplies, food; the outrages of deserters and thieves and maddening clouds of biting insects; constant attention securing his precious cargo from ravenous insects and fungi that seemed to sprout and entomb specimens in the blink of an eye. He had nearly lost his life not once but three times between severe disease and shipwreck, yes, but what about all those lower-grade bouts of fever, infection, dysentery, nausea? The heaviest burden was surely the death of his brother; the Wallace siblings had now dwindled from nine to three, a heavy burden, too, for their mother.

His collections were to make his name, but now what? He was a collector without a collection. In the stratified society of the time, his prospects must have seemed dim indeed: materially little better off than he had been four years ago, he had not much by way of social connections. Bedraggled and nearly killed for his efforts, he was down and out. Or was he? On the other hand, his name was no longer unknown to London naturalists, thanks to Stevens's constant promotion: four of his previous consignments of collections *had* made it to London, where the savvy Stevens, a fine naturalist himself, displayed the choicest rarities at the learned societies and sold his collections one after another to the museum curators and wealthy armchair collectors of the London scene for good money. What's more, Stevens periodically published extracts of Wallace's letters from the field in the *Annals and Magazine of Natural History* and communicated Wallace's paper on the umbrella bird to the Zoological Society.[9] All of this helped introduce the relative unknown to a diverse and admiring audience as an intrepid explorer and first-rate collector.

Wallace himself made an appearance as Stevens's guest at a meeting of the Entomological Society of London a mere *three days* after his return. The gaunt and limping Wallace must have been quite the pitiful sight. John O. Westwood, president of the society, noted to the assembled fellows that Wallace had lost all of his valuable collections in the tragic fire and "narrowly escaped death in an open boat, from which, after long privation and suspense, and while yet in the midst of the Atlantic Ocean," he and his crewmates were rescued.[10] Wallace published a brief but vivid account of his ordeal in the *Zoologist* the following month, expressing the magnitude of his losses that included an entire fifty-foot leaf of the *jupaté*, or royal palm (now *Roystonea regia*), which, he noted wistfully, would have made for a splendid display in the Botanical Room of the British Museum.[11] He conveyed the immediacy of it all with a postscript commenting that he had left Spruce up the Rio Negro at São Gabriel ("hard at work and in good health") and had just missed Bates at Santarem as his friend departed for the Rio Tapajos.

His health, strength, and name recognition growing, Wallace was soon making the rounds of the Zoological and Royal Geographical Societies too, attending meetings and gaining access to their libraries, thanks to Stevens. His talents clear—and perseverance in the name of science even clearer—Wallace was also granted access to the libraries of the Linnean Society and Kew Gardens and

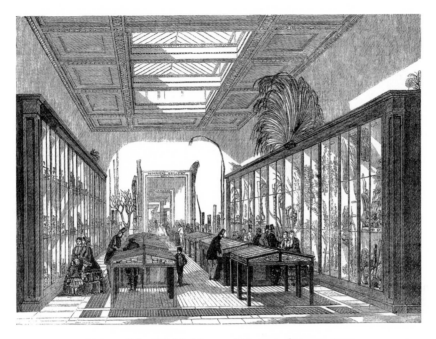

A view of the British Museum's Botanical Room, 1858.

given a pass to the Zoological Gardens (then still only open to fellows of the Zoological Society). Membership in these societies had long been limited to the wealthy and well connected of British science, but beginning in 1851 the new category of "associate" member was created at the Entomological Society admitting "working entomologists"—collectors and agents like Wallace and Stevens. This was in no small part owing to the efforts of Edward Newman, son of a wealthy manufacturer, cofounder of the Entomological Society, and founding editor of the *Zoologist*, whose "big-tent" philosophy in promoting science had opened that journal to dispatches from such aspiring young enthusiasts as Bates and Wallace in the 1840s, their very first publications, such as they were. Stevens was soon a member of the Entomological Society's council, then treasurer, a sign of the slow democratization of British science—well, "meritocratization," if not outright democratization. Members were elected based on their talents and contributions, but membership still entailed dues, and not everyone could afford to be a member. For now, Wallace was content to attend meetings and consult the libraries as Stevens's guest, saving on the membership fees.[12]

Wallace was soon joined with his family at last; he never wrote about his reunion with his mother and sister Fanny, but it had to have been an emotional roller-coaster given the tragic death of Edward and his own harrowing ordeal. Determined to bring them all back together, he rented a house for himself, his mother, and his sister and her husband, Thomas Sims, an aspiring (but not yet successful) commercial photographer. They were moved in by Christmas. The house was near Regent's Park at 44 Upper Albany Street—just up the road from Mr. Webster's carpentry shop where he and John worked when they first came to London. Having gotten word of Alfred's tribulations, John wrote their mother that although it was surely very hard to "lose at one 'fell swoop' what has taken him years to collect and what no money can replace," he had faith that "the name of Sir Alfred Wallace may shine forth as an enterprising collector & author."[13] For his part Alfred missed his only surviving brother, and it would be many years yet before the two would meet again.

Wallace found their new home conveniently located near the Zoological Gardens and not far from Stevens's office near the British Museum, with the Zoological Society's headquarters a bit further south in Hanover Square (the society occupied that building until 1910, when it was forced to move, it is said, because the weight of the books in its prodigious library threatened to collapse the building). It was there at the Zoological Society, on 14 December 1852, that Wallace presented a remarkable paper: "On the Monkeys of the Amazon."[14] Considering his losses, it is surprising in its detail, a paper that seems on the surface to be a straightforward report on primate natural history but in fact reveals a profound insight. In this paper Wallace discusses the geographical distribution of twenty-one monkey species in nine genera, his main point being that the great rivers of the Amazon Basin seem to delineate the primates' range limits to a striking degree. Known as the *riverine barrier hypothesis* today, this paper reveals Wallace's underlying and ongoing research interest: not merely documenting the diversity and natural history of Amazonia but seeking *pattern* as a critical step toward understanding *process*.[15] He lamented that naturalists are too often woefully imprecise in recording the localities of their collections, thinking "South America," "Brazil," or even "River Amazon" sufficient. No, no, no, says Wallace—altogether different (albeit related) species are encountered from one side of a great river to the other, so such vague locality information is next to useless for really understanding species and their multitudinous relationships to each other and their environment. Accurate determination of species' ranges helps answer a host of interesting questions, he pointed out. Are closely related species ever separated by great distances? What landscape features are important in determining the boundaries of species and genera, and why? And for that matter, why do certain rivers and mountain ranges seem to delimit many species and not others? "None of these questions can be satisfactorily answered till we have the range of numerous species accurately determined."[16]

In his travels he took, he said, "every opportunity" to determine the range limits of species and soon found that miles-wide rivers like the Amazon, Negro, and Madeira formed boundaries dividing related species—not just the monkeys but many birds and insects as well. Giving example after example, Wallace took pains to mention not once but twice that the native people know this very well—a nod to the knowledge of the Indigenous people that foreign naturalists would do well to learn from, even as, perhaps audaciously, he dinged famous and esteemed figures like Johann Spix for failing to grasp this fact. He even corrected the august Humboldt, who had maintained that the deafeningly loud howler monkey calls could only be achieved by troops of monkeys calling together. Wallace flatly rejected this, citing Indian accounts and the results of his own observations, which included anatomical dissection. Confidence in his own powers of observation and reasoning continued to grow, and with them his willingness to engage even the most illustrious authorities if he believed he was in the right. Another of his key points was that toward the headwaters where these rivers dwindle in depth and breadth they cease to be barriers, and the same species are often found on both banks. Rivers appear to separate related species only when sufficiently broad—but why?

The patterns that Wallace pointed out in this paper came to be understood in the context of speciation by isolation—the formation of new species by splitting or separating populations, or *allopatric speciation*, in the parlance of modern evolutionary biology. But this was 1852, when the idea of species change was anathema to most naturalists, not a subject worthy of investigation. Not that Wallace was investigating *precisely* this but recall back in 1847 the desire he expressed to Bates that he wanted to travel and really study some group minutely in view of "the theory of the origin of species." "By that means," he declared at the time, "I am strongly of opinion that some definite results might be arrived at."[17] By "results" he meant insights into process. His intuition told him that there is some fundamental connection between the environment—geology, climate, and so on—and how species and varieties arise. If you think about it, the geographical distribution of species and varieties is where the rubber meets the road: mapping distribution is to map interface with environment writ large.

We can thus understand Wallace's point about recording precise locality data, thereby discerning distributional patterns as part of a larger, ambitious endeavor.

Map locally, infer globally. That ambition was as far ranging as his curiosity. Recall his letter from the field to his brother-in-law, Thomas Sims, dreaming of his book plans: a travel narrative, palms, fish, and a comprehensive physical history of the Amazon Basin and its people.[18] Anyone might reasonably assume that these plans were completely derailed by the tragic loss of so much of his material, but in fact between his prodigious memory, the information in the letters, reports, and collections he had previously sent home, and the material he did manage to save from the *Helen*, Wallace was able to realize a significant portion of these goals—and, even more remarkably, consider that he did so largely in the space of a single year, between December 1852 and December 1853!

———

Wallace, now thirty years old, was hard at work in 1853. He was fast becoming very much a part of the scientific scene, regularly attending the meetings of several scientific societies, consulting the collections at Kew and the British Museum (where he briefly met Darwin one day), and visiting the Zoological Gardens. He presented some short but insightful papers at the Zoological and Entomological Societies. In May he had one reporting observations on skippers (a group of insects sort of intermediate between butterflies and moths), in June another discussing entomophagy (the consumption of insects) by the Indigenous people of the Amazon Basin, and in July a third paper on "curious fishes allied to the Electrical Eel." There were also two more substantial papers, again reflecting his keen interest in species distribution. One bookended that productive year that had started with the Amazonian monkeys paper. This time it was insects: "On the Habits of the Butterflies of the Amazon Valley" was read over two meetings of the Entomological Society of London, 7 November and 5 December 1853.[19] Its detail underscored just how closely Wallace observed and recorded the species he collected. Opening with the physical geography of the basin, a "vast alluvial valley" of dense tropical forest through which "the Amazon rolls its mighty flood," bordered by the lofty Andes to the west and the great uplifted granitic tablelands to the north, he pointed out that the conditions were very favorable to the "development and increase"—read: astounding diversity—of butterflies and moths. "Where else in a single locality can 600 species of butterflies be obtained?" This can be done just strolling around Belém! The paper was pretty impressive, treating the distribution and behavior of a dozen butterfly families in some ten pages and peppered with superlatives that speak of his love of these forest gems of the *richest* hues, stunning in *variety* and *brilliancy*, by turns *ornamented, magnificent, unsur-passed, singular, exquisite* . . . you get the idea. But it was Wallace's "Heliconi-dae," the "most elegant" longwing butterflies, that inspired an idea well worth our notice here.

Now classified as a tribe (Heliconiini) of about one hundred species in a subfamily of the great brush-footed butterfly family Nymphalidae, the long-wings are distributed from the southern United States to central South America, with the richest species diversity in the Andean and Amazonian regions where they are as bewilderingly complex as they are beautiful. With elongated wings sporting stripes or blotches of deep orange red, lemon yellow, and downy white in various combinations on a velvet-black background, these butterflies were considered so delicate and lovely by Linnaeus that he was inspired to name them for the muses and graces of Greek mythology—as opposed to the squabbling and bellicose warriors of Homer's *Iliad* and *Odyssey*, whose names he and his students liberally applied to many other butterfly groups. The cat-erpillars of heliconiines specialize on passionflower vines, Passifloreaceae (giv-ing rise to their other common name, passionflower butterflies), and the adults are unique among butterflies in that they feed on pollen, specializing on certain members of the squash family, a relationship that entails a suite of remarkable foraging adaptations.[20]

It was thought in Wallace's time that the longwings were the most "advanced" of butterflies, based mainly on aspects of anatomy. To Wallace, Bates, and others this was measured by how dissimilar a group was reckoned to be from generalized moth anatomy, a kind of reference group. (Without getting into the details, let us just say that modern biologists would not agree; looks can be deceiving, after all, and it helps to have additional data like DNA.) But the important thing here is that, in the thinking of Wallace (and Bates), most "ad-vanced" means youngest, most recently *arisen* . . . however that happens. In the conventional view, this was the day the species appeared by special cre-ation, with debate over whether this was by Divine fiat all at once in the six-day burst of activity in Genesis or at intervals over the eons according to some plan. A related issue was whether Creation took place at just one place on Earth initially or at multiple so-called centers of Creation. European natural-ists of the day may have been devout and rather doctrinaire in their religious beliefs, but the Bible was not to be taken literally: those six "days" could rep-resent vast periods of time. Just how *much* time, and how to reckon it in relative and absolute terms, was a topic of lively discussion, making the up-and-coming science of geology a hot area that promised profound new insights into Earth history and the physical processes involved—and, by extension, the history and physical processes involved in regulating the life upon it.

Hence Wallace's keen interest in the subject, especially, as we have seen, regarding how geology intersects with diversity and geographical distribution. This helps explain the significance of a fascinating passage in his "Butterflies of the Amazon Valley" paper. Wondering how a group like the longwings— thought to be the most "advanced" of butterflies, full of closely related and geographically localized species and varieties—might relate to geology, he noted that since "there is every reason to believe that the banks of the lower Amazon are among the most recently formed parts of South America, we may fairly regard those insects, which are peculiar to that district, as among the youngest of species, the latest in the long series of modifications which the forms of animal life have undergone."[21] Putting together geology and biology, Wallace's idea here is that the geologically most recent parts of the Amazon Basin will also have the "youngest" or most recently arisen species, the latest in a succession of such species no less, underscoring a profound link. Although one can fit phrases like "youngest of species" and "latest in the long series of modifications" into the special creation model of the time—the latest in a long series of successive and *specially created* species of longwings, say—we know that was not Wallace's mindset. His was proto-evolutionary thinking, the "modifications" transmutational.

In the modern view, Wallace was closer in some ways than others to the mark. To begin with, the heliconiine subfamily is not as a whole especially recent or the most evolutionarily derived among nymphalid butterflies—in fact, most other subfamilies and several tribes within Nymphalidae appear to have arisen more recently than the longwings.[22] Certain lineages within it do, however, seem to be fairly recent indeed, evolving at so rapid a rate as to make it difficult to reconstruct their ancestral coloration patterns. What drives this diversification? The geological link Wallace intuited is part of the story, but there is more to it than he could have known and perhaps imagined at the time. First, he was correct that the geographical (and therefore geological) template of their diversification is young, but the locale was not just the banks of the lower Amazon but central and especially western Amazonia, in the Andean uplands and Amazonian slope. A youthful, dynamic mountain range, the central and northern Andes were uplifting in earnest by the Oligocene period, approximately 35 million to 23 million years ago, and the heights reached in the eastern flank of the cordillera by late Oligocene times fundamentally reorganized Amazonian river systems while creating both barriers and a vast environmental mosaic of elevation, slope, and aspect.[23] Surely not by coincidence, the Oligocene is also when the Heliconiini clade (evolutionary lineage) began to diversify in that very region, diversification that increased as Andean uplift proceeded, reaching its most intense phase from the mid-Miocene to the early Pliocene, approximately 12 million to 4.5 million years ago. But another driver of that diversification stems from a fascinating ecological and evolutionary dynamic.

The diversity of colors and patterning of *Heliconius* butterflies, which are mostly unpalatable, vary geographically, and *within* a given geographic area different species, even of other butterfly families, have converged on the same color and pattern as *Heliconius* so closely that at first glance it is hard to tell them apart—they are mimics. It was Wallace's friend Bates who first cottoned onto this nearly a decade later in a paper read to the Entomological Society, when he realized that certain butterflies he called "Leptalides" (family Pieridae, now subfamily Dismorphinae) closely resembled heliconiine species in the same area—a form of protective resemblance now called Batesian mimicry in his honor.[24] "On these expanded membranes, nature writes, as on a tablet, the story of the modifications of species," Bates later evocatively wrote in his travel memoir.[25] Wallace was tickled, heaping praise on Bates's paper—"in every respect an admirable one"—and pointed out that it also reinforced his own argument about river barriers delimiting Amazonian species.[26] It is thus both the dynamically changing geological scene (including the formation of the great river systems we have today) and the predator-prey and mating interactions behind the mimicry complexes that fostered the evolution of this incredible panoply of heliconiine varieties and species.[27]

In that year, bracketed by his insightful monkeys and Amazonian butterflies papers, Wallace also managed to produce two books. The first, *Palm Trees of the Amazon and Their Uses*, was published in October 1853. A slender volume of 129 pages, it featured forty-eight plates based on his drawings saved from the *Helen*—his first book! More labor of love than financially rewarding—having to lay out the money for an artist to engrave his sketches and produce a print run of 250 copies, barely recovering his investment—it was the long-planned homage to his favorite trees, that "graceful" group so emblematic of the tropics, combined with his ethnological interests. It was homage, too, to Martius, whose "magnificent" three-volume work on palms Wallace acknowledged.[28] He trusted that his "little book may be of some use" to the botanist and "not uninteresting" to general readers.[29] It was well received by those general readers among naturalists, many of whom surely picked up a copy after reading the nice review in the *Annals and Magazine of Natural History*, taking a cue from Wallace's preface for the opening: "We beg more strongly to recommend this book, as one that will not interest the botanist alone, but give pleasure to unbotanical readers." The botanists, however, were unimpressed; it was panned by the dean of botanists, W. J. Hooker, and even Wallace's friend Spruce, who, perhaps ingratiating himself to his boss (Hooker), alternated between faint praise ("The figures are very pretty . . . accounts of the uses are good") and damning critique ("The descriptions are worse than nothing, in many cases not mentioning a single circumstance that a botanist would most desire to know").[30] He had certainly made some errors; Wallace never claimed to be a botanist, after all, and you can't always please everyone. Still, it seems uncharitable to focus on the book's weaknesses and not on what Wallace had managed to accomplish.[31]

He fared a bit better (at least financially, barely) with his travelogue, the *second* long-planned book he produced that year: *A Narrative of Travels on the Amazon and Rio Negro, with an Account of the Native Tribes, and Observations on the Climate, Geology, and Natural History of the Amazon Valley*. Darwin was a bit dismissive of it as a work of natural history (complaining that it had "hardly facts enough" to Bates and about its "extreme poverty of observation" on the nature of species to Hooker). For his part Hooker replied that he didn't recall the book interesting him at all when it appeared.[32] All right, fair enough, it was light on analysis and synthesis, but give the guy a break. Considering all that Wallace had lost, this book was nothing short of astonishing, a concise and entertaining travelogue that was hardly devoid of facts or incisive observations. Drawing on his memory, a rescued diary, odd notes, and surviving letters he had sent home, he produced a chronological memoir of some four hundred pages with a general map of northern South America as the frontispiece, followed by overviews of the physical geography, geology, climate, vegetation, zoology of major groups (mammals, birds, reptiles, fishes, and insects, with a six-page discussion of species distribution) *and* an account of the Indigenous peoples, from the diversity of groups to cultural practices to a discussion of petroglyphs, with illustrations (his few that survived). If all this were not enough, he concluded with a twenty-page appendix on the "Vocabularies of the Amazonian Languages," complete with a foldout chart of ninety-nine common words and phrases for comparison between English, the *Língua Geral*, and ten Indian tongues![33] Just astonishing.

Note that this was published in December 1853, about the time he delivered the Amazonian butterflies paper and two months after *Palms* was published.

His map of the Rio Negro and the "almost unknown" Rio Uaupés was being readied for the press. Wallace's *Narrative* may not have been quite the tour de force of Bates's fine book to appear a decade later, as Darwin then pointed out, but to be fair Bates spent eleven, not four, years in the Amazon, did not tragically lose half of his collections and most of his notebooks on his way home, and, in 1863, had the benefit of retrospectively seeing his species data in the light of natural selection. It is no wonder he received every encouragement and high praise from Darwin. In Hooker's letter to Darwin, where he dissed Wallace's *Travels on the Amazon*, he twisted the shiv, asking rhetorically why "Wallace does not fructify as Bates does." I would argue that Bates was to rest on his well-earned laurels after publishing his travel memoir, writing little or doing much else of consequence afterward, and it was Wallace who "fructified"; before his own travel memoir was even hot off the press, the naturalist was already making preparations for his next great expedition.[34] His eyes were now on the Far East.

———

There was a moment of despair in December 1850 when Bates decided to give up and return home. He would travel downriver and catch a ship home the following April or May, he wrote Stevens. Wallace, he supposed wistfully, would follow up the profession: "He is now in a glorious country [up the Uaupés at the time], and you must expect great things from him. In perseverance and real knowledge of the subject, he goes ahead of me, and is worthy of all success."[35] It was a rare and frank acknowledgment from Bates of his friend's talents. But letters of credit and praise soon arrived—including the exhilarating news that lepidopterist William Hewitson had named a beauty of an Amazonian butterfly after him (now *Asterope batesii*). His spirits lifted, Bates bucked up and decided to carry on. Had he not, it is possible that his friend Alfred Russel Wallace may have returned to South America instead of heading east.

That is what Wallace said, anyway; he was itching to travel despite writing Spruce from the *Jordeson*, freshly plucked from his lifeboat bobbing on the high seas, that he had vowed fifty times since leaving Brazil that he would *never again* trust himself to the ocean after his narrow escape. "But good resolutions soon fade"—he knew even as he wrote those words that he would travel again. How he envied Spruce, he opined, way off "in that glorious country where 'the sun shines for ever unchangeably bright,' where farinha abounds, and of bananas and plantains there is no lack."[36] The question was not *whether* he would

sail for distant shores again but where and when. This time, though, his growing reputation helped him.

Working up his compendium of Indian vocabularies, he sought a consulta-
tion with the well-regarded ethnologist and philologist Robert Gordon
Latham, who was glad to help and contributed remarks to accompany the
vocabularies in Wallace's *Narrative*. Wallace helped Latham too. Appointed to
oversee the Natural History Department of the new Crystal Palace, just relocated from the original Hyde Park exhibition site to its permanent home in
Sydenham, South London, Latham included statues of several Indigenous
Amazonian Indians in his "museum of man" exhibition.[37] He asked Wallace to
help correct the tendencies of his classically trained Italian sculptors to make
Greco-Romans of the Indigenous South Americans—probably no one in London at that time knew the peoples of Amazonia better. (Wallace did his best
but was not altogether satisfied with the somewhat incongruous group: picture Arekuna-Romans reclining as if at banquet aside a vaguely Turkanoan
paterfamilias greeting passersby, while just beyond a pair of Italic-Arawakan
sagittarii fished with bows and arrows.)[38]

Wallace consulted the collections and curators of the British Museum to
get a sense of the taxonomic and geographical gaps in the collection, part of
the calculus of collectors: gaps needed filling, and their very existence in the
heyday of natural history collecting underscored how this was no mean feat.
The rarest of the rare from far-flung and storied locales fetched awfully good
money, and being little known also presented opportunities for ambitious
naturalists to make their mark by finding new species and describing natural
history.

Thanks to his paper and remarkable map of the Rios Negro and Uaupés,
Wallace also earned the respect of the Royal Geographical Society. He called
upon long-standing and congenial President Sir Roderick Murchison, asking
for help in securing free passage to Singapore on a government ship and in
obtaining the necessary permissions from the Dutch and Spanish colonial
governments, as Spain then controlled the Philippines and Holland much of
what is now Indonesia. Happy to help, Murchison encouraged Wallace to sub-
mit a formal application. His request, submitted in late June 1853 to the president
and council of the Royal Geographical Society, was literally formal: written in
third person, he begged leave to lay before the council his proposal to investigate the natural history and geography of the Eastern Archipelago, using Singapore as a base and collecting in Borneo, the Philippines, Celebes (modern
Sulawesi), Timor, the Moluccas, and distant New Guinea in turn. He pointed

to his successes in South America but also reminded the council of his recent losses, "render[ing] necessary the present application." What is more, having been promised "every assistance" from none other than Sir James Brooke—Royal Geographical medalist and the newly reigning "white Rajah" of Sarawak, just then visiting London—"he has little doubt of success in exploring the great Island of Borneo."[39]

The easy part proved to be the green light from the Royal Geographical and securing permissions and an agreement from the Admiralty to provide passage. An actual departure, however, was far more elusive. The first miscue came in late August 1853. Hearing no news regarding permissions or sailing orders, he headed off to Switzerland for a two-week hiking trip with his old friend George Silk, now working as secretary to the archdeacon of Middlesex. On the way a letter from the Royal Geographical's secretary, Henry Norton Shaw, caught up with him in Paris, informing him that permissions were in hand from the Foreign Office—good news—and a ship to take him as far as Ceylon (modern day Sri Lanka) was departing imminently. This was not so good news and in fact was a triple inconvenience. He was on holiday, after all, plus working away to finish his travel memoir (no publisher yet lined up), and anyway, he could not afford to pay for passage from Sri Lanka to Borneo out of pocket. He would await another opportunity, preferably later in the autumn, and puzzlingly flirted with the idea of going to the mountains of Africa instead.[40] There were further miscues the following January and February 1854, when Wallace got word from the Foreign Office that he could sail to Sydney aboard the HMS *Juno* and make his way from there to Singapore. By then desperate to get underway, even to Australia first if he had to, he headed to Portsmouth and shipped off his equipment and instruments to Singapore—only to find the *Juno*'s sailing orders had changed. He might have been relieved, having heard that the captain, Stephen Grenville Fremantle, had a poor reputation, but in any case was told that the sixteen-gun sloop HMS *Frolic* would take him to Australia instead.[41] His departure seemed certain now; Edward Newman, in his presidential address to the Entomological Society in late January, wished Wallace well from his entomological colleagues: "His face is familiar to us here; his writings are known to most of us, and some of them are on the point of publication in our 'Transactions.' I am sure that there is not one member of the Society but will wish him God speed!"[42]

He soon moved aboard, slinging a hammock in the cabin of Captain Matthew Nolloth, a congenial companion with a good library and scientific tastes—a bit like Darwin and Fitzroy. But the ship did not budge beyond bobbing with the waves and oscillating with the tides. Weeks later he was *still* swinging in his hammock in port when the *Frolic*'s sailing orders changed: she would be heading to the Crimea, where war between the Ottomans and Russia, which had broken out the previous October, had now escalated to the point where Britain and France were on the verge of going to war with Russia in defense of the Ottomans. He almost accepted a second offer of passage on the *Juno* but thought the better of it and schlepped back to London instead, where, exasperated, he showed up on Roderick Murchison's doorstep.[43]

The Admiralty had earlier suggested that he travel by mail packet—on small ships designed to carry the post and similar cargo—instead of Royal Navy vessel, and now he jumped at that option: the helpful Sir Roderick soon procured a first-class ticket for him aboard the *Euxine* of the Peninsular & Oriental Company (now the P&O family of companies), departing in days. In the back-and-forth of letters settling the travel arrangements between the Foreign Office and Henry Shaw at the Royal Geographical Society, we get an indication that Wallace was not traveling alone: "Mr. Wallace," Shaw wrote, "will at once accept Lord Clarendon's kind offer" of passage on a mail packet, and by the way, he hopes he can include "his servant lad, who was by the permission of the Lords of the Admiralty allowed to accompany him on board HMS 'Frolic' and 'Juno.'"[44] The "servant lad" was Charles Martin (Charley) Allen, a youth of fourteen years later described by Wallace as "a London boy, the son of a carpenter who had done a little work for my sister, and whose parents were willing for him to go with me to learn to be a collector." Charley eventually did become an adept collector and valued assistant, but it took awhile: he was not exactly a quick study, at least not quick enough in Wallace's view, as we will see.[45] But for now, both travelers were eager and excited to be underway at last!

———

Wallace was thirty-one now, tall, serious, congenial, with a thick mop of dark hair parted on the left and a short beard stretching from ear to ear in the Victorian fashion, outlining his full, bespectacled face. His health had long since been regained from the disease-wracked and malnourished state he was in at the time of his Atlantic rescue. It must have been bittersweet, leaving his family again just a year and a half after (barely) making it home. Among the portraits made in that familial interlude by his brother-in-law Thomas Sims, the aspiring photographer, is one of Alfred with his mother and sister Fanny. Alfred and his mother are seated with Fanny standing between and just behind them, leaning

Reunited: Alfred with his mother, Mary Ann Wallace, and sister
Fanny Sims, 1853.

forward a bit with one elbow on the back of Alfred's chair, her other hand
extending to the back of her mother's. At first glance they look like a serious
lot, decked out prim and proper in the usual Victorian sartorial excess, with a
certain somber note perhaps created by the image's black-and-white tone. A
closer look reveals faint smiles—a departure from that other Victorian standard, the ever-so-serious portrait pose. They seem like they are about to crack
up, and probably were—who knows how many "takes" they ruined before
managing to stifle their laughing long enough for a keeper? We know this was
a family with a sense of humor that loved wordplay and puzzles. You can imagine Thomas cajoling them to be still—"Keep your eye on the toucan!" We are
fortunate to have this portrait and four others that survive from around this time,
ambrotypes and calotypes that were likely taken as Thomas's photographic
exercises.[46] They say something about Wallace and family. Tellingly, in three
of them he is *doing, thinking*: playing chess with Fanny in one; pensive, with

book in hand in another, his finger holding his place; intent on something he is fiddling with in his hands in a third. They convey a certain restlessness, even in those in which he is just posing—a faraway look in his eyes, as if he sees through the lens and beyond, to the infinite tropics of his dreams. The tropics beckoned; no wonder he had no sooner planted both feet on English soil before he had one foot off the island already. And now the other one, as he boarded that steamer.

Think about how far he had come literally and figuratively since his last departure from England in 1848, just six years earlier: he had traveled thousands of miles in that time, collected a multitude of impeccably preserved and documented rarities, authored two books, delivered several well-regarded papers at the learned societies of London, and published a fine map that garnered him election as a Fellow of the Royal Geographical Society. He also became, more and more, a philosophical naturalist. No mere collector, he was intent on understanding the nature of species and varieties and their distribution, relationships, and transformations. As we will see, those motivating questions were ever present, picking up in his travels to the East where he had left off deep in Amazonia.

The SS *Euxine* was a sturdy iron paddle steamer about 220 feet (67 m) long and 25 feet (7.5 m) broad, with a nearly 16 foot (4.8 m) draft. It accommodated ninety-eight passengers—none of whom Wallace felt simpatico with, unsurprisingly, though he found a few "amusing." At least the trip was uneventful, in the way his last oceangoing voyage had *not* been. But it was plenty eventful in the novelty of the sights and peoples he encountered. The travelers departed the port of Southampton on England's southern coast for Alexandria, Egypt, on 4 March 1854, taking in grand vistas of the Mediterranean along the way: south along the Iberian Peninsula and through the Pillars of Hercules to Gibraltar (where they were quarantined due to fear of cholera, unfortunately), then on through the Alboran Sea, admiring the grand snow-capped Sierra Nevadas of Spain. East along the North African coast they chugged, rounding Tunisia, where they plied the strait crisscrossed by battling Roman legions and Carthaginians in classical antiquity. Stopping at tiny Malta, then still a British colony, there was a day of sightseeing ("where the town and the tombs of the knights were inspected"). Continuing east, they passed to the south of Crete, landing at storied Alexandria—the grandest of the cities founded by Alexander the Great—in time for the spring equinox, 20 March 1854. He was dazzled. "Of all the eventful days in my life (so far), my first in Alexandria was . . . the most exciting," he wrote George Silk. That was saying something. Naively thinking he and Charley would go on a quiet stroll and explore, he found himself thronged by donkeys and their drivers, all raucously insisting on being his guide—"The pertinacity, vigour, and screams of the Alexandrian donkey-drivers cannot be exaggerated." Declining the service was out of the question, so he eventually selected one: "Now, then, behold your long-legged friend mounted upon a jackass in the streets of Alexandria," he wrote Silk, "a boy behind, holding by his tail and whipping him up." Off they went, self-consciously, amid jostling crowds of Jews, Greeks, Turks, Arabs, everywhere veiled women, "donkey-boys" yelling their commands, and irregular bands of Turkish soldiers. It was a head-spinning feast of sights and sounds: the bazaar, the slave market, the pasha's new palace, beautiful mosques "with their grace-ful minarets."[47]

In those days before the construction of the Suez Canal (which opened in November 1869), there was no direct connection by water from the Mediter-ranean to the Red Sea and ultimately the Indian Ocean, so it was necessary to travel overland. The shortest route was from Cairo, just inland from Alexandria at the head of the Nile Delta. The following day they headed there via boat, paddling up the Nile, dazzled anew. Wallace beheld, awestruck, the fabulous soaring triumphal column honoring the emperor Diocletian, dating to 302 CE, and then the ancient and fabled landscape of the Nile delta: "Mud villages, palm-trees, camels, and irrigating wheels turned by buffaloes,—a perfectly flat country, beautifully green with crops of corn and lentils; endless boats with immense triangular sails.[48] Then the Pyramids came in sight, looking huge and solemn; then a handsome castellated bridge for the Alexandria and Cairo rail-way; and then Cairo—Grand Cairo!"[49] Refreshed with "splendid tea, brown bread, and fresh butter" at the small hotel where they stopped for the night, he was overwhelmed with delight: "I could hardly realize my situation. I longed for you to enjoy it with me," he wrote Silk.

The party followed the Cairo to Suez road established in the 1830s by Briton Thomas Fletcher Waghorn with the backing of the pasha of Egypt, Muham-mad Ali. Waghorn promoted this overland desert route as a means of greatly reducing the travel time from Britain to India for both post and passengers, saving some six thousand miles circumnavigating the entire African continent, down the west coast to the Cape of Good Hope at the southern tip and back up the west. Traveling by small horse-drawn carriage, Wallace found the road itself excellent, but the hundreds of camel skeletons littering the shoulders were a grim reminder of the journey's dangers, especially in the withering heat of summer, despite the rest houses erected along the way. It is a reminder, too, of how travelers like Wallace benefited from the infrastructure of empire—the very existence of the Cairo-to-Suez road, not to mention the network of government-backed P&O mail steamers, was made possible by Britain's global domination at the time.

Wallace took in the natural history along the way, marveling at the undulating desert of sand and coarse volcanic-looking gravel, shimmering mirages in the heat of day, endless camel trains, fragrant desert plants, improbable land snails, and curious birds, including vultures (probably both the Egyptian vulture, *Neophron percnopterus*, and the Bearded vulture, *Gypaetus barbatus*), flocks of plump, pigeon-like sandgrouse (genus *Pterocles*), and the drab Desert lark (*Ammomanes deserti*). From Suez they boarded another steamer, the *Bengal*, on 26 March and headed down the Gulf of Suez to the Red Sea. He would have been fascinated to know that the long, narrow shape of these water bodies is a clue to some interesting geology. He was traveling a rift zone, a spreading center where the African and Arabian plates are pulling apart, and rather rapidly, too, in geological terms: nearly half an inch (approximately one centimeter) annually on either side of the rift. That is a good sixty-six feet, or twenty meters, in just a millennium, a geological blink of an eye. (At that rate a natural canal will open in no time, relatively speaking, rendering the Suez Canal superfluous.) The *Bengal* paused a day at Aden, near the tip of the Arabian Peninsula, a landscape Wallace found "desolate" and "volcanic." Indeed, lying at the interface of three rift zones, this region is one of the world's most active plate boundaries, dotted with extensive volcanic fields.

They landed at Pointe de Galle (now just Galle) on Sri Lanka's southwest coast on 9 April, just two weeks after leaving Suez. Known as Gimhathiththa prior to contact with the Portuguese and Dutch beginning in the sixteenth century, the town struck Wallace with its coconut groves and crowded markets where the locals hawked precious stones. Although he did not comment on the old Dutch fort, now a UNESCO World Heritage Site, he surely admired the broad fortification overlooking the harbor, with ramparts enclosing a town of whitewashed, red-roofed buildings—most prominently the Dutch *kerk* and Meera Mosque—with a gleaming white lighthouse flanked by palm trees standing sentinel at the point. Here they changed ships for the final leg of the journey, trading the *Bengal* for the SS *Pottinger*, steaming off again on 10 April, and stopping briefly one week later at Penang, along the west coast of the Malaysian Peninsula. They stayed just a day, long enough to admire the "picturesque mountain [likely Penang Hill], its spice-trees, and its waterfall." Excitement was mounting as they steamed south through the Strait of Malacca "with its richly-wooded shores," arriving at their destination the next day, 18 April 1854.

At last, the remarkable city-state of Singapore, his gateway to the east! Thus began a fourteen-thousand-mile odyssey of exploration and discovery. For the next eight years, Wallace crisscrossed the vast Malay Archipelago, collecting rarities, observing patterns, intuiting process: it was to be "the central and controlling incident" of his life.[50]

CHAPTER 7

SARAWAK AND
THE LAW

THE HARBOR of Singapore was quite a sight to see, teeming with vessels great and small in the bright tropical sun: imposing white-sailed men-of-war, paddle steamers, sleek trading vessels with flags from around the world, and myriad praus, junks, fishing boats, and passenger sampans crisscrossing the harbor, dexterously navigating among the great ships. In the town he had never seen such diversity before, not even in London—English, Chinese, Portuguese, and Arabs, as well as West Indians, Parsees, Bengalis, Bugis, and native Malays, all busily coming and going while trading, delivering, bartering, fixing, building, eating, praying. It was electric with energy for a smallish town, bustling with shops and bazaars, handsome public buildings, temples, mosques, and churches, with massive warehouses at the waterfront overlooking the crowded harbor. "Few places are more interesting to the traveler . . . than the town and island of Singapore," Wallace later wrote on his first impressions of the city.[1]

Interesting indeed: it was, and is, a remarkable place, the world's only island city-state, small at some 280 square miles (728 square km) yet now home to about 5.7 million people—nearly 23,000 people per square mile, modern construction materials and technology permitting growth vertically, in the form of skyscrapers. The quality and advantages of this harbor on the island's southern coast had been recognized since at least the early twelfth century, when the island was known as Temasek. Perhaps one thousand or so native Malays then lived in the interior, while the harbor area was the lair of pirates and the sometime base of Chinese traders and fishermen. The island's position at the choke point between the Strait of Malacca and the South China Sea was enviably strategic, and the site boasted such advantages as fresh water (the Singapore River) and topography: a series of low hills dotting the harbor area afforded a

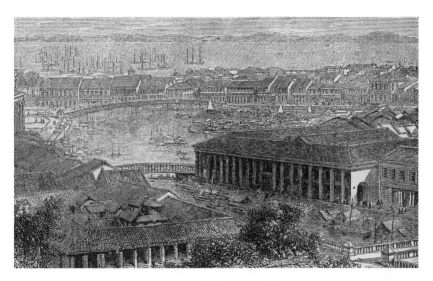

Bustling Singapore, 1860s.

commanding vantage point. In the early fourteenth century, Prince San Nila Utama of the Malayan Srivijayan Empire founded a city on the site, Kuala Temasek, which he dubbed Singapora, the "Lion City," reportedly inspired by the sight of a lion.[2] (More likely a tiger, but more on that later.) Sir Stamford Raffles of the British East India Company certainly saw great potential on his arrival in 1819, at a time when the European maritime powers of Britain, Portugal, and Holland were vying for sea and trade route supremacy in the East. Raffles struck a bargain with the local ruler, the sultan of Johor, first for the right to establish a free trade port and then governance of the entire island—with the perhaps reluctant acquiescence of the Dutch. By 1826 Singapore was the regional capital of the Straits Settlements, a string of strategically located towns under British rule from Penang and Malacca on the west coast of the Malay Peninsula northwest of Singapore to the island of Labuan to the east.

The bustling Singapore that Wallace encountered in 1854 was laid out according to a plan drawn up by Lt. Philip Jackson in 1822 at Raffles's request. It consisted of a series of districts along a six-or-seven-mile stretch of coast centered on the river, laid out largely in a grid that reflected a kind of Enlightenment orderliness but also a paternalistic apartheid that saw the populace ethnically segregated as far as housing: Malay and Chinese communities on the west side of the river, Bugis and Arab on the east, with European Town and the seat of government at the center—originally Raffles's residence, later renovated into the handsome Government House on Government Hill.[3] Four years after Wallace's arrival, the house would be demolished to make way for Fort Canning, the hill renamed Fort Canning Hill; the site is preserved today as lovely Fort Canning Park, complete with a reconstruction of the Raffles residence, a greenspace amid a modern metropolis, albeit one pierced beneath by subway lines. Raffles would surely be awestruck at his own prescience in seeing the tremendous potential of his fledgling town, given its location and his (relatively) enlightened governance.

It is worth our notice that Jackson's plan for the city included a "Botanical and Experimental Garden." Often today, "botanical garden" conjures up images of pleasure gardens, with rich displays of botanical beauties to delight the senses. Not so this garden, as the "experimental" bit of the name reveals. This was the heyday of botanical imperialism, a time when economic and military supremacy rested in large part on cultivating and ideally monopolizing certain plants that yielded coveted products: exotic spices like nutmeg, mace, cinnamon, and clove; rubber trees; nutritious fruit-bearers like breadfruit trees; and of course countless medicinals.[4] To realize just *how* coveted, think of the strategic importance of being able to cultivate quantities of quinine-producing cinchona trees, staving off deadly malaria for troops at the frontiers of empire,[5] or recall that in the 1667 Treaty of Breda the Dutch traded the island of New Amsterdam (Manhattan) for tiny Pulau Run, the smallest of the already tiny Banda Islands of eastern Indonesia, in order to regain their monopoly over the highly lucrative trade in nutmeg and mace from the fruit of the tree *Myristica fragrans*. The wily British dug up young nutmeg trees on their way out the door and transplanted them in tropical territories under their control, including Sri Lanka, Grenada, and Raffles's Botanical and Experimental Garden in Singapore.[6]

The burgeoning spice trade created headaches for Wallace. He estimated the population of Singapore at about sixty-five thousand at the time, more than half being Chinese working the ever-spreading plantations of nutmeg, black pepper (*Piper nigrum*), areca palm (*Areca catechu*, the source of the universally chewed betel nut, a mild stimulant), and versatile gambier (*Uncaria gambir*), the leaves of which yield an extract with uses from medicine and flavoring to tanning and dyeing. And the population was still rapidly growing—it reached some eighty thousand by the end of the decade, giving the town something of a raw up-and-coming feel, Wallace thought, a bit like Belém. And like the Brazilian port town, burgeoning growth had its downside: the forest was by then cleared for four or five miles all around Singapore town, and Wallace realized he and Charley needed to head to the interior if they expected to find much by way of insects.

The pair soon moved to rented rooms at the mission of Bukit Timah, named for the nearby prominent hill (the highest point in Singapore at nearly 535 ft, or 163 m) and where in 1846 the energetic and talented French missionary Anatole Mauduit had established the Church of Saint Joseph. Mauduit's mainly Chinese congregation had outgrown the original thatch church he built, so he raised funds and oversaw the construction of a new one completed just months before Wallace's arrival, a handsome neoclassical structure featuring gingerbread, decorative gables, and a Palladian portico supported by six Doric columns. Wallace found the mission an ideal base for collecting forays, surrounded as it was "by such patches of the virgin forest or jungle as the rapid increase of cultivation has suffered to remain."[7] The forest was restricted to hilltop redoubts, ever shrinking as bands of sawyers took their daily toll cutting timber for the growing town. He saw where that was going and more: ahead of his time, Wallace understood the feedback loop between forested landscape and climate, observing in one letter that before long the whole island of Singapore would be deforested and when that happened "its climate will no doubt be materially altered (probably for the worse), and countless tribes of interesting insects [will] become extinct."[8] Now reforested and preserved as the Bukit Timah Nature Reserve and Dairy Farm Nature Park (the latter a tract used as a dairy farm between 1929 and 1970), today it is possible not only to walk in Wallace's footsteps atop Singapore's highest hill but to do so on the approximately 1.8 mile (3 km) Wallace Trail, complete with natural history interpretive stations drawing on Wallace's writings.[9]

For Wallace the only silver lining to the forest's destruction was the bounty of insects found in and among the resulting piles of sawdust and slash—especially the sought-after long-horned beetles, family Cerambycidae, the larvae of which consume wood. Highly prized for their large size, fantastic colors, and long, elegantly curving antennae, long-horned beetles are in a close race with showy butterflies for "most coveted tropical insect" title. If the butterflies tend to win that race, it is only by a proboscis (which these beetles lack). In a letter to Stevens sent just three weeks after his arrival, Wallace reported great collecting success: some eighty butterfly species already and a bevy of "exceedingly beautiful and interesting" beetles, including—"Mirabile dictu!"—an amazing fifty species of long-horns. Figuring that Stevens would publish his letter—which he did[10]—Wallace also played to his audience, commenting on the one big drawback to all this feasting at the "entomological banquet" of rural Singapore:

"Suspended by a hair over the head of the unfortunate flycatcher" like a Damoclean sword was the ever-present possibility of the insect hunter becoming the hunted and made a meal by a tiger. Tiger attack was in fact a fairly regular occurrence in the Singaporean interior at that time,[11] and hearing distant roars one evening only made Wallace and Charley more apprehensive in their collecting. But the exaggeration is perhaps understandable—not just another impressively big cat, tigers have long struck terror as deep as the sense of wonder they inspire in the human heart. Romantic poet William Blake's 1794 masterpiece "The Tyger" expresses these mixed feelings with its celebrated fiery imagery of fear, awe, and astonishment that a Creator would have fashioned such a creature:

> Tyger Tyger burning bright | In the forests of the night,
> What immortal hand or eye | Could frame thy fearful symmetry?
> In what distant deeps or skies | Burnt the fire of thine eyes?
> On what wings dare he aspire? | What the hand dare seize the fire? . . .

Probably with mixed disappointment and relief, Wallace didn't catch even a glimpse of a tiger in his time in Singapore, though he had a couple of close calls falling into potentially deadly tiger pits—deep and well-camouflaged pitfall traps, once outfitted with a long, sharp upright spike at the bottom, outlawed after one too many hapless humans fell in.

His collecting proceeded apace. By late May, just about a month after arriving at Bukit Timah, Wallace shipped off a whopping one thousand specimens to Stevens: "A very valuable" consignment, he thought. He and Charley had a routine now, worth quoting here in full from a letter to George Silk and passed to Wallace's mother and sister:

> I will tell you how my day is now occupied. Get up at half past 5. Bath & coffee. Sit down to arrange & put away my insects of the day before, & set them safe out to dry. Charles mending nets, filling pincushions & getting ready for the day. Breakfast at 8. Out to the jungle at 9. We have to walk up a steep hill to get to it & always arrive dripping with perspiration—Then we wander about till two or three generally returning with about 50–60 beetles, some very rare & beautiful. Bathe, change clothes & sit down to kill & pin insects. Charles with flies bugs & wasps, I do not trust him yet with beetles—Dinner at 4. Then at work again till six. Coffee—Read—if very numerous work at insects till 8–9 then to bed.[12]

He was so busy with insects, he said, that he had no time for anything else, but fortunately for us he did make time for one other activity: travel writing.

Wallace had apparently arranged with an eclectic magazine called the *Literary Gazette and Journal of the Belles Lettres, Arts, Sciences, &c.* to contribute occasional essays from the field: "We have pleasure in presenting to our readers a communication, which we hope is the first of a series, from Mr. Wallace, the South American traveller." This venture was likely facilitated by the magazine's publisher, conchologist Lovell Reeve, who had also published Wallace's *Travels on the Amazon* and thought highly of the naturalist. For Wallace's part this would have been a great opportunity, earning him a bit of income while further helping get his name out there in British intellectual circles. He sent in four pieces over the next two years.[13]

It is perhaps surprising that Wallace did not say much about the local geology, though. For now he seemed to be single-minded in getting to those remaining jungle patches teeming with insects, understandably, and the landscape may have seemed uninteresting from a geological perspective in any case, being mainly low-lying and dotted with granitic hills here and there. But he surely would have been fascinated to know that, once again, he had managed to find himself in a borderland of space and time: he and Charley were residing practically atop the Bukit Timah Fault, a major line of demarcation between mid-Triassic Bukit Timah granites on the east side of the line and younger, late-Triassic fossiliferous sedimentary rocks of the Jurong Formation on the west, full of corals and a host of curious bivalves—*Cassianella, Spondylus, Cuspidaria, Plicatus, Myophoria*, and more, dating to between 227 million and 247 million years old. In a borderland bonus, Wallace's base was even a stone's throw, if you will, from the Gombak gabbro formation, an exposure of an even older, Permian-age, igneous rock forming the nearby hill Bukit Gombak.[14] Wallace likely did visit this well-defined formation, long the site of quarries that yielded handsome building stones.

———

As productive as Bukit Timah was for insects, it was poor in birds, something Wallace decided to remedy with an excursion to Malacca, northwest of Singapore on the Malaccan Strait. They made the two-day journey in a trading schooner, the *Kim Soon Hin*, arriving on 15 July 1854. The layers of history of the old city were immediately apparent: variously under the authority of Malay sultans, Ming Chinese, Dutch, Portuguese, and then the British by Wallace's time, it was once the most important trading center in the region but now well past its prime. The population reflected that history, and the economy too. He admired the industrious Chinese as the most efficient and hardworking farmers of the widespread spice and gambir plantations and the miners of tin found in the alluvial sands of the floodplains, separated by laborious washing and melting into ingots. He hired two Malaccan Portuguese assistants, a cook and a hunter, and in a case study in language evolution noticed that their Portuguese differed strikingly from that of the homeland or Brazil, being "woefully mutilated in grammar" and simplified with the loss of feminine and plural forms of adverbs and a single form for all tenses.

He was impressed with the beauty of the place, picturesque with many fruit and forest trees supporting an abundance of monkeys, insects, and especially birds: in short order he captured a trove of Malacca's "rich ornithological treasures," each more exquisite than the last. The jewels included "handsome woodpeckers and gay kingfishers" galore, wonderful green-and-brown cuckoos sporting velvety red faces and green beaks, red-breasted doves, and improbable metallic honeysuckers. But the real prizes were the "remarkably striking and beautiful" Blue-billed gaper (*Cymbirhynchus macrorhynchus*), with a cobalt-and-orange bill complemented by plumage of black and deep wine and white shoulder stripes; lovely trogons of rich chestnut brown with crimson breasts and "beautifully pencilled wings"; large green frugivorous Barbets (*Megalaema versicolor*), small toucan-like birds sporting a short, straight, and bristly bill and vivid patches of blue and crimson on the head and neck; vivid Green gapers (*Calyptomena viridis*) with wings marked with delicate black bars—all of which kept him "in a continual state of pleasurable excitement."[15]

After two weeks of collecting in the hamlet of Gading about thirteen miles from town, where the little band boarded with a group of Chinese plantation workers, illness struck. First one and then the other of his assistants fell ill with fever, probably malaria, and left him. Then Wallace had an attack too. He self-medicated with liberal doses of quinine and relocated to a government rest house in the village (kampong) of Ayer Panas outside Malacca, a crossroads now near the modern Jasin Hot Springs, where he befriended a young man named George Rappa, son of a local natural history specimen dealer. There he recovered and did a bit of collecting, including the fortuitous bagging of a prize butterfly. It was the one and only time he ever saw this particular species, a male with upper wings of creamy yellow and delicate blue framed in velvet black and a dazzling canvas of salmon and black spots with splashes of yellow and green below. He came upon the beauty as it lapped juices from fresh dung, as butterflies are wont to do—an incongruous sight. Naturalist William Hewitson was rhapsodic when he received the specimen from Stevens back in

England: dubbing it *Nymphalis* (now *Agatasa*) *calydonia* the following year, he noted that "this glorious butterfly is beyond description. . . . It is one of the many beautiful new species sent home by Mr. Wallace, who, after suffering shipwreck, and seeing his South American collections burned on board, is now exploring the Indian islands."[16]

The landscape beckoned; Wallace resolved to climb massive Mount Ophir, now properly known by its original Malay name Gunung Ledang, at the heart of Gunung Ledang National Park in the state of Johor. Although far from the loftiest mountain in peninsular Malaysia at a mere 4,186 feet (1,276 m), the mountain rises abruptly from the lowland rain forest, making its 4,000 feet imposing indeed, inspiring lore and legend. It is even referenced in traditional pantun quatrain Malay verse:

> The scent of the bread flower is fragrant | As dawn approaches,
> the aroma will grow
> How high is Gunung Ledang? | Once you climb it you will know
> How tall does the Areca palm grow? | The smoke of the fire wafts higher
> How high is Gunung Ledang? | The hopes of the heart are greater.

Climbing the legendary mountain was no mean feat, as there were no paths or trails. Wallace put together a party of six Malays as porters and hunters, plus Charley and George Rappa, and off they went trekking thirty miles through the jungle from Ayer Panas, well provisioned with food and equipped with blankets, clothing, insect and bird boxes, nets, guns, ammunition, and scientific instruments—a boiling-point thermometer and sympiesometer, a type of barometer to measure altitude. It took several days of slogging through progressively hilly and untracked forest, braving knee-deep mud holes and myriad terrestrial leeches that constantly infested them. Picking off these engorged blood sausages was a nightly ritual—Alfred may have found them merely "annoying," including the one that narrowly missed his jugular vein, but it was typical of him to admire the more attractive ones, "beautifully marked with stripes of bright yellow."[17]

They steadily ascended, through open forest and over broad, steep *padang-batu* (stone fields) on the flanks of the mountain, then thickets of tall ferns: graceful *Dipteris horsfieldii* (now *D. conjugata*), with dark green leaves cloven into two deeply lobed fan-shaped halves atop a long, smooth stem, and elegant *Matonia pectinate*, sporting striking spirals of toothed (*pectinate*) leaflets at the end of a stem so thin they appear suspended in the air, like frozen firework displays. Parched from the exertion and unable to locate water, they reluctantly turned to drinking from the *Nepenthes* pitcher plants. Although full of drowned and half-decayed insects—"uninviting," said Wallace in understatement—thirst overcame revulsion, and he was pleasantly surprised: "On tasting it, however, we found it very palatable, though rather warm, and we all quenched our thirst from these natural jugs."[18] On they went, as the towering hardwoods of lower elevations—notably lofty *Dipterocarpus*, so emblematic of the Asian tropics with their dual-winged fruits—gave way to shorter trees and shrubs of the tea and myrtle families and then the gnarly trees and heathy scrub of the summit, where montane-ericaceous forest prevails—*Rhododendron, Vaccinium*, and other heaths, family Ericaceae. From hiking to clambering to hand climbing, over hard rocks and deep cushiony mosses, they finally beheld the "noble prospect" from the summit and bedded down for the night on blankets draped over twigs and branches, not unlike the "nests" of certain other primates.

Descending to their base camp, where he stayed for another week of collecting, he came away with hundreds of prized insects, including some fine butterflies that he was sure were new species. He was rapturous, in fact, about the butterflies, later commenting that in diversity of form, color, size, and beauty, "no country can surpass" Malacca. His greatest treasure, he declared, was "a magnificent green and gold powdered species" of swallowtail, which he hoped was something new: a butterfly of inexpressible beauty, as if the wings, each with a broad green stripe on a deep black background, were heavily dusted with minute yellow pollen grains.[19] But what he most longed for here was a specimen of the fabulous great Argus pheasant (*Argusianus argus*)—so named by Linnaeus in reference to the rows of startlingly realistic eyespots along its long wing feathers, reminding him of Argos Panoptes, the hundred-eyed guardian of Io in Greek mythology.[20] Wallace was tantalized—or should we say tortured—by their frequent calls: they clearly abounded, but the elusive bird remained completely out of sight. Even his experienced hunters said the Malayan equivalent of "Fuhgeddaboudit!" This was true, too, of other rare specimens he sought: rhinos, tigers, and elephants. Apart from dung and tracks, no luck, though two of his men claimed to have glimpsed a rhino.[21] He was to see elephants and rhinos later in his travels, but all three of these impressive mammals were already declining; as he later put it, they seemed to "retire rapidly before the spread of cultivation"—no surprise there.[22]

———

They were back in Singapore by the end of September, where a welcome pile of mail awaited: letters from friends and family, newspapers and books, his watch, instruments, insect pins from Stevens, and more. He was glad to catch up on news from home and got to work preparing his prodigious Malaccan specimen haul for shipment to Stevens in London. But there was a fly in the ointment: Wallace's patience with Charley was wearing thin. The lad was pleasant and generally capable enough but was no self-starter, and the exacting Wallace found him sloppy in both dress and work habits. Just as he was heading to Malacca, he had commented in a letter home that Charley could now shoot pretty well and would actually be useful if only Wallace could "cure him of his incorrigible carelessness." He had to constantly supervise the kid, he reported frustratingly, and if he wanted something done right he had to do it himself. Charley apparently did not improve in their time in Malacca. "If it were not for the expense I would send Charles home; I think I could not have chanced upon a more untidy or careless boy," he complained in his next letter to his mother.[23] As we will see, things with Charley went from bad to worse before they got better.

But he also reported exciting news: Sir James Brooke, who had arrived in Singapore from Britain, invited Wallace to visit him in Sarawak, a province he ruled along the north coast of Borneo. A sometime soldier with the British East India Company turned adventurer, in 1835 Brooke had used a sizable inheritance to purchase and crew a 142-ton schooner. Originally planning on trading, he had arrived in Borneo in 1838 to find the powerful sultan of Brunei facing a range of woes, from rampant piracy to an uprising in Kuching, the nominal capital of the neighboring province of Sarawak. Brooke came to the rescue on both fronts and was rewarded first with the governorship of Sarawak and then, a few years later in 1841, was named hereditary rajah of Sarawak for helping restore the sultan to the throne after further turmoil and political intrigue within the royal family. (He was also rewarded with a knighthood back home for arranging for the sultan to cede the island of Labuan in Brunei Bay, becoming one of Britain's Straits Settlements.) Thus began the storied dynasty of the white rajahs of Sarawak, which ended only with the Japanese invasion in World War II and the subsequent cession of the province to Britain in 1946, where it remained a colony until 1963.[24] Brooke was a benign dictator and surprisingly efficient administrator, vigorously putting down piracy—and headhunting by the Indigenous Dyaks—restoring order, and implementing civil and legal reforms. He was, however, charged with using excessive force against the native people in the name of fighting piracy, paying bounties handsome enough for each pirate head that those cashing in likely did not pay too close attention to whether the heads they gleefully collected belonged to pirates or not. Brooke went to Singapore in 1854 to testify before a Commission of Inquiry appointed by the crown. (The charges were later dismissed, though his reputation remained blemished.)

Wallace was appreciative of Brooke's warmth and hospitality: "He received me most cordially, & offered me every assistance at Sarawak."[25] Wallace had been flirting with the idea of a collecting trip to Cambodia with his Jesuit friend Father Mauduit of Bukit Timah, but now decided to take Sir James up on his offer. The rajah had a taste for literature, science, and stimulating intellectual conversation and was happy to invite Wallace to join his regular dinner salons— the intrepid young explorer-naturalist was a deep thinker with a growing reputation in British scientific society, and fellow Europeans with intellectual interests were scarce. Needless to say, the paternalistic and frankly racist attitudes of the rajah and most other Westerners in the region did not exactly encourage or recognize intellectual pursuits by the native Malays and other nonwhites. For his part Wallace looked forward to enjoying a bit of "pleasant society" and an opportunity to improve his Malay. But mainly there was the irresistible allure of collecting in a remote and exotic locale hardly visited by Europeans: the land of orangutans, fascinating Indigenous peoples, and biological treasures yet to be discovered.

Wallace and his assistant Charley arrived in Kuching at the end of October 1854, the start of the east Indo-Australian monsoon, a season of near-incessant rainfall that ran from November through February. The Kuching of that time was a modest Malay kampong, a settlement of a few thousand people sited several miles from the sea on the Sarawak River, with dwellings, a pasar, or market (the Malay root of the modern word "bazaar"), and temples. Today the site of that early kampong is the heart of a dynamic city of half a million people straddling the river, the north and south banks joined by the sinuous Darul Hana pedestrian bridge. Near the bridge on the north bank is the Astana, or Government House, the third, largest, and last residence of the Brooke dynasty, built on the same site as the original residence Sir James had constructed in 1842. That had been a modest wood and thatch dwelling built on raised posts, which served as both home and office. Wallace fondly recalled many a pleasant evening in the rajah's company, between the spirited philosophical discussions and carte blanche access to his excellent library. By all accounts Sir James thoroughly enjoyed Wallace's conversation too—which not infrequently included debates over the scandalous *Vestiges of the Natural*

History of Creation, the implications of which intrigued the rajah. Spenser St. John, Brooke's private secretary and biographer, later remarked that if Wallace "could not convince us that our ugly neighbours, the orang-outangs, were our ancestors, he pleased, delighted, and instructed us by his clever and inexhaustible flow of talk—really good talk. The rajah was pleased to have so clever a man with him. . . . Our discussions were always either philosophical or religious. Fast and furious would flow the argument."[26] In the morning, St. John added, the guests would find one another in the library, fact checking each other and arming themselves for that evening's debate.

Sir James helped Wallace in another way, too, lending him the use of a small bungalow, a government rest house sited at the mouth of the Santubong River at the foot of Gunung Santubong, an imposing sandstone massif rising abruptly from the sea to over 2,600 feet (810 m) at its highest. The site, called Wallace Point today, is now owned by the Sarawak Museum, with plans to develop a museum or research center dedicated to Wallace (and Ali, a local who, as we shall see, was to become Wallace's most trusted and capable companion) as part of a larger archaeological park.[27] The broad base of iconic Santubong forms the bulk of the peninsula it sits on, and the long north-sloping spine of the massif projects fingerlike into the South China Sea. The almost jagged appearance of its peaks and ridges speaks to the relative youth of the mountain, formed of riverine and shallow sea sediments laid down in the very late Cretaceous to early Paleogene (approximately eighty to sixty-three million years ago) and later consolidated, uplifted, tilted, and now dubbed by geologists the Kayan sandstone: massive cross-bedded sandstones and conglomerates admired by today's hardy trekkers clambering to Santubong's summit by rope, ladder, and bridge. Wallace could not have known that he took up residence on another profound boundary: Mount Santubong lies along the Lupar Line, a major fault representing an ancient continental margin, a former subduction zone where the oceanic plate of the proto–South China Sea was carried beneath the continental Sundaland plate, consumed in a slow tectonic gulp. Santubong's geology differs strikingly from that just across the broad bay not sixty miles to the east. It is a geological suture zone: the intervening Lupar Line separates the Kuching Zone, with its distinctive shallow-water sandstones and limestones intruded in many places by igneous rocks, from the uplifted deep-marine rocks of the Sibu Zone immediately to the north and east—a divide even manifested in the geographical distribution and genetics of many species, especially plants.[28] But Wallace did not know: he was not collecting plants, and the geology is not all that easy to observe with all that exuberant tropical vegetation on top. Plus, it was the rainy season: the collecting was lousy, though they did what they could over the next few months, plying the Sarawak River in hired canoes and sampans—traditional Chinese and Malay flat-bottomed wooden boats—exploring far upriver to the southwest as far as the gold-mining districts of Bau and Bidi, to use their modern names. But mostly there was little else to do but wait out the rains indoors, reading and pondering. It turns out this was a most fortuitous circumstance for science.

Besides having access to the rajah's fine library, Wallace traveled with a small library of his own and periodically received journals and newspapers from Stevens from back home.[29] His traveling library included reference works useful for identifying specimens—books like Lucien Bonaparte's encyclopedic *Conspectus Generum Avium*, the authority on birds, and Jean Baptiste Boisduval's comprehensive treatment of sulphur and swallowtail butterflies (Pieridae and Papilionidae), as well as works bearing on his larger scientific and philosophical interests, such as Charles Lyell's monumental *Principles of Geology*. Wallace traveled with the fourth edition of 1835, issued in four volumes. At first glance a geological treatise would seem to have little relevance for his interests in species, but remember that naturalists of Wallace's time took a more holistic view of the natural world, and for them the study of Earth and the life upon it was all of a piece. This was (and is) most obvious in the field of paleontology, where fossils themselves give testimony to the former world, helping delineate geological periods and epochs as well as infer the slow evolution of climate and landscapes . . . and, of course, *species*, ever since Wallace and Darwin's seismic insights, as we shall see.

But to Lyell, fossils may have spoken eloquently of the state of the former world and its climate, but they said nothing about species change. Well, that's not quite accurate: far from mute on the subject, he argued that fossils spoke *against* the possibility of species change. Not that the fossils found in successive layers of the geological cake did not exhibit changes: they did; it's just that he thought each fossil group, each layer of the geological record, represented a separate round of creation, one after the other, paving the way for you and me. He was not even sure the changes over time were all that directional, or progressive, even entertaining the idea that as Earth's climate cycles on, each set of climatic conditions brings with it a certain set of divinely created and well-adapted species. If the fossil record of the British Isles revealed a tropical landscape in the deep antediluvian past, with lumbering iguanodons and pterosaurs flitting among the tree ferns, just wait around long enough and these conditions, with precisely these specially designed species, will reappear.[30]

This all involved some mental gymnastics, and the lawyer-turned-geologist devoted quite a lot of space in *Principles* to discussing species in a range of contexts, from the nature of the fossil record to the variability and migrations of living organisms, all aimed at scuttling the idea that species change over time. As a young man studying in France, he flirted ever so briefly with Lamarckian notions of transmutation, the idea that species are slowly transformed, giving rise to new ones. But the deeply devout Lyell soon thought the better of it: the religious implications made him shudder, and he soon became the most devastating critic of transmutational notions.[31] Wallace—and Darwin— could see Lyell's inconsistency, how his eloquent arguments for a gradually changing Earth logically extended to the biological realm, the evidence of the fossil record clear. It took time to convince Lyell, and in the 1850s both Wallace and Darwin were, in their own ways and unbeknownst to one another, working to do just that, constructing a pro-transmutation case so compelling, so watertight, that the barrister-trained Lyell, scientific luminary and most eloquent of anti-transmutationists, would have to capitulate. And thus everyone else too.[32]

Not that either one said this was his plan; no, not overtly. Rather, it is their actions that speak: what they were working on, their motivation and objects. On the other side of the world in 1854, Darwin was just publishing the last two volumes of his epic four-volume treatment of the fossil and living barnacles of the world, even while constantly making observations, conducting experiments, devouring literature bearing in myriad ways on the species question, and building on the case for transmutation that he had written out in a private essay in 1844, sealed in an envelope with instructions to his wife, Emma, to publish it immediately should he unexpectedly die.[33] He was a man on a mission: having had a personal epiphany of the reality of species change seventeen years earlier, in 1837, and then a eureka moment of insight into the process he dubbed natural selection a year later, Darwin had ever since steadily amassed *data* to back up his claims and had shared his "heterodox" notions with only one person by that time: his trusted friend the botanist Joseph Dalton Hooker, Sir William's son. As we have seen, for his part Wallace had been a transmutation convert for some nine years by then, since reading *Vestiges* in 1845. He had been collecting to fund his travels ever since so that he could collect data bearing on the question too. It is worth recalling his efforts to relate the distributions of groups like palms and long-wing butterflies in South America to geology and other factors, his riverine barrier observation, and that for all of the insect specimens he shipped to Stevens for sale, he designated a significant proportion of them to be kept for his private study later. His approach, his MO, was quite different from Darwin's—he liked to start with first principles and come to thoroughly understand the matter at hand—*pattern*—in order to draw logical inferences and see if he could gain some insight into *process*.

The earliest indication of Wallace in this analytical frame of mind is found in his specimen notebook from 1854. This small marble-board notebook, now in the collection of the Linnean Society of London,[34] is mainly a record of specimens and consignments, but the first dozen and a half pages include notes on eclectic topics of interest to him: ethnology, comparative morphology, paleontology, geographical distribution, and more. Some were probably jotted down while studying in Sir James's library, while others came from his own books and journals. On the very second page is a fascinating note: "Geoffroy St. Hilaire believes mutability of species," citing as the source William Whewell's influential 1837 *History of the Inductive Sciences*. The anti-transmutation Reverend Whewell—one of Darwin's Cambridge professors—discussed the French zoologist St. Hilaire in a section titled "Hypothesis of Progressive Tendencies," only to dismiss him: "Not only, then, is the doctrine of transmutation of species in itself disproved by the best physiological reasonings, but the additional assumptions which are requisite, to enable its advocates to apply it to the explanation of the geological and other phenomena of the earth, are altogether gratuitous and fantastical."[35] Wallace begged to differ. The note on St. Hilaire was followed on the next page by comments on a passage from Joseph Beete Jukes's *A Sketch of the Physical Structure of Australia* (1850), speculating on the effect of cycles of geological uplift and subsidence, where hills are made islands as land subsides and the sea rushes in, only to be raised and rendered dry land again. Such cycles can result in a distinctive species distribution pattern, yielding a group of different yet related species in adjacent areas belonging to the same genus. The unspoken implication is that species somehow change when marooned on so many islands, the one original or parental species producing, or at least *succeeded* by, a bunch of separate but related species once dry land is restored. Darwin once jokingly referred to his studies of geographical distribution as "a grand game of chess, with the world as a board."[36] Wallace's version here was three-dimensional chess: geographical distribution in space and time.

These two forms of distribution were fully complementary, and Wallace was sure you could not understand the one without understanding the other, so it is not surprising that observations of species distribution both in modern geographical terms and historical-geological terms are the subject of several other tantalizing entries in his 1854 notebook. Upon reading, for example, R. C. Tytler's report on the birds of Barrackpore (north Kolkata, India) in the *Annals and Magazine of Natural History*, Wallace wondered, "How many of the land birds are found in the countries on both sides [of] the Ganges?"— reminiscent of the theme of his "Monkeys of the Amazon" paper. And then there are his notes on Swiss paleontologist François Jules Pictet's monumental *Traité de Paléontologie*. Wallace summarized ten "laws of geological development" from Pictet, observations on the relative order of occurrence, distribution, and duration of fossil species in the geological record. He then added a comment on each: "undisputed," "doubtful," "important," "true, with specific exceptions," "very important," and so on.[37]

Wallace was in synthesis mode, thanks to the enforced downtime of the incessant rains. Largely alone in the cottage at the foot of Santubong Mountain that rainy season, with only Charley and a hired Malay cook for company, he had "nothing to do but look over my books and ponder over the problem which was rarely absent from my thoughts."[38] It may have been Pictet's list that inspired him to put two and two together: pairing a list of what is known about the distribution of species in time—the fossil record—with what is known about their distribution in space—geographical distribution.[39] What do the two sets of patterns reveal? Wow! The result was a tour de force, and he quickly composed an essay, "On the Law Which Has Regulated the Introduction of New Species." He said that his immediate impetus for writing it was a paper by marine biologist Edward Forbes, newly appointed Regius Professor of Natural History at the University of Edinburgh and president of the Geological Society. Wallace likely came across Forbes's paper in the rajah's library—it was reprinted in the *Literary Gazette* of 19 August 1854, just two pages after Wallace's own first writings in the magazine appeared. Forbes was an adherent of *polarity theory*, a quasi-mystical oscillating model of the ebb and flow of life through the geological ages according to a divine plan—precisely the sort of nonscientific thinking masquerading as science that Wallace had little patience for; you can imagine him rolling his eyes and shaking his head, then deciding he was going to set the well-meaning but misguided professor straight.[40]

Yes, Forbes was surely important in spurring Wallace to write this paper, and Pictet's laws of geological development may have inspired a structure. But the *real* object of Wallace's essay was Lyell. About the very time Wallace penned this epochal paper, he was also immersed in an analysis of the *Principles of Geology*, and his first substantive notes on Lyell, under the heading "Notes on Lyell's Principles," were made in another and truly remarkable notebook now dubbed the Species Notebook.[41] Interestingly, the entry relates to Lyell's argument that scientific advancement in geology was finally made when the nature of fossils and landforms was understood to result from natural law—material processes, or what Lyell called "secondary causes." Wallace agreed that this was a giant leap forward in scientific understanding, but Lyell was clearly not, however, inclined to apply natural law to understanding species origins, an inconsistency Wallace was determined to correct. A key theme for Lyell in *Principles* was just how the "introduction of new species" was realized—it was a phrase he used repeatedly, including as the heading for a long section summing up his chapter on the "Extinction and Creation of Species." But there Lyell set up an argument for a succession of species "introduced" via special creation, in accordance with needs demanded by environment: it was a vision of sets of species tailor-made by a creator to fit their environment. As the environment changes (climatically or geologically), he argued, earlier species are rendered extinct and replaced with new ones suited to the latest conditions. As suites of species are thus swapped out over time, the *relationship* of new to old species is irrelevant—there *is* no relationship per se. To Lyell it was all about fit to environment, wherein each species and species group is created de novo to fit the bill. Wallace disagreed with the great geologist about as strongly as he disagreed with Forbes. But while he found Forbes's model little more than metaphysical nonsense, Lyell was a more formidable opponent: given his stature, his word was authoritative, definitive. Yet, Wallace was convinced, it was also incorrect.

Thus, we can plainly see why the title as well as the central arguments of Wallace's Sarawak essay refer directly to Lyell. Contrary to what we might expect, however, Wallace does not explicitly analyze Lyell's model in this paper—not yet. That would come soon enough as he contemplated what he found in the amazing Aru Islands, but here and now, in Sarawak, Wallace looked to first principles and cut to the chase: his essay laid out nine numbered "propositions in Organic Geography and Geology," culminating logically in his *Sarawak Law*, as it came to be known: "The following law may be deduced from these facts: *Every species has come into existence coincident both in space and time with a pre-existing closely allied species*" (his emphasis). That is, every species arises in immediate proximity to a preexisting and closely related species—indeed, the inference is that every species is somehow materially *derived from* that preexisting species, but although we can plainly see a transmutational implication, he did not say as much.[42] Nonetheless, it had astonishing explanatory power: this simple fact "connects together and renders intelligible a vast number of independent and hitherto unexplained facts," Wallace declared. Why? Because material links from one species to another constitute *evolutionary* lineages, lineages that can also branch, ramify. It thus makes sense of classification, geographical distribution, the geological sequence, comparative anatomy, and more.

It's no wonder that clear and unambiguous classification of species is so difficult, Wallace says, when the branching chains of relationships extending back through generations of ancestral species are so often obscured. He put his finger on two key reasons: first, consider that species relationships are often (especially in Wallace's day) identified by anatomical features. Greater similarity suggests a closer relationship, or "affinities," to Wallace, which bears on just how the species are placed taxonomically. But "similarity" can be superficial, owing to adaptive convergence (what are termed "analogies" in structure) and is also a matter of scale: two groups may be *analogous* at one level (for example, the "jumping mouse" morphology convergently hit upon by distantly related marsupial and placental small-mammal groups) but have close *affinities* at another level (both are mammals closely related to each other relative to, say, worms). For groups where you have a great fossil record and can identify lots of representatives in the chain of relationship between these groups—that is, their genealogy—you would have a better chance of identifying adaptive convergences, or analogies, and more clearly identifying the actual relationships, or affinities. But we do not have a great fossil record for most groups, innumerable links are inconveniently missing, and this is Wallace's second reason for why the chains of affinities are obscured: "We have only fragments of this vast system," he points out, "the stem and main branches being represented by extinct species of which we have no knowledge, while a vast mass of limbs and boughs and minute twigs and scattered leaves is what we have to place in order [classify], and determine the true position each originally occupied with regard to the others." Thus, "the whole difficulty of the true Natural System of classification becomes apparent to us." Here Wallace independently articulated two key interrelated insights that Darwin had also come to: first, all species are linked through an infinitude of ramifying chains of relationship—a treelike system of irregular branches upon branches and "a complicated branching of the lines of affinity," Wallace evocatively writes, "as intricate as the twigs of a gnarled oak or the vascular system of the human body." Second, our system of classification should ideally reflect this great tree: it would be a natural system, "a true classification."[43]

It is just astonishing, really. Clearly, these ideas had been percolating in Wallace's mind for some time, and here they all came together in one scintillating paper: an evolutionary vision of ramifying lines of relationship coursing through space and time. Of course! With this model it all makes sense—the nested hierarchy of higher classification, the family resemblance of body plans in disparate groups, the mapping of relationship onto geography, the dramatic succession of extinct beings and their essential bond with the living. . . . All of these are beautifully explained; in fact, he declared, more than merely explained, it *necessitates* the patterns we see, rendering nonsensical hypotheses like that of Forbes moot. "Granted the law, and many of the most important facts in Nature could not have been otherwise, but are almost as necessary deductions from it, as are the elliptic orbits of the planets from the law of gravitation." It was a breathtaking insight, elegant in its simplicity yet powerful in its implications.[44]

———

He wrote out a fair copy of the essay in February 1855 and sent it off in one of the several shipments to Stevens that month containing well over five thousand insects (reserving a few thousand for his personal study), as well as assorted snails, crabs, and several boxes of orchids, including *Vanda* (*Dimorphorchis*) *lowii*, collected at Mount Serumbu, a spectacular species bearing long cascading strings of flowers, some yellow and some orange (dimorphic, as the genus suggests)—a botanical Rapunzel in rainforest tree towers. Stevens duly forwarded the paper to the *Annals and Magazine of Natural History*, where it was published in the September 1855 issue. Wallace hoped and expected that Forbes and his supporters would engage him in discussion and debate over the issue, but before it was published, Wallace received the sad news of Forbes's untimely illness and death at age thirty-nine that past November, just a month after his polarity paper came out. Perhaps feeling a bit badly about his treatment of the late naturalist under the circumstances, he was able to write the editor in time to insert a respectful footnote in the printed version: "Since the above was written, the author has heard with sincere regret of the death of this eminent naturalist, from whom so much important work was expected. His remarks on the present paper—a subject on which no man was more competent to decide—were looked for with the greatest interest. Who shall supply his place?"[45]

The trouble with writing pithy papers on the other side of the world was the excruciating wait. He sent letters and smaller (or more valuable) shipments home via the faster and more expensive overland route via Suez and larger

shipments via the longer but cheaper sea route around the Cape of Good
Hope; this shipment may have gone the longer route, considering it was published some seven months later. (Even after it came out, there seemed to be
only resounding silence, though as we will see the paper hit its intended target:
Lyell himself was sufficiently shaken to start a new notebook on the species
question, opening with a detailed summary of Wallace's "Sarawak Law" paper.)
But that was in months to come; for now, with the prospect of the wet season
soon ending, Wallace decided it was time to head inland, to the storied realms
of the orangutan.

———

The world's third-largest island, at nearly three hundred thousand square
miles, with a vast densely forested and mountainous interior, Borneo was then
one of the great landmasses little explored by Europeans and little affected by
its many Indigenous peoples. The island's ecology is tragically imperiled today,
with well over half the primary forest gone, between demand for tropical hardwoods (an estimated half of the world's tropical timber annually comes from
Borneo alone), sprawling palm oil plantations, quarries, and coal mining[46]—
the latter an industry that was just starting up in earnest in Wallace's day and
from which he benefited. Coal had been discovered up the Si Munjon (Simunjan)
River, a small tributary of the Sadong, a year or so before Wallace's arrival,
outcropping in seams about three feet wide at the isolated little peak of Gunung
Ngeli—more *bukit* (hill) than *gunung* (mountain) at just 650 feet (200 m). Sir
James and the British Borneo Company starting mining operations there with
mostly Chinese labor soon after. Mining ceased only in the early twentieth
century, though it was temporarily resumed during World War II by Japanese
occupying forces.[47] With his collecting successes at Bukit Timah no doubt in
mind—where the slash and sawdust from timber clearing created bounteous
conditions for insects—the new coal works promised similarly good collecting. He and Charley meandered their way fifteen or so miles south up the
Sadong through a vast level and swampy plain, then a few miles more up the
somewhat smaller Simunjan, equally meandering and closed in by lofty forest
canopy on either side. They arrived in mid-March, making their way from the
boat landing to the mine site by unsteadily walking the "Dyak road" of tree
trunks laid end to end. Robert Coulson, the mine engineer, hospitably put
them up at first, but it was soon clear that Wallace's hunch was right on. The
place was teeming with insects, thanks to all the timber cleared for the mine

and a railroad line to serve it, so he had a small bungalow built where he and Charley spent the next nine months in collecting heaven.

It was here that Wallace started another set of entries in his Species Notebook, "Entomological Notes—Sadong River, Borneo." Besides the prospect of great insect collecting, he was mainly motivated to come here "to learn something of the Geology of the district," as well as an opportunity to study "the great Orang-utan (which here abounds) in its native haunts."[48] The rains having abated, day after day he and Charley returned with dozens of new species, and the workmen, paid a penny an insect, brought in an additional bounty. But while the collecting was going well in regard to specimens, all was not well in his estimation of Charley. Having evidently read Wallace's vented frustrations more than once in letters home, Stevens and his mother and sister helpfully conferred about recruiting another helper, "a very nice young man," to replace Charley. This provoked an irritable missive from Wallace. Being merely a "very nice young man" was useless: Is he quiet or shy? Talkative or silent? Sensible or frivolous? Can he live on rice and salt fish for a week at a time? Do without wine or beer or sometimes tea or coffee? Can he sleep on a board? Does he like hot weather? Is he too delicate to skin a stinking animal? Can he walk twenty miles a day? Can he work? Can he *draw*? Can he make anything or even saw a board straight?[49] It goes on and on, but you get the idea—hapless Charley was evidently lacking in several departments. Though Wallace acknowledged that the kid was finally getting tolerably good at insect collecting and skinning, he declared that another helper "with a similar incapacity would drive me mad." A replacement was not forthcoming—Stevens, Fanny, and his mother wisely realized that no one they recommended was likely to live up to Alfred's exacting standards.

For all the frustrations, Wallace later wrote that in his twelve years in the tropics of Amazonia and the Malay Archipelago, he never had such good collecting as at Simunjan, in an area not more than a square mile in extent. His collections racked up in a predictable way, like a classic species accumulation curve: steeply increasing at first, then slowing until eventually the curve flattens as most available species are collected. In little more than a month, he had over a thousand species, then new ones came in at a slower and slower pace, but it is unlikely the collecting would have plateaued completely in so biorich a locale. He later reckoned that of the two thousand or more beetle species he had collected on Borneo, all but about a hundred came from Simunjan— including some three hundred species of fabulous cerambycid beetles (most new!) and five hundred different weevils and their relatives, such as the spectacular Spider weevils of the genus *Mecopus*—some with snouts greatly elongated to match their long spindly legs, others with slender antennae four times the length of their body (including one later named for Wallace: *M. wallacei*).[50] The butterflies were less numerous, and if the ones he caught were jewels, the *crown* jewel was surely *Ornithoptera* (now *Trogonoptera*) *brookeana*, the stunning birdwing butterfly he named in honor of Rajah Brooke. This exquisite butterfly, which Wallace described as "one of the most elegant species known," flutters on wings spanning some seven inches (seventeen centimeters), the handsome males in fine attire: a velvet-black body and wings, each adorned with a row of large iridescent-green triangles like the tips of so many gleaming ferns, and a vivid red collar about the neck like a stylish scarf. Indeed, the Rajah of Butterflies.

Other rarities abounded. One day a Chinese worker brought him a curious green-and-yellow frog with greatly enlarged toes connected by thin, expandable membranes, each ending in a small adhesive disk. The worker described the creature as sailing through the air, which Wallace confirmed—the photogenic frog is known today as Wallace's flying frog (*Rhacophorus nigropalmatus*). Among the mammals he collected were Bornean squirrels, odd shrew-like moonrats (*Echinosorex gymnura*), semiaquatic otter civets (*Cynogale bennettii*), and, rarest of all, the secretive bay cat, *Catopuma badia*—live specimens of which did not turn up until 1992![51] But the prize mammal he sought was the orangutan, a Malay name meaning "man of the forest" but known to the Dyak people as *mias* and to modern biologists as genus *Pongo*, a name bestowed by French naturalist Bernard Germain de Lacépède in the late eighteenth century. Three orang or mias species are recognized today, one in Borneo (*P. pygmaeus*) and two in Sumatra (*P. abelii* and *P. tapanuliensis*, the latter just described in 2017). All are listed as critically endangered.[52]

From a modern perspective, one might reasonably argue that the likes of Wallace contributed to the decline of these magnificent creatures—it is certainly true that Wallace hunted orangs at every opportunity for profit and study, killing perhaps fifteen or more: specimens were sought-after rarities back in Europe.[53] In fairness, however, two points should be borne in mind: first, the biology of orangs was little known, and Wallace's studies did contribute significantly to the understanding of these great apes. He commented that he had "probably seen more of these animals in a state of Nature than any other European," reporting in several papers his extensive observations and as much information as he could obtain from his Dyak neighbors, trying to determine if there was one or several species, among other things.[54] Second, it is fair to say that the catastrophic decline in orang populations stems not so much from overzealous collectors like Wallace in the nineteenth century but from rapidly accelerating habitat destruction in the twentieth, stemming from explosive growth in timber harvesting and palm oil plantations for global markets, exacerbated by human conflict that was evident even in Wallace's day. The Dyaks, he remarked, considered him a great benefactor in killing the orangs, owing to their destructive foraging in their durian trees.

Nonetheless, it is painful to consider these gentle giants of the forest mercilessly killed; even Wallace was moved when he realized that a female orang he had just shot high in a tree was nursing a baby, the pitiful infant still clinging desperately to her lifeless fallen body. Wallace took it home to try to raise it, ultimately unsuccessfully but making careful observations in the process. He was struck by how human, how "baby-like," the infant orang was, its emotional expression and behavior ranging from inquisitive to playful to content to having a meltdown. As an aside, it is noteworthy that Darwin, too, had spent time with an infant orang, closely observing her behavior and emotional expression. That was the famous Jenny, acquired by the Zoological Society in November 1837 and soon had London abuzz. Darwin spent time with Jenny periodically over the next year and a half, until her death in May 1839. Like Wallace later, he, too, was struck with the humanlike qualities of the little primate: "Let man visit Ourang-outang in domestication," read one notebook entry on Jenny, "hear expressive whine, see its intelligence when spoken; as if it understood every word said—see its affection.—to those it knew.—see its passion & rage, sulkiness, & very actions of despair."[55] It is no coincidence that both Wallace and Darwin took advantage of opportunities to better understand the orang mind and emotion, given their transmutational convictions. At that time orangutans were considered the most humanlike of the great apes, and while anthropomorphizing animals was (and is) extremely common, for all sorts of reasons, Wallace and Darwin saw the human qualities in these orangs through the lens of ancestry and common descent.[56]

Wallace was sad when the poor creature died—malnourished, no doubt, despite his efforts to feed it, deprived as it was of its mother's milk. Writing home to Fanny and his own mother, he reported that he had adopted an orphaned baby, one that he feared they would find rather ugly. He eventually revealed the baby's identity. His was no common baby, he declared, "and I am sure nobody ever had such a dear little duck of a darling of a little brown hairy baby before!"[57] Yet whatever pangs of sadness or even regret he felt did not prevent him from making a specimen of the deceased infant orang—he was a

Wallace's "dear little duck of a darling" orangutan.

collector and a naturalist first, after all, but there were surely mixed emotions between seeing the young orang as human relation, pet-like companion, and scientific specimen.[58]

Wallace kept up his correspondence while in residence at Simunjan, welcoming news from the family. His brother John returned to England in January 1855 to marry Mary Elizabeth Webster, the daughter of their old boss and landlord the master carpenter, and the newlyweds returned to the States soon after. Now a modestly prosperous mining engineer and co-founder of a water company, John and his bride planned to remain in the United States. Alfred wished his brother and his wife "unalloyed domestic happiness" and only hoped he might draw "as high a prize in the matrimonial lottery" one day. Fanny and her husband had moved their home and photography business to Conduit Street in central London, but Alfred fretted over their mother needing to move from Albany Street and hoped they might find a small cottage for her not far from Fanny and Thomas. He was impatient for more news, but his sister sent him a small music box (a "never ending delight to the Dyaks" who called it a bird) and new shoes, as well as some cured bacon, less successfully shipped ("eatable, *just!*").[59]

He also kept up with his literary pursuits, sending off two articles via Stevens from Simunjan. In one, published under the heading "Proceedings of Natural-History Collectors in Foreign Countries" in the August 1855 issue of the *Zoologist,* he gave a detailed account of his insect collecting in the vicinity of Simunjan, with its frustrations and successes—"I daily got numbers of species, and many genera which I had not met with before." In describing his haul, the word "elegant" appears three times, "beautiful," five—the *most* beautiful catch was the aptly named *Belionota sumptuosa,* a nearly two-inch (five centimeters) wood-boring beetle, sumptuous indeed in glittering metallic green-and-copper tones. Then there were lucky catches, like the time at breakfast when a "magnificent" black-and-yellow-spotted long-horned beetle flew practically into his hand.[60]

His second Sarawak memoir, appearing in the 27 October 1855 issue of the *Literary Gazette,* was more journalistic in tone, with vivid accounts of the natural history and people.[61] Of the former he reported on the fragrant and beautiful *Caelogyne* orchids, "magnificent" scarlet-flowered *Aeschynanthus* vines, and curious eastern oaks with red, brown, and black acorns; colorful hornbills (the toucans of Southeast Asia, he thought); those "most strange and interesting animals" the orangs and the immense flights of flying fox fruit bats (*Pteropus vampyris edulis*) extending as far as the eye can see—with their impressive five-foot wingspan, such a sight might evoke visions of the flying monkeys from *The Wizard of Oz* today. But his commentary on the Malays, Chinese, and Dyaks he encountered is interesting in another way, once again helping us understand Wallace in the context of the colonial enterprise, of which he was after all a product and from which he benefited. Wallace has sometimes been portrayed as an almost uniquely nonracist, egalitarian Victorian, in a sense an observer who was "woke" a century and a half before many in Western society realized there was anything to awaken to. Reality is more nuanced, and instructive. It is true that he saw people as *people* and valued their humanity. "The more I see of uncivilized people, the better I think of human nature on the whole, and the essential differences between so-called civilized and savage man seem to disappear," he famously wrote in this memoir from Simunjan. But it is important to consider the context: marveling at how orderly and prosperous this mixed population of Chinese, Malays, and Dyaks was under the firm, paternalistic hand of the ruling Europeans, he implies they are being saved from themselves by the beneficent and wise whites. He mentions the Chinese, both the lying and thieving "lowest and least educated" class and the greater proportion who were "quiet honest decent sort of men"; the

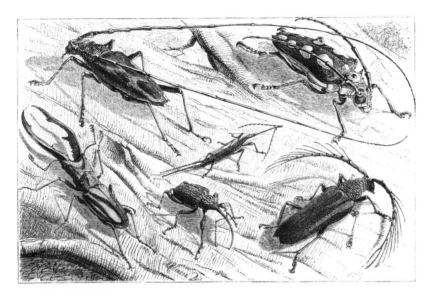

Some of Wallace's fabulous Bornean beetles.

"treacherous and bloodthirsty" Malays all armed with their kris (*keris*), the traditional wavy-bladed dagger of southeast Asia; and the Dyaks, only lately come round to the view that headhunting is not an absolute necessity. Sure, the old men missed their head-hunting days, proudly swapping stories of conquests and tallies of heads taken in their youth. They were fond of the rajah but still thought the harvests would improve if only they could take a few heads now and then, we are told. But now all three groups gave their consent and support to the rajah and all was well: crime was low, there were few instances of the murderous rampages the Malays called "running-a-muck," people slept with unlocked doors, and most went about unarmed (notwithstanding all those krisses). While Wallace the Owenite socialist was real, and it is true that he was remarkably respectful and humane when it came to other cultures, it is equally true that at times he saw these cultures through colonialist, rose-colored glasses. There was never any question in his mind of what society was where on the ladder of "social progress" and that European civilization, despite its many ills (which he also readily and pointedly called out), was intrinsically superior when at its best.[62]

In keeping with his perennial geological and anthropological interests, he was keen to see more of the country and its people. Toward the end of the year, as the rainy season approached, he decided to pack up shop at Simunjan, and sending Charley back to Santubong the way they came in a boat laden with their treasures, he took the scenic route to Kuching, traveling up the Sadong to its headwaters and then overland to descend to the Sarawak River. There was a distant peak visible from the lower Sadong that beckoned, Gunung Ampungan, and he wondered what marvels of natural history and humanity might greet him there and beyond. He pared down his gear and hired a Malay youth named Bujon as his guide and interpreter, knowing the language of the Land Dyaks they would encounter, along with some Malay boatmen.

The Dyak people, it should be mentioned, are the Indigenous people of Borneo. The name is thought to have originally been derived from a derogatory Malay term for "savage" but became the common appellation now used, even among themselves. Seven main Dyak ethnic groups are commonly recognized today, mostly in the Bornean interior, but Wallace mainly encountered two general groups: the coastal Sea Dyaks (Iban or Hivan)—greatly feared marauders, sort of Southeast Asian Vikings of legendary ferocity—and the upland Land Dyaks (Klemantan or Bidayuh). Most Dyak groups, including these, had a tradition of headhunting (*ngayau*), believing that the heads of enemies conferred special powers and were a tangible sign of fighting prowess. Defeated by Brooke, the Dyaks became subjects and allies, and although Brooke and later colonial administrators, like the Dutch who controlled much of Borneo (Kalimantan) before them, officially banned head-hunting, it tended to be permitted (or ignored) by authorities in times of war—even well into the twentieth century. While ngayau was by Wallace's time not often practiced, many other elements of traditional Dyak culture were—and are— strong, including the communal longhouse. Wallace likely found these communal structures, easily hundreds of feet in length and housing a hundred or more families, reminiscent of the malocas of the upper Uaupés, although unlike the South American structures they are typically raised high on posts for protection.

His entire journey took just nine days, necessitating changing boats and crew as the river became narrower and shallower, then hiking overland. Traveling southwest up the Sadong and then the Kayan, he did not have to ascend endless cataracts as in Brazil, but the geology and landscape around them, an increasingly mountainous terrain of tilted slates, conglomerates, and sandstones with myriad valleys and ravines running this way and that, were interesting, one vista reminding him of the "Himalayas in miniature." It was slow going by boat up to the headwaters, then overland through precipitous terrain, where they forded ravines and snaked along the rocky faces of gorges on

A Dyak bamboo bridge spanning a Bornean river.

narrow bamboo bridges and walkways, making their way along slick bamboo
trunks a mere four inches wide with slender handrails so shaky that he dared
not use any for support. He had a grand view of beautiful Gunung Penrissen,
the loftiest peak of a lofty massif at just over 4,000 feet (1,300 m). Penrissen

forms a high point along a great semicircular escarpment enclosing an elevated sandstone basin—the mid-Paleogene Penrissen Basin, flanked by two even larger sandstone basins that Wallace would have found interesting: the Semuti Basin just to its southeast and the yawning Bengoh Basin to its northeast, now partially flooded thanks to the Bengoh Dam.[63] He was close to a borderland twice over from his vantage point, if only he knew: practically on the boundary of the Penrissen sandstone of the basins and the older Kayan sandstone that dominates the area to the north, including Mount Santubong, and within sight of the international border that runs along that high escarpment separating the Sarawak of Wallace's day from Dutch Borneo (today Indonesian Borneo, with Sarawak now a state of Malaysian Borneo, along with Sabah and Labuan). That border today is not nearly as remote as it was for Wallace: the Penrissen Basin is now home to the Borneo Highlands Resort and golf course, with a short drive to a lookout point on the escarpment at 3,281 feet that overlooks the mountains of southwest Kalimantan, Indonesia.

Wallace and his companions stayed at several Dyak villages along the way, hospitably received by the people they encountered. He found himself guest of honor by the *orang kaya* (headman), of each village in turn, with public audiences, feasts, gifts of food and rice wine (*tuach*), demonstrations of dances, kids' games, and contests of strength. He admired their form and dress—diminutive people, the men and women were typically bare-chested and ornamented, the men wearing a *chawat* or waistcloth, kerchief about the head, large moon-shaped brass earrings and brass arm and leg rings, beads, and armlets of shell; the women were equally decorated, with the addition of corset rings around their body. He was put up in "headhouses"—multiuse circular buildings on long posts adjacent to the communal longhouse, typically adorned, as the name suggests, with the heads of vanquished enemies. He did not give a head count, so to speak, but Dyak dwellings could contain hundreds of skulls. Wallace and the Dyaks shared a mutual curiosity: everywhere eyes were on him, and he thought this must be how zoo animals feel with throngs of the curious watching every move. In one village a girl of ten or twelve years old screamed in horror at the sight of him, dropping the water vessel she was carrying and jumping into a stream to escape. In another the curious residents wondered about his color—could he really be white all over? Obligingly pulling up a pant leg was as far as Victorian propriety would allow him to go to show them.

Descending the Sarawak River watershed, the geology changed, and Wallace admired the sometimes oddly eroded limestone formations they passed.

He learned the hard way that the Sea Dyaks were far surer boatmen than the Land Dyaks he had hired in Sennah, the last Dyak village he had stayed at, when they repeatedly ran the boat aground and nearly swamped in rapids. They managed not to capsize, fortunately, and proceeded with the aid of some helpful Malays who came to the rescue. The small party was soon serenely floating down the left (*kiri*) fork of the Sarawak River to Kuching. In Sennah and environs, Wallace was struck by the abundance of fruit-bearing trees. While mangosteen, langsat, rambutan, jackfruit, jambu, and bilimbi fruits were all abundant, the highly esteemed durian (*Durio* species, in the mallow family, Malvaceae) was the most abundant of all. A large fruit with a thick, thorny rind enclosing chambers of custard-like pulp, ripe durian smells (reeks?) to many of some unholy witch's brew of skunk, sewage, and vomit— which is to say, it smells horrible. Horrible enough that today it is banned from airplanes and many hotels, and riders of mass transit in some eastern cities, such as Singapore, are prohibited from carrying it aboard. But while it is understandably repulsive to the uninitiated, for those who can get past the smell it becomes immediately apparent why this is considered the "king of fruits," prized throughout Southeast Asia, where hundreds of varieties are cultivated. Wallace was put off at first but soon became a convert, and in an article written for William Hooker's *Journal of Botany*, he sang the praises of the durian—"a new sensation worth a voyage to the East to experience"—giving perhaps the most detailed yet evocative description of this remarkable fruit ever penned.[64] He also impishly pointed out that durians present a valuable lesson to boot, in a barb aimed at certain moralists seeking sermons in stones, or at least fruits. Falling from tall trees, heavy, spiky durian fruits can cause serious injury, even death—contrary to the belief of natural theology enthusiasts who, seeing beneficent design in nature, claimed that only small fruits were produced by lofty trees, while large and potentially dangerous ones were providentially bestowed upon low-growing plants. "From this we may learn two things," Wallace states matter-of-factly. "First, not to draw general conclusions from a very partial view of nature, and secondly, that trees and fruits, no less than the varied productions of the animal kingdom, do not appear to be organized with exclusive reference to the use and convenience of man."[65] Ouch, in more ways than one.

He was back in Kuching in early December 1854, and at Sir James's invitation he, Charley, and a Malay lad named Ali traveled to the rajah's small country cottage at Peninjau, just southwest of Kuching on picturesque Gunung Serumbu, for some relaxation and collecting.[66] It was no mean feat getting up the

Durian: as prized for its taste as reviled for its odor.

steep-sided mountain: it required climbing a succession of ladders up cliff faces, negotiating tree trunks with notched steps, more of those slightly concerning bamboo bridges over chasms, and clambering over "house-sized" boulders (and Wallace meant longhouse sized). But once there they found themselves in moth heaven: on good nights they collected 100 to 250 moths, a half to two-thirds of them new species each night! They returned to Kuching to spend Christmas with the ever-hospitable rajah, then hoofed it back to their lepidopteran Shangri-la, where they stayed through the middle of January. Wallace later tallied up his moth data—a total of 1,386 specimens over twenty-six nights, more than 800 of which were caught in just four of those nights under, he noted, particular conditions: dark (no moon) and wet. And a veranda. He concluded that perhaps the best thing going for that collection locality was a nice whitewashed veranda with a lamp—easy for the moths to get in, not so easy to get out before captured. Thus, he prescribed such a setup for aspiring moth collectors in the tropics: "I feel sure that . . . it would well repay them to carry a small-framed veranda, or a veranda-shaped tent of white canvas, to set up in every favorable situation."[67] Modern entomologists actually employ a version of this, using lights with white sheets to provide a convenient and perhaps irresistible surface for the insects to alight on. It may have been here that Wallace penned yet another interesting paper, a report for the Royal Geographical Society, with map, describing his inland Sarawak journey in great detail. It included separate sections on three of his favorite subjects: *geography* (including corrections to the latest maps), *geology* (accurately noting that he traversed a dividing line of sorts, with ancient limestones to the east and younger sandstones and conglomerates to the west), and *ethnology* (among other things noting the kinship between the Dyaks and the Indigenous people of Amazonia). This last bears notice: that insightful observation reflects how Wallace tried to understand human societies in the same context of geographical

View of Santubong mountain from near Rajah Brooke's cottage on
Mt. Serumbu.

distribution and descent that he applied to other species. Dyaks, he noted, are
most closely related to the Malay, then to the Chinese, and next to the Indians of
South America. These groups have so many physical and even cultural traits in
common that he realized "we must consider them as branches of one great division of mankind."[68] His interest piqued in such divisions and branches of "mankind," this was to become a major line of investigation for Wallace in his subsequent travels east.

———

His thoughts were certainly trending in that direction. Wallace prepared to
depart Sarawak, shipping off his sixth and final Borneo consignment to Stevens: orangutan skins and skeletons, a box of ferns, skulls and skins of various mammals and birds, bottles of reptiles, and nearly 4,000 insects, 1,690
of which were marked "private," to be kept for his personal study. The shipment would arrive safe and sound in London by the summer solstice. He
also finished up two more papers—that article for Hooker praising the durian
(and bamboo too, by the way) and "Observations on the Zoology of Borneo"
for the *Zoologist*. In that one he made a remarkable inference, "strong pre-
sumptive evidence" that Borneo, Sumatra, and the Malay Peninsula had

been closely connected in the not-so-distant geological past. Attuned as a collector would be, he found that their *shared* avifauna rendered them redundant and therefore "uninteresting" to him from a collecting perspective. Which meant they were uninteresting for an interesting reason: past geological events shape modern geographical distributions, an idea that was to become a key theme of some of Wallace's greatest scientific contributions. He also revealed his next steps in the paper: his time in Singapore and Sarawak were but a prelude, time spent "in a great measure, preliminary or preparatory" to the main object of his journey, to "investigate the less known islands of the Eastern part of the Archipelago"—the "rich and almost unexplored" Spice Islands, remote Timor and even more remote New Guinea, where he was as keen to revel in the anthropology as the natural history. Actually, to him they were largely the same.[69]

He would first head to the trading town of Makassar, near the tip of the southwestern arm of the tentacular island of Celebes (now Sulawesi), for passage to the far east of the archipelago. Timing was everything, though: ships sailed east and back on the monsoon winds, and to miss the window of opportunity either way was to wait months for the next one. He had to return to Singapore to catch a boat to Makassar, but in Singapore he could restock on clothes, ammunition, and other "necessaries" he could not get in Sarawak, send off his broken spectacles for repair, and hopefully get some desperately needed funds from Stevens. He and Ali departed on 10 February 1856 on the barque *Santubong*, arriving in Singapore a week later. The rajah was sorry to see Wallace go,[70] but Charley was not. His hapless assistant decided to stay behind—it was doubtless mutual, Wallace fed up with what he saw as the boy's sloppy work and Charley tired of his perfectionist boss's incessant complaints. Though now Wallace half complained about losing him: "I hardly know whether to be glad or sorry he has left," he wrote to Fanny. "It saves me a great deal of trouble & annoyance & I feel it quite a relief to be without him. On the other hand it is a considerable loss for me, as he had just begun to be valuable in collecting."[71] Charley remained with the bishop of Sarawak to become a teacher, initially, but it was not the last they would see of each other: four years later Charley came to work with Wallace again, having grown in both stature and ability.[72] In the meantime the talents of the Malay youth, Ali, whom Wallace had engaged in his inland journey, were more and more in evidence, and he went with Wallace in Charley's stead. A quick study, Ali was to become Wallace's most trusted and respected assistant for the rest of his travels in the East.[73]

His arrival in Singapore was lousy timing: the very day he got there, a ship to Makassar had departed, and he lamented in a letter to his sister Fanny that it would be some time before there would be another. It ended up being rather more time than he had hoped, remarking in his next letter to her, two months later, that he was "still a prisoner in Singapore."[74] Staying with his friend Father Mauduit, the Jesuit missionary, he tried to make the most of his time there, practicing Malay with Ali, preparing for his voyage east, catching up on news and correspondence, and writing. Species were on his mind: distribution, relationships, varieties, changes. He fired off a fascinating paper titled "Attempts at a Natural Arrangement of Birds," published the following September in the *Annals and Magazine of Natural History*.[75] In it Wallace drew upon his intimate experience with the birds of Amazonia and now Southeast Asia, together with Charles Lucien Bonaparte's fine treatise *Conspectus Generum Avium*, and offered a novel approach to understanding the relationships of bird families using a kind of "whole data" approach including internal and external anatomy as well as behavior and physiology. We can see his attempt as overtly evolutionary today, even recommending a method for drawing unrooted trees of relationship, with varying branch lengths reflecting degree of divergence, lines of relationship that represent transmutation, and evolutionary transitions (though certainly not put in these terms by Wallace). In so doing, from a modern perspective his approach even sets up testable hypotheses about relationship and common descent: insofar as it is an "article of our zoological faith," as he put it, "that all gaps between species, genera, or larger groups are the result of the extinction of species during former epochs of the world's history," we can look to the fossil record to identify forms linking related groups. "We believe this view will enable us more justly to appreciate the correctness of our arrangement."

All the while he fretted, too, about the reception of his Sarawak Law paper. Recall that it had been published the previous September, but he had heard nary a word about it except for an irritating comment from Stevens about grumblings coming from some in the learned societies of London, wishing for "less theorizing" and more facts—specimens—from Mr. Wallace. You can imagine how well that went over with him! He wrote Sir James and sent him a copy of the paper, evidently explaining his ideas about the "mutation of species" and, keen to hear how scientific men back in London viewed his hypothesis but hearing nothing, interpreted their silence as rejection and lamented the resistance he assumed the idea was being met with. The rajah wrote back encouragingly—he himself could only see a role for a Creator regardless of whether there was a "successive development of species" or not but was surprised there would be such "bigotry & intolerance" at views or facts, like Wallace's, that were opposed to received wisdom. "What harm can truth do us?—What good can it not do us?"[76] Little did Wallace know that his paper had in fact caused quite a stir back in London: Charles Lyell was sufficiently impressed that within days of reading it he opened the first of what grew into a series of seven notebooks on the species question. Lyell kept many such notebooks, but this was a fresh one dedicated to the vexing "species question," and the timing is surely no coincidence given that notes on Wallace's paper fill its very first pages: "Of innumerable ways in which Omnipotence might fit a new species to all the present and future conditions of its existence, there may be one which is preferable to all others, and if so this will cause the new species to be in all probability allied to preexisting and extinct or with many coexisting species of the same genus." In another notebook he noted that Wallace's paper "goes far toward Lamarck's doctrine [of transmutation]."[77] Right about the time that Wallace wrote to Brooke, Lyell was in fact paying Charles Darwin a visit to discuss Wallace's paper, among other things. It turned into a momentous occasion where Darwin revealed his theory of transmutation by natural selection to Lyell, only the second person after Hooker he had confided in. Lyell was astounded and, clearly seeing the implications of Wallace's paper, soon urged his friend to publish his theory immediately.[78] Darwin at first resisted, convinced that Wallace's paper, while interesting, had nothing on him. As mentioned earlier he was likely misled by Wallace's repeated use of the word "created," but Wallace did not mean that in the sense of divine fiat. He used the word casually, neutrally; it could have been "arisen" or "formed," but, perhaps consciously, he refrained from using "developed from." It was likely difficult enough for the paper to get a hearing, after all. Darwin, after some discussion with Lyell and Hooker, relented, noting in his diary for 14 May 1856 that he "began by Lyells advice writing species sketch."[79]

Barely one week later, Wallace, on the other side of the world, departed Singapore at long last. He had resigned himself to traveling to Makassar via the islands of Bali and Lombock, small islands to the southwest of Makassar and just to the east of Java where the Dutch colonial capital of Batavia (now Jakarta) was located. It was irritating—the winds were against them, so it would take over a month to get there and then who knows how long to get another boat north to Makassar and then arrange for another east. @#%!&! He had not planned on visiting those islands, and the detour was sure to be a waste of time. He reckoned he would have wasted six whole months by the time he got to Makassar, precious time "utterly lost & at great expense," he lamented to Stevens. But, he trusted, when he finally made it to the eastern reaches of the archipelago, his efforts would be amply repaid.[80] Little did he know that he would be repaid sooner than that: he was unknowingly heading to another borderland of time and space, on a detour that was to lead him to one of his greatest discoveries yet.

CHAPTER 8

CROSSING THE LINE(S)

"13TH JUNE. Arrived at Bileling on the N. side of Baly 20 days from Singapore in 'Kembang Djepoon.'" So reads the first entry in the earliest of Wallace's surviving journals from the Malay Archipelago, a document that seems as weathered as the ancient geological formations Wallace so admired. Barely bound by its disintegrated cover, nibbled on, torn, worn, and water stained (long-ago monsoon rains or errant splashes of the Indonesian seas?), it has traveled long distances in space and time, albeit on a human scale. Precious cargo once carried while island-hopping across the vast Malay Archipelago and then around the world on sailing ships and steamers, nearly a century later the notebook found a home with its three surviving companions at the venerable Linnean Society of London, on the last island to which it had hopped.[1]

The word "JOURNAL" on the opening page in Wallace's neat script declares it a chronicle, its pages like so many stacked strata—but in reverse to geological ones, with the oldest records at the top. But like geological memoirs of the ancient world, each record, each entry, can give only limited insight into the time and place it was made but hint at more. "Bileling," "Baly," "Singapore," "Kembang Djepoon"—these names speak of the rise and fall of kingdoms and empires, vying for control of the region over centuries. Singapore, we have seen, was a British territory, one of the Straits Settlements. The name of the island he arrived at, Bali, is likely derived from the Sanskrit *bali-dvīpa* (बलि-द्वीप), "island of offerings," reflecting the Indian connection: Hindu Buddhist kingdoms ruled the island from about the tenth century BCE. Some nine warring kingdoms had divided up the island by the time Europeans, first the Portuguese and then the Dutch, arrived in the sixteenth century. The Vereenigde Oostindische Compagnie (VOC), or Dutch East India Company, had greatly expanded Dutch colonial control by the early nineteenth century in its quest to dominate the spice trade. The northern Balinese kingdom of Buleleng (as it is now spelled), founded in the late seventeenth century, was the last of the Balinese kingdoms to come under Dutch control, conquered just half a dozen years before Wallace's arrival in the East and thus incorporating Bali fully into the sprawling Dutch colonial system ruled from the administrative capital of Batavia (now Jakarta) on neighboring Java. Wallace's vessel, the *Kembang Djepoon—Blossom of Japan*—was a Dutch barque, a three-masted sailing ship, one of hundreds of trading ships plying the eastern seas under the Dutch flag. Although the old Dutch East India Company had been defunct for over half a century by then, taken over by the Dutch government, it was the private armies and warships of that former company-state that wrested control of territories from Sumatra to New Guinea from the local sultans and kings, controlling the highly lucrative trade in spice, coffee, and other commodities from the East. As in Brazil, the administrative apparatus of the Dutch colonial system—local administrators, traders and trading vessels, and a clockwork mail delivery network extending from Europe to these remote island outposts—greatly facilitated the work of naturalists like Wallace.[2]

Born of volcanoes, Bali is dominated by a lofty mountainous district where two great peaks stand out: Batur and Agung. Of the two, Gunung Agung is perhaps the more impressive, rising dramatically from the plain to nearly 10,000 feet (3,000 m), taller and younger than neighboring Gunung Batur but lacking the charm of that volcano's wide caldera and lake. Disembarking from the *Kembang Djepoon* at Buleleng that bright June day, Wallace would have seen the slopes of distant Gunung Batur rising to the southeast—but was perhaps not impressed by it since, lofty though it is, the compound stratovolcano is far broader than it is tall. He was more impressed with the prospect closer at hand: the countryside around Buleleng amazed and delighted him, not because of deep, lush forests but the opposite, a well-cultivated district of fruit trees and fields worked by the curious native ox, the Banteng, *Bos javanicus*. Bantengs are wild oxen of Southeast Asia, a somewhat diminutive but sturdy species with short, curved horns, light to deep chestnut in color except for white on the lower half of their legs and a distinctive large white oval marking their hindquarters.[3] The fields give way to "luxuriant rice-grounds," a series of green stairstep rice paddies extending up every slope, fed by myriad streams coursing from the volcanic uplands. The ancient paddy fields represent an elaborate irrigation system (*subak*), dating to the ninth century—but more than simply an irrigation works, subak is both a marvel of landscape engineering and a profound merging of a people with their land. Now a UNESCO World Heritage Site, the cultural and environmental are one in this spectacular fifty thousand-acre (twenty thousand hectare) complex of terraced paddies and water temples, where ancient earthworks and channels regulate a thriving artificial ecosystem. Or is it artificial? That depends on how you view the nature of humanity: Are we not organisms, too, part of the planet's teeming biosphere?

He was fascinated with the culture—together with Java and Lombok, the region's last remnant of Hinduism—but did not have high hopes for collecting on either Bali or Lombok. This island was extensively cultivated, which never bodes well for biodiversity, but more importantly he figured from geography that the fauna would differ little from his Singaporean and Malaccan collections, these islands being more or less strung in the same chain. Sure enough, the birds he collected in Bali, though lovely enough, were well known from points west and north: pretty black-and-white magpie robins (*Copsychus saularis*), pied starlings (*Sturnopastor jalla*), rosy barbets (*Megalaema rosea*), spectacular flame-back three-toed woodpeckers (*Dinopium javanense*), and the vibrant-yellow Asian golden weaver (*Ploceus hypoxanthus*), brightening the dark lava-sand beaches as they came and went from their pendant nests along the waterfront. He did a bit better with butterflies, bagging among them a new species he dubbed *Pieris tamar*, now called *Cepora temena tamar*, with wings of soft black and white with a bit of yellow above and hind wings suffused beneath with a deep orange glow.

He would have found more novelties but was impatient to get on with his travels farther east—but to go east he had to first go north, to Makassar, and now he found that there was no direct boat from Bali. So, first a zig, then a zag—he needed to hop over to Lombok to catch a lift to Makassar. He left for Lombok three days later, "a pleasant sail of two days" that took him around the Amed coast of eastern Bali, with its string of quiet fishing villages, each marked by a flotilla of sleek outriggers, and then across a strait of perhaps fifteen miles to the town of Ampanam (now a district of the city of Mataram) on Lombok's west coast. What a prospect, sailing between the islands! Twin mist-wreathed volcanoes seemed to guard the strait, one behind and one before him, "magnificent objects" standing high above the clouds where over the course of the day sun, shadow, and mist painted an ever-shifting picture in rich tones. Two lush, green volcanic islands, about the same size, within plain sight of each other, meant that the collecting was bound to be more of the same. He could not have imagined that he was in yet another double borderland in space and time.

———

Wallace was surely clued in to one kind of borderland here, without of course the fullness of a twenty-first-century understanding of what was going on. He recognized that one of the earth's "chief volcanic belts" ran through the archipelago and that it was no coincidence that the volcanic islands he was traveling along were aligned: the Lesser Sunda (Nusa Tenggara) chain—Lombok, Sumbawa, Komodo, Flores, Alor, and many satellite islands—in a line with the larger Sunda Islands of Sumatra and Java just to the west, mark a long, linear zone of volcanic and seismic activity that signifies geological uplift, much like the Andes in South America. He knew his Lyell, whose oscillatory model of slow and steady regional uplift and subsidence were central to his vision of a dynamically cycling planet. In *Principles* Lyell wrote of the "continued series of elevatory movements" caused by volcanic activity, where masses of rock are brought to the surface from deep within the earth—a surface continuing slowly up and up to become lofty mountains where, Lyell marveled, geologists "may then, perhaps, study in some mountain chain the very rocks produced at the depth of several miles beneath the Andes, Iceland, or Java."[4] Yes, Wallace knew he was traveling an uplift zone but could not know why. With the discovery of plate tectonics a century later came new insights into volcanic chains like this one. To modern geologists this is the Sunda Arc, the zone where the Indian and Australian plates are inexorably pushing beneath the Sunda Shelf, part of the Eurasian Plate, at the remarkable rate of 2.8 inches (7 cm) per year—very slow on a human scale (a bit faster than the rate fingernails grow but slower than the growth of hair, which averages about 6 inches per year) but quite fast in geological time: twenty-three feet in a century, some forty-four miles in a million years, a geological blink of an eye![5]

The long, deep oceanic trench that marks the boundary where the oceanic crust of the Indian and Australian Plates begin their slow-motion dive beneath the continental Sunda Shelf lies several miles offshore of the island arc itself. Dragged down—subducted, as it is termed—the oceanic crust is remelted and eventually recycled into terra firma: the islands are amalgamated masses of upraised sediments mixed with igneous rock, where vast magmatic plumes of that melted crust have made their way to the surface, often explosively via the plumbing of the ever-growing volcanoes. Indeed, volcanic activity is almost continuous here, and the most powerful known volcanic eruptions occur along this island arc, including the massive August 1883 explosion of Krakatau (Krakatoa), in the Sunda Strait between Sumatra and Java. It is perhaps the most violent volcanic event in recorded history, estimated to have released energy equivalent to some four times that of the most powerful nuclear bombs detonated.[6] Attendant on such violent volcanism are deadly earthquakes and tsunamis, the most recent of which struck this immediate area in 2018, with a massive earthquake and tsunami off the Lombok coast in July, then a catastrophic tsunami generated by a partial collapse of Krakatau in December.[7] The earthquakes are sometimes directly related to volcanism, but subduction itself generates them on a regular basis in a process known as the *seismic cycle*. As the oceanic crust is slowly but surely subducted beneath the continental crust, its downward movement is not necessarily smooth and continuous. Friction between the plates can "lock" them in places, and where they get stuck, the subducted plate slowly pulls the edge of the overlying plate down with it. Owing to the elastic properties of the earth's crust, this results in its stretching, with a simultaneous sinking of the seafloor near the edge of the overlying plate and an upward bowing or bulging just inland. Stress builds and builds over time until a critical point is reached, and the overlying plate suddenly springs free, generating an earthquake and tsunami. This also raises the near-coast seafloor while lowering the adjacent land. It is typically easier to see the effects of the coastal uplift than the inland subsidence, as raising the shallow seafloor often leaves marine life high and dry—precisely what happened along the coast of the small island of Nias, just west of Sumatra, following the massive earthquakes of December 2004 and March 2005, when long strands of fringing coral reef were uplifted and left bleaching in the tropical sun.[8] In the previous century, during the *Beagle* voyage, Charles Darwin investigated this phenomenon along the Chilean coast where, following the devastating earthquake at Concepción in February 1835, he and the captain found the seafloor uplifted, killing all the exposed marine life. In keeping with "the principles laid down by Mr. Lyell," Darwin declared, we "may fearlessly maintain that the problem of the raised shells . . . is explained."[9]

Wallace, who read Darwin's *Voyage of the Beagle*, agreed—it was the kind of volcanically induced uplift he was to observe some years later on the south coast of Java, an island that, as he later noted, contains more volcanoes active and extinct than virtually any other region of comparable area yet also beautiful karst (limestone) topography and extensive cliffs of coralline limestone—all marine in origin.[10] But he also perceptively observed that tsunamis arise from upheavals of the ocean floor. During his stay on Lombok, he experienced a relatively minor earthquake and soon after learned that there was an unusually high tide, resulting in flooding. Putting two and two together, Wallace concluded that "the sudden heavy surfs and high tides that occur occasionally in perfectly calm weather may be due to slight upheavals of the ocean-bed in this eminently volcanic region."[11] Yes, Wallace was alert to the volcanic arc even if he did not understand the deep boundary responsible for it, the meeting of tectonic plates. But there was another line, an invisible one that runs more or less perpendicular to that plate boundary, that he was now serendipitously crossing.

Wallace and his assistants Ali and Manuel landed at Mataram on 17 June 1856. For centuries this island was ruled by various *datus* or chiefs of the Indigenous Sasak people, who had converted to Islam in the sixteenth century. Conquered by the Hindu Buddhist Balinese in the early seventeenth century, the Sasaks were now subjects of the Balinese rajah of Lombock.[12] The sprawling town was, and is, the capital of this smallish island dominated by the 12,000-foot (3,700 m) volcano Gunung Rinjani, now the focal point of the Rinjani-Lombok UNESCO Global Geopark.[13] Wallace soon understood why they had anchored a quarter-mile offshore: at that distance the bay was wide and calm, dotted with Western ships and native praus, small twin-hulled out-riggers with triangular "crab-claw" sails. But he was alarmed to discover that the placid water suddenly gave way to huge rollers at the beach. It was no small relief when he—and his boxes and collecting equipment—was safely delivered ashore, having escaped the "devouring surf," as the natives put it, telling Wallace with pride that "their sea is always hungry, and eats up every thing it can catch."[14]

They hurried out of reach of that hungry surf, making their way up a steep beach of dark volcanic sand, where they were greeted by a local *bandar*, or licensed trader, a fellow Briton named Joseph Carter who generously offered the use of his house, office, and storerooms. The next day Carter even loaned a horse to facilitate Wallace's trip down the coast a little way to visit a Mr. S., for whom Wallace had letters of introduction. He was accompanied by another obliging expat, a Dutchman who offered to guide him. Following their guide, Wallace and his assistants wended their way through a richly cultivated landscape of rice paddies and pastures along the bay, lovely tropical agrarian vistas marred only by the shocking sight of bamboo cages bearing human skeletons— executed unfortunates, in all likelihood, left as examples or warnings to would-be transgressors, his first introduction to the severity of laws under the rajah of Lombok. The very day before, he had noted in his journal how the rajah, who was an absolute ruler, "appears to be tempered with more moderation & wisdom than usual with the rulers of the Malayan races."[15] He may have spoken too soon, given this gruesome punishment. In subsequent entries Wallace noted other examples of draconian Lombokian laws: death by kris (knife) for theft and, for married women, accepting any token or gift from a stranger, and the practice of suttee (*sati*), the ritual burning of the widowed wives of the rajah. In another entry a local Malay man inspired fear in Wallace's assistants by warning them that the rajah had decreed that a certain number of heads should be collected as an offering for a good rice crop—putatively an annual event, corroborated by others. Wallace had to accompany a terrified Manuel to get him to go collecting, and Ali would not collect water or firewood without arming himself with a large spear.[16] It was around that time that a somewhat superstitious Manuel matter-of-factly related that "gools" (ghouls) or *hantus* (spirits) were scarce in Lombok. He reached this conclusion based on having to avoid, back home, places where people died, especially at night, for fear of the terrible noises made by the ghouls. But conferring with a Malay on Lombok, he learned that on Lombok "there are numbers of men killed & their bodies lie in the fields & by the wayside and you can walk by them at night & never hear any noises at all." This surely elicited an eye roll from Wallace, who jotted in his notebook: "Note. In Lombock *hantus* are scarce!!"[17]

This initial exploratory jaunt was perhaps also marred—in a less gruesome way—by the paucity of birds and insects. He tried north of town too: "Similar country, numerous rivers—Pretty place—A few birds & insects."[18] He watched with interest as kids caught dragonflies with long sticks smeared with a sticky homemade glue—frying up dewinged dragonflies with onion and shrimp was a popular local dish—and managed to add a few birds to his collection from around town: the fig trees lining the open-air market were tenanted by a beautiful subspecies of black-naped oriole, *Oriolus chinensis broderipi*, a striking yellow bird with black wings and mask and an orange bill. He also collected the curious helmeted friarbird he called *Tropidorynchus* (now *Philemon*), a drab streaky-brown bird with a naked black mask and knobby bill known locally, he was told, by the onomatopoeic name *quaich-quaich*—now usually rendered *koak-kaok*. Wallace was to soon discover that throughout the region the orioles are dead ringers for the friarbird species they co-occur with—a celebrated example of mimicry in birds that he first described in 1863.[19]

But at the time, the friarbird was significant in another way. Friarbirds belong to the family Meliphagidae, honeyeaters and their relatives—a characteristic group of eastern Indonesia, Australia, Papua New Guinea, and New Caledonia. Lombok, he realized, was the westernmost limit of this group—or was it? It is significant that the first entry in Wallace's Species Notebook after his departure from Singapore in March (aside from a brief note on a book about dodos) is a "note to self" regarding this friarbird: "*Tropidorynchus* sp. Common in vicinity of Ampanam in the Island of Lombock but said not to occur in *Baly*." Squeezed between the lines is a superscript qualifier, "except on the side next Lombock."[20] He did not visit western Bali and thus did not know for sure, so this information probably came from Mr. Carter and his other contacts in Mataram. But they were not naturalists or collectors. Were they correct? Was Lombok the western limit of friarbirds? Or might they be found in western Bali too? The next entry on this page of the Species Notebook reveals Wallace wondering about what it means for a bird to be considered "resident" in some location, part of the fauna. It is one thing to be found there year-round and breed there, or even migrate to and from the locale regularly, but "occasional visitors . . . can not be so considered." He was surely thinking about that friarbird and perhaps having second thoughts of its purported occurrence in eastern Bali. What about other birds? It was beginning to dawn on him that Lombok might differ greatly from its near neighbor to the west.

He heard that birds abounded a bit further south, around the beautiful harbor of Labuan Tring (now Lembar Bay). He hired a boat and decamped there, where he presented a letter of introduction to one *Inchi* (Mr.) Daud, who offered him space to work, though did not have much to spare. It was here, collecting in that lovely landscape of rounded bamboo-covered volcanic hills bordering valleys and plains teeming with soaring gebang palms (*Corypha utan*), that he began to notice more and more how decidedly un-Bali-like the avifauna was: "Birds very interesting, Australian forms appear. These do not pass further West to Baly & Java & many Javaneese birds are found in Baly but do not reach here."[21] Reports of abundant birds proved true: here he found elegant white cockatoos feasting on golden fruits, little brown *Lichmera* honeyeaters, large metallic-green pigeons, handsome Australian bee-eaters (*Merops ornatus*), shy green-and-buff ornate pittas (*Pitta concinna*), and no fewer than eight species of kingfisher, including the diminutive *Ceyx rufidorsa*, a violet-and-orange beauty that "darts rapidly along like a flame of fire," and a gorgeous new species, dubbed *Halcyon* (now *Caridonax*) *fulgidus* by the distinguished British ornithologist John Gould, sporting a cobalt-blue back, white breast, black head with glittering red eye, and bright-orange bill and feet. The list went on—lovely doves of grass-green, crimson-and-black flower-peckers, black cuckoos, metallic king crows, golden orioles, and even a fine specimen of jungle cock, "the origin of all our domestic breeds of poultry."[22] But his most unexpected find must have been the great megapode, *Megapodius gouldii* (now *M. reinwardt*).[23] These large chicken-like birds, dubbed orange-footed scrub fowl, use their long, curved claws to rake and scratch together leaves, sticks, and other debris into enormous mounds some 6 feet (1.8 m) high and 12 feet (3.6 m) across, into which they deposit their brick-red eggs to be incubated by the heat of the decaying organic matter. Dozens of birds may converge on an area, leaving it studded all over with great mounds, a breathtaking sight—and mouthwatering, for the locals: both eggs and hen-sized megapodes were considered good eating, Wallace noted.

Truly a bounty of birds, Labuan Tring—birds mostly unknown just to the west. As he put it later, "I now saw for the first time many Australian forms that are quite absent from the islands westward."[24] What did it mean? Think about that: the Bali-Lombok disconnect is completely counterintuitive given the proximity (just fifteen miles apart) and environmental and geological similarity of these islands. For two side-by-side islands to differ in their fauna so radically defied common sense. An old canon of geographical distribution, Buffon's Law (named for the eighteenth-century French polymath Georges-Louis Leclerc, the Compte de Buffon), held that environmentally similar but widely separated regions are tenanted by similar but different groupings of organisms. Buffon had an Old World/New World comparison in mind: two great north-temperate land masses with *different yet similar* groups of species of bears, deer, big cats, rodents, canines, and so on, by and large the same groups east and west but each region having distinct species representing these groups. Buffon's interest lay in the faunal disconnect: similar environment, different species, and insofar as these different species belonged to the same classes and genera, this was a provocative observation that could support the possibility of species change, if only in the form of degeneration from a supposed ancestral ideal.[25] Buffon's scheme made sense, in its way, when thinking about related groups on complementary land masses on opposite sides of the globe. But here we have two essentially identical little islands side by side, with distinct assemblages of birds that are not even of the same families, let alone genera.

This profound discontinuity in space says something about distribution in time, Wallace came to realize. His was part of a growing awareness among naturalists that anomalous patterns of distribution might provide unique insights into the history of the planet, geologically and climatologically. At almost the exact moment that Wallace was cottoning on to the invisible divide represented by the Bali-Lombok Strait, botanist Asa Gray, a world away in Cambridge, Massachusetts, was discovering another profound pattern, albeit a bit closer to Buffonian expectations: the high congruence between the flora of eastern North

America and eastern Asia. A corollary to Buffon's law spoke to similarity of groups of species on the same continent, but now Gray found that the eastern North American flora was far more similar to that of East Asia than to the flora of the western parts of its own continent. Gray was prompted to investigate the distribution of families and genera across continents by his friend Charles Darwin, who was equally bowled over by Gray's findings: "Is it one of the many utterly inexplicable problems in Botanical Geography?"[26] Wallace was to later marvel at this floral discontinuity too—it was all of a piece.

———

He was now more alert than ever to distribution. He returned to Mataram in hopes of catching that ride to Makassar, but no luck. He waited, and waited, frustrated but keeping busy: there were letters to write, specimens to ship off, patterns to ponder. Then some opportunities for a diversion and decent collecting arose: first an excursion to Kopang, on the lower slopes of Gunung Rinjani in central Lombok, in the company of a Mr. Ross of the Cocos/Keeling Islands, who was acquainted with the *pumbuckle* or *perbekel*, the village chief.[27] It was a bit of a misadventure. But his next excursion was more successful: he joined a large party to see the country *puri* (palace) and pleasure gardens (Indraloka Gardens) of Gunung Sari, north of the town, where he got permission to collect. Fierce Hindu deities carved into the gateway belied a serene and beautiful scene within, a verdant mixed landscape of manicured gardens and jungle, with a large pond fed by a rivulet directed through an elaborate brick-and-stone crocodile's mouth. In the center of the pond sat a highly ornamented pavilion flanked by carved statues. The serenity was disturbed only by the retort of his and Ali's guns as they collected birds, bagging more of the beautiful new blue, white, and black *Caridonax* kingfishers—unusual for a member of this group in frequenting woodlands and thickets and preying on ground-dwelling arthropods—and succeeding in collecting the "curious and handsome" Sunda thrush *Zoothera andromeda*. Unlike the kingfisher, this thrush was true to its group (family Turdidae) in displaying the brown-and-white plumage and breast spots so typical of thrushes, as well as the characteristic ground-hunting behavior.

These diversions came at a cost: Wallace discovered he had missed a ship heading to Makassar while he was away, and he wanted to kick himself. No more excursions! He had been on Lombok over a month, and now it would be more weeks before he could leave the island. He packed up a consignment of specimens to send to Stevens, lamenting both missing the boat and the poor collecting. It was the dry season, after all, and so pretty lousy for insects, with only 250 beetle specimens (constituting eighty species) and some 150 butterfly specimens (thirty-eight species) to show for two months of effort—though the latter included the consolation prize of the beautiful swift peacock swallowtail *Papilio peranthus*, a dusky species suffused with a glowing nucleus of metallic green at the base of all four wings. He fared better with birds, packing up sixty-nine species and three hundred prepared specimens all told. As with a choice selection of insects, he marked with a red stripe on the label which birds Stevens should hold back for his private study. Together with a few dozen miscellaneous insects, shells, and a few bats and skulls, it was not a very small consignment, but he felt that it was for the tropics: "It is really astonishing & will be almost incredible to many persons at home that a tropical country when cultivated, should produce so little for the collector." The worst collecting spots in species-poor England would produce ten times the number of beetles, he lamented to Stevens, and even the butterflies of Britain are finer and more abundant than Lombok in the dry season.[28] Tellingly, he also related to Stevens that the birds, proportionally much more numerous than the insects, interested him greatly in that they "throw great light on the laws of Geographical distribution of Animals in the East." He went on to describe his discovery—how Bali and Lombok were nearly identical in every way yet differed dramatically in their avifauna and that they in fact belonged to "two quite distinct Zoological provinces," he declared, "of which they form the extreme limits." Many species illustrated this odd fact, he found, and he was now preparing a paper for publication.[29] That paper was not to be completed for several years, during which he had time to further collect and reflect, pondering the significance of this puzzling discontinuity. As we will see, he was not far off the mark in the explanation he came to—which was more complex and interesting than anyone could have known then.

At long last, on 30 August, he sailed from Mataram on the little Dutch schooner *Alma*, reaching his destination on the south arm of storied Celebes (now Sulawesi) three days later: "It was with great satisfaction that I stepped on a shore which I had been vainly trying to reach since February." The geologically minded Wallace would have noticed that Sulawesi is a most curiously shaped island, with four "arms" extending south, southeast, and east and a long curving northern one that gave the island the appearance of a misshapen starfish, or perhaps a stylized letter *K* with a long, playful flourish atop the upright backbone of the letter. Together with its myriad islets great and small, it gives the impression of an island that has just been exploded or the product of an island mash-up. In fact it is all of those things, at once an amalgam and in a state of slow-motion deformation and dismemberment. The island is the meeting place of three tectonic plates: suture zones upon suture zones, where the north-moving Australian plate is met by the Eurasia plate moving south–southeast and the Pacific plate moving westward. The resulting five tectonic provinces represent bits of times and places: Paleozoic chunks of the Australian continental plate mixed with Mesozoic metamorphics and Cenozoic volcanics, with a liberal sprinkling of geologically recent clastic sediments throughout. With all those suture zones, one might say it is something of a Frankenstein island, if it were not so beautiful. Sulawesi Island is—for now—the largest island of the region now known as Wallacea, one with geology and biology so bizarre that Wallace later suggested it was the most remarkable island on the planet. But that was a bit later, after having had time to make sense of his collections over the course of two stays on the island. We will get to that, but first, it started here, with his arrival in the old trading port town of Macassar (Makassar) on the west coast of the south arm of Sulawesi (okay, technically the south*west* arm, paired with the neighboring southeast arm just to the east, but by convention referred to as the south arm).

Austronesian peoples arrived at south Sulawesi in deep prehistory, ancestors of the modern Makassarese, Bajau, and Bugis, or Buginese, who might be considered the Indigenous people of the island, or this part of it. Makassar later became part of the prosperous Hindu-Buddhist kingdom of Gowa in the early 1300s, one of several kingdoms of the island, and then an Islamic sultanate in the early 1600s, by which time it had long been a hub of traders and merchants from far and wide—Portuguese, Arab, Malay, British, Dutch, Chinese, and others. Hostilities between the island's kingdoms provided an opening for intervention by the ever-opportunistic Dutch East India Company (VOC), for which commerce and bayonet-and-bullet-backed coercion went hand in hand. The VOC ultimately took control of Makassar and its environs in 1667 in its quest to dominate the nutmeg and clove trade. In that year fort Ujung Pandang, built in the 1630s against Dutch incursions, was ceded and soon rebuilt and renamed Fort Rotterdam, with twenty-foot stone walls and six bulwarks in the classic "turtle" fort design of the time (today a tourist destination and home to a museum, library, music conservatory, and archaeological center). But the town's role as an important trading center never ceased, and in Wallace's day such valuable commodities as pearls, rattan, copra, trepang, sandalwood, and exotic spices moved briskly through the port to and from all points of the compass—but not the famous "Macassar oil," then all the rage in Europe. The popularity of this vegetable oil blend as a hair treatment for the well-groomed gentleman was owed to the marketing efforts of London barber Alexander Rowland, who began peddling his curative and restorative Rowland's Macassar Oil beginning in 1783, claiming it to be the product of an exotic fruit grown on the "Island of Macassar" and endorsed by the royal houses of Europe—a kind of snake oil, then, but harmless enough. Rowland succeeded beyond his wildest oily dreams, and Macassar oil became a veritable cultural institution, lampooned in satirical cartoons of the day and celebrated in verse (Rowland's garners mention in Lewis Carroll's *Through the Looking Glass*, and Lord Byron, a fan, opened his poem *Don Juan* with a reference to "thine incomparable oil, Macassar"). The oil even inspired a minor industry in doily-like embroidered *antimacassars* to protect the backs and arms of fine upholstered chairs from oily heads and hands.

Wallace surely knew of Macassar oil from home, but if he puzzled over the fact that there did not seem to be any in Makassar itself he did not say. What he did comment upon was, in a blush of first impressions, how pretty, orderly, and clean he found the town—in particular the European district, unsurprisingly, with its familiar white-washed houses, churches, and clean streets that the residents were required to tend to each afternoon.[30] This is now a narrow district of today's Makassar, grown from a town dominated by a single *passarstraat* (market street) to a sprawling city of 1.5 million. One feature of the area that Wallace would recognize is the extensive rice paddies and farms. The city has since grown around the enormous district of paddies nourished by the sinuous *Sungai* (river) Tallo, and there were farms and orchards of fruit trees, ever-useful bamboo ("that necessary of life"), and stately feathery-leaved sugar palm trees, *Arenga saccharifera* (now *A. pinnata*), the sap of which was used to make blocks of brown palm sugar (still for sale in every market in Indonesia) and mildly alcoholic palm wine, or toddy. But as much as he admired the great agricultural productivity these fields and paddies represented—and the hard work of the farmers with their buffalo-drawn wooden plows—he knew that they also meant poor collecting. Compounding this problem was his timing: he arrived in the middle of the dry season, a time when the otherwise verdant paddies resemble the brown fallow fields of an English winter. So he did what he always did in this circumstance: with permission from the local rajah and assistance from locals, in this case prosperous (and slave-owning) Dutch farmer and merchant Willem L. Mesman, he and his assistants headed inland—but not before the helpful Mr. Mesman arranged for Wallace and his assistants to use a small bamboo house as a base in Mamajang, then on the outskirts of town. In a letter dashed off to Samuel Stevens, he lamented the expansive cultivated ground ("absolutely barren for the Naturalist") and hoped to arrange to use a small house near forest with better collecting prospects. Still, he had not done badly thus far: in just three weeks he already had a host of insects and some forty bird species, several new.[31]

Over the next couple of months, he made two extended collecting trips farther into the interior. It didn't go too well initially—Ali came down with a fever, then another hired hand, who abandoned the trip as a result, and then Wallace himself. Liberal doses of quinine helped get them back on their feet, and he was able to hire new assistants: a young man named Baderoon, who could cook and shoot, and an "impudent little rascal" of twelve or fourteen named Baso, who would help carry his gun and net and, he hoped, "make himself generally useful." The collecting got better—fine pigeons, rollers, cuckoos, drongos, and more among his bird catches and loads of butterflies, including striking new species like *Pareronia tritaea* ("beautiful pale blue & black") and *Appias ithome* ("a rich orange band across the wings on a blackish ground"). But the real prize, on one of his first collecting forays no less, was a fabulous birdwing butterfly with a nearly eight-inch (twenty centimeters) wingspan. "Grand" and "superb," he cried, its coloration was stunning: the upper wings were a solid "rich shimmering bronzy black" and the lower were black flecked with white and large spots of "the most brilliant satiny yellow" along the margins. The undersides of the lower wings were satiny white, the marginal spots half black and half yellow. The butterfly's head and thorax were solid black, while the abdomen was a beacon of yellow and fiery orange. Declaring it "the largest, the most perfect and the most beautiful of butterflies," it is no wonder that the catch practically gave him heart palpitations.[32] He excitedly returned to his little house in Mamajang with this choicest of choice catches secure in his collecting box, which he hung by a cord from a bamboo rafter out of harm's way while he first worked on processing his bird specimens. Except that they were not out of harm's way . . . at one point he looked up to see a long column of voracious ants marching up and down the cord—the "little red rascals" had already invaded the butterfly box and had started to dismember his prized specimens! He stopped them just in time, cleaning the precious cargo one at a time, then safeguarded the box with a moat by setting it on a stand in the middle of a large basin of water, about the only barrier capable of stymying ants (for a while, anyway). At the time he was sure the prized birdwing was new, but alas! It turned out to be a known species, *Ornithoptera remus*, a name later synonymized with Rippon's birdwing, *Troides hypolitus*, described by seventeenth-century Dutch merchant and entomologist Pieter Cramer in 1775.[33] *Troides, Ornithoptera*, and *Trogonoptera* are the three bird-wing genera recognized today, a distinct group within the swallowtail family Papilionidae and, as we will later see, an important case study in evolutionary diversification for Wallace and subsequent investigators.

The collecting was good but had its challenges beyond pesky ants, including the perils of staying in a house atop spindly posts perilously leaning in the wind, recurring fever, and cultural miscues and faux pas. Where to begin? With his house. He asked the rajah for a bamboo house he could use far inland, closer to the forest, only to find that in granting his request the ruler had turned out a family from their home. They left grumblingly, eyes shooting darts at him, and he tried to make amends by asking them to a *bitchára*, or talk, where he apologized profusely and handed out tobacco and paid five silver rupees for rent. He promised to pay more, for provisions as well as specimens, giving the village kids a copper for each snail shell and insect they brought him. They seemed satisfied and all was well, except for a couple of issues. First, he was in constant fear that the house, perched high on posts leaning far out of perpendicular in the strong prevailing winds, was going to tumble over at any moment. He was a bit disparaging of the "mechanical geniuses" of the country who had not discovered the use of diagonal struts to stabilize raised structures but instead tethered the houses with a rattan cable on the windward side or built them on posts that were crooked to begin with, ostensibly lending stability in fierce winds. He was at first dismissive but had to concede that it worked—the sometime surveyor and builder sketched it out and checked the math: "A true square changes its figure readily into a rhomboid or oblique figure; but when one or two of the uprights are bent or sloping, and placed so as to oppose each other, the effect of a strut is produced, though in a rude and clumsy manner."[34] Maybe they *were* mechanical geniuses after all! But this house evidently had neither tether nor crooked posts, and the laws of physics likely weighed heavily on his mind every time he climbed up to the listing structure.

The other issue was location: good forest was farther away than he realized, requiring long collecting jaunts—good in its way, reducing time in that risky house, but increasingly difficult as his fever returned, draining his energy. He might have abandoned the spot, but after all the trouble procuring the house and with the rainy season right around the corner, he was determined to make the best of it—and hope that his temporary domicile did not blow over (at least not with him in it). It was on one of their long collecting trips that he experienced another mishap. Tired, hungry, and far afield, he, Ali, and their two younger assistants blundered into a house seeking provisions, only to beat a hasty retreat with apologies as they realized with horror their mistake: the occupant was a young mother in seclusion with her newborn baby, as per the custom of the country. Her own horror may have been compounded by one of the trespassers being a sweaty, bearded, bespectacled white man. His presence was a constant reminder that few of the locals had ever seen a European before, and the sight of him seemed to inspire terror everywhere: "Wherever I went, dogs barked, children screamed, women ran away, and men stared as though I were some strange and terrible cannibal monster. . . . If I came suddenly upon a well where women were drawing water or children bathing, a sudden flight was the certain result." It was all a bit tedious and depressing, "very unpleasant to a person who does not like to be disliked, and who had never been accustomed to be treated as an ogre," he lamented.[35] Could it have colored his perception of the women? While he admired the "quiet and dignified" manner of the rajah, he was lukewarm on the queen and princesses. He found them pretty good-looking and wished he could wax lyrical about their elegant costumes and gold and silver ornaments, he said, not to mention their "sparkling eyes," "jetty tresses," and "heaving bosom" beneath gauzy bodices, but on the whole he was underwhelmed.

————

With growing signs of the approaching rainy season, the collectors returned to Mamajang in mid-November, where Wallace packed up a consignment for London and prepared for his next move—a voyage a thousand miles east, to the very fringes of the archipelago, the storied Aru Islands at last! He hoped to depart in December, when traders took advantage of the monsoon winds. In the meantime, serenaded by farmers' rhythmic cries as they plowed by day and the bass tones of chorusing frogs by night, he enjoyed little luxuries after his weeks in the field—a proper table instead of a box to sit at, a glass of fresh milk every morning, sweet bread made with toddy, good coffee and tea, fresh Dutch butter, and cut flowers on his table were among the comforts of Mr. Mesman's estate, lands tended by servants from Makassar and enslaved Timorans. The latter were likely the victims of native slaver pirates or raiders feeding the enormous Dutch and Portuguese demand for forced labor; indeed, Timor was the second-largest slave market in Dutch possessions, after

Makassar itself.[36] If Wallace was troubled by the nature of the "employment" here, he did not mention it—we have seen that while he personally took a dim view of slavery, he was not very vocal in his criticism, especially when it came to friends.

During his Makassar interlude at the Mesman estate, Wallace also caught up on news from friends and family. Letters awaited him, including ones from his mother and sister that brought glad tidings: he was now an uncle! His brother John and wife, Mary, in California, were the proud parents of a son, John Herbert Wallace. He dashed off hearty congratulations, thanking his brother and sister-in-law for their kindness in giving him a new title, "that of Uncle, which I trust to repay in kind, some of these fair days."[37] He also wrote a long letter to accompany his next consignment, which Samuel Stevens duly communicated to the *Zoologist*.[38] The island was a bit odd: "Whole families and genera are altogether absent." There were none of the groups found to the west, like barbets, broadbills, trogons, bulbuls, or thrushes, except maybe one measly flycatcher, yet there was not much to take their place. His collections were disappointing as a result, he warned Stevens, but it was impressive enough given his time there, with a nice array of birds (267 specimens), snails (410), mammals (14), and insects (a whopping 2,774 specimens, of which 470 were marked for his private collection), plus a bundle of plants from Lombok he finally packed up. He was sure there were rare and new species among the lot (and there were), but for now he was off again ahead of the coming rains, he informed readers: to Aru, one of the great objects of his journey to the East. What did he hope to find there? The rarest of the rare, certainly—fabulous birds of paradise—but also perhaps some of the "strange and beautiful natural productions" of storied New Guinea, so close to Aru and yet so far, being dangerous country to visit owing to the deadly Papuan people.[39]

And speaking of which, what *of* the peoples across that vast archipelago— Malay, Timorese, Buginese, Aru Islander, Papuan, and countless others? What of their similarities, differences, and distribution? Here his investigations would continue, several lines interweaving like a rattan mat: natural history, geography, geology, ethnology. He was leaving behind one enigmatic island and heading to another, but he would be back. He could learn only so much from a limited stay in a limited district of Sulawesi, in an area heavily cultivated and with little geology in evidence, but already he recognized far more of an Australian than Asian affinity in the fauna, such as in Lombok, and the wheels were turning. He was beginning to connect some dots—to draw a line—in his mind. On his return he would travel to the north of that tortured island, and his insight would deepen. But first, a grand six-month expedition, one that the traders made just once each year, riding the western monsoon winds in December or January and returning on the eastern monsoon in July or August. Now was the time, as the long months of sunshine gave way to dark, leaden clouds and the winds picked up. He was introduced to the Javanese Dutch captain, Abraham van Waasbergen, who gave him a snug thatch cabin before the mainmast. The ship was a great prau with two triangle-shaped masts and seventy-foot lateen sails, with a crew of thirty plus twenty traders and passengers. After one false start when the fierce winds turned, they finally got underway in the wee hours of 18 December 1856 to prayers and wishes of *selamat jalan*, safe journey.

———

Wallace marked their progress by the islands great and small sighted along the way: Selayar, Kabaema, Buton, and WangiWangi; and then a long expanse of the Banda sea before Buru, Ambon, and great Seram appeared in the distance to port; and then the tiny Banda group to starboard, where he had his first view of an active, smoking volcano, rising up like a great Egyptian pyramid topped by a cloud of its own making. The voyage had its ups and downs, both literally—with stomach-churning storm-driven swells and squalls that snapped the boom at one point—and figuratively, as Wallace delighted in the knowledge that every one of those islands probably teemed with rarities: brilliant birds and insects, he imagined, and veritable *hosts* of unknown species, yet he also became impatient for the time when he might "explore these '*terrae incognitae*' of the Naturalist." In the coming years, he would indeed visit several of them. But now, Aru beckoned. As he cruised the border of sea and sky, the two at times seemed to merge into one: by day, elegant flying fish rose and fell over the sea swells, looking for all the world like gracefully flying swallows, and by night whirling eddies of "phosphoric light gemmed with whirling sparks of fire" streamed in the rudder's wake, reminding him of the beautiful star clusters he so admired with his telescope, with the added bonus of "dancing motion & ever changing form"—not unlike the stars' own eons-long dance in galactic currents and eddies.[40]

Contemplating the Bali-Lombok disconnect he had discovered and anticipating the character of the fauna awaiting him at Aru, he was taken by surprise by another "faunal boundary" he happened upon: that of people. Nearing Aru, the prau stopped at the long, narrow, and heavily forested island of Ké (Kai)-Besar for a week, where Wallace encountered native Papuan-descended peoples for the first time. Fifty or more men came alongside in canoes, and in microseconds he was bowled over by the contrast with the Malays in both physical features and personality: the animated, boisterous, in-your-face manner of the Papuans—singing, shouting over one another, darting about—was like night and day compared with the reserved and unfailingly polite Malays. "Had I been blind, I could have been certain that these islanders were not Malays." It struck him that here he had an opportunity to compare, side by side, "two of the most distinct and strongly marked races that the earth contains."[41]

As in Amazonia, Wallace was keenly interested in human diversity and its origins and was aware of the debates of the day back home between the polygenist "anthropologicals" (members of the Anthropological Society of London) and monogenist "ethnologicals" (members of the breakaway Ethnological Society). The former were convinced that human races were separate and distinct entities—different species, even, something often used at the time to justify the enslavement of certain races, in the manner that domesticated animals are subjugated. As such they did not put much stock in transitions between peoples or their culture but held to a more typological treatment of human variation with so-called races and ethnicities each in their own category. The ethnologicals, in contrast, held that all races were simply variants of a single human species and found the enslavement of fellow humans abhorrent.[42] For them, intermediates and transitions were the norm, reflected in their preferred research method of philology, which is essentially linguistics, the study of languages and their families, which show patterns of relationship that indicate pathways for deriving one from another. Although Wallace's drawing of boundary lines might seem to imply the delineation of permanent, well-marked racial differences aligned with the polygenist agenda of the anthropologicals, as opposed to a more monogenist-oriented sussing out of transitional forms, Wallace came at this from a fresh perspective.[43] Although a monogenist, certainly, he was also a surveyor and mapper who realized that present-day distributions give us insight into the past: insofar as species, varieties, and races have a discernable geographical distribution, this can teach us much about relationships and migrations of the past. He also emphasized "mental and moral" characteristics in identifying relationships.[44] The approach Wallace intuited, then, was nothing less than a natural history of humans, understanding peoples in all of their diversity in the same historical context he tried to understand the *other* animals that he so assiduously studied.

———

They departed beautiful Kai-Besar on 6 January 1857, arriving at the trading center of Dobbo in the Aru Islands late the next day. Known as Dobo today, the town is no longer the strictly seasonal trading center of Wallace's time. Then it was something of a pop-up town of three crowded streets of "rude thatched houses," dozens of tall A-frames of bamboo, rattan, and thatch amid which there was a confusion of little lean-tos, cooking sheds, shops, and makeshift pens for chickens and hogs, all opening up and shutting down with the coming and going of the western monsoon. By the end of the century, it was a perennially bustling town where steamships called regularly, and large shipping companies like the Dutch Koninklijke Paketvaart-Maatschappij had a permanent presence, steam power having ended the age-old dependence on the monsoon winds.[45] It was the mercantile impetus that brought traders from all points of the archipelago—Javanese and Dutch, Ceramese and Malay, Makassarians and Papuans from Timor, Aru, and Babar. While he admired the Chinese merchants neatly dressed in blue trousers and white jackets, with long red silk-plaited queues characteristic of the Qing dynasty reaching down their backs, and the dignified Bugis traders in flowing robes of green silk and brightly colored turbans, often attended by boys dutifully bearing sirih- and betel-filled boxes, he was also fascinated with the "half-wild Papuans," natives of Aru in waist-cloths, dark complected with great masses of Afro-textured hair. Wallace marveled that this "mixed lawless bloodthirsty thievish population" of Dobo managed to self-organize into a more or less self-governed community that conducted business peaceably, for the most part, without any real governmental apparatus such as police, courts, and lawyers. Chalk it up to the "genius of Commerce," he thought, where "trade is the magic that keeps all at peace." Compared with the layer upon layer of governmental administration back home, with its vast legal apparatus and lawyerly priesthood interpreting the hundreds of acts of Parliament, "one would be led to infer that if Dobbo has too little law England has too much."[46]

Commerce. They all came to buy and sell precious commodities: mother-of-pearl, turtle shell, and pearls; delicacies of birds' nests, dried shark fins, and trepang, or *bêche de mer*, those smoked sea cucumbers that repulsed him ("like sausages which have been rolled in mud and then thrown up the chimney"). But most prized of all were the birds of paradise—*burung cenderawasih*—spectacular creatures that came from *belakang tana*, "the land beyond." Wallace was tremendously excited: to him Aru was the back of beyond, an almost mythic place thanks to those almost mythic birds bedecked with plumage of such beauty and elegance that they could only be creatures of some heavenly

Dobo, Aru Islands, in the busy trading season (*note Wallace at center*).

realm. Or so went the legend: birds so rare that hardly any Europeans had ever seen one alive—indeed, with some irony they were known as *burung mati*, dead birds, by the Malay traders, who themselves only procured them lifeless and stuffed from the Aru Islanders in belakang tana. To Europeans they were the "footless" birds of heaven, *Paradisaea apoda*, so named by Linnaeus in a nod to the legend that these ethereal birds had no need for feet, perpetually sailing about the heavenly vault where they subsisted on rain and dew—a legend that arose from the practice of the native collectors who routinely lopped off the feet as they prepared the specimens, perhaps so as not to detract from the gorgeous and lucrative plumage.[47] For his part Wallace was interested in them alive *and* dead—burung mati for his collection, certainly, but he was also keen on *burung hidup*—live birds—for natural history and behavioral study.[48] If only he could find them.

The Aru archipelago consists of several great low-lying islands dissected by channels, almost river-like in character, but waterways of salt water. Dobo was situated on an islet Wallace knew as Wamma, now Warmar, off the west coast of much larger Wokam Island. The thick forests reminded him of Amazonia, with trees heavy with epiphytic orchids and ferns and stands of stately palms soaring to a hundred feet or more topped with great drooping leaves. But the "greatest novelty and most striking feature" of this forest was a species of supremely beautiful thirty-foot tree fern, topped by elegant fronds: "There is nothing in tropical vegetation so perfectly beautiful."[49] Collecting, too, was up to Amazonian standards, yielding the most species he had thus far captured in a single day since South America: lovely beetles; a "superb" bug; and some thirty butterflies, including glittering blues (Lycaenidae), elegant black-and-white *Idea durvillei*, the silky owl butterfly *Taenaris catops* sporting great eyespots on the hind wings, and an astonishing clearwing moth (*Cocytia durvillii*) with a metallic-blue body and sleek wings of rice paper, each vein neatly traced in India ink.[50] But the real prize was the stunning green birdwing butterfly, now dubbed *Ornithoptera priamus poseidon*. He could hardly believe his eyes as the giant butterfly languidly fluttered toward him; trembling with excitement, he swung his net. He was at first afraid to look, but there it was, caught, body golden with crimson thorax, radiant metallic emerald wings slowly opening and closing. . . . He was lost in admiration,

Sixteenth-century engraving of a "footless" paradise bird subsisting on rain.

"gazing upon its fresh and living beauty, a bright gem shining out amid the silent gloom of a dark and tangled forest." "The village of Dobbo," he declared, "held that evening at least one contented man."[51]

This *more* than whetted his appetite for the interior, that "promised land," as he put it, where he hoped to bag more entomological rarities as well as birds unknown in parts west: burung cenderawasih, of course, but also black cockatoos, great brush turkeys, and those large ratites, the cassowaries. Getting there was half the battle. Having written the governor at Ambon, then the Dutch administrative center for the region, to request permission and assistance for collecting in Aru, Wallace was initially delighted to receive a prompt and favorable response. Everything was falling into place! But then the pirates struck. The first indication that all was not well came on a February day when a small ransacked prau limped into the harbor at Dobo, raising the alarm of pirate attack. There were several groups of raiders in the region, but the most feared were from the modern-day Philippines. Variously called the Moro, Sulu, and Mindanao pirates, these Muslim outlaws were murderous in the extreme, plundering and burning villages, capturing hapless men, women, and children for the slave market or as galley slaves, and killing all who defied them. It was one such group that James Brooke subdued in a series of campaigns around Borneo in the 1840s, his title of white rajah of Sarawak later bestowed as a reward. The pirates typically swooped in on swift outrigger sailing galleys (*lanong*) supported by a fleet of smaller praus armed with guns, cannon, and the ubiquitous deadly kris. As Wallace's luck would have it, it had been eleven years since their last attack on the Aru islands—a long interval that was part of the pirates' strategy, allowing enough time to pass between raids for everyone to let their guard down. Now word arrived of a devastated settlement just east of Ceram and two praus plundered near Aru, the unfortunate crew of one of them murdered to a man. Everyone was on high alert, and for several weeks no price could entice the locals to take him to the mainland.

"March 13th. 1857. My boat being at last ready & having, with as much difficulty as usual obtained two men besides my own Malay & Macassar boys, we left Dobbo for the main land of Aru." So opens Wallace's second Malay journal— it took nearly a month, but he finally made the journey, assisted by no less a personage than the orang kaya (headman) of Warmar. They sailed and paddled up the channel that runs in a sinuous line between the Wokam and Kobroor Islands, deep into the interior, where they came upon a tiny outpost consisting of a small communal house of about a dozen people and some sheds or pens. For the price of a *parang*, or chopping knife, he was given working and living space and promptly set up camp and got to work—the first European to ever reside on a Papuan island, he reflected. Over the next two months, he pursued three main lines of activity: collecting and natural history research, Aru Islander ethnology, and making sense of the landscape.

The collecting was certainly a spectacular success. The Aru Islanders, expert archers, hunted the paradise birds from blinds constructed high in trees where the birds were known to gather.[52] They ascended before dawn with a bit of refreshment to last some hours, lying in wait and stunning the birds with blunt arrows when they assembled. But the first specimen Wallace received was downed not by them but by one of his assistants, Baderoon—and here it is worth noting that Wallace actually credits the young Malay boy for this in his journals and other writings, a reminder both of the fact that Wallace did not achieve all he did single-handedly and that he readily acknowledged this in a departure from the common practice of the day wherein the white explorer, as head of the expedition, typically took all the credit. Baderoon's specimen was the *burung rajah,* or *goby-goby* of the Aru islanders: the king bird of paradise, dubbed by Linnaeus *Paradisaea regia* but soon recognized as so special it was put in its own genus, *Cicinnurus regius.* Wallace could not believe his eyes: plumage of scarlet with a gloss like spun glass, a belt of metallic green across the belly, a yellow bill, and feet of cobalt blue—the Gaudi of birds. But there was more: springing from either side of the breast were tufts of erectile feathers tipped in emerald green, which the bird could raise into paired fans. And the tail feathers! The two middle feathers of the tail formed slender glossy wires about five inches long, diverging in a graceful double curve ending in a tight spiral disk of metallic green that looked like suspended glittering buttons. He was speechless: it was "one of the most perfectly lovely of the many lovely productions of nature," he rhapsodized. Only poetry could adequately celebrate a creature of such unsurpassed rarity and beauty, perhaps making him think of his long-lost brother Edward, the one member of the family gifted with poesy. But short of waxing poetic, he waxed philosophical. Beholding this gem of a bird, he pondered the countless generations of these exquisite birds over long ages, "year by year being born, and living and dying amid these dark and gloomy woods." The locals seemed to take them for granted, exciting no more wonder than a common house sparrow might back home. Were they incapable of appreciating these living gems? He suspected so, unfairly. But if he was dismissive of the aesthetics of the natives, he was also clear about the toll "civilization" would take on these birds. He put his finger on a tragic irony: while only "civilized" people can ostensibly appreciate the sublime beauty of such beings, the advent of civilization there would "so disturb the nicely-balanced relations of organic and inorganic nature" as to render these creatures extinct. What does it all mean? To Wallace it was clear: "This consideration must surely tell us that all living things were *not* made for man."[53]

He was soon treated to a view of great birds of paradise, too, young male *P. apoda* not yet in full plumage but practicing their courtship display hopping and flying about their lekking trees: "Wawk–wawk–wawk, wŏk–wŏk–wŏk," their calls resonated through the forest.[54] Wallace reveled in being surrounded by such beauty, the first Westerner to observe the feeding and courtship behavior of the almost mythic birds of paradise![55] He had to pinch himself, thinking how many before him had longed to reach "these almost fairy realms," to see with their own eyes the wondrous beauty he was so fortunate to behold. But he snapped out of his reverie as he smelled the coffee young Baso was preparing, heard Ali and Baderoon readying their guns for the hunt, and remembered with a start that he had a beautiful black cockatoo to skin.

Determined to further explore the interior and make observations of the channels dissecting the islands, he next moved the party to the interior village of Wanumbai—not without difficulty, as usual, this time owing to a false alarm about more pirates. Wallace haggled over renting a portion of a large house, shared with four or five families. Solidly built on seven-foot posts with a bamboo floor, it had a pitched thatch roof with a shutter that could be opened for light and air and a cooking area separated by a thatch partition. With mosquito netting hung and boxes arranged to serve as tables, he was all set—except that he was finding it increasingly difficult to walk. Ever since leaving Dobo, Wallace had been beset with bloodthirsty sand flies and mosquitoes, and now his feet and ankles were ulcerated so horribly that he could hardly get up. He may have joked that these biting insects "seemed here bent upon revenging my long-continued persecution of their race," but it was no laughing matter: he could not go out collecting and had to crawl down to the river to bathe where, adding insult to injury, beautiful swallowtail butterflies and other rarities tormented him by fluttering just out of reach.[56] But his faithful assistants kept up a steady stream of specimens—besides varied insects and snail shells there were lovely birds of paradise, brightly colored pigeons, radiant kingfishers, and beautiful little parakeets, and their mammalian catches included a cuscus, a curious Australasian possum. But he did, eventually, get to witness the raucous lekking of the great bird of paradise, described as *sácaleli*, or dancing parties, not altogether inaccurately.

He later reckoned that he was laid up about half the time he was in the interior; pretty much all he could do was help with the skinning and pinning and ensure that all specimens were secured from pests great (marauding dogs) and small (ravenous ants). But there was more: whenever Wallace had enforced downtime his mind turned to more philosophical questions. He recorded observations of the people and his interactions with them, noting in his journal that "the human inhabitants of these forests are not less interesting to me than the feathered tribes." People, then, were his second line of research. He was more attuned than ever to what he saw as striking racial differences between the Papuans and the Malay and made observations on their modes of dress, ornaments, implements, and tools. There were long group conversations, with the aid of Malay interpreters, in which the Western naturalist and the eastern denizens of this remote village of remote Aru struggled to better understand

Aru Islanders hunting king birds of paradise from their tree blinds.

one another and one another's beliefs. They begged him to name his country, as they could not accept that "Unglung" or "N-glung" was a proper name. Who ever heard of so unpronounceable a place name? "Do tell us the real name of your country," the *orang-wanumbai* said, "and then when you are gone we shall know how to talk about you." They also related a legend of "lost people," men, women, and children long ago taken over the sea, and implored him for information on their whereabouts, as he had come from over the sea. He thought this legend might have its origin in Portuguese raiders taking a group captive decades ago, and he tried to convey that they were likely taken to another island and were probably no longer even alive, it had been so long ago. But the Aru Islanders had a very different conception of life and death; they were sure the captives were still alive and would return, like the animals he collected—they, too, would come back to life. They thought him some kind of conjurer, and he imagined that if he dazzled them with some simple demonstration of Western science—magnetism, say—magical properties would be ascribed, and in a couple of generations he, too, would become mythologized, part of their lore.[57]

Such, he thought, is the "savage" mind; it was all part of his pigeon-holing at the time. Since first encountering the Papuan "race" at the Kai Islands, his observations were directed at further characterizing them and distinguishing them from the Malay "race"—convinced as he was of a sharp distinction of both "type" and geography despite ambiguities he observed in Aru, where there were signs of intermixture. His research plan was clear—although he was sure that the "Malay & the Papuan appear to be as widely separated as any two human races can be," the latter with clear affinities "physical and moral" to what he called the "true negro races," the ambiguities pointed to the need for careful fieldwork: "It is a most interesting question & one to which I shall direct my attention in all the islands of the Archipelago I may be enabled to visit."[58]

Wallace's account of his time living among the Aru Islanders at first seemed to lack the sense of wonder and admiration so evident in his descriptions of the Uaupés Indians of Amazonia in their great malocas or the Dyaks of Borneo in their longhouses. Recall there was a certain primitive romanticism in those earlier accounts, a picture painted of noble savages uncorrupted by "civilizing" forces. Harkening back to Amazonia, it was a place where he could live contented, he wrote, watching his children grow up like wild fawns. Here the Aru "savages" he had first encountered, in March 1857, seemed a miserable lot to him—idle, pock-marked from their poor diet, living in equally poor houses. He had seen (and taken part in) Indigenous communal living before and been admiring of the sense of community, but his choice of words to describe the first Aru houses he encountered seems to say something about his frame of mind at the time: describing the house's numerous partitions that formed separate sleeping places, he noted that each accommodated "the two or three separate families that usually herd together."[59] Herd? He seemed to view these people as barely human. The negative vibes may simply reflect the rough patch he was going through at the time, or more likely, he was reacting to the corrupting influence of the leading edge of "civilization"—notably, in the form of the rotgut spirit arrack, of which "the traders bring great quantities & sell very cheap." Wallace used it for pickling specimens and the occasional nightcap. For the Aru islanders, a day's fishing or rattan cutting fetched a half-gallon bottle, and for a season's worth of trepang or birds' nests, the traders paid in boxes that each contained fifteen such half-gallon bottles, all consumed in a matter of days of straight drinking. The islanders themselves told Wallace that in their drunken stupor they often tore their bamboo and rattan houses to pieces.[60] Just as he had deplored the corrupting influence of the Portuguese traders in Brazil while admiring the pure and unadulterated Indian peoples he had encountered, here it was the same. His attitude improved within a few weeks, at Wanumbai, "among the genuine natives of Arru tolerably free from foreign admixture," where the diet was very good, the people healthy, and the houses fine. He drew a direct comparison with the Dyaks and Indians of South America: as with them, here he was now delighted with the beauty and nobility of the human form, "a beauty of which stay at home civilised people can never have any conception."[61]

Wallace the zoologist and ethnologist was also joined by Wallace the geologist in the Aru interior. Following one of the curious saltwater channels to the eastern side of the islands, he began to connect some dots in a distinctly Wallacean manner. These low-lying islands sit in shallow water, on the edge of the continental shelf about a hundred or so miles from New Guinea. They consist of coralline limestone, as many islands do, but he could think of none in the world that were divided by saltwater channels with all the characteristics of rivers. Could they have *been* rivers? How else to explain their apparent meanders and uniform channel width? The Aru archipelago, he realized, offered an object lesson in Lyellian geology, with evidence of stages of slow uplift and subsidence. The limestone would have initially formed in a marine depositional environment, then uplifted to such a degree that the Arafura Sea receded, rendering what is now Aru a bit of New Guinean real estate, maybe a promontory. Rivers coursing west from the highlands of New Guinea—perhaps ancestors of today's Pulau, Lorentz, Agimuga, Muras Besar, or any of the dozens of other rivers along New Guinea's Aru-facing coast—could have meandered along and cut into what are now Aru lands, until subsequent subsidence flooded the intervening lower-lying lands and, filling the river channels with saltwater, rendered Aru an archipelago.

It was a breathtaking vision, one that at once helped explain the odd topography and river-like channels as well as the zoology of Aru, with its distinctly New Guinea-esque fauna, dovetailing with his growing vision of how geological changes of the deep past leave their stamp on the modern geographical distribution of species. He sketched the theory out in his journal soon after leaving Wanumbai, then fleshed it out in a paper for the Royal Geographical Society, where it was duly read at the 22 February 1858 meeting and published later that year.[62] Right about the time this paper was read, he had already made another discovery that was to eclipse all he had done to date. But first things first.

———

He had been six weeks at Wanumbai, and it was time to get going. The birds had become scarcer of late, his provisions were running low, and his feet were still swollen and ulcerated—on top of which he came down with a fever. He bid farewell to the "simple and good-natured" people of the village and distributed among them his remaining stock of salt and tobacco. With a parting gift of a flask of arrack to his host, shortly before dawn on 9 May 1857 Wallace and his three young assistants set sail for Dobo, arriving that evening.

The town was livelier than ever. Crews of the hundreds of trading praus busily loaded up newly acquired cargo as sailmakers and carpenters made repairs. The streets were animated with music, cockfights, rounds of a game resembling hacky-sack, and the cooing and squawking of exotic birds—parrots, lories, cockatoos, and pigeons, tethered to bamboo perches outside the houses. People came and went, young cassowaries gamboled in the streets, and the sweet scent of smoke from cooking fires and drying trepang wafted through the village. Yes, it was lively, yet there was something in the air, a growing sense that it was soon time to depart. As changes in the winds and rains signaled the coming eastern monsoon, things began to wind down in the trading town, bringing a hint of melancholy like the end of an endless summer or the murmur of emptying fairgrounds. Many had made their fortune this season, or soon would with their holds packed with precious commodities for sale in the West. Others had lost fortunes and even their lives. There were twenty or more dead this season, buried in a grove of casuarina trees behind the house where Wallace stayed.

His thoughts, too, were trending west, back to Makassar, where he would launch his next expedition. He appended a postscript to a short letter to Stevens that was supposed to have been sent back in March on a ship that got delayed: "Rejoice with me, for I have found what I sought; one grand hope in my visit to Arru is realized: I have got the birds of Paradise."[63] Two species, in fact! He would depart soon, he said, and was so pleased with his Aru successes that he now planned on returning to the East as soon as he could, specifically New Guinea. He learned which parts of the great island were safe and which were dangerous and was determined to go. To get there he would hopscotch from Makassar to the northern arm of Celebes (Sulawesi), then to the islands of Ternate, Gilolo (Halmahera), Ceram (Seram), and others in between, thence to the north coast of New Guinea. He hoped to procure a host of additional bird of paradise species along the way. Just think, he said, he was the only European to have shot, skinned, and even eaten paradise birds, and he was hungry for more (metaphorically speaking). Although he did not say so explicitly, he sensed, correctly, that many of the islands dotting the far eastern Malay Archipelago were probably home to these spectacular birds . . . perhaps even some hitherto undescribed![64] He would be departing with one fewer assistant, as Baderoon took his wages and left him in a huff after getting scolded for laziness. The luckless kid promptly lost all his money gambling, then became indebted by gambling away borrowed money after that. As was the custom, he was now effectively enslaved by his creditor and probably would remain so for life. Industrious and dependable Ali, in contrast, was sent back to Wanumbai to procure more burung mati, returning with sixteen "glorious" specimens despite developing a fever himself. Wallace was impressed with the honesty and hospitality of the people of the village, who assisted Ali in every way they could.

2 July 1857. Back on board Captain Waasbergen's vessel, they sailed for Makassar that morning in the company of a flotilla of fifteen praus, coursing a thousand miles of deep blue seas in an astonishing nine and a half days. As soon as Wallace made arrangements for the transport of his collections and equipment, he made his way to Mamajang in the company of his friend Mr. Mesman. He stayed up all night devouring the seven months' worth of letters and news that awaited him, then got to work ticketing and packing up his "treasures from the forests of Aru" for shipment to London. It was an unqualified success—he reckoned that the collection exceeded a whopping nine thousand specimens, sixteen hundred of which were distinct species—many new! Besides the fabulous new birds and spectacular new insects, he had been able to make detailed observations on the Papuan peoples and their geographical distribution and, as icing on the cake, elucidate a fascinating object lesson in reading geological history in landscape.

If these accomplishments were not illustration enough of the interconnected lines of research Wallace's fertile mind pursued, some of the letters in that pile of mail that awaited him underscored his overarching interest: the nature of species and their origin. In one chatty letter, his friend Rajah Brooke asked Wallace if perchance he had read the recent collection of essays by the clergyman and mathematician Baden Powell, of Oxford, commenting that the liberal theologian "adopts your view of the transmutation of species"—a telling comment.[65] A newsy letter from Bates, written the previous November in faraway Amazonia, was sent in response to Wallace's of April 1856. Bates had read Wallace's Sarawak Law paper with great interest: "I was startled at first to see you already ripe for the enunciation of the theory. . . . I must say that it is perfectly well done." Heaping on the kudos and accolades, Bates found the case admirably made, the idea like truth itself: closely reasoned, perfectly original, and quite complete, it "embraces the whole difficulty & anticipates & annihilates all objections." Bates pointed out that he, too, had a share in its formulation: "The theory I quite assent to & you know was conceived by me also, but I confess that I could not have propounded it with so much force & completeness." He waxed lyrical on the implications and in so doing revealed the trend of thought of the two philosophical naturalists as they had swung gently in their redés on the Upper Amazon what must have seemed a lifetime ago: "A great deal remains to be done to illustrate & confirm the theory—a new method of investigating & propounding Zoology & Botany inductively is necessitated, & new libraries will have to be written." Imagine, Bates said, how amazing it would be to write a monograph of flora and fauna of a region with differing but related species in each province, "tracing the laws which connect together the modifications of forms & colors with the local circumstances of a province or station—tracing as far as possible the actual affiliation of the species."[66] Together they could do this, combining their unparalleled collections once back in England.

A third letter worthy of our notice is from Darwin: "I can plainly see that we have thought much alike & to a certain extent have come to similar conclusions. In regard to the Paper in Annals [the Sarawak Law paper], I agree to the truth of almost every word." Perhaps in a bid to signal his priority, Darwin further informed Wallace that it had been some twenty years since he had opened his first notebook on the question of the nature of species and varieties and was making steady progress on a book on the subject. Wallace wrote Darwin back, delighted that they concurred in their views—he had been rather disappointed, he confided, that his paper had not gotten any response. Then

Wallace revealed his next steps: his paper was just an opening salvo "prelimi-nary to an attempt at a detailed proof of it, the plan of which I have arranged, & in part written."[67] When he got around to replying to Bates a few months later, he made the same statement: "To persons who have not thought much on the subject I fear my Paper 'On the Succession of Species' will not appear so clear as it does to you. That paper is of course merely the announcement of the theory, not its development. I have prepared the plan & written portions of an extensive work embracing the subject in all its bearings & endeavouring to prove what in the paper I have only indicated."[68] He told Bates about Darwin's letter and that the elder naturalist agreed too and was then working on a great work on "species and varieties." Darwin might save him the trouble of writing the second part of his hypothesis, he thought. How? "By proving that there is no difference in nature between the origin of species & varieties." This, then, is a key element of Wallace's project: to show that there is no real difference between species and varieties or, put another way, that varieties are but incipient species, so whatever the process is that yields new varieties is also responsible for yielding new species. This is why the Sarawak Law holds: new species *must* originate in close association with a previously existing, closely related species insofar as they are *derived from* those preexisting species—through the inter-mediate step of varieties. Wallace's "law" is also why *affinities*—relationships—often map onto geography, which is what made invisible boundaries like that between Bali and Lombok so utterly intriguing to him. In a way such patterns seem a departure from the Sarawak Law. Or does the discontinuity reveal something about how the chains of affinities of different groups play out in time and space?

Wallace did not know *how* new varieties and species arise, but that was part of his research agenda. The easy part, for him, was to make a solid case for the reality of species change to begin with. *That* is the plan he referred to in his letters to Darwin and Bates—the plan he said he had prepared and partially written. Wallace was alluding to a book he had in the works: a blockbuster of a book, no less than an explicit rebuttal of the great Charles Lyell's anti-transmutationism. He had already written out a series of arguments both *against* Lyell and *for* transmutation in his Species Notebook, systematically copying out Lyell's arguments from the *Principles* and rebutting them in turn under the heading "Note for Organic Law of Change."[69] In one revealing "note to self" squeezed in on one page of this most important of notebooks, Wallace flags an argument of Lyell's as a useful starting point for the conclu-sion to his own book: "Introduce this and disprove all Lyells arguments first

Note for Organic law of change.

We must at the outset endeavour to ascertain if the present condition of the organic world, is now undergoing any changes. of what nature & to what amount, & we must in the first place assume that the regular course of nature from early Geological Epochs to the present time has produced the present state of things & still continues to act in still further changing it — While the inorganic world has been strictly shown to be the result of a series of changes from the earliest periods produced by causes still acting, it would be most unphilosophical to conclude without the strongest evidence that the organic world so intimately connected with it, had been subject to other laws which have now ceased to act, & that the extinction & production of species and genera had at some late period suddenly ceased. The change is so perfectly gradual from the latest Geological to the modern epoch, that we cannot help believing the present condition of the Earth & its inhabitants

Page from Wallace's Species Notebook. His planned transmutation book might have been titled *On the Organic Law of Change*.

at the commencement of my last chapter."[70] The reference to a "last chapter" implies a *series* of chapters: Wallace was writing an *Origin of Species* before *Origin of Species*![71]

————

You can imagine Wallace working away on his project—contemplating pattern and process over those long hours at sea, writing in his bamboo huts or around cooking fires, and composing his thoughts in between suppressed urges to scratch his red and oozing ankles and chasing off scavenging dogs looking to snatch his prized specimens. Darwin was encouraging without revealing much of his own thinking. To Wallace's expressed disappointment that his 1855 Sarawak Law paper seemed to be all but ignored, Darwin assured him that "two very good men," none other than Charles Lyell and zoologist Edward Blyth, in India, had specially called his attention to it. He didn't mention the urgency with which Lyell had called his attention to the paper. The astute Blyth, himself something of a closet transmutationist and a longtime correspondent of Darwin's, wrote an enthusiastic letter to Darwin soon after reading it: "What think you of Wallace's paper. . . . ? Good! Upon the whole! . . . Wallace has, I think, put the matter well."[72] Darwin replied that he supposed he went much further than Wallace and was about halfway done with his species book—another FYI.[73] But despite Lyell's warning, Darwin could not believe that Wallace was anywhere close to his own seminal insight: natural selection. The clever specimen collector may dabble in philosophical questions and may even trend toward the heterodox on the matter of species change. But natural selection? The idea is deceptively simple, its majesty and explanatory power counterintuitive. No, Wallace had nothing on him. Or did he?

CHAPTER 9

EUREKA

Wallace Triumphant

IF DARWIN had a plan, it backfired. His casual remarks to Wallace about how long he had been working on the species question, and how far along he was with his book, were likely intended to warn the junior naturalist off and mark his territory. But the enthusiastic Wallace seemed only to come away with encouragement: the self-taught field collector had an irrepressible philosophical bent, and to hear that his ideas in the Sarawak Law paper had been so well received by the likes of Darwin—well, that was a real shot in the arm. He expected his old buddy Bates to be supportive, but even *his* emphatic praise went above and beyond: Wallace was tickled and later confessed to Bates that he had read and reread his letter more than twenty times.[1] But on top of this, the esteemed Darwin "agreed to the truth of almost every word" of the paper and was just then studying "in what way species and varieties differ from each other."[2] Exciting!

Back at Makassar, working on packing up his extensive Aru collections for shipment, the wheels were turning . . . species, varieties, degrees of affinity, permanence, transitions, distribution in time, space. . . . Thinking about Aru, his collections, the letters, and the patterns, he had further bursts of insight. In the space of just a few months at Makassar, before sailing off on 19 November 1857 to Amboyna (Ambon), he dispatched no fewer than six papers to London. Three were essentially zoological, rewrites of his extensive field notes. They included a detailed account of the behavior and appearance of the great bird of paradise, an overview of his insect collecting successes at Aru, and observations on the caterpillar and pupa of that "extraordinary and unique" green birdwing butterfly of Aru, *Ornithoptera priamus poseidon*—"absolutely luminous with a brilliancy which nothing in animated nature can surpass."[3] The other papers were more astonishing. If the first three papers reflect Wallace the diligent field naturalist, the next reflect Wallace the astute philosophical naturalist.

They were all of a piece, these three—a short (two-page) note on the nature of species and two much longer papers, one on the physical geography and the other on the natural history of the Aru Islands.[4] His hypothesis of a former connection of Aru with New Guinea is central to both of these papers, but the natural history paper stands out: in a stunning application of his Sarawak Law, Wallace makes the clearest case yet for his evolutionary vision. His approach mirrors that in the Species Notebook: his foil was Charles Lyell. Look, says Wallace, we know that it was not that long ago, geologically speaking, that currently living species did not exist. "How do we account for the places where they came into existence? Why are not the same species found in the same climates all over the world?" The widely accepted explanation is that over time the slow revolutions of the earth—the uplift of mountains, sinking of continents, shifts in climatic conditions—render species existing under the original conditions extinct, while rounds of special creation populate the landscape under the new conditions with new species. "Sir C. Lyell, who has written more fully, and with more ability, on this subject than most naturalists, adopts this view," says Wallace. It is telling that the example from the *Principles* that Wallace gives next in his paper was also written out verbatim in the Species Notebook. Lyell imagined the effects that uplift of a hypothetical mountain chain in North Africa might have: "Then," [Lyell] says, "the animals and plants of Northern Africa would disappear, and the region would gradually become fitted for the reception of a population of species *perfectly dissimilar in their forms, habits, and organization*."[5] Wallace's emphasis here takes aim at the heart of the matter. In his paper he constructs a detailed step-by-step refutation, but we need only consider his concise follow-up in the Species Notebook:

> But have we not reason to believe they would be modified forms of the previously existing Northern African species. The climate might then more resemble that of the W. Indies, but we know the productions would not resemble them. It would be an extraordinary thing if while the modification of the surface took by natural causes now in operation & the extinction of species was the natural result of the same causes, yet the reproduction & introduction of new species required special acts of creation, or some process which does not present itself in the ordinary course of nature.[6]

This insightful passage is a dig at Lyell's lack of consistency—note the passage recalling the subtitle of Lyell's *Principles*: "Being an Attempt to Explain the

Former Changes of the Earth's Surface, by Reference to Causes Now in Operation." Ouch! Wallace then pointed out that according to his "law," *contra* Lyell, each new species arising would be "closely allied" to its preexisting counterpart (Wallace awkwardly called this an "anti-type"), implying it was a modification of that counterpart—he just didn't know *how*. If true, each successive round of faunal change as geological time unfolds will be more or less *incrementally* different from what came before. This is consistent with the seemingly progressive or directional change we see in the fossil record and consistent with transmutation.

Applied to the Aru Islands, Wallace did something of a thought experiment—an explicit application of his Sarawak Law. Given that most of the species of Aru are found in New Guinea, he surmised that the separation of the two islands was pretty recent. Looking to the future, with continued separation of the islands and slow geological change in each, he imagined a progressive divergence in their fauna: some groups that might go extinct or flourish on one island will likely not be the same that go extinct or flourish on the other. Successive new species will be similar but not identical to those that came before, and this occurs independently on each island. In time the two islands might first present groups of different species of the same genera, perhaps like we see with North Australia and New Guinea today. Allowing for even more time and change, "the faunas will come to differ not in species only, but in generic groups"—rather like the Caribbean Islands in comparison with Mexico, he suggests. As yet more time passes, even greater divergence occurs: "Then we should have an exact counterpart of what we see now in Madagascar, where the families and some of the genera are African, but where there are many extensive groups of species forming peculiar genera, or even families, but still with a general resemblance to African forms." Wallace's evolutionary vision was clear. But note the coup de grâce for the Lyellian model inherent in his thought experiment, the simple observation that completely undermined Lyell's view of environment as the main determinant of species relationships over the globe. Compare Borneo and New Guinea: two great islands virtually sitting on the equator in the same region, comparable in size, geology, topography, and environmental conditions. Yet they lie on either side of that invisible dividing line he had stumbled upon at Bali and Lombok; the two were dramatically different in their fauna, Borneo decidedly Asian in its birds and mammals, New Guinea decidedly Australian. But then run that transect a different way, comparing New Guinea with the island-continent of Australia. The two differ in every way that New Guinea was similar to Borneo. They differ dramatically in size, geology, topography, and environmental conditions: lush rainforest-to-alpine New Guinea, mostly flat and bone-dry desert Australia. But their bird and mammal groups are virtually identical! There was a certain elegance in the simplicity of this argument: in one fell swoop, Wallace swept away a central tenet of Lyell's whole vision for changes in the earth and the life upon it. Wallace never shied away from taking on any scientific authority if he thought he was in the right; how would this argument, this scholarly potshot at Lyell, be received back in the scientific salons of London? And transmutation, the idea his Sarawak Law compels?

––––––

Papers written, collections sorted and packed—over nine thousand specimens consisting of some sixteen hundred species, including a set of more than eight hundred species for his private study—he shipped the prodigious consignment to London via Singapore, from where the schooner *Maori*, under the command of Capt. Charles Petherbridge, departed on 4 September 1857 on the first leg of that multistage journey. Almost exactly one month later, the ever-efficient Stevens read Wallace's letter from Aru at the Entomological Society of London, whetting the appetite of the London collectors and museum men for the rarities to come.[7] Wallace later noted that sales through June produced a whopping £1,000, not even counting the separate set of insects he had collected on commission for British entomologist William Wilson Saunders, just then serving his second stint as president of the Entomological Society.[8] Wallace's letters and papers went by mail rather than cargo, arriving well ahead of his specimens. The papers were all read and some even published by December—you can well imagine the high hopes he had for his more philosophical papers after Darwin's encouraging words.

It being mid-September, with no ships heading east soon, Wallace figured he could get in a bit more collecting while the dry season lasted. Ali was laid up with a fever, so he hired two new assistants for a short trip just north of Makassar to the small kampong of Maros where his friend Mr. Mesman's brother ran a prosperous little farm nestled against limestone mountains— paddies and fields worked with water buffalo, surrounded by a forest of breadfruit trees and the stately *Arenga* feather palms so prized for making sugar and palm wine. With the rajah's permission for collecting, Wallace had a small bamboo bungalow built in the forest at the foot of a mountain, his base for the next month or so. The geology of that part of Sulawesi is an interesting mix of limestone intruded by volcanics, eroded into bizarre and beautiful formations and precipitous mountains—highly scenic and one of the world's largest karst regions. Wallace followed the Sungai (river) Pute inland near modern-day Rammang-Rammang, ascending the river's twists and turns to the falls, impressed with the increasingly spectacular landscape: "Such gorges chasms and precipices as here abound I have nowhere seen—a slope is scarcely anywhere to be found, huge walls & rugged masses of rock bounding all the mountains & enclosing the vallies."[9] Sheer cliffs and overhanging precipices five hundred feet high were draped with "a tapestry of vegetation"—*Pandanus* trees, ferns, creepers, and trees interwoven into an "evergreen network," here and there revealing white limestone, sinkholes, and openings to caverns. Indeed, the Maros-Pangkep karst is famous today, renowned for its complex of caverns full of prehistoric cave art. Wallace would surely have been impressed to learn that the world's oldest known cave art is found in this very area, dated by an Australian and Indonesian team to nearly forty-four thousand years old.[10] But he had entomology on his mind, and as he admired the geology with one eye, the other was on the insects. Reaching the falls, he was met with a breathtaking sight: the moist sand beach of the pool below the waterfall was a dazzling canvas dotted with butterflies of vivid oranges, yellows, whites, blues, and greens in a Jackson Pollock-meets-Georges Seurat kind of way—a canvas that would suddenly erupt into a literal action painting with hundreds of airborne flecks of color, a painting come alive like a spectacular rainbow storm. That would have been reward enough, but he was also treated to not one but *three* birdwing butterfly species, awed as they "wheel through the thickets with a strong sailing flight" on seven-to-eight-inch wings gorgeously colored with spots or masses of deep satiny yellow on a jet-black canvas.

The rains came a bit early, in mid-October, bringing with them fever, dysentery, mysteriously swollen feet, snakes, and a plague of thick millipedes eight or ten inches long, crawling absolutely everywhere ("I even found one in my bed!"). It was time to go, so he packed and shipped off yet another prodigious consignment—this time just over 8,000 insects all told, plus 166 bird species, 140 shells, 3 bats, 2 wild pig skulls, and 1 squirrel. He later got word that all arrived safely in London the following July.[11] While his collections made their way west, he headed east for Ambon, departing aboard the mail steamer *Padang* on 19 November. Or rather, he headed southeast and then north, calling at the towns of Kupang and Dili on Timor and then north to Banda Neira island and on to Ambon, the usual zigging and zagging of a mail route. The steamer was slow but more than made up for that with roominess and

Dutch ships in De Reede van Banda (the Road of Banda).

comfort—not to mention good food and libations: the shipboard routine was pretty nice, with tea or coffee and a light breakfast served each morning at eight o'clock followed by a glass of Madeira or gin and bitters at ten, *déjeuner à la fourchette* at eleven, tea or coffee brought round again at three, at five another aperitif followed by dinner at six, then tea or coffee again at eight to cap the evening: "There is no lack of gastronomical excitement to while away the tedium of the days on board."[12]

The brief stops at Timor and Banda allowed for observing but not collecting. He noted that the Timorese were of Papuan descent and made geological as well as what we could today call ecological observations. He could not know that mountainous Timor is a geological borderland—a complex product of the subduction of the Australian plate inexorably moving northeast—but his Lyellian eye noted the tale of uplift and subsidence past told by the raised and broken coral reefs on which the town of Kupang was built, as well as the walls of an old house now sunken into the tidal zone. Timor was remarkably dry, he noted, dominated by scrubby vegetation, and neighboring Wetar Island even more so. The contrast with Banda Neira and adjacent islands in the Banda Arc just a couple hundred miles north, all "clothed with a bright robe of ever verdant forests," was striking, and he put his finger on why: the dry monsoon winds blowing from the southeast over the parched Australian continent dried out Timor, rendering it more tropical savannah than tropical rainforest in today's terms. He would return to Timor a couple of years later, staying a few months to collect, but now Ternate was his destination by way of Ambon, where he planned to stay a month or so. On 30 November, twenty hours after departing Banda Neira, they landed at Ambon, where the indispensable letter of introduction got Wallace the immediate assistance of a pair of kind expatriate doctors in the service of the Dutch East India Company who were also amateur entomologists: German Otto Gottlieb Mohnike and Hungarian Carl Ludwig Doleschall. He was also taken to see the Batavian-born Dutch governor, Carel Frederik Goldman, who proved to be equally helpful.

Ambon is today the capital of the Indonesian province of Maluku. Known in Wallace's time as Amboyna or Amboina, the name was applied to both the town and the entire small, rugged island on which it sits, in the chain known as the Banda Arc. It is almost two islands, in fact, with larger and smaller oblong lobes connected by a narrow isthmus—if islands could reproduce by amoeboid budding, Ambon might be the type specimen. The great shallow bay between the two dazzled Wallace with the "magnificent dimensions varied forms & brilliant colours" of the teeming coral and other marine invertebrates covering the bottom of the crystal-clear bay. "It was a sight to gaze at for hours, & the beauty & interest of which no description can do justice to."[13] Nestled at one end of far larger Seram, Ambon's small size belied its historical importance as the capital of the Moluccas and the Dutch spice trade. Since their arrival around 1512, the Portuguese had a tenuous hold on the island, between conflicts with the Muslim Malay Melanesian Ambonese and growing tensions with the English and Dutch. They never managed to fully control the spice trade and a century later gave up the island without a fight and turned over the keys of their stronghold, Forte de Nossa Senhora da Anunciada, to Dutch VOC admiral Steven van der Hagen, who renamed it Fort Victoria (parts of which can be visited today). Except for brief periods, the island remained largely under Dutch control over the next two hundred-plus years, the center of not only the spice trade generally but of clove production in particular. At one time the Dutch decreed that cloves, the wonderfully aromatic flower buds of the tree *Syzygium aromaticum* (family Myrtaceae), prized for its culinary and medicinal uses, could be grown only on Ambon.

Wallace was given the use of a small cottage situated in the northern part of the island in a recently cleared section of plantation planted with cacao—just the kind of site he loved, like the Simunjan coal works in Borneo, where the combination of downed trees, brush piles, and forest-edge habitat attracted wood-boring beetles of all kinds. He was not disappointed, with loads

"Ejecting an Intruder"—easier said than done.

of fine weevils (Curculionidae), long-horn beetles (Cerambycidae), metallic
wood-boring beetles (Buprestidae), and more—insects "remarkable for their
elegant forms or brilliant colours & almost the whole entirely new to me."[14]
The shady paths into the forest yielded lovely butterflies, too, including the
luminous metallic-blue Ulysses or blue mountain swallowtail, *Papilio ulysses*,
"the prince of lepidopterous insects." This wandering sapphire of a butterfly
is widely distributed throughout the region, but Ambon is the type locality,
the place from where *the* representative specimen made its way to Uppsala,
Sweden, where Carl Linnaeus bestowed its Latin name in 1758. But the site was
perhaps a bit too productive: the brush-filled clearings and forest edge that
attract a diversity of insects also attract a diversity of insect predators—and
predators of those predators, and predators of *those* predators. . . . Hence one
evening Wallace discovered that a twelve-foot python, thick as a thigh, had
taken up residence in the roof of his little bamboo and thatch cottage, and he
had been sleeping directly beneath it for some time! His panicked assistants
ran for the hills, and nothing could induce them to return, let alone help. A
local snake wrangler eventually came to the rescue, subduing the great snake
with a rattan noose. But extricating it from the cottage proved to be an ordeal
as the infuriated serpent thrashed about, coiling around chairs and posts and
(nearly) people.[15]

Wallace stayed from Christmas through the new year in the company of Dr. Mohnike, suffering again from a touch of fever but well enough to help his friend arrange his collection and attend the annual New Year's soiree at the governor's residence. Not one for such social occasions, Wallace was mainly impressed with the rare black, or rajah, lory of New Guinea, *Chalcopsitta atra*, kept by the family as a pet—a regal-looking bird indeed, covered in feathers of black just tinged with gray at the tips, with a bright-red eye and spectacular yellow and crimson undertail feathers. He may have longed to add it to his collection, but even the socially awkward Wallace would not be so gauche as to ask his hosts if he could purchase their pet lory . . . at least not at a New Year's party. Instead, it may have reminded him that he longed to visit New Guinea—that time was wasting. A few days later, he boarded the mail steamer *Ambon* for the volcanic island of Ternate, which was to be his base for expeditions to storied regions east. He arrived on his thirty-fifth birthday.

———

The small island of Ternate is a textbook circular island dominated by a single conical 5,600-foot (1,700 m) volcano, *Gunung api* Gamalama, in a textbook string of such volcanic islands—Hiri, Tidore, Mare, Moti, Makian, Kajoa, and more—a geometry that reveals geological tumult below. In fact, that string continues to the northeast along North Maluku, part of the neighboring and much larger island of Halmahera, or Gilolo as it was known in Wallace's day, where some of the region's most active volcanoes lie (some still in a state of continuous eruption). Here Wallace entered a singular borderland in a region rife with fascinating borderlands—the Moluccan Sea Collision Zone. The small Molucca Sea Plate, a microtectonic plate between Halmahera to the east and Sulawesi to the west, is the only known example of *divergent double subduction*, with seafloor spreading east and west from a midplate ridge and new seafloor simultaneously butting into and subducting under neighboring plates (in this case the great Eurasian plate to the west and the Philippine Sea plate to the east).[16] It is no coincidence that Sulawesi and Halmahera bear strikingly similar tortured morphology—both are born of a complex geological mashup, creating their long volcanic, earthquake-wracked arms that seem to be in a state of slow-motion flail. If the northern peninsula of Halmahera seems very different from the rest of the island, that is because it is: they are separate geological terrains altogether, sutured together at the narrow isthmus opposite Ternate.

Ternate was at the center of the Moluccan spice trade and therefore was the center of a centuries-long tug-of-war of kings, sultans, and European powers vying to control it—a history reflected in the assortment of *benteng*, or forts, dotting the island: Ternatese Malay, Portuguese, Spanish, and Dutch, some still standing and others marked only by the odd wall, gate, or rampart. The largest of the forts was Oranje, the thick-walled and moated bastion of the Dutch East India Company. It is largely intact, if dilapidated in places today, the rusted and graffiti-covered cannon atop its old battlements belying the immense wealth, power, and strategic importance the site once commanded. Wallace presented letters of introduction to the "king of Ternate," Maarten Dirk van Duivenbode, a wealthy merchant who owned, Wallace said, half the town, a fleet of ships, and "above a hundred slaves," mainly Papuans. Slavery was rife in the region, the trade plied with large traditional outrigger canoes, *kora-kora*. Raiding and trading went hand in hand. Large kora-kora, heavily armed and with rowers on the outriggers as well as the central boat, often featured a tall pole at the stern: not a mast, it was used to display the heads of resisters.

Through Mr. Duivenbode's help, Wallace soon found a low ramshackle house to rent—the location only fairly recently identified, in fact, thanks to historical sleuthing by talented and persistent Wallaceophiles.[17] Abutting the southwest bastion of Fort Oranje, on the corner of what is now Jalal Pipit and Jl. Merdeka, the stone, wood, and palm-thatch cottage was more than spacious enough for Wallace and his assistants, with both front and back verandas and several rooms, conveniently located just minutes from the market and the waterfront, and surrounded by a "wilderness" of fruit trees. The icing on the cake for Wallace was the well—deep enough to provide plenty of clear, cold water, a precious commodity on any tropical island. Structures are perishable in that torrid (and earthquake-prone) region, and so, unsurprisingly, the original house no longer stands, but the City of Ternate and the Department of Culture plan to acquire the site to rebuild the house as a museum.[18]

But he did not stick around to enjoy the place for long—collecting beckoned, and Ternate itself was an unpromising hunting ground. The town and surrounding agricultural fields and orchards of the compact island crowded the limited level land between the shore and the steeply rising Gunung Gamalama, and he knew pickings would be slim. No, just twenty or so miles across the narrow strait lay great Halmahera, barely explored. Wallace, Ali, and a few companions along for the ride headed across in a borrowed boat—rowed by enslaved Papuans—landing at Sidangoli, a small village on a headland just

north of the isthmus and still the site of a ferry port today. The group was probably greeted by locals in rickshaws for transport, much like today's fleet of
waiting motorbike *tuk-tuks* and Bajaj autorickshaws. But Wallace was not
going anywhere: while the distant hills looked promising, the immediate prospect was an expanse of tall, coarse reed-like grass—tough walking, and he
knew it would be a bird and butterfly desert. He stuck around not two days
before heading out, relocating about ten miles southeast to the island's isthmus
where the tiny village of Dodinga lay in a bend of a sluggish river. The village
was overlooked by an old Portuguese fort on a low hill defended with four
cannons, two of which can be seen today ornamentally "defending" the local
town hall. Its turrets toppled and walls rent by earthquakes by the time of
Wallace's visit, it seemed appropriate that the ruinous fort was garrisoned by
a ragtag band of soldiers.

Wallace soon found a promising hut to rent, in disrepair but serviceable,
and comfortable enough once his landlord was cajoled into fixing the leaking
roof. He planned on staying about a month and had high hopes for his new
collecting locality, situated in the midst of a rugged and picturesque landscape
of abrupt limestone hills and valleys with great outcrops, and a lofty and luxuriant green forest enlivened by swathes of crimson *Ixora coccinea* shrubs just
then everywhere blooming. The flat valleys, however, were very grassy, not
forested. He got to thinking about that and those expansive grasslands in the
vicinity of Sidangoli, just ten miles away—why so much grass in a region
where trees grow perfectly well and indeed cover most of the landscape
all around? He had seen such puzzling grasslands in the tropics before, in
Sulawesi, and knew of the great llano grasslands of the Orinoco River in South
America. The tropical grasslands cannot be explained by climate or soil, he
thought, and hit upon an intriguing explanation: the vast grasslands result
from a competition dynamic, set up by geological uplift. As rapid elevation
creates a muddy plain where there was formerly a shallow sea, what plants will
most likely colonize the newly available land first? Most likely grasses, he
thought, given the sheer number and ease of dispersal of their seeds compared
to those of forest trees. The grasses outcompete the trees in this scenario: they
are faster to arrive and thrive, and once they get established, trees and other
plants have a tough time getting so much as a toe- (rootlet?) hold. "Ground
once taken possession of by grasses cannot be reconquered by forest even if
surrounded by it," he declared. "A clearing for a few years only, will if left become forest, from roots & seeds left in the earth, but if once covered with grass
all woody growth is kept down."[19]

These ruminations are noteworthy for their timing: such a competition dynamic lies at the heart of selection, and Wallace's notebook entry on grass-versus-tree competition was made tantalizingly close to his greatest of insights: *natural selection*. It came to him here, in a malarial fit, he later related. Indeed, he commented that his collections from the Dodinga area were pretty limited owing to his being ill most of the time, and his recurring fever and chills certainly fit malarial symptoms. Here is Wallace's recollection of his epiphany, written years later:

> After writing [the Sarawak Law paper] the question of how changes of species could have been brought about was rarely out of my mind, but no satisfactory conclusion was reached till February 1858. At that time I was suffering from a rather severe attack of intermittent fever . . . and one day while lying on my bed during the cold fit, wrapped in blankets, though the thermometer was at 88° F., the problem again presented itself to me, and something led me to think of the "positive checks" described by Malthus in his "Essay on Population," a work I had read several years before, and which had made a deep and permanent impression on my mind. These checks—war, disease, famine and the like—must, it occurred to me, act on animals as well as on man. Then I thought of the enormously rapid multiplication of animals, causing these checks to be much more effective in them than in the case of man; and while pondering vaguely on this fact there suddenly flashed upon me the idea of the survival of the fittest—that the individuals removed by these checks must be on the whole inferior to those that survived. In the two hours that elapsed before my ague fit was over I had thought out almost the whole of the theory, and in the same evening I sketched the draft of my paper, and in the two succeeding evenings wrote it out in full, and sent it by the next post to Mr. Darwin.[20]

It was a stupendous insight! Wallace's immediately prior reflections on "rapid multiplication" in another context, that of grasses in competition with trees, may well have sparked a conscious or unconscious transfer of the concept of struggle and competition to animals. Indeed, this is a more likely catalyst than Malthus—in fact, it is not clear to what extent Malthus really played a key role in Wallace's evolutionary insights despite this and other later recollections to the contrary. Wallace is, of course, likely to have read Malthus's important *Essay on the Principle of Population* years earlier and maybe even the 1820 *Principles of Political Economy*. After all, these works were found in just about any well-stocked library, including those of the mechanics' institutes, and were very much part of the early nineteenth-century social and political discourse over issues he had been awakened to—the Poor Law reforms of 1834 and the misery they caused, for example, and the vigorous disagreements between Malthus and the Owenites over the relevance of Malthusian diminishing returns and unsustainable population growth to the cooperative communities envisioned by Owen.[21] He would have come across Humboldt's brief mentions of Malthus in the *Personal Narrative* too. In these instances the context was human societies and the causes of their population increases or declines, certainly also something Wallace thought about in connection with the peoples he encountered in the Malay Archipelago.[22] But when might Wallace have connected Malthus with *natural* populations and struggle?

The "struggle for existence" itself was a concept he was well familiar with—Lyell used the phrase in *Principles* (but did not mention Malthus), and Wallace himself used it in his 1853 Amazon book (and the phrase appears twice in the Dodinga paper). But Malthus per se in this context? Not so much. It has sometimes been suggested that Wallace later falsely claimed an inspiration from Malthus to more closely align his process of discovery with that of Darwin, who clearly had been inspired by Malthus on population. I disagree—it is more likely that Wallace was somehow led to connect Malthus with his "struggle for existence" thinking and thereafter merged them completely in his mind. It's a natural linkage, after all, and how many of us have not had the experience of clearly and vividly remembering some long-past event or detail, only to learn that it was not as we remembered at all? Malthusian struggle could indeed have made a "deep and permanent" early impression on Wallace in connection with the social issues he was passionate about, then easily linked with the struggle discussed by Lyell and others in a biological context. How might that link have been catalyzed in Wallace's mind? Well, it could have just occurred to him, as he suggests in the passage quoted. Or perhaps it was through Darwin—consider that Malthus is mentioned but once in Wallace's Species Notebook, and as we shall see in chapter 10, that mention is connected with Darwin.[23] What's more, Malthus is not mentioned in any of Wallace's known correspondence until 1859, in a letter from Darwin.[24] However his insight was catalyzed, whether through subliminal fever-wracked images of competing grasses or the struggles of individuals and the ebb and flow of populations, the idea was a seed fallen on the fertile soil of his mind. The words of the great American philosopher-naturalist Henry David Thoreau resonate here: "I have great faith in a seed. Convince me that you have a seed there, and I am prepared to expect wonders."[25]

Wonders indeed: Wallace's long pondering over the nature of species and varieties had come to fruition at last! Consider the backdrop of his pro-transmutation book project: remember that he had been spending his evenings writing out arguments in the Species Notebook refuting Lyell and that he had lately been contemplating a competition dynamic between plant species. Wallace's idea, so elegant in its simplicity, was not so much a bolt from the blue as a sudden sprouting—the sprouting of a latent seed full of potential. Here, then, was a mechanism that explained the Sarawak Law beautifully and indeed all of the compelling facts he argued that the Sarawak Law neatly tied together: facts of geological succession, geographical distribution, classification, comparative morphology, and more. It was a triumph, made all the more puzzling, then, by his silence on the discovery in his journal and notebooks at the time. Putting it in perspective, although the paper was a triumph in retrospect, was he a bit tentative about it at the time? Was it as important as he thought it might be? Was he *really* on to something? It sure seemed compelling; perhaps write it up and run it by others for comment? "I know, I'll send it to Mr. Darwin," Wallace thought to himself. "He's quite keen on the question of species and varieties too. I'm sure he would give some good feedback. Besides, Darwin is a good friend of Sir Charles—I'll ask Darwin to show it to him if he thinks my argument holds water." Or so I imagine him musing . . . not unreasonably, since that is what he did.

Fleshing out his idea when his fever broke, Wallace titled the resulting essay "On the Tendency of Varieties to Depart Indefinitely from the Original Type." It may as well have been called "On the Organic Law of Change" and subtitled "A Reply to Sir Charles Lyell."[26] Practically every paragraph contains direct or indirect references to *Principles*, including such key topics as lessons from domestic varieties and the struggle for existence and what we now call ecological interactions, biotic (competition and predation) and abiotic (slow environmental change), and how these cause population growth or decline and thus the extinction of some species and the continued success and divergence of others. The title Wallace formulated concisely expresses the key argument of the paper: a process by which varieties depart further and further from the parental form, tackling head-on Lyell's central anti-transmutation claim that varieties can only change so much, that they inevitably snap back or "revert" to their parental form. Limited change around a parental "type" means the transformation of one species into another species is impossible. Wallace begged to differ.

Lyell's Exhibit No. 1 for such "reversion" was domesticated varieties run wild—think of a bunch of purebred dogs of several breeds gone renegade, interbreeding over several generations. It will not take long before they are all mongrelized, by and large converging on a generalized canine morphology not too far removed from the wolf-like ancestor of these dogs—"reversion to parental type." Thus, domestic breeds could never become so many new species, Lyell asserted. Wallace's strategy was to shoot down this argument by pointing out that since domestic varieties are human inventions, wholly unnatural, a bunch of domestic breeds run wild are irrelevant to the question of the persistence and continued divergence of natural varieties in the natural world.[27] In fact, Wallace says, I propose a mechanism that can explain the "reversion" of domestic varieties *and* promote the continued divergence of natural ones. He then built an argument for the process we know as natural selection (Darwin's term) as a mechanism by which varieties arising can not only persist, but, over time, become increasingly divergent from their original form, eventually so much so that they might be considered new species altogether. Keep iterating this process and get further (and irregular) branching, and rebranching, and re-rebranching . . . an ever-ramifying tree.

It is worth briefly considering the paper's central arguments. After dismissing Lyell's domestic varieties argument, Wallace opens with cases for two foundational points: First, owing to a variety of checks, populations are generally static despite a tendency to grow rapidly—population pressure and checks on growth are real. Second, population size, a direct measure of success, must be related to the "organization and resulting habits" of its individuals—in other words, how well adapted individuals are to their local conditions. He next considers the effect of naturally occurring slight variants among individuals of a population bearing minute differences in, say, color or physiology or some anatomical trait. Such differences may determine how these individuals fare in their environment: insofar as some variations are more favorable and others less favorable to survival, health, or reproduction, we might expect individuals (and so populations) to proliferate or decline according to these traits. But then throw into the mix slow and inexorable environmental change—progressively drier or wetter, warmer or colder. One of those variants rating as "meh" under the original conditions just may have the edge under changed conditions. The balance shifts and individuals with this trait spread, their population growing, while those lacking the trait ebb and perhaps disappear. Iterate this process over and over—variations constantly arise, so it is easy to envision variants of variants of variants—each new and successful one "might itself, in course of time, give rise to new varieties, exhibiting several diverging modifications of form." As conditions continue to change, the population does, too, as long as there is variation that enables at least some individuals to prosper. At some point "the *variety* would now have replaced the *species*," Wallace writes (his emphases), "of which it would be a more perfectly developed and more highly organized form." The net result, taking the long view: "Here, then, we have *progression and continued divergence* deduced from the general laws which regulate the existence of animals in a state of nature" (again, Wallace's emphasis). Wallace concludes his paper by stepping back and painting the big picture: the "continued progression of certain classes of varieties further and further from the original type"—a progression made by minute steps in various directions and with no definite limits—may "be followed out so as to agree with all the phenomena presented by organized beings, their extinction and succession in past ages, and all the extraordinary modifications of form, instinct, and habits which they exhibit."[28] It is telling how the finale of this paper echoes that of the Sarawak Law—that essential insight that, Wallace declared, "connects together and renders intelligible" the same phenomena as those he mentions here.

Wallace may have written out a fair copy of his essay once back at his home base on Ternate before mailing it off to Darwin, Halmahera being too much off the beaten track to have a post office.[29] As for dispatching his essay to Darwin, that choice, too, is clear. Darwin was not only a senior and well-regarded (indeed, famous) naturalist who had shown Wallace they had a shared interest in the nature of species and varieties, he had given Wallace encouragement, no less, saying how he agreed with nearly every word of the junior naturalist's earlier essay and was working on a book on the subject. Just as importantly, the definitive factor behind sending his essay to Darwin and not to Stevens or some journal directly was Wallace's knowledge that Darwin was close to Lyell. Indeed, the essay was written to refute Lyell, and in his cover letter, Wallace asked Darwin to forward it to the great geologist.[30]

Wallace's essay coursed its way to London, where it eventually landed in the study of Down House like a bombshell. Darwin was devastated. He duly forwarded the essay to Lyell—maybe right away or maybe after fretting a bit about what to do, but he did forward it: "Your words have come true with a vengeance that I should be forestalled. . . . I never saw a more striking coincidence. . . . So all my originality, whatever it may amount to, will be smashed." He was staggered: if Wallace had Darwin's own private sketch of his theory in front of him, he could not have made a better abstract of it. "Even his terms now stand as Heads of my Chapters," he wailed to Lyell. What to do, what to do? He was scooped. Lyell offered reassurance, and Darwin followed with another letter: if he could publish honorably, he would. But were not his hands tied now? Or maybe he could publish and send Wallace a copy of a letter he had previously written to Asa Gray in the United States, in which he had described the theory, proving he had not stolen Wallace's ideas? But he had not *planned* to publish, is the thing, so doing so now would be "base & paltry" and downright dishonest, would it not? "My good dear friend forgive me—this is a trumpery letter influenced by trumpery feelings." This letter was followed immediately with yet another, fretting how if he published he would surely be seen as taking advantage of Wallace being away in the field, of Wallace having unwittingly put in Darwin's hands the very paper he would have scooped him with, giving Darwin the opportunity to preempt it.[31]

Complicating Darwin's "trumpery feelings" was surely the sickness just then striking his family: two of his children, fifteen-year-old Etty and two-year-old Charles, were sick with scarlet fever, and it was getting worse—three children in the village had died already. Lyell contacted their mutual friend Joseph Hooker, the botanist, and together they swung into action: Wallace may have had priority in one sense, they reasoned, but they knew that Darwin had been working on this for years and had priority in another. They took matters into their own hands and decided to present Wallace's luminous essay together with excerpts from several of Darwin's private writings. Opportunistically, there was an opening at the Linnean Society of London: a special meeting that had already been called for 1 July, days away. They stood before the gathered fellows and read a preamble explaining the circumstances, then read Darwin's material followed by Wallace's.[32] Was the order of presentation alphabetical? Partisan? We will never know.

What we do know is that the papers were received with interest by the thirty-two members present, a number that did not include either Darwin or (of course) Wallace. Darwin's infant son had tragically succumbed to scarlet fever three days earlier, dying on 28 June. Darwin was despondent, attending the funeral on the very day his and Wallace's papers were read. As for Wallace, he was a world away in New Guinea and, in an eerie correspondence, had just buried one of his young assistants, who had also tragically succumbed to disease two days before Darwin's child. The sad loss was not the first tribulation of his New Guinea expedition, and it would not be the last.

———

Rewind a few months, to March. Wallace had been longing to visit New Guinea, all the more so since the wonders of the Aru Islands. A trip there was not to be taken lightly, however—the hostility of the headhunting native peoples to outsiders was well known, and very few places on the mainland were safe. The village of Manokwari (Dorey in Wallace's day) on the northeast coast of the vast western Papuan "bird's head" peninsula was one, and one of Mr. Duivenbode's trading vessels, the schooner *Esther Helena*, was heading there in a few weeks. Preparations made, the ship set sail on 25 March 1858 with Wallace and four assistants aboard: the ever-dependable Ali, his main assistant; a lad in his late teens named Jumaat as shooter; a Javanese cook named Loisa; and middle-aged Lahagi, a reliable all-around helper. It was an inauspicious beginning, though, as getting there was no mean feat: when they weren't buffeted by squalls or becalmed, they drifted at the mercy of the rapid currents. Slowly progressing east through the Straits of Dampier, they were hailed by Papuan traders in outriggers from the rugged islands of Waigeo and Batanta. Wallace turned down a "miserable specimen" of the rare red bird of paradise (*Paradisaea rubra*) but traded some calico and a copper ring for a beautiful turtle-spear float carved in the form of a bird in flight and a palm-leaf box—the first of several New Guinean artifacts he was to acquire, reflecting his ethnographic interests.[33]

As the schooner rounded the north coast of the peninsula and entered Dorey Bay, Wallace marveled at the vast mountainous interior, range upon range receding in the misty distance. The great Arfak Mountains rose abruptly from the coastal plain, the highest, Gunung Arfak, reaching nearly 9,700 feet (2,960 m). The geologist in Wallace noted that the islands and shore around the bay were uplifted, and recently, the strand strewn with masses of white coral. He later found that the low, flat-topped ridge forming the point along the north side of the bay, Gunung Meja, or Table Mountain, showed similar evidence of uplift, but far earlier, consisting of compact crystalline limestone (today preserved as Taman Wisata Alam Gunung Meja, or Gunung Meja Nature Park, with limestone caves popular with exploring hikers). Those great mountains in the distance, he surmised, were very different: "Primitive" rock, he suggested. "It would be very interesting to visit them," if only.[34] He was right on, and indeed the very abruptness with which this rugged mountain range rises is a clue to some interesting geology. It is another line where the metamorphic Paleozoic-era Arfak Mountains meet the far younger sedimentary coastal plain: the range is a borderland marking the long east–west trending Sorong Fault Zone, a bit of the Pacific plate sutured onto the Australian plate. The two plates grind against one another in the kind of fault known as *strike-slip* by geologists, where the terrains are mostly moving horizontally past one another, periodically generating great earthquakes. What rarities must make those mountains their abode! It was the storied land of the cassowary and tree kangaroo, mountains whose "dark forests gave birth to the most extraordinary & the most beautiful of the feathered inhabitants of the earth, the birds of paradise, the no less brilliant Epinachidae, the glittering *Astrapia* & the golden necked oriole."[35] It was also a land where "the foot of civilized man had never trod"—"civilized" and "savage" being the stock terms du jour delineating European and non-European cultures. He was right that Europeans like himself had never trod there—or if any did, they surely did not live to tell the tale. The aggression of the Arfaki, as the lowland Papuans referred to the mountain peoples, was legendary even against fellow Papuans. The Arfaki launched regular headhunting raids against their coastal-dwelling relatives, not infrequently leaving behind heads of their own in the process, their skulls displayed, Wallace observed, as trophies in the abodes of the otherwise peaceable coastal Papuans. In fact, not all of the Arfaki were hostile—at least not all the time—and Wallace paid a visit to one village in the hills above Manokwari, perhaps on or near Gunung Meja, bearing gifts for the chief. Between drags on an elaborate long-handled pipe carved from a single piece of wood, the chief expressed, through an interpreter, thanks for the gifts and promised specimens and protection for Wallace and his assistants when they came collecting. Whew!

Agriculturists, these hill people grew yams, rice, plantains, and breadfruit, while their coastal brethren subsisted mainly on fish, and traded in turtle and trepang, or sea cucumber. Both groups lived in houses raised high on poles, those of the coastal people lined up in the intertidal, reached by ramps at high tide, along with a large boat-shaped council house with great supporting posts carved with anatomically correct human figures ("obscene caryatides," as Wallace described them), decorated with trophy skulls. The houses of the hill people were scattered in and among their fields, raised some fifteen feet high. The dwellings were communal, laid out much like the communal houses he had seen in South America and Sarawak, though more dilapidated, with a central passage on either side of which a series of side passages led to paired rooms, each accommodating a single family. Fascinated, he took every opportunity to observe the people, making notes on their language, culture, health, and physical appearance. He was still trying to delineate the boundary between Malayan and Papuan peoples, much like that remarkable boundary he had found running between Bali and Lombok. Western Papua was not as clear-cut as he had thought it would be. "We see here many signs," he wrote, "of being in the debatable ground between the Malayan Archipelago & the

Pacific Islands," where elements of each culture mix—for example, rice from the west was a staple, grown with taro, a small bean ("which makes a very fair vegetable"), and breadfruit from the Pacific east.[36] At times he sounds like a dispassionate anthropological observer, while at others the judgmental yardstick of European tastes and mores intrude. He was a bit quick to dismiss the "miserable" houses, perhaps, and while admiring the "elegant tracery" of the chest tattoos of the women, "following the curves of the bosom," he seemed obliged to disavow any attraction: "The females, however, are, without exception, the least engaging specimens of the fair sex it has yet been my fortune to meet"—![37] Beauty is in the eye of the beholder, certainly, but in fairness, chances are he was not viewed by the locals as the hottest specimen of white European manhood either. But at least the Papuan women and children did not run screaming at the sight of him like they did in Sulawesi, as far as we know.

The help and protection of the chief assured and now ensconced in a little "jungle house" built with his assistants and others, Wallace was delighted to be "the first English & 2nd European resident on the main land of New Guinea," following the remarkable French physician and naturalist René Primevère Lesson, likely the first naturalist to see birds of paradise in the wild during the 1822–1825 'round-the-world voyage of *La Coquille* under Capt. Louis Isidore Duperrey.

The intrepid collectors were all set to explore and procure the rarities that awaited, but despite the return of good weather, the rare birds and insects did not follow suit. And it got worse: butterflies were scarce, and the few he found he had collected already. Ditto the tree kangaroos and birds—even the lovely lesser bird of paradise (*Paradisaea minor*) got old, as the only species around. Greatly adding to his frustration, he discovered that the Prince of Tidore accompanied by a functionary from Banda had swanned into town, dispatching a small army of collectors and spreading word that they were paying top dollar for paradise birds. Aarrgg! He could hardly compete with a wealthy prince. In fact, when a Dutch war steamer such as this arrived from Ambon he learned that, as a rule, the locals would take any precious specimens—and fresh fish—straight to the visiting ships first. And all the more so when royalty came to town.[38]

If this were not bad enough, Wallace managed to hurt his ankle clambering among the branches and trunks of fallen trees, an injury that soon turned into a nasty open ulcer that laid him up. It was no sooner getting better when his heel became infected, and he could not walk at all. The doctor (and a "brother naturalist") from the prince's steamer, tried to help, but poulticing did not work, and it got worse and worse until he had to resort to lancing and leeches followed by incessant poultices and ointments, incapacitating him for agonizing days and weeks—what a waste of precious time!! Then his men got sick, too, dysentery striking some, others coming down with fever, and often the whole lot sick at the same time. "I was almost in despair," he wrote. In desperation Wallace decided to send two of his assistants, when well enough, to distant Amberbaki (now Ambubaki) along the north coast of the Bird's Head, with instructions to collect insects and buy as many bird of paradise skins as they could. They were to remain a month, and he was sure they would have better luck over there and bring him a fine collection. "They left May 27th. I still remaining a prisoner in the house with my unfortunate lame foot."[39] Alas, he was disappointed yet again when in late June his assistants returned empty-handed, while Hermann von Rosenberg, an amateur naturalist from Darmstadt, Germany, who worked as a draftsman aboard the Dutch steamer (and who had two collecting assistants), astonished Wallace with his collection of precious rarities from the interior: new birds of paradise, a rare crowned pigeon (*Goura victoria*)—a large and regal species with crimson eyes and a spray of elegant head feathers for a crown—and a pair of stunning Arfak astrapias, *Astrapia nigra*, a "magnificent bird" he had never seen before. Live tree kangaroos were on board, too—yet another group he longed for but could never find. Frustration upon frustration . . .

It's worth stepping back to note that all of this activity—the Dutch steamer and its surveying party, the collectors amateur and professional, the networks of trade in commodities like birds of paradise from natives to Westerners—represented so many ripples of interaction following on the tsunami of colonialism. The coveted currents of the spice trade were at the heart of it, driving the competing Portuguese, Dutch, and English to vie for control of territory and sea-lanes, but trade extended to anything salable. The survey ship crew were looking for local sources of coal to fuel their steamers and could do so freely up to the 141st meridian: the Dutch line, marking the extent of their claims on the great island of New Guinea, which still marks the mid-island border delineating Papua New Guinea from Western Papua, part of Indonesia today. As in South America, Wallace very much benefited from this Southeast Asian colonial network as a working naturalist procuring rarities whose very presence and movements were facilitated by colonial infrastructure: the mail steamers, ever-helpful expats, letters of introduction to colonial representatives or local potentates bought by the Dutch—typically kept happy by receiving an annual annuity and retaining royal privileges—native traders or hired hands. Overall the network tended to work well for Wallace, but in this case collectors higher up the collecting food chain had dibs on the choicest morsels.

Just when things could not seem to get worse, they did. He came down with fever again himself, followed by a bizarre inflammation of the whole inside of his mouth, his tongue, gums and all so sore that he could only slurp liquified food. While he contended with this strange affliction, two of his assistants were again struck by dysentery and fever, falling gravely ill. It was bad: he tried administering calomel and other medication, but nothing seemed to help them. One fortunately recovered, but the other, the lad Jumaat, succumbed to dysentery and died on 26 June. Wallace provided a bolt of white cotton cloth for his burial and lamented the whole misadventure. "This is a terrible country," read one journal entry, and another: "Nothing to be done, nothing to eat & all of us ill. Fevers & cold succeed one another & make me long to get away from New Guinea."

They waited and waited for Duivenbode's schooner, expected any day. The collecting still as terrible as the weather—the two in fact linked—Wallace took time to observe the natives instead, marveling at the seeming contradiction of a people who struck him, with his European standards, as living in "complete barbarism" and dwelling in "the most miserable crazy & filthy hovels" atop their rickety poles in the intertidal, strangers to either furniture or even a stitch of clothing ("*in puris naturalibus*"), and yet masterful wood-carvers demonstrating a refined artistic sense. Wallace asked rhetorically in his journal: If these are not savages, who are? Yet they possess "a decided taste for the fine arts" and spend their leisure time creating works "which could not but be admired in our schools of design, & might perhaps not often be surpassed there."[40] Such reflections got Wallace thinking: Are these "savages" really so different from us? They are *people*, fellow humans after all, despite what their complexion and physical features and (comparatively) rudimentary technology suggested to Europeans (and helped justify the policy of subjugation, exploitation). Here in the East, as in deepest Amazonia, Wallace was again struck with commonalities over differences: the essential unity of humanity. That was duly filed away in his mind, to emerge again and again, as we will see.

In late July he got his wish, and they gladly quit New Guinea at last. It had been a disaster, and he didn't regret leaving a bit: few places he had visited were more disagreeable, he wrote, being cursed with incessant rain, a plague of pesty ants and flies, and debilitating—indeed fatal—illness. Adding further insult to injury, what should have been fairly quick sailing back to Ternate on the monsoon turned into another ordeal, with nearly three weeks of winds too weak and currents too strong. The only silver lining was a one-day stop at Bacan (Batchian in Wallace's day) off the long southern peninsula of Halmahera. It was an odd-shaped island of four joined lobes, rugged in places—long-ridged Gunung Sibela in the south rising to over 6,500 feet (2,000 m)—and clothed in beautiful forest. In no time Wallace found some fine insects and birds, but more interestingly, primates: handsome black macaques similar to some he saw in Sulawesi. (They are considered the same species today, likely introduced by humans to Bacan: the endangered Sulawesi crested black macaque, *Macaca nigra*.) Wallace the mapper of biogeographic boundaries took note: on this island he had found, he thought, the "extreme limit" of primates— "the most easterly spot on the globe on which any monkey is found in a wild state." (He was close: monkeys are actually found a bit further east, in Japan.) And what of the *human* primates? Wallace thought they were a handsome people, noting that they were light skinned with Papuan features and hair, resembling what he imagined a Dyak Papuan mix would look like. Bacan was fascinating; he would be back.

———

"Pretty well knocked up" is how Wallace described his mental and physical state on arriving back in Ternate in mid-August 1858. Read: weak and exhausted, leavened with frustration and disappointment. He recovered both health and spirits pretty quickly, though, thanks to small creature comforts like milk for his tea and a bit of society, attending a ball celebrating the marriage of one of Duivenbode's sons, complete with waltzes, cards, and plenty of beer, claret, and the ever-popular *jenever*, a Dutch gin that Wallace knew as geneva. He also experienced his first earthquake, which proved to be something of a disappointment, and wished for "something a little stronger." He should have been careful of his wishes—in his eight years in the archipelago, Wallace was fortunate to have been spared a sizable and cataclysmic earthquake, often with attendant tsunami, regular occurrences in that tectonically wracked region.

The consignment that he packed up for Stevens suggested more success in Manokwari than he may have realized: three cases, totaling about 7,400 insects alone, including 6,400 beetles, 413 butterflies and moths, and a miscellany of 635 specimens that included a nice collection of those most charming of flies, the "horned" flies of the genus *Elaphomyia* (now *Phytalmia*), bearing curious "antlers" that make them look for all the world like diminutive moose of the fly world.[41] He caught up on his mail and decided to head back to Halmahera with Ali and one other assistant. He went to a different part this time, coastal Djilolo (Jailolo), the traditional seat of the island's sultans on the long northern peninsula. There he was hoping for good collecting in lofty rainforest, but the highest vegetation he found was ten-foot-high reedy grass in endless expanses ("utterly barren of interest in a zoological point of view"). It was not altogether destitute of interesting natural history, though: among the nice birds they collected was a red-flanked lorikeet, *Charmosyna* (now *Hypocharmosyna) placentis*, the "smallest and most elegant" of the brush-tongued lories. Insect rarities were discovered too—most notably, the spectacular clear-winged moth *Cocytia d'Urvillei*, family Erebidae, a monotypic species, meaning the only one in its genus, with a beautiful metallic green-blue body and a bright-orange patch at the base of each of its clear forewings.

Despite these successes, the grass did not bode well for continued decent collecting, so he pivoted to plan B: Sahoe (Sahu), a small village about twelve miles away, where he heard he could access forest. It had the added benefit of being home to the Alfur people, thought

Phytalmia fly species sport antler-like protuberances on their head.

to be Indigenous to Halmahera.[42] The anthropological and zoological were never far apart in Wallace's mind, and there he would have an opportunity to acquaint himself with yet another interesting group of people. He was kindly given the use of a large cottage on the beach, and it seemed like a serviceable if not ideal site as he settled in for a couple months' collecting and observing. Unfortunately, that pesky tall grass turned out to be more widespread than he had realized. He did manage to collect some neat birds, insects, and even the most beautiful gastropod he had yet seen, *Helix* (now *Anixa*) *pyrostoma*, known as the fire-mouth tree snail. But, the collecting underwhelming, he decided to cut his losses after just a week, thinking now of beating it back to Ternate and heading south to return to beguiling Bacan—but not before taking full advantage of the people watching, which was better than he had hoped for.

Recall that all this time Wallace was keenly interested in identifying the boundary between the western-derived Malay peoples and the eastern-derived Papuans, an interest that may well have been sparked by his reading of *Vestiges of the Natural History of Creation*, the book that had convinced him of transmutation back in 1845. Speculating on the center of origin of humanity—one of the book's heresies, positing that humans have a single origin from which all principal so-called races derive—the anonymous author of *Vestiges* suggested working backward along their supposed lines of migration. These, he maintained, pointed to the greater Indian region and what is more could be narrowed down further: "We should expect man to have originated where the highest species of the quadrumana are to be found," he wrote, referring to orangutans, considered at that time the most humanlike of the great apes. Where might this be? "These are unquestionably found in the Indian Archipelago," Mr. Vestiges asserted.[43] It is worth noting that this argument built upon a principle in *Vestiges* resonant with Wallace's Sarawak Law—namely, that the simplest or most primitive "type" (species) gives rise to the "type next above it," which in turn produces the next-higher type, "and so on to the very highest" over time.[44] In other words, species are derived from preexisting species, a principle that means clues to the origins of current species can be found in geographical distribution. Precisely Wallace's interest! The biogeographer in him was sure that despite some intermixture, the stamp of separate geographical origins in the distant past was evident in an enduring racial and cultural contrast between these broad groups, the Malay and the Papuan.[45]

Note that "distant past" was very distant for Wallace: a very deep antiquity of humans, bordering on the geological in timescale. This dovetailed with his conviction that the lands of the Papuan/Polynesian peoples, broadly considered, represent the remains of a former Pacific continent, now sunken and broken up into the myriad islands great and small we see from the south Pacific to the Malay Archipelago. If the dark-skinned Papuans and Africans shared a common origin, their current distribution suggested that it was ancient indeed. In contrast, such leading lights of early anthropology and philology as Wilhelm von Humboldt (brother of the great explorer-naturalist), Robert Gordon Latham, and James Cowles Pritchard held that Papuans were somehow derived from Malays more recently, the two constituting a great "Oceanic race." If these thinkers were correct, peoples intermediate between the Malayan and Papuan should be found somewhere in the great archipelago. Were the Alfuros such intermediates? Was there a gradation between the Malay and Papuan? Monogenists like Latham and Pritchard thought so, while their polygenist opponents preferred to see things in black and white—a set of well-defined races, each a separate and independently created species, making the enslavement of some of them by others more palatable to most polygenists in the perverse morality of the day.

Contrarian of contrarians, Wallace ultimately satisfied both camps in some ways and failed to satisfy them in others. For example, although he initially vacillated, Wallace ultimately decided that Alfuros were not intermediates but a "mixed race," and the more he thought about it, the more he was convinced of a sharp delineation between these and the Papuan and Malay—even, it must be said, if that meant glossing over some linguistic and physical evidence to the contrary.[46] He commented in a letter to his childhood friend George Silk, "If I live to return I shall come out strong on Malay and Papuan races, and shall astonish Latham, [Joseph Barnard] Davis, & Co.!," and later declared that it was there, at Halmahera, that he had "discovered the exact boundary-line between the Malay and Papuan races, and at a spot where no other writer had expected it."[47] This was music to the ears of the polygenists—except he was no polygenist. He was simply pushing common origins far back in time as he tried to link human with geological history. Wallace's insistence on this putative human boundary line was surely influenced by his success in identifying the great Asian/Indo-Australian faunal boundary, not so much in terms of the allure of drawing satisfyingly distinct boundaries, but in his belief that humans, like these other animals, have a history, too, a history that should be equally amenable to the insights of biogeography. But his commitment to that human boundary line required some mental gymnastics, particularly when it came to the idea of uniting Papuans and Polynesians, all groups east of his line, into a single "Oceanic or Polynesian race" scattered among the island remains of a supposed disintegrated Pacific landmass. His work in this arena was both a product of colonialism and a contributor to colonialist thinking, and as we will see Wallace continued to write about race and distribution for many years following his travels in the East, engaging in the debates and controversies of the day . . . and generating his own controversies. But for now, suffice it to say that while Wallace's vision for the origin and migrations of the Malayan and (especially) Papuan peoples did not stand the test of time, even in his own time, he helped make field observation de rigueur for subsequent investigators.[48]

Back on Ternate by the end of the month, Wallace found a fresh batch of letters and packages awaiting him. I can see him flipping through the envelopes, scanning names and postmarks: any given stack typically had mail from friends, family, his agent, and far-flung correspondents posted months ago, slowly wending their way to find him wherever his current residence might be in those remote longitudes, thanks to the efficiencies of colonial mail service. Wait—hullo, what is this one? An envelope bearing postmarks from the Isle of Wight and London, posted last July. His interest piqued as he recognized the handwriting: Charles Darwin. Immediately opening it, he found not one letter but two: a letter from Joseph Dalton Hooker was enclosed within Darwin's.

———

Wallace had no idea how his "eureka" of a paper sent from Ternate the previous March was received. All this time—throughout his months-long New Guinean ordeal, his recuperation once safely back in Ternate, and his second albeit brief return to Halmahera—there had been no word. He could not know what tumult he had caused back in England. What did the letters say? We do not know, as the letters have apparently been lost, but we have a pretty good idea.

A triply devastated Darwin, prostrated by the death of his infant son, ashamed at being concerned about priority over Wallace at such a time, and fretting over the honorable course of action—while not wanting to fall on his sword—had left matters in the hands of his friends Hooker and Lyell, as we have seen. Within a few days of the baby's funeral, the family seemed to be out of the woods: Etty was on the mend, as were the two nurses who became ill treating her and her now deceased little brother, and the rest of the children had been sent out of harm's way to stay with Emma's sister Sarah Wedgwood at Hartfield, in Sussex. Charles and Emma joined them on 9 July, and from there they proceeded to the Isle of Wight for some seaside R, R, & R—rest, relaxation, and recuperation in their grief. Darwin wrote a flurry of letters: one to Asa Gray, asking him to send a copy of Darwin's letter from the previous year describing his theory of natural selection—added proof of priority—and others to Hooker and Lyell, thanking them profusely for coming to his aid, agreeing that he must write a longer abstract of his theory, and wondering how long it should be and how best to structure it.[49] But first, Wallace must be notified. Hooker offered to write him, an idea that Darwin jumped at: "I certainly should much like this, as it would quite exonerate me: if you would send me your note, sealed up, I would forward it with my own."[50] A week later, Darwin had Hooker's letter in hand: it seemed to him "perfect, quite clear & most courteous." He duly forwarded it to Wallace along with a letter of his own on 13 July 1858.[51]

The heat of the moment behind him, priority seemingly assured, Darwin reflected that having Hooker and Lyell introduce his writings was a very good thing: their involvement would, he was sure, "have the most important bearing in leading people to consider the subject without prejudice." In fact, he was "almost glad of Wallace's paper for having led to this."[52] The savvy Darwin was quite right: years later Hooker recalled that Darwin's and Wallace's papers were discussed "with bated breath" after they were read but that "Lyell's approval, and perhaps in a small way mine, as his lieutenant in the affair, rather overawed the Fellows, who would otherwise have flown out against the doctrine."[53] In the meantime Hooker shepherded the papers into publication, prefaced by an introduction by Lyell and himself, and they finally appeared in the 20 August 1858 issue of the *Journal of the Proceedings of the Linnean Society*.[54]

For his part, Wallace could only have been astounded at the letters he received from Hooker and Darwin. The only record of his response at the time comes from letters written just before his departure for Bacan: one to Hooker, sent care of Darwin (an accompanying letter to Darwin himself has been lost), and another to his mother. Written on the same day, they reveal a Wallace generous of spirit: to Hooker he expressed gratitude and gratification at both the favorable opinion of his essay and the course of action he and Charles Lyell took. He regarded himself "a favoured party in this Matter" since it was all-too-common to give full credit to the first discoverer of a new fact or theory and none to others who may have independently made the same discovery. (In the debate over proper priority, the first to make a discovery vs. the first to publish it, it is clear that Wallace fell in the former camp.) Indeed, Wallace told Hooker, it was fortuitous that he had just a short time before started up a correspondence with Darwin over the nature of species and varieties, suggesting that it was lucky he had mailed his Ternate essay to Darwin rather than sending it directly to some journal. As a result, Wallace wrote, this helped Darwin secure a claim to priority that would have been complicated had Wallace's essay appeared by itself, out of the blue, something that would have caused Wallace "much pain & regret."[55] Wow—that is the very definition of magnanimity: Wallace was delighted that he did not accidentally scoop Darwin, helping assure his priority! His generosity is evident in the letter to his mother too. Indeed, anyone inclined to suggest that Wallace's gracious deference to Darwin in the letter to Hooker is not genuine, reflecting, say, only stiff-upper-lip class-conscious politeness, should consider that any ill-feeling or sense of injury would surely have been expressed in private letters to his family. On the contrary:

> I have received letters from Mr Darwin and Dr Hooker, two of the most eminent Naturalists in England, which have highly gratified me. I sent Mr Darwin an essay on a subject in which he is now writing a great work. He shewed it to Dr Hooker & Sir Charles Lyell, who thought so highly of it that they immediately read it before the Linnean Society. This insures me the acquaintance and assistance of these eminent men on my return home.[56]

In a letter to George Silk the following month, Wallace expressed his pride in what he had accomplished, encouraging him to borrow a copy of the August 1858 *Journal of Proceedings of the Linnean Society*: "In the last article you will find some of my latest lucubrations with some complimentary remarks thereon by Sir. C. Lyell & Dr. Hooker which (as I know neither of them) I must say I am a little proud of."[57] A few months after they were posted from Ternate, in January 1859, Wallace's letters were anxiously opened by Darwin. He may have had good reason for concern but need not have fretted. Forwarding them to Hooker, he admired "extremely" the spirit of Wallace's letters: "He must be an amiable man."[58]

———

Think of it: a mere ten years after launching his first expedition, to South America, in pursuit of the mystery of the origin of species, the impecunious, self-educated Wallace succeeded against all odds in uncovering the mechanism behind transmutation! It was surely exhilarating to realize that he had not only succeeded in this quest but in so doing was immediately catapulted into the very center of the British scientific establishment: collector extraordinaire, yes, but so much more. Beyond mere field naturalist, he was a *philosophical* naturalist, making original, insightful contributions to some of the

Beautiful Comet Donati by William Turner of Oxford, ca. 1858–1859.

most profound scientific questions of the day! He was riding high when, just three days after he penned those letters to Darwin, Hooker, and his mother, he was off again. Hooker had evidently wished Wallace a speedy return to England, but Wallace was far from ready to come home. As Hooker surely knew, he said, it would take an awful lot "to induce a Naturalist to quit his re-searches at their most interesting point." There were fresh discoveries to be made, rare and precious species to find, exotic peoples to get to know.

Traveling with trusty Ali and four additional assistants—steady Lahagi, the Ternate native who had traveled to New Guinea with him; Lahi, from Halma-hera, as woodcutter; a lad named Garo as cook; and Latchi, an enslaved Papuan, "very civil and careful," who came with the rented boat as pilot—the intrepid party shoved off on 9 October 1858. That night, stopping on a sandy beach in the shadow of Tidore's towering volcanic cone, they admired brilliant Venus as they ate their supper, just then an "evening star" so bright it cast shadows on the beach. But then, as they launched again that evening, they noticed a bright light emerging from behind the volcano. It was no eruption but a mag-nificent comet with a luminous, curved tail—a rare *bintang ber-ekor*, or "tailed-star" in Malay. It was known as Comet Donati back home, one of the brightest comets of the nineteenth century, first sighted by Italian astronomer Giovanni

Battista Donati at the Florence Observatory in early June 1858. Wallace and his crew happened to observe the great comet just as it reached perigee, its closest approach to Earth. Marveling, he sketched the spectacular visitor from deep space in his journal, describing it in terms that included "magnificent," "brilliant," and "singular"—terms that could as easily be applied to Wallace and his luminous ideas just then. Comets may have once portended doom and disaster, but this one augured great things to come.

CHAPTER 10

ISLAND HOPPER

I have got here a new Bird of Paradise!! of a new genus!!! quite unlike any thing yet known, very curious & very handsome!!![1]

WALLACE CONSIDERED it the greatest discovery his voyage had yet yielded—and that's saying something, having just learned of the splash his Ternate essay had generated back in London. The fabulous find came just days after their arrival, by Ali: "Look here, sir, what a curious bird!" Curious indeed: about the size of a crow, with orange legs and feet, viewed from above the male bird was a somber brown save for a faintly metallic violet sheen atop its flat head, along with a dark eye and a short brush of feathers at the base of the beak. Nothing too remarkable there. But wait: a pair of long, slender white plumes as long as the body projected from the shoulders, the first indication of oddness. Below, the bird was not only gilded in feathers of metallic green mixed with brown but sported two great pointed metallic-green wing coverts extending from the throat, like epaulettes on either side of the body—the most "elegant cravat in shape or colour worn by any bird," Sir David Attenborough would (much) later declare. In life the bird can raise the white feathers above the body and raise up the epaulettes. In full display the effect is spectacular: vibrating wings stretch out to reveal white below, green wing coverts rise to the horizontal like a long, sleek Elizabethan neckpiece, and those long white feathers are held high overhead, waving and quivering with an unmistakably sexual urgency.[2] An astounding find! In 1859 George Robert Gray, at the British Museum, dubbed the new species *Semioptera wallacii*, Wallace's standardwing, the genus derived from the Greek *sema* (signal) and *pteros* (wing), in reference to the semaphore-like white plumes the birds wave about.[3] Gray was right on: those plumes do function in signaling, as part of the bird's elaborate, frenzied

Statue of Wallace and Ali in the field.

courtship dance to attract mates. The wonderful discovery of this new bird of paradise—and Ali's leading role in it—is memorialized at Singapore's Lee Kong Chian Natural History Museum with a fine bronze statue of Wallace and Ali, the tall, bearded, and bespectacled Wallace pointing at something in the distance while the young, turbaned Ali, with gun in hand, is looking up, intent on the object—a lovely standardwing in bronze, perched on a bronze snag mounted on a tall white column.[4] At Ali's foot is a subtle nod to another of their fabulous Bacan finds—a new birdwing butterfly.

If you thought Wallace was excited about the new paradise bird, that was nothing compared to his reaction when he finally succeeded in capturing that birdwing. He had spotted it only a few times in as many months, always sailing along frustratingly out of reach on its whopping seven-inch golden wings. Then, at last: "I was nearer fainting with delight and excitement than I have ever been in my life," he declared. Upon carefully extricating the netted butterfly, his heart beat violently, and the blood rushed to his head, giving him a headache the rest of the day. It was a small price to pay: this fabulous butterfly surpassed all expectation, with wings of velvet black and scintillating fiery orange changing to gold-flecked green as it turned in the light. It was "perfectly new, distinct, and of a most gorgeous and unique colour." Indeed, he thought it the finest *Ornithoptera* yet discovered and thus, he said, the finest butterfly in the world! He thought of a name for it: *Croesus*. What name could be better for this jewel of a butterfly than that of the ancient Greek king of Lydia renowned for his fabulous wealth? He dashed off an excited letter to Stevens and separately sent him a special package bearing the prized finds by the expedited overland route via Suez and Alexandria.[5] It contained several carefully packed specimens of his fabulous bird of paradise and golden birdwing butterfly, along with another butterfly he thought was new: a splendid black-and-metallic-blue swallowtail similar, he saw, to the widely distributed blue mountain butterfly *Papilio odysseus* but deeper in color. If new, what to call it? How about son of Odysseus, he suggested, *P. telemachus*?[6]

Despite these initial successes, Bacan proved to be a bit of a disappointment—butterflies and birds were scarcer than he had first thought, and changing locations a couple of times did not help much. He stuck it out for a while, but his stay had its ups and downs—in some cases literally, rocked by a couple of sharp earthquakes. But the biggest downer was being burglarized, his lockbox stolen one day and nearly all of his money taken along with other items—most inconveniently, all of his keys. He had a good idea who was responsible and went to the authorities, but they found the culprit's defense—ludicrous in Wallace's view—to be compelling and let him off. They told Wallace he could shoot the perpetrator if caught in the act again, but otherwise it was case closed, to his great irritation.

On 25 January 1859, just three days before Wallace penned his excited letter to Stevens, halfway around the world Darwin put pen to paper in his study at Down House. A few days before, he had received Wallace's letter of the last October, from Ternate. It gave the first indication of how Wallace took the news of the events at the Linnean Society. Darwin must have been anxious indeed when he saw the address and carefully opened the envelope . . . but then: "How heartily I admire the spirit," a relieved Darwin wrote, assuring Wallace that although he had "absolutely nothing to do" (*well* . . .) with what Hooker and Lyell took as the fair course of action, he was of course anxious about how Wallace would take it. He complimented the younger naturalist on his tour-de-force Ternate essay but could not resist inserting a note reasserting priority in a "by the way" fashion: "Everyone whom I have seen has thought your paper well written & interesting. It puts my extracts [written in 1839, now just twenty years ago!], which I must say in apology were never for an instant intended for publication, in the shade."[7] And what of Lyell's reaction to all this? Should we be surprised that Wallace asked specifically about this in his letter to Darwin? No, given that it was with Lyell that he wanted to argue all along. "You ask about Lyell's frame of mind," responded Darwin. "I think he is some-what staggered, but does not give in." Lyell said, perhaps tongue-in-cheek, that he was horrified at being "perverted." "But," Darwin assured Wallace, "he is most candid & honest & I think will end by being perverted"—and anyway Hooker was fully on their side, and he was "*by far* the most capable judge in Europe." Darwin was in the home stretch of his book, he also said, on the last chapter, and would of course send Wallace a copy when it came out. For his part, Lyell never did directly engage with Wallace over the issue, no doubt to Wallace's disappointment, but in years to come they met often.

It may have been in this letter that Darwin sent Wallace the table of contents of his book, then to be titled *Natural Selection*. Wallace neatly copied the contents out in his Species Notebook and quietly dropped his own book plans—a real pity![8] It would have been tremendous to have Wallace's and Darwin's fresh and independent takes on evolution in book form—just imagine having, in essence, twin *Origins*! In that alternate universe, graduate theses would be written and courses taught on their comparative analysis! They would later be issued in elegant boxed sets! How fascinating it would have been to have not one but *two* founding books of evolutionary biology. But alas, in our universe the timing was all wrong, the plans and prerogatives of our joint protagonists otherwise: Wallace did not plan on writing his book in earnest until his (as yet undetermined) return home, with years to go; Darwin was already far ahead with his. It would have been impossible, perhaps, for Wallace to later write his book uninfluenced by Darwin's. But still, I cannot help but think that *On the Organic Law of Change* would have made for a compelling and powerful companion volume to *On the Origin of Species*, underscoring and amplifying as it does, in the privacy of Wallace's notebook, many of the very same key lines of evidence for transmutation treated by Darwin in his epochal book.[9]

———

Wallace would not receive Darwin's letter for some months, when he returned to Ternate. In the meantime he plugged away at collecting, observing, and exploring Bacan and its adjacent islets. It was a beautiful and varied landscape, heavily forested with rushing pebbly streams. In typical fashion he noted the geology, observed the people—mainly different groups of Malays with varying degrees of Portuguese influence—and made a prodigious collection of insects, birds, and mammals. Each group had its highlights.[10] Second to the fabulous standardwing to Wallace was surely the beautiful Nicobar pigeon (*Caloenas nicobarica*), a large regal-looking bird of slate gray and shimmering metallic green-bronze with a ruff of long, slender neck feathers. There were several curious mammals too, from both west—Indian civets (*Viverra zibetha*), rusa deer (*Rusa timorensis moluccensis*), and babirusas (*Babyrousa*)—and east: marsupials including the sugar glider (*Petaurus breviceps*) and Molluccan cuscus (*Phalanger ornatus*).[11]

Of the 10,400 insects Wallace collected on Bacan, the rarest must be the two represented by but a single specimen each: a beautiful swallowtail later named in his honor (*Graphium wallacei*) and an immense black bee the size of a human thumb, with formidable-looking jaws like a stag beetle.[12] It was a *Raja ofu*, "king bee"—in fact the world's largest, the king of bees indeed. In 1860 Frederick Smith of the British Museum described this species along with 131 other ants, bees, and wasps first collected by Wallace, dubbing it *Megachile* (now *Chalicodoma*) *pluto*—king bee, but of the underworld, the specific epithet *pluto* resonating with the mythological names he bestowed on some of Wallace's other *Megachile* species, such as two named for the Fates *Clotho* and *Lachesis* and one for the Fury *Alecto*. The giant bee was the "gem of the collections," Smith declared, and the "grandest addition which Mr. Wallace has made to our knowledge of the family *Apidae*."[13] In between the collecting, Wallace admired the botany: elegant tree ferns and stately palms dotting the forest, kanary trees (perhaps *Canarium vulgare*, Java almond) alive with the "hoarse cooings & heavy flutterings" of great green wood pigeons feasting on the nuts, and great candelabras of Dr. Seuss-ish screw pine, *Pandanus*, lining the beaches. He was in awe over a giant fig with an aerial root lattice soaring one hundred feet or more, the tree that once supported it long gone. Having given up plant collecting to focus on zoology, he could only comment, plaintively, that "Batchian is an island that would perhaps repay the researches of a botanist better than any other in the whole Archipelago."[14]

Nearly six months had passed; he was getting restless. When a government kora-kora delivering rice for the local troops called at Bacan en route to Ternate, he jumped at the opportunity to head back to his base. The weeklong trip north was uneventful except for the venomous snake coiled up next to his cot one night—heart-racingly discovered by reaching for what he thought was his handkerchief in the dim light. Fortunately, he was not bitten, and the snake was carefully dispatched, but he was plagued with snake-filled dreams and kept unusually still the rest of the night, fearing to roll over. That was the second time he nearly had a snake for a bedfellow—that he knew of.

He already had his next moves mapped out by the time they landed at Ternate: back to Sulawesi but to the northern arm this time, and he would get there via Ambon and Timor on the mail steamer. But this was only the beginning: having solved the puzzle of how new species are formed, his thoughts turned to the biogeographical puzzles of the great archipelago and how this related to species origins. Wallace's game plan is revealed in a letter to his brother-in-law, Thomas Sims. Back at Ternate there was, as usual, a pile of mail awaiting him, among them one from Sims bearing news of the family and evidently informing him that his mother and sister very much wanted him to come home. The original letter has not been found, but we can surmise that Sims made the case at length, speaking for the family. They felt enough was enough: Wallace was thirty-six years old now and had spent the better part of the past eleven years abroad. His collections had sold well, and he could be more than comfortable selling some of his other rarities too. He could live off that in style, and in his private collection he had material enough for a life-time's work in entomology—work that could as easily be done in London as in remotest Southeast Asia, if not more so. In London his scientific successes and discoveries gave him an entree to scientific society. Don't be angry, Sims evidently soothed, but really, every enthusiast eventually needs to rein in the enthusiasm and look to the future, settle down. Come home and reap the fruit of your labors—enough of this collecting junket, toiling in one pest-infested, fever-wracked, steaming backwater after another with its Russian roulette existence: On a given day, will you be done in by shipwreck and drowning? Deadly snakes? Dysentery or fever? Volcanic eruption, earthquake, or tsunami? Or will it be at the hands of pirates, thieves, kris-slashing locals running amok, or headhunting cannibals?

An exaggeration, this depiction of Sims's pleas, but not by much I'm sure. Predictably, the letter went over like a lead balloon, and worse—he sure hit a nerve! "Your ingenious arguments to persuade me to come home are quite unconvincing," Wallace flatly informed his brother-in-law. What he was doing was not work but pleasure, and he was good at it—why should he not follow his vocation? Sims was wrong, too, that he had made enough money to live on, and besides, who cared about getting rich? In fact, retorted Wallace, "it strikes me that the power or capability of getting rich is in an inverse proportion to a man's reflective powers & in direct proportion to his impudence." And, he declared, "It may be good to be rich but not to get rich or to try to get rich." He, an enthusiast? Yes! He wore the label with pride: "Who ever did any thing good or great who was not an enthusiast?" What's more, most people seemed to be enthusiasts at only one thing: "Money-getting." They have some nerve calling others "enthusiasts" as if it were a bad thing because they cannot imagine there is anything in the world better than making money. As for health and life, how about peace and happiness? Happiness, he informed his brother-in-law, can be defined as "work with a purpose, & the nobler the purpose the greater the happiness." And his purpose was noble indeed. Here Wallace gives insight into his overarching goal, a "wider & more general study—that of the relations of animals to time & space" and nothing less than the geographical and geological distribution of species through time and its causes. "I have set myself to work out this problem in the Indo-Australian Archipelago & I must visit & explore the largest number of islands possible & collect materials."[15] That was it—that was always it! The relationships of animals in time and space—transmutational change over time, distribution in space—these lay at the heart of so many of Wallace's key papers, and he knew he was on to something profound there in the great Malay Archipelago, land of invisible lines drawn by the earth itself. "I know not exactly where I am going next," he told his brother-in-law in closing—but definitely not home. Not yet.

———

Over the next three years, Wallace did precisely what he said he would: "Visit & explore the largest number of islands possible," collecting and observing in the service of a grand quest to chart the interrelationships of species and Earth. He became quite the island hopper, crisscrossing the archipelago visiting dozens of islands great and small in six more or less discrete journeys, traveling by mail steamer, local boat, and even his own prau. He always traveled with Ali, his steadfast and talented assistant, and additional hired hands who varied from trip to trip. He eventually even hired his old assistant Charley, who had since matured into a tall and more capable and self-assured young man, to undertake separate collecting expeditions. Wallace met with ready assistance everywhere he went, whether from expatriate Europeans making their living on the fringes of empire or local rulers in their service. That he had continued to ponder geographical distribution was evident in one of the last letters he sent off before departing Bacan. It was more of a paper, actually—at least it was published as such. It was to ornithologist Philip Lutley Sclater, founding editor of the journal *Ibis*, commenting on a paper Sclater wrote on the global distribution of birds. Sclater, thinking big picture, had divided the earth into six great "zoological provinces" based on avifauna—the foundation for the modern system of biogeographical realms recognized today. After offering more or less minor corrections to the boundaries of Sclater's six provinces, Wallace zeroed in on the special puzzle posed by the Malay Archipelago: there is no more seemingly inexplicable fact about geographical distribution than the division of that great archipelago, so homogeneous looking, into two provinces that "have less in common than any other two upon the earth." To the eye of the geographer or geologist, "absolutely nothing" marks the division between the two regions. It could only be a vestige of landscapes past, a world of landscapes altered and lost by inexorable geological processes that nevertheless left their stamp upon living species. "Here," Wallace enthused to Sclater, "is a wide and most interesting field of research, in which I have long been working, and which I hope by the assistance of my collections to do much to elucidate."[16]

This was a consuming interest, and Wallace was sure that careful analysis of his collections would help shed light on distribution present and past. He resisted requests from entomologists back home to study and describe selected specimens from his private collection—usually the most beautiful and charismatic groups, of course. He insisted that they agree to catalog *complete* series, ideally from a given island or taxonomic group, or forget it. As he put it to one correspondent, entomologist Francis Polkinghorne Pascoe: "You are no doubt aware that my hobby is 'Geographical distribution'—which I am glad to hear you are also interested in—but it is for that very reason that I am so anxious that my collection should be worked at, either in classificational or geographic *groups*."[17] Indeed, not long before penning this letter Wallace wrote a lengthy paper on this very subject, one of his greatest: "On the Zoological Geography of the Malay Archipelago." It was a tour de force detailing for the first time the faunal discontinuity that would later be known as the *Wallace Line*. He had Sclater's paper on his mind, referring to it in the very opening sentence. Sclater referenced the east–west divide for birds in the Malay Archipelago, but Wallace argued that this was a more general phenomenon, with a deep meaning. His object, he said, was to precisely map out the limits of each region and "to call attention to some inferences of great general importance as regards the study of the laws of organic distribution." He pointed out that given the proximity and physical similarity of the islands in the region no disparity in their fauna should occur, yet the invisible divide he had first noted at Bali and Lombok and had since extended north was rather general and striking: "The most anomalous yet known, and in fact altogether unique." The only possible explanation lay in Earth history, he maintained: no less than a "bold acceptance" that the earth's surface has undergone dramatic change. His model, it should be said at the outset, did not stand the test of time in detail—specifically, his belief that Australia, New Guinea, and several islands on the eastern side of the archipelago all the way into the South Pacific represent the vestiges of a great Pacific continent now broken up and largely sunken. But he correctly put his finger on the significance of the shallowness of the seas around the islands in the eastern and the western halves of the archipelago: sitting on the Sunda and Sahul Shelves, respectively, islands on each side can be (re)united into a common landmass with their nearby mainland when sea level drops low enough (or, he thought, land levels are raised high enough). At those times the joined islands become extensions of Asia and Australia/New Guinea, respectively. He rightly recognized that sea depth correlates with time since isolation, and hence affinity as well as endemism—the uniqueness of their fauna. He intuited the difference between oceanic and land-bridge islands and was, at that time, rather taken with the idea known as *continental extensionism*—the extensions of continents, which in some formulations meant land bridges on steroids. It was an old idea championed by the late Edward Forbes, the naturalist whose benighted concept of "polarity" describing species relationships in time had provoked an exasperated Wallace to write his famous Sarawak Law paper in 1855. But unlike absurdly idealized polarity, continental extensionism gained traction among naturalists. Many thought that the only way to explain how far-flung species managed to reach out-of-the-way places was to posit that they walked via land bridges extending here, there, everywhere, even right across entire ocean basins. How to explain the similar species of, say, Madeira and mainland Europe? Perhaps a land bridge, now broken up, once connected them. The problem with this ad hoc hypothesis is that it was easy to invoke while being generally devoid of evidence, at least when it came to the abyssal seas. But land-bridge islands are real. Islands sitting on continental shelves can and do unite with the mainland on a local scale—think of Tasmania, Sri Lanka, Majorca. The problem was that the continental extensionist school extrapolated the local to the global—an extension of extensionism that was, shall we say, a bridge too far for some. Like one Charles Darwin—the idea drove him nuts, as he was sure that in the fullness of time normal chance events of wind and wing, plus flotation, were more than sufficient to explain colonization of even the most remote island outpost.[18] Darwin was certainly not opposed to the Lyellian model of regional uplift and subsidence—a key concept behind his model of coral reef and atoll formation—but he argued there was no evidence at all for vast sunken continents and thought it lunacy that some respected naturalists even flirted with myths and legends like the lost continent of Atlantis to give credence to extensionism.[19] Not exactly a parsimonious explanation.

Wallace eventually, years later, saw the light, realizing there was a world of difference between local (continental-shelf) land bridges and the kind of ocean basin–spanning bridges that Forbes and his acolytes promulgated, but as we will later see, the idea took on a bizarre life of its own.[20] For now, however, here in this watershed paper, Wallace was not so much concerned with championing extensionism as invoking it as he made a case for the profound insights that the geographical distribution of species can provide: "Geology can detect but a portion of the changes the surface of the earth has undergone. It can reveal the past history and mutations of what is now dry land; but the ocean tells nothing of her bygone history. Zoology and Botany here come to the aid of their sister science." He built to a masterful conclusion in his paper as he stressed how "humble weeds and despised insects" of distant shores reveal much about geological changes past and answer key questions such as where and when former continents must have existed, to what continental areas islands must have been joined, and how long ago their separation occurred. Think about it: species, by their mere distribution, can reveal secrets of earth history and enable us to reconstruct the rise and fall of continents and oceans!

Wallace fired off his latest brainchild to Darwin—again!—asking him to convey it to the Linnean Society. Darwin found the paper impressive and was glad to oblige, though he told Wallace he rejected the idea of lost continents. The paper was read at the next available meeting, 3 November 1859, and published in the society's *Zoological Proceedings* the following year.[21]

————

By then Wallace had already traveled hundreds of miles, visiting a dozen or more islands. The first of his island-hopping journeys took him to the town of Manado near the end of the sinuous northern arm of Sulawesi in the usual zigzagging way on the Dutch mail steamer, first zigging south some 570 miles to Timor (via Ambon and Banda), where he planned to do a bit of collecting, before zagging north again. At Ambon he saw his old friend Dr. Mohnike, the amateur entomologist, who showed him the latest enviable additions to his beetle collection—which perhaps made the next stop, at Banda, all the more painful as he mostly struck out in the insect and bird departments. The geology was interesting, though, showing clear evidence of dramatic uplift, with coral shelves stranded three hundred to four hundred feet above sea level, capped by volcanic basalt as if to underscore how volcanic activity and uplift often occurred in tandem and were causally related. Completing the saga of geological upheaval told by these formations was the remarkable sight, in another district, of swathes of standing tree skeletons extending far inland, the telltale reach of a deadly tsunami that had followed a great earthquake two years prior. He continued his zig to Coupang (now Kupang) on Timor's west coast, the main Dutch town of handsome, low, white, red-tiled buildings on sheltered mangrove-lined Kupang Bay. The collecting was poorer than he

Wallace's parents, Mary Ann and Thomas Vere Wallace

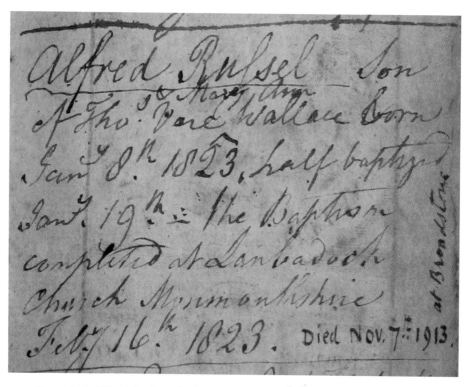

Entry for Wallace's birth in the family prayer book, showing his "half-baptism"

View of the River Usk and town from Llanbadoc Rock

The bee beetle *Trichius fasciatus*, Wallace's prize capture in Wales

Falls of the Afon Hepste, Vale of Neath, Wales

Samuel Stevens, agent par excellence, ca. early 1850s

A few striking longwing butterflies (*Heliconius* spp.)

The umbrella bird, *Cephalopterus ornatus*

The spectacular Guianan cock-of-the-rock, *Rupicola rupicola*

Howler monkeys (*Alouatta* spp.), among the largest and certainly the loudest of South American primates

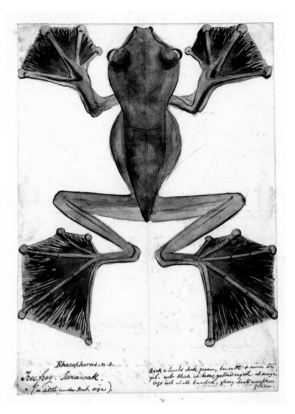

Wallace's flying frog (*Rhacophorus nigropalmatus*), painted by Wallace

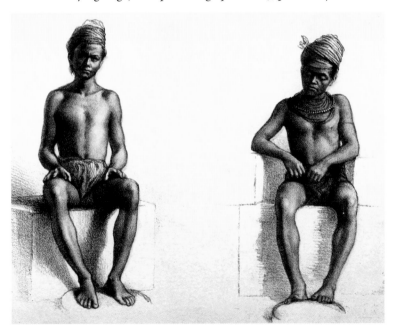

Land Dyaks of Borneo

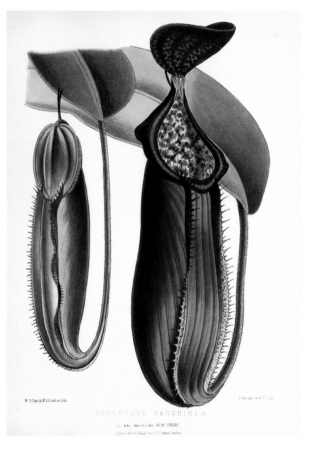

Nepenthes sanguinia pitcher plant, thirst quenching in a pinch

Wallace's painting of the rhinoceros hornbill of Borneo (*Buceros rhinoceros*)

CYMBIRHYNCHUS AFFINIS. *Blyth*

The black-and-red broadbill, *Cymbirhynchus macrorhynchus*

The rufous-backed kingfisher (*Ceyx rufidorsa*), among the
most diminutive of kingfishers

Three fabulous birdwing butterflies: *Ornithoptera brookeana* (*top*), *O. croesus*
(*bottom right*), and *O. poseidon* (*bottom left*)

The South Moluccan pitta, *Erythropitta rubrinucha*

Skin of the python unwise enough to try to share Wallace's Ambon hut

Wallace's giant bee (*Megachile pluto*), the world's largest bee, dwarfing the common honeybee

Red bird of paradise, *Paradisaea rubra*   Lesser bird of paradise, *Paradisaea minor*

A prize find: *Semioptera wallacii*, Wallace's
standardwing bird of paradise

Page from Wallace's field notebook describing the king bird of paradise (*note the tail feather drawing*)

King bird of paradise, *Cicinnurus regius*

PLATE LXXXIII.

EVOLUTION OF THE EYES ON THE WING-FEATHERS OF THE
ARGUS PHEASANT.

Gradations that may reflect the stepwise evolution of the remarkable
trompe l'oeil eyespots of the great Argus pheasant (*Argusianus argus*)

A harbor scene with Malay praus

Plate from Wallace's Malay swallowtail butterfly paper, showing mimetic forms

A few of the tens of thousands of beetles collected by Wallace

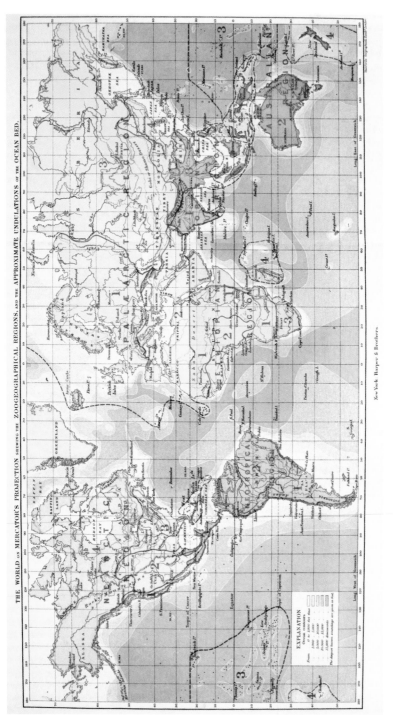

New York, Harper & Brothers.

Wallace's 1876 map codifying the Sclaterian biogeographic realms of the world

Wallace's childhood friend George Silk

Portrait of Wallace's fellow explorer
Henry Walter Bates, in his mid-forties

Richard Spruce, botanist extraordinaire

The remarkable Alice Eastwood

Annie Wallace's painting of Corfe View

The plant-loving Wallace in his greenhouse, ca. 1890s

anticipated—it was the dry season, and the island was even more parched than usual. The vegetation all around was scrubby, used to dry, baking heat, a rather desolate scene relieved only by tall doub palms (*Borassus flabellifer*) dotting the landscape—an ever-useful species, a favorite for toddy palm wine and yielding edible nuts and sturdy workable wood. Once again geology provided an interesting pastime: Kupang was built on a raised coral platform, he noted, and he traveled to the tiny village of Uiasa on the neighboring islet of Semau to see the hot springs of bubbling "soapy" water but was nearly swamped on the return across the narrow strait when his small hired prau sprung a gushing leak. That was fixed readily enough, but heavy seas then nearly capsized them. "I vowed not to go to sea in such small & miserable vessels for the future." He decided to cut his losses and leave Timor early, though as a going-away present he and his assistants collected a veritable flock of fine birds as they waited for the mail steamer, most notably the endemic Timor figbird (*Sphecotheres viridis*), a beautiful yellow-and-olive-green oriole with a brilliant red mask.

Two weeks after departing Kupang, the steamer landed at Manado (Wallace's Menado), in the district of Minahasa. He found it "one of the prettiest in the East," a veritable garden town with rustic cottages lining a neat grid of broad streets while everywhere gardens and thriving plantations interspersed with fruit tree orchards. Over the next three months, the collectors launched a series of expeditions into the interior, beginning with an ascent through rolling mountains to the village of Tomohon, where the local chief treated Wallace to a fine meal of roasted wild pig and fricasseed bat, washed down with fine claret and beer. Continuing on to Rurukan at about 3,000 feet (1,000 m), he was keen to see if any high-elevation rarities presented themselves (they did not) but enjoyed the refreshingly cool climate, at least initially: overnight the thermometer fell to sixty-three degrees (seventeen degrees Celsius), "which produced an effect on my tropicalized constitution equal to a sharp frost in England." An even bigger jolt awaited him a few days later when the cry went up: "*Tanah goyang*!! *tanah goyang*!!—earthquake!! earthquake!!" Everyone scrambled outdoors amid screams and wailing babies. He finally got what he wished for, a quake strong enough to inspire fear and awe, in seconds making the house "visibly rock about & creak & crack as if it would fall to pieces." It was a unique sensation, never to be forgotten, he wrote: "We feel ourselves in the grasp of a power to which the wildest fury of the winds & waves are as nothing in comparison." But he realized that this was a tame one, really, and that he didn't truly want to experience a really severe earthquake, "the most destructive & fearful catastrophe to which human beings can be exposed."[22]

Collecting in the mountains proved a letdown except for the not inconsiderable consolation of a fabulous green swallowtail, *Papilio blumei*, one of the loveliest insects he had ever seen. Clothed in metallic green and gold with elegant spoon-shaped tails of azure on the hind wings, modern researchers regard recreating the spectacular coloration of this butterfly with nanotechnology to be the holy grail of biomimicry.[23] But fine birds and insects generally wanting in this area, as he proceeded south, crossing Lake Tondano, his attention shifted to geology for now. The distal end of the northern arm of Sulawesi is dotted with active and extinct volcanoes as well as steaming, sulfurous hot springs, mini-geysers, and lakes of boiling mud—part of the Minahasa-Sangihe Volcanic Arc, one of no fewer than five tectonic provinces of that curiously amalgamated island, the tortured product of three converging plates: the Australian plate moving north, the Eurasian plate moving south–southeast, and the Pacific plate moving west, all converging, diverging, shearing. Wallace was traveling borderlands within borderlands here, one of the most geologically complex regions on the planet. Reading the landscape, stratified sandstones alternating with volcanic basalts told him that Sulawesi was very old, "the remains of a country older than any of the islands now surrounding it." He was not quite correct but on to something: insofar as the island is a product of a complex collision of microcontinents from east and west, rocks of which now mix with later sedimentary and volcanic additions, it consists of terrain old and new.

Pushing further south still, he went up and over the sinuous ridge that marks the southern rim of the great Tondano caldera, in the shadow of dormant Gunung Manimporok and its highly active neighbor Gunung Soputan—the 5,800-foot (1,780 m) peak completely devoid of vegetation, testament to its regular explosive eruptions. A bit lower on the slopes, at the tiny village of Pangu, collecting got better despite rain and an illness that struck his assistants. He bagged a few birds, notably, imperial pigeons, honeysuckers, and flower-peckers, and a handful of lovely beetles—the prize being *Cicindela* (now *Thopeutica*) *gloriosa*, a glorious beetle indeed of "deep velvety green," wonderfully camouflaged on the wet mossy stones of a stream bed.

Coming down with fever himself, he backtracked and recuperated for two weeks at Manado, then headed north. Skirting the towering Gunung Klabat, the island's highest volcano at over 6,500 feet (1,990 m), he traveled through Lumpias to Likupang at the tip of the long north arm, his considerable baggage—collecting equipment, boxes, books, clothing, gear—following by porters. This was the realm of some of Sulawesi's most interesting zoological

Babirusa, or deer-pig,
Babyrousa celebensis.

oddities, including the endemic *sapi-utan*, or anoa (*Bubalus depressicornis*), a diminutive bovid whose Malay name translates as "forest ox," the endemic babirusa or deer-pig (*Babyrousa celebensis*) sporting elaborate curved tusks, and the celebrated maleo bird (*Macrocephalon maleo*), a large, handsome megapode, black with a pink belly, red and yellow about the eye and beak, sporting a curious black casque, or bonnet, on the head.[24] In the dry season, maleos gathered by the hundreds on the beaches, mated pairs excavating holes in the dark volcanic sand in which to lay their large pale-brown eggs to be incubated by the tropical sun. One might think that the eggs would be more poached than incubated in that fierce heat, but the only poaching they suffered was at the hands of humans. Wallace noted that each year the locals came from miles around to extract the eggs, "esteemed a great delicacy & indeed when quite fresh are delicious—as well flavoured as a hen's egg & much richer."[25] Wallace, who was of course quite the poacher himself, collected twenty-two of the birds, and he and his assistants ate several more.

Back at Manado an opportunity arose to catch a mail steamer southwest to Ambon, a base from which he planned to launch expeditions to the much larger neighboring island of Seram. The stay would last months, so he made a quick stop at Ternate for a few days to pick up some supplies and his latest stack of mail. Among them was a letter from Darwin, written the previous April in answer to Wallace's of November 1858.[26] His book was nearly finished and in the hands of his publisher, he reported, a "small volume" of five hundred pages. He hoped Wallace approved of the notice he gave in the introduction of the younger naturalist's "excellent memoir" presented to the Linnean Society, apparently enclosing a copy of the paragraph, and mentioned that he also cited Wallace's Sarawak Law paper, explaining that they both came to the same explanation for the "law" Wallace described there.[27] Wallace must have discussed in his letter how Darwin came to transmutation and the discovery of natural selection, and here Darwin confirmed he was correct: domestication provided insight into the process of selection, and Malthus provided the impetus for seeing how this could apply in nature. But it was geographical distribution and paleontology in the context of his South American explorations that first opened his eyes to transmutation. Species relationships in space and time were key—just like for Wallace. There was a P.S.: Darwin reiterated how he admired Wallace's generous spirit and, perhaps propelled by lingering pangs of guilt, said that after receiving the Ternate paper he had actually written a letter to say he would not publish in favor of Wallace but had not yet posted it when the one from Lyell and Hooker arrived "urging me to send some M.S. to them, & allow them to act as they thought fair & honourably to both of us. & I did so." Well, there was a bit more to it than that, but Wallace certainly seemed to take it in stride.

He spent a month getting situated on Ambon, packing up an impressive consignment for Stevens, writing up his ornithological observations from Manado, and planning his next move. The consignment, including material from Ternate, Halmahera, and Manado, included over 5,400 insect specimens, 664 birds, and 35 mammals, many, as always, tagged for his private collection.[28] It went out on the 10 November mail steamer to Batavia (Jakarta), then on to Singapore and home. He fired off a letter to Stevens giving him a heads-up about the shipment and, likely knowing his remarks would be published, took the opportunity to complain about so-called authorities who cannot seem to provide accurate locality information and weighed in again on the nature of species and varieties and when you should call a species a species.[29] The idea of a "permanent local variety" is an absurdity and contradiction, he insisted; where differences are constant and permanent, they are species, period. At the same time, Wallace also wrote up a paper on the ornithology of Sulawesi, focusing especially on the curious maleo birds—a paper worth our notice because in it he makes his first public reference to natural selection after the Linnean Society papers. Discussing the biology of maleo birds, Wallace considered the interrelationship of adaptive structure and behavior (what he termed instinct, or habit) in relation to environment. Rather than assume instinct is fixed and structure is adapted to accommodate it, as so many naturalists did, flip this on its head, he urged, and consider that instinct adaptively follows from the constraints of structure and demands of environment—indeed, certain instincts could be seen as the "inevitable and logical result" of structure and environment. Then a telling statement followed: the subject being too large to fully discuss here, "for a perfect solution of the problem we must . . . have recourse to Mr. Darwin's principle of 'natural selection,' and need not then despair of arriving at a complete and true 'theory of instinct.'"[30] Not "Mr. Darwin's and *my*" principle but Mr. Darwin's, solely. Consider: in Wallace's first published application of natural selection is also his first complete public deference to Darwin, declining to claim any credit as the co-discoverer of natural selection. Consciously or not, could this be connected with the letter Wallace had just received from Darwin, informing him that Darwin's book was soon coming out? That may have given him some feeling of finality, Darwin's priority cemented. We will never know.

Just days after posting his letters and paper, he was off to Seram by local boat, arriving on 31 October 1859. The great island of Seram (spelled Ceram in Wallace's day) is a rugged and geologically complex island in the outer Banda Arc, sitting on its own microplate that has been spun about some eighty degrees in the past eight or so million years, dizzyingly fast in geological terms. Its three geological provinces, spanning the past half-billion years, Paleozoic to very recent, speak of repeated bouts of uplift and subsidence, with intrusives and extrusives thrown into the mix. The island is dominated by a belt of great Triassic-age limestone mountains, the highest of which, Gunung Binaia, rises to 9,931 feet (3,027 m) amid a montane karst landscape of deep caves and underground rivers. Rarities were said to abound, and Wallace was finally answering the exhortations of naturalists back home: "If you want fine birds, go to Ceram."[31] Indeed, this island is an ecoregion unto itself, home to sixteen endemic birds (including the monotypic rufescent darkeye, *Tephrozosterops stalkeri*) and nine endemic or near-endemic mammals, the greatest number of any island in the Moluccas. Unfortunately, however, Wallace found few novel birds or mammals and fared little better in the insect department. In fact, his entire eight-or-nine-month stay was something of a mixed bag, mainly because the price he paid in exertion and privation was hardly commensurate with the rewards.

One thing after another went wrong. In his first foray, he stopped at villages along Elpaputih Bay on the southwest coast but was not impressed: "I have nothing particular to say now, except that Ceram is a *wretched place* for birds," he wrote to Stevens from the village of Awaiya.[32] Next, following the Sungai (river) Ruatan inland from tiny Makariki on the bay to the center of the island, he endured trial by water: drenching by storms, with two pairs of shoes completely disintegrating between being constantly soaked through from fording rivers and torn from clambering over rocks. On the final day of the return trip, he was reduced to schlepping in his stocking feet, the skin of his toes rubbed raw by the sand. He made it back to Makariki limping and scratching, covered with intensely itchy welts from chiggers, minute parasitic mite nymphs that feast on skin cells that they dissolve by biting and injecting proteolytic enzymes. Wallace found them "worse than mosquitoes ants & every other pest because it is impossible to guard against them & their effects are more lasting & dangerous," going nearly mad between the itching and the ulcerations made by his incessant scratching. Declaring that part of Seram a "forest desert," he was rather deflated by the time he got back to Ambon in late December. Finding his letter from Awaiya returned for insufficient postage (naturally . . .) surely didn't help; he appended a weary line: "I have just arrived here, being quite tired of the barrenness of Ceram."[33]

But another package awaiting him surely lifted his spirits: *On the Origin of Species*! Ever since their papers had been read at the Linnean Society back in July 1858, Darwin knew he had to bring out the big book he had been working on, in no small part to establish his priority and show he really *had* been working on transmutation for twenty years and in fact was well into writing a lengthy book on the subject. He fretted: the "big species book," *Natural Selection*, was just that—BIG, and far from done, so Lyell and Hooker urged him to come out with a synopsis, a paper detailing the theory. Darwin tried, but after several false starts, he threw his hands up in despair. Given the sensitivity of the subject and the reams of supporting evidence needed to make a convincing public case, he just couldn't do it justice in a mere paper. He went back and forth on the best course of action and eventually settled on writing an "abstract" of his big book. His original title, "An Abstract of an Essay on the Origin of Species and Varieties through Natural Selection," was nixed by his publisher, John Murray, in favor of *On the Origin of Species by Means of Natural Selection, or the Preservation of Favoured Races in the Struggle for Life*. Published in November 1859, Darwin asked Murray to send Wallace a copy. It was evident from the accompanying letter that Darwin awaited its publication with trepidation.

He was just then at Ilkley, West Yorkshire, undergoing treatment at the Wells House Hydropathic Establishment for his nerves. He wrote plaintive letters to fellow naturalists and old mentors, bracing himself: "I fear . . . that you will not approve of your pupil," he fretted to his revered Cambridge professor John Stevens Henslow. "My book will horrify & disgust you," he warned his friend Thomas Eyton. "I know that there will be much in it, which you will object to. . . . I am very far from expecting to convert you to many of my heresies," he ventured to T. H. Huxley. But he may have been feeling especially low the day he wrote naturalist Hugh Falconer: "Lord how savage you will be . . . and how you will long to crucify me alive! I fear it will produce no other effect on you."[34] He was anxious for Wallace's reaction, hoped Wallace would find at least some of it new, and urged him to remember it was only an abstract, much condensed. "God knows what the public will think."[35] Darwin need not have worried: consuming the book practically overnight, Wallace was delighted and quickly wrote Darwin from Ambon to tell him so. His letter is lost, but Darwin's response, written the following May, says it all: "I received this morning your letter from Amboyna dated Feb. 16th, containing some remarks & your too high approbation of my book. Your letter has pleased me very much, & I most completely agree with you on the parts which are strongest & which are weakest."[36] He told Wallace again how he admired his generosity of spirit on the whole matter and filled him in on how the book was being received: who was staggered, who was a convert, who was on the fence, who was a foe. The most venomous in this camp, he reported, was surely Richard Owen, in a vicious attack in the *Edinburgh Review*. But he had gotten "case-hardened," and "all these attacks will make me only more determinately fight," he declared. Wallace would prove to be a devastatingly effective defender of the Darwinian-Wallacean faith.

––––––

At distant Ambon, a weary Wallace hoped to recuperate a few weeks in a little house he rented in Paso, on the narrow isthmus between the island's two "lobes." He had high hopes of finding the magnificent birdwing butterfly *Ornithoptera priamus* there, as well as prized racquet-tailed kingfishers, yellow-necked lories, and other birds. But all he managed to catch was a bizarre disease, an eruption of boils so severe that he was laid up for a month. What the suffering Wallace really needed to catch was a break. That proved elusive, but ever hopeful, he forged on ahead planning his second Seram expedition and was able to hire his old assistant Charles (Charley) Allen, now twenty-two years old, to do some supplemental collecting for him.[37] Charley came over from Borneo, where he had been working for a mining company, and was to travel north to Mysol (Misool) island with two assistants in hopes of procuring birds of paradise while Wallace and his other assistants headed east along the south coast of Seram—this time in style, as he had a letter from the Dutch governor requesting the local rajah provide assistance for Wallace. And assistance they got: the rajah kindly provided four boats, and one bright morning Wallace found himself sailing with an escort across a deep blue bay to the village of Telutih, framed against the backdrop of the magnificent mountains of central Seram. The boats were rowed by sixty men, surely quite a sight with flags flying and drums beating, their song and shouting reverberating across the bay as they plied the crystalline waters. Received by the orang kaya and secondary chiefs garbed in bright silk jackets, Wallace was conducted to a specially prepared house. He found the people helpful and fascinating, recording their customs and some vocabulary, but he found few novel specimens and so in a few days pressed on. Or tried to—with difficulty he made his way from village to village all the way to Kissalaut (modern Kisalaut), where he tried to procure men and a prau to take him further east to the islet of Goram (now Gorong). The rajah promised to assist, but of course that was to take place in "local time," not "European time," much to Wallace's endless frustration. As beautiful as the tropical site was, he was stuck in a "perfect desert in zoology."[38]

Four weeks later they were off, hopscotching along the string of islands extending southeast from Seram. Between April and June 1860, Wallace and his men visited Manawoka, the Watubela archipelago, Kisyui, Baam, Gorong, and diminutive Kilwaru—an islet like a Malayan Venice, he thought, "most extraordinary." The town appeared to float upon the sea like a vision out of Calvino's *Invisible Cities*. Kilwaru was the "metropolis of the Bugis traders in the far east," a bustling crossroads on pilings where a brisk trade in sago, opium, trepang, nutmeg, tortoiseshell, rice, massoia-tree essential oil, and Papuan slaves was carried on constantly. Cultural and geological observations dominate Wallace's account of this journey, reflecting the fact that it yielded precious little by way of collections. He puzzled over the people—a confusing mixture of Alfuros, Goramese and Ceramese, Bugis, Papuan, and Malay. He recorded what vocabulary he could of their languages as well as their customs and beliefs, but they tended not to compare very favorably in his opinion with the native peoples of Borneo or South America.

Wallace gained more insight with the geology, noting with fascination the uplifted coral—in some cases dramatically so, as at Manawoka Island, where he marveled at the perpendicular coral cliffs soaring 100 to 200 feet. Indeed, the heights of this little island reach to well over 650 feet (200 m). At Watubela (Wallace's Matabello), he was impressed with the reefs lining the shores—his first example, he said, of a true barrier reef formed through the process of subsidence that Darwin had demonstrated so clearly in his 1842 book. And at Gorong a classic fringing coral reef encircled the island about a quarter mile from the shore, just visible as a thin strip of emerald-green water, coral rock protruding only at the lowest ebb tides. Setting out to explore inland from the village of Ondur, he described a stairstep topography that suggested cycles of sinking and uplift in Gorong's past, comparatively recent history since even far inland the coral and seashells seemed fairly fresh and not yet bleached by the elements. He speculated that before there was coral the island was terra firma that gradually sank at intervals. During periods of stasis, the island became encircled by reefs—first one circle, then another closer in as it subsided further, and so on. When the whole island was later uplifted, evidence of the start-and-stop subsidence would be seen in the undulating concentric pattern of coral on the island's surface, another exciting example of reading history in the landscape. This process is consistent, he realized, with why the island seemed species-poor: if it had reemerged from the sea relatively recently, its fauna would be wholly attributable to chance recolonization and subject to the vagaries of wind and currents.

Both wind and currents were certainly major controlling factors in Wallace's *own* movements, with several close calls. It was often a waiting game, lying low in some sheltered spot until a favorable wind arose and then making a break for it and hoping the wind did not fail or shift—the Scylla and Charybdis of being becalmed and at the mercy of currents versus being blown far off course, the two equally deadly. It was the latter that befell them among the islands of the tiny Watubela archipelago, where owing to fierce winds they found themselves in the open sea at nightfall, miles from a safe haven. Now there was yet another Scylla and Charybdis confronting them: "My men were now all very much frightened, for if we went on we might be a week at sea in our little open boat, laden almost to the water's edge; or we might drift on to the coast of New Guinea, in which case we should most likely all be murdered." They could not return, and almost certain death awaited if they went on, but they got lucky that time and came upon a tiny coral islet, hugging it for dear life until sunup, hoping for a change in the wind. The threat from New Guinea was very real, as they were reminded when two trading praus limped into harbor at Gorong one day with six ragtag starved men, the survivors of a crew of twenty. The rest had been massacred by the Papuans, among them the rajah's son, and these six were lucky to have escaped with their lives. Chilling shrieks and wails of grief went up for the lost sons, brothers, and husbands, reverberating all night throughout the island.

It was May 1860, and he was looking ahead to his next expedition—a return to storied New Guinea where paradise birds beckoned, despite the dangers.

Seram and the outlying islands had been a bust—not completely, but largely.[39] He outfitted his own prau for the trip this time, paying the equivalent of nine pounds for the boat, then worked with a group of carpenters that included expert boatbuilders from Kai-besar to renovate it to his specifications. The rest of the group predictably failed to live up to his expectations: "Their ideas of work were however very different from mine." After the equally predictable hassles with hired hands disappearing for various reasons on the eve of the expedition, despite (or because of?) having paid them in advance, an exasperated Wallace finally managed to assemble a crew: three men and an enslaved youth from Gorong, two lads who had accompanied him from Ambon, and trusty Ali. Stopping at Kilwaru to ship off some specimens to Ternate and stock up on supplies—knives, knickknacks for trading, and two additional muskets for protection—a Bugis trader brought him news of Charley Allen on Misool: he was doing well on the whole, though no birds of paradise yet, but he was getting low on supplies. Wallace decided they would stop at Misool to restock Charley en route to New Guinea.

Easier said than done. On 1 June 1860 they headed for Misool, making their way along the northeast coast of Seram. But just a day into the expedition, he awakened to find that his crew from Gorong had absconded, taking with them some critical items such as oars. Dragging anchor in the unrelenting wind, the remaining party had to fire their muskets in a distress signal and were fortunately rescued by a boat dispatched from Seram. He fumed about the deserters. Had he not treated them with the greatest kindness, met nearly all their requests? Running away could only be attributed to their being unaccustomed to the "restraint of a European master," he declared.[40] Perhaps "restraint" and "master" are the operative words here—a reminder that Wallace, however sympathetic and kind he might have been, was, like all European scientific travelers, very much part of the colonialist system, which locals often had little motivation to respect. In that light an absconding crew is no surprise. Indeed, perhaps the only surprise is that it did not happen more often out there on the fringes of empire, and that more locals, chaffing under the yoke of paternalistic despotism, did not exercise what little autonomy they had and thumb their nose (or worse) at the white guy with his bird and bug obsession. That is not a state of mind Wallace could have understood, perhaps, but he did try to understand where they might be coming from in his own way, speculating that they may have been uneasy about his "ultimate intentions" toward them. By this he may have been referring to their unease with the dangers of sailing near deadly New Guinea, even though he had determined they would head to the island of Waigeo, just to the northwest of the "bird's head" of western New Guinea. It was safer there, sure, but raiders and pirates were rampant, and the all-too-real prospect of one's head ending up dangling outside a Papuan long-house was surely enough to give pause to most. But not Wallace.

————

If the desertion of his crew a day into the journey was not inauspicious enough, it didn't get better.[41] The next forty days and forty nights saw trials and tribulations of biblical proportions, appropriately enough, as they fought fierce currents and fiercer winds—when the wind blew, which it often and unpredictably did not. Wallace suffered like a voyaging Job, to continue the biblical metaphor, as they strived to reach Charley, camped on Misool's southern coast at Silinta (modern Lelintah). To no avail—they snapped rattan anchor cables, foundered on coral reefs, and were alternately becalmed at the mercy of the currents or driven by raging squalls. The eastern monsoon roaring through the sixty miles of open ocean between Seram and Misool proved too much for the little prau, and they were constantly blown northwest of their mark.

A terribly seasick Wallace was relieved when they made it to Pulo Kanary, today's Pulau Nampale, a tiny archipelago just eight miles northwest of Misool. The plan was to wait for a favorable wind to jump over to Misool, then hug the coast, creeping their way around to Silinta where Charley awaited. That evening they got their chance and off they went, getting so close that they were congratulating themselves on their success—prematurely, as the fickle winds abruptly quit, and a current drove them out to sea despite their frenzied rowing. The men threw down their oars in despair. Morning found them where they had started, miles from Misool.

Another miserable day and night passed fighting the squalls, and they had to give up on Misool for now. The furious winds took them further north, and when another day dawned, they found themselves tantalizingly close to an island Wallace identified as Poppa (Kofiau) on his chart; the desperate little band scrambled to make for the island and safety, only to miss again. Wallace fretted about not being able to reach Charley and his dwindling supplies. The lad would have found it unthinkable that Wallace and his crew could miss a forty-mile-long island, so what would he conclude? Most likely that they were lost at sea, Wallace imagined, or that the crew had murdered him and made off with the boat. All he could do was try to get to safety and somehow send word and provisions to Charley.

Driven farther and farther away, they tried mightily to reach specks of islands about twenty-five miles north of Kofiau, probably the Fam Islands—a tiny group among the more than fifteen hundred islands of the Raja Ampat archipelago, a world of deep-green islands dotting seas of blue and turquoise. After four harrowing days, they made it to the lee of one of them and dropped anchor just inside its fringing reef, mercifully able to sleep. The next morning Wallace saw that the reef was high and dry at low tide, a dangerous spot to anchor, hemmed in by sharp coral. To move they had to get around a rocky point, and Wallace thought it safest to get there by towing from shore—all they needed was a length of "jungle rope," strong rattan lianas easily found on shore. But his men scoffed—why tow when they could just row around the point in a few minutes, easily? If this were a Greek tragedy, the chorus would be intoning a warning.

Against his better judgment, Wallace agreed. The mistake was evident as soon as they left the safety of the fringing reef: the still-furious winds and currents quickly drove the boat away. They scrambled to drop anchor and managed to halt their seaward drift. Now two of his men, a Papuan and a Malay, dove in, making for shore to get those ropes after all. They had disappeared into the jungle, leaving Wallace and the rest of the crew anxiously awaiting their return, when suddenly the anchor started to drag . . . Carried into deeper water, they quickly ran out all available anchor line, halting their drift again. At the end of their tether, the unrelenting winds strained the anchor cable. As they fired their muskets to get the attention of the men onshore, the anchor gave way again; they desperately took to the oars but already knew it was futile. The shore men either did not hear the muskets or were blithely unconcerned as they poked about for shellfish. The increasingly frantic waving in the boat failed to get their attention until it was too late: "They stared at us," Wallace later wrote, "and in a few minutes seemed to comprehend their situation; for they rushed down into the water, as if to swim off, but again returned on shore, as if afraid to make the attempt"—undoubtedly the wiser decision. Wallace desperately hoped the men would quickly fashion a raft or even just cut down one of the multitude of softwood trees and use it to paddle out to them, but rather than remaining cool, calm, and collected, the two ran along the beach, "gesticulating wildly to us."[42]

The prau drifted more than a mile to a neighboring islet, where Wallace determined to wait in hopes that the marooned men would try to reach them. Low on water, he and his remaining men scavenged from unpromising depressions scraped amid sago palms, scooping out rotten leaves and detritus to fill jars with the turbid semiliquid remaining, optimistically thinking that perhaps the depressions were springs. They were not. As they searched for water, the parched island taunted them with dry stream beds, but eventually, their luck turned and they found deep potholes in one bed still holding gallons of water. "When the cup came we enjoyed a good drink of the cool pure water, and before we left had carried away, I believe, every drop on the island."[43] But that was not the extent of their luck: they were back in the boat when the anchor line snapped from abrading against the sharp coral, the wind pushing them to and fro. Thank heavens it did not happen at night! They would have been blown out to sea anchorless, all but doomed, Wallace soberly realized.

They could not take that chance. The anchor retrieved and the line repaired, Wallace resolved to sail on; they would have to leave their marooned men behind and try to send help later. Distant smoke rose from their direction, so between shellfish and sago palms they should have food, he figured, and would hopefully find water by digging. Shoving off into the winds that evening, by morning they neared the west coast of Waigeo—and blundered full on into a coral reef. Whew, no damage. But they were lost. The marooned men being both the strongest rowers and the only ones who knew the region, the party could only gingerly pick their way amid the maze of islets, reefs, and mangrove thickets off the coast of Waigeo, searching for a village—any village. The islets of limestone were everywhere eroded into bizarre and whimsical shapes beautifully surrounded by coral and emerald-green sea. They were also sharp, deadly, and completely porous, with not a drop of water to be had. For three days they wandered: "The shores seemed all desert; not a house, or boat, or human being, or a puff of smoke was to be seen; and as we could only go on the course that the ever-changing wind would allow us (our hands being too few to row any distance), our prospects of getting to our destination seemed rather remote and precarious." Then their luck changed again: as they happened upon a small outpost, the orang kaya came to their rescue. They were still days from their destination, the headman said, and pointed them in the direction of the nearest proper village. Limping into the coastal settlement of Muka, also called Umka, they were saved at last after eight days in the islet maze. He urgently dispatched a boat to rescue his marooned men, but after ten days it returned without them. The castaways had been spotted, but foul weather and illness had prevented the rescuers from landing. With the inducement of more pay and provisions, he sent them off again, successfully this time. His resourceful men had indeed found water and subsisted on shellfish, turtle eggs, and the roots and flower stalks of a bromeliad-like plant. They returned "in tolerable health, though thin and weak," Wallace recorded, grateful for their rescue.

He spent about a month there in Muka, building a long, narrow shed-like house using the prau's sails for makeshift walls and cadjan, woven palm-leaf mats, for roofing. He was pleased with the site except for the food—or lack thereof. He noticed that here, like most places where the sago palm abounded, the local people tended to live in an abject state of poverty. Rather than plant gardens or orchards, they were content to live on sago cakes and whatever else they might scrape together, and they hardly did even that themselves, relying instead on the ubiquitous enslaved Papuans. Their greatest exertions were to secure the annual tribute that was due the sultan of Tidore, in whose realm they lived.

His house was near a nice fig tree that attracted his chief quarry: the rare red bird of paradise, *Paradisaea rubra*—a gorgeous endemic of cinnamon brown with a velvet-green-and-black head, a yellow bill and mantle about the shoulders, and signature bright-red tail plumes with two additional long black ribbon-like tail feathers resembling glossy twisted wires. Their function is evident at the height of the spectacular courtship dance of the male, when, inverting his body with outstretched wings, the sinuous wires fall to either side, forming an elegant heart-shaped frame. Wallace's admiration for such gems of the forest can seem at odds with his ruthless insistence on collecting as many as possible, but for him this was as much a financial as scientific imperative. When the birds became wary of the fig tree, Wallace decided to relocate to Bessir (now Besir), on rugged Gam Island just southwest of Waigeo, where he heard that the Papuans—all enslaved—had an ingenious way of catching them. Rather than shooting the birds with blunt arrows as in Aru, the bird-catchers set a trap baited with red-ripe arum fruit, a favorite of the birds, positioning it in a tree above a well-hidden cord snare ready to be closed on their legs when they came to take a bite. With the bird catcher's help, he ultimately collected twenty-four "fine specimens," sketching and describing them in detail in his Species Notebook.[44]

He was given a small but serviceable hut to use in Besir. Raised up on posts, there was just enough clearance beneath for him to sit and work at a little table while using the thatch structure above to sleep. He found the arrangement fairly comfortable despite occasionally banging his head getting in and out of his workspace. The collecting was not bad overall—a range of lovely butterflies and birds, including several new species, thanks to the help of local men and boys compensated in such goods as handkerchiefs, beads, and hatchets for their efforts. There was much else to contemplate besides: the local geology (more coral and limestone mixed with granitic formations) and certainly the local people. He concluded there were no Indigenous Besirans, just intermixed Malays and Papuans, reinforcing his view that there was no "true transition" between these two groups, with one "race" derived from the other, only varying degrees of intermixing. He based this distinction on an absence of a large *homogeneous* population intermediate between the extremes, as opposed to different groups exhibiting different degrees of intermediate features, which would be more suggestive of intermarriage. Recall, though, that Wallace did not deny the ultimate relationship of all peoples. He just thought such racial evolution occurred deep in human history, on an almost geological timescale, a belief that dovetailed with his sunken Pacific continent idea. His thinking here has not quite stood the test of time: modern genetic and linguistic analyses show that Papuan- and Malayan-affinity peoples do indeed share a common ancestor and likely within only the past six thousand years or so.[45]

As September waned it was time to depart. The eastern monsoon would soon cease, winds he needed (if they would only cooperate) to get back to Ternate. Besides, he was getting increasingly malnourished and ill with each passing day. There was little by way of nutritious food to be had, and he was reduced to consuming meager rice and sago complemented with the occasional tough cockatoo or pigeon, scavenging wild fruits, and boiling ferns for greens (generally not a good idea unless young and tender). He developed an acute pain in his right temple, which had no sooner resolved itself when he came down with a fever. As the intrepid band of collectors was about to leave the island, Wallace was impressed with the honesty of the men he had engaged when one unsuccessful hunter showed up to return the axe he was paid in advance, and another ran up at the last moment, out of breath, to hand him the sixth and final bird specimen he had been paid to collect. "Now I owe you nothing," the collector said with evident satisfaction. Wallace was touched; to him these were "remarkable and quite unexpected instances of honesty among savages" in a situation where it would have been very easy for them to have just shrugged him off.[46] He filed this away; it was further evidence of the essential goodness of humankind and the universality of a moral compass.

Back they went to Muka, where they got everything packed up, replaced the rat-gnawed sails of his prau, stocked up on water and what food they could, and headed off—or tried to. Battling the wind and currents for four days, they made it to Gagie (Gag) Island, anchoring by moonlight in a sheltered harbor. Two days (and two anchors . . .) later they finally found themselves

A cramped home office: Wallace working beneath his hut at Besir, Gam Island.

approaching the southern extremity of Halmahera, Wallace's Gilolo. That night, 4 October, they experienced the remarkable phenomenon of an earthquake-spawned tsunami at sea. It started with a deep roaring sound to the south and then a white line of foam moving rapidly toward them followed by a series of ten or twelve long cresting waves passing rapidly through, raising and lowering the prau and leaving the sea as perfectly calm as before. The next morning they made for the point, to no avail. They tried repeatedly to round it and were each time blown or carried north. Days passed; they sometimes rowed mightily for hours only to find themselves where they started. Then they lost their anchor yet *again* to the sharp coral and drifted out to sea. With but three days of provisions remaining, they realized they needed help rounding the point, so they decided to run north to the village of Canidiluar—except the Anemoi, gods of the winds, were in a fickle mood that day, and the fierce south wind that had kept them from making progress in that direction suddenly abated when they turned their bowsprit north. They cautiously rowed to shore, making a temporary anchor of a sack filled with stones, then north to the village. It was just their luck that the orang kaya was on the opposite side of the peninsula at the village of Gani, so while word was sent to fetch him, Wallace's men set about making new anchors: taking a thick forked branch with one arm shorter than the other, the two arms were bound with rattan for strength while a flat stone was lashed to the end of the longer arm.[47] In due time the headman arrived and graciously provided rowers and food for Wallace and his men.

16 October 1860. They departed Canidiluar at dawn and by noon the next day had rounded the point (losing only one anchor in the process) and headed north again. By 18 October they had reached Gani on the west coast and by 22 October the narrow strait between Bacan and Halmahera: Paçiençia Straits or Fretum Patientia, so named by the early Portuguese navigators because great patience is needed to get through, so fierce are the currents. Fighting every nautical mile as they slowly rowed and sailed their way north to Ternate, they rested at a village on Makian island, where a letter from Charles Allen managed to reach him. Charley was just then at Ternate, following his collecting in the Sula Islands, awaiting his boss after running out of supplies on Misool when Wallace and crew could not land months earlier. Wallace was keener than ever to get to Ternate now, but *paçiença* was still the watchword of the voyage. The crew grew more and more convinced that Wallace's prau was unlucky. Apparently it had not been properly blessed, a ceremony that entailed pouring a kind of holy oil through a hole bored in the keel. When the fiercest squalls yet struck them off the coast of Makian, "a regular little hurricane," they scrambled to get the mainsail down, and the Bugis helmsman implored "Allah! il Allah!" into the teeth of the storm—the eventual abatement of which he credited to his pious entreaties. At last, on 5 November, they arrived at Ternate. Whew! Looking back at the voyage from the time he had left Gorong the previous May, Wallace observed that his experiences traveling in a native prau "have not been encouraging."[48] Indeed.

————

There was no rest for the weary. He was two months at Ternate, but he had hardly any time to recuperate. As he put it in a letter to his old friend George Silk, "For two months I was stupified" [*sic*]—a year's worth of letters and accounts to sort, papers, magazines, and books to go through, some sixteen thousand specimens to clean, arrange, and pack for shipment, all on top of preparing for his next journey—with the usual hassles of finding reliable men to hire and trying to get necessary supplies in town ("None of them could be had for love or money").[49] No, no rest for the weary, but by late December he was more or less caught up and put pen to paper. On Christmas Eve he wrote to Bates, who had, after eleven years in Amazonia, arrived back in England a year earlier, in November 1859. By then Wallace had read *On the Origin of Species* several times over, each time more and more deeply impressed. He did not know how or to whom to express his admiration, he confided to Bates. To Darwin it would come off as flattery, to others self-praise. But his hat was off to Darwin, so sure was he that he himself could never have managed so complete and masterful a treatment of the subject, with such a vast accumulation of evidence, such an overwhelming argument, and such an admirable tone and spirit. "Mr. Darwin has created a new science & a new Philosophy," he declared, "and I believe that never has such a complete illustration of a new branch of human knowledge, been due to the labours & researches of a single man." He was awed: "Never have such vast masses of widely scattered & hitherto utterly disconnected facts been combined into a system, & brought to bear upon the establishment of such a grand & new & simple philosophy!"[50] He expressed the same sentiment in his letter to George Silk in even stronger terms, urging his friend to read the book: "It is the 'Principia' of Natural History. It will live as long as the 'Principia' of Newton." Wallace saw the deeper implications of the concept of evolution by natural selection like no one save Darwin himself—a vision of "grandeur & immensity," of the burgeoning tree of life growing through myriad mutual interactions, intricate relationships, epic struggle unfolding over untold eons. The laws of physics were simplicity itself in comparison, he told George: "Mr. Darwin has given the world a new science & his name should in my opinion stand above that of every philosopher of ancient or modern times. The force of admiration can no further go!!!"[51] There is no better proof of Wallace's utter respect and admiration for Darwin and his accomplishment—while giving his own contribution a back seat—than such praise from the heart written in private letters to friends.

While he may have deferred to Darwin on transmutation by natural selection, geographical distribution was still front and center of his research. He related his latest thinking to Bates, sure now that birds and mammals give a better indication of historical geological processes than insects, for a variety of reasons. Therefore, he thought, a study of first mammals and then birds can give insight into the geographical and physical changes that help understand current insect distributions. That was one of his motivations for stopping at Buru after visiting Timor: he knew the babirusa occurred there and wanted to determine if the fauna was overall more eastern or western in character. Looking to the future, he looked forward to exchanging specimens with Bates when they were reunited in England, which "according to my present plans will not be delayed beyond a year & a half from this date." A year and a half would be the summer of 1862, but as we shall see, thoughts of home were growing, and in fact Wallace departed the East little more than a year after penning that letter.

There was much to see and do in the meantime. Not least, he had to make up for the debacle, as he saw it, of his Seram and Waigeo expedition. Although his collections were immense and there were some real gems in there, he was disappointed overall. In a letter to Stevens, the words "poor," "miserable," and "nothing good" are perhaps the most salient, and as for the places: Seram, he declared, was "a wretched country" and Waigeo "not worth visiting" except for the paradise bird found there.[52] Speaking of which . . . so close and yet so far. How, he asked himself, could he have struck out not once but *twice* in obtaining more bird of paradise species in the very lands where earlier naturalists had gotten them by the dozen?! Charley had better luck, but not by much—his six months on Misool yielded a few specimens of the king bird of paradise (*Cicinnurus regius*), one of the lesser bird of paradise (*P. minor*), and but a single traded skin of the magnificent bird of paradise (*Cicinnurus magnificus*). But he missed the *Paradisaea* breeding season when Wallace failed to show up and was forced to return to Seram for supplies. Wallace desperately wanted more fabulous *Paradisaeas* but did not think he was up to a third expedition himself. He dispatched Charley instead, sending him in his outfitted prau from Gorong. Whether the vessel was now appropriately "blessed" he did not say, but thanks to the support of the sultan of Tidore, he did send three soldiers to assist and protect Charley from the Papuans. "If he does not succeed this time, I must give up the attempt in despair," wrote Wallace in a paper for the *Ibis* on the birds of Seram and Waigeo.[53] For his part, he said, he was now off to Timor on the next steamer. After Timor the plan was to meet up with Charley at Buru after his New Guinea collecting—Buru being the largest remaining island of the Moluccas he had yet to visit—after which his young assistant would head on to the Sula Islands. They would then meet up again at Ternate in September, where they would pack up their collections and leave the region for good.

He had long wanted to return to Timor in the wet season, and although that time of year posed its own challenges, at least there should be far more bird and insect activity compared with the desert conditions of the dry season. He was right, procuring quite a few nice specimens—notably, butterflies and birds. He arrived at Dili on 12 January 1861—at the age of thirty-eight, having celebrated his birthday a few days before—and was immediately assisted by two friendly expats, Capt. Alfred Hart, a trader, and a mining engineer named Frederick Geach, who invited Wallace and his two assistants—Ali and another hired hand—to stay with him at his home in a fertile valley a couple of miles from town. Dili itself, on the north coast of East Timor, occurs on a flat plain bisected by the seasonal Rio Comoro, surrounded by rugged mountains. Part of the outer Banda Arc, the island sits on the very boundary between the Indo-Australian and Eurasian plates, resulting in another kind of island amalgam, with crustal plate fragments from each. Wallace found the fauna decidedly Australian but astutely speculated based on a few notable *absences* that the island was populated mainly by chance colonization from the surrounding region. In that he was essentially correct, with the preponderance of incoming species unsurprisingly Australian in ancestry, given the strong seasonal eastern monsoon that blows from that continent.

East Timor came under Portuguese rule in the sixteenth century, continuing until the mid-1970s, a fraught time when an independence movement was crushed by opportunistic Indonesian forces that already controlled the former Dutch territory of West Timor. (East Timor is independent today, however, and officially known as the Democratic Republic of Timor-Leste, or just Timor-Leste—a curious redundant name insofar as "Timor" is derived from the Malay *timur*, "east," and "leste" means "east" in Portuguese.) The region has a long history of unrest, as Wallace witnessed firsthand in the Indigenous insurrection of 1861 when Timorese warriors from two kingdoms, or *reinos*, marched on Dili, cutting off food and other supplies—and cutting short Wallace's planned collecting excursions to the interior. Without connecting the two, Wallace put his finger on the source of the conflict: gross mismanagement, incompetence, and exploitation on the part of the Portuguese colonial authorities.[54] He was able to make just a few excursions during his three-and-a-half-month stay, including one on horseback to Balibar, a mountain hamlet situated at nearly 2,000 feet (600 m) just south of Dili. Taking stock just before his departure, he wrote a paper later published in the *Ibis* noting that he had managed to collect about one hundred bird species, a remarkable two-thirds to three-fourths of which were unique to Timor, although closely related to those of the surrounding islands (especially Australia).[55]

The Dutch mail steamer *Macassar* departed Dili on 25 April and, after brief stops at Banda Neira and Ambon, called at the town of Kayeli (Wallace's

Cajeli) on 4 May. Situated in a bay on the eastern side of Buru, at the mouth of the wide plain of the meandering Waeapo River, Kayeli was the site of a seventeenth-century Portuguese fort that still had a commandant in Wallace's day. After the usual formalities, stopping to see the town *opziener* (overseer), and the rajah, he and Ali went separate ways to explore a bit and see if they could find good collecting grounds. Wallace went upriver about five miles in the company of the rajah to a tiny Alfuro village but encountered a veritable specimen desert: a plague of tall, coarse *kusu-kusu* grass dotted with broad white-trunked cajuput trees (*Melaleuca cajuputi*), the source of pungent cajuput oil, a medicinal still used today. Ali did not fare much better, but then they were directed to a spot to the east with nice forest and a rushing stream, Waypoti. They arrived on 19 May and promptly set up in a little shed for a house, with a large bamboo platform at one side. Wallace got busy, setting up his mosquito netting; building a rough-and-ready work table where he set out his books and labels, penknives and scissors, and pliers and pins; stringing up a clothes line; and devising hanging shelves in an effort to evade the ubiquitous ants. He marveled at how the barest and most unpromising little structures can soon be made comfortable and functional. They made the most of it, launching collecting forays daily with no mishaps for a change—save his forgetting his good boots on the mail steamer and so once again eventually reduced to walking barefoot or in stocking feet through the forest. As planned, Charley joined them briefly from New Guinea, stopping for about ten days to deliver his impressive haul on his way to the Sula Islands (Sanana and Magole islands).[56]

As for Ali, he was fortunate to come upon an enormous python before it came upon him. He was stepping up and over what he thought was a fallen tree across the trail, but then it moved. He jumped aside in fright and watched in awe as the huge animal slid on into the forest, looking for all the world like a tree being dragged through the grass. Thankfully, the collector was not collected that day . . .

By the end of their stay, they could call the journey a success. The small band had collected over seventeen hundred insects, richest in beetles and butterflies, but their most spectacular finds were among the birds: over sixty-six species, of which seventeen were altogether new to science, among them two spectacular kingfishers—a local variant of the racquet-tailed kingfisher (*Tanysiptera galatea acis*) and Buru dwarf kingfisher (*Ceyx cajeli*)—as well as a black sunbird (*Leptocoma aspasia proserpina*) sporting a cap of metallic green-gold with a deep metallic purple throat gorget, and the black-tipped monarch flycatcher

Wallace, age thirty-nine, shortly before departing for England.

Symposiachrus loricatus in elegant black and white.[57] Then there was a new species of *Pitta*, hard-won by Ali, now known as the South Moluccan pitta, *Erythropitta rubrinucha*.[58] The very day after Ali bagged his prize bird, they packed up and headed back to Kayeli, then booked passage on the mail steamer *Ambon* bound for Java, with brief scheduled stops in Ternate, Manado, and Makassar. On 3 July 1861 they departed Buru, landing at Ternate on 6 July. Wallace had two days to pack up his remaining belongings and say his goodbyes: "During our two days' stay at Ternate I took on board what baggage I had left there and bade adieu to all my friends. We then crossed over to Menado, on our way to Macassar and Java, and I finally quitted the Moluccas, among whose luxuriant and beautiful islands I had wandered for more than three years."[59]

It was time. Home and family beckoned, certainly, but so, too, did scientific society. His heart was surely full as he and his faithful companion Ali made their way west. Enduring years of endless frustrations, privations, afflictions, and near-fatal disasters on land and sea, he had persevered. His collections were rich beyond measure, his scientific contributions richer still. He was a seer of time and space, perhaps the only person on the planet besides one Charles Darwin who understood, who was gifted with the vision of Earth and life in a dynamic embrace, dancing through the grand sweep of deep time "from so simple a beginning"— cycling climate, an infinitude of struggling, striving, multiplying individuals, an endless ebb and flow of species, the rise and fall of continents, an eternally branching and rebranching tree.[60] "Urge and urge and urge, Always the procreant urge of the world," wrote Walt Whitman half a world away just a few years earlier. Multitudinous and singular, typical and novel, population and individual, species and varieties. What did it all mean? "Immense have been the preparations for me," declared Whitman. "Before I was born out of my mother generations guided me, My embryo has never been torpid, nothing could overlay it. . . .

For it the nebula cohered to an orb,
The long slow strata piled to rest it on,
Vast vegetables gave it sustenance,
Monstrous sauroids transported it in their mouths
and deposited it with care.
All forces have been steadily employ'd to complete and delight me.[61]

Now Alfred Russel Wallace stood on this spot with his robust soul, going home.

CHAPTER 11

FIRST DARWINIAN

AS THE *AMBON* steamed across the Java Sea from Makassar that July of 1861, he knew he was in that most remarkable of faunistic borderlands, passing from the Indo-Australian to the Asian realm. Did he nerdily track their progress, pestering the ship's navigator for updates on their longitude? Right around 115 degrees he may have exclaimed to Ali, "Now we've crossed it!!" "No, wait . . . *now*!" Perhaps he fancied he could almost *feel* that mysterious line as they passed near the Kangean Islands north of Bali and Lombok, a fleeting sign like that tsunami swell, suddenly upon them and as quickly gone. Crossing, he was back in what is now called the Indo-Malayan Realm.

"Going home" from so remote an outpost is a process, especially if you are Alfred Russel Wallace. There were localities in the west of the archipelago he had yet to see, after all, and many a loose end to tie, so his final six months in Southeast Asia—Java, Bangka, Sumatra—were spent exploring, sightseeing, and collecting as he made his way back to Singapore, the gateway to home. His and Ali's first destination was Surabaya, on the northeastern coast of Java, the old seat of the Hindu kingdom of Janggala, dating to 1045, which grew into the powerful Majapahit Empire. After coming under the control of a succession of sultanates beginning in the mid-sixteenth century, power shifted again, to the expanding Dutch East India Company beginning in the 1740s. The town was doubly strategic, on sheltered Lamong Bay dominating the Madura Strait, at the mouth of the *Kali* (River) Mas, "Golden River," for the riches that flowed through. The site thus became the home of the main Dutch naval base in the East Indies (Maritiem Etablissement), with an enormous floating dry dock. Wallace found a prosperous, attractive, highly Dutch-influenced town, Java itself "the garden of the East, and probably without exception the finest in the world," he wrote his mother from his hotel room.[1] Mary Ann Wallace was surely delighted to learn that her youngest surviving son was returning home at last. She had since moved into her own cottage and hopefully was equally delighted to learn that he planned on living with her on his return, if she felt herself "equal to housekeeping for us both," he ventured (perhaps with equal parts sexism and anticipation of parental doting?). All he needed was a room or two to organize and study his collection, he said, and a workshop for "rough & dirty work" such as preparing skins and building cases. He had asked his brother-in-law to measure the rooms of the cottage so he could see if there would be space enough, and if not he hoped to find a place in the suburbs, "in a quiet neighbourhood & with a garden but near an omnibus route" so he could easily get into town.[2]

He planned to arrive home in the spring. Traveling lighter now, he was no longer lugging his miscellaneous furniture and cooking supplies, that mobile household of "bed, blankets, pots, kettles & frying pan," "plates, dishes & wash basin," "coffee pots & coffee, tea, sugar," "butter, salt, pickles, rice, bread & wine," "pepper & curry powder, & half a hundred more odds & ends," everywhere he went. Just think, that was on top of his collecting gear, books, and clothing and his prodigious collections themselves (which, when things were going well, tended to accumulate as he went). Reckoning that he had made about eighty trips, averaging one per month over the past seven years, it is no wonder that packing and repacking all that stuff was "the constant standing plague" of his life.[3]

After shipping off the latest sizable consignment to Stevens—further relieving his burden—he took time to be a tourist, visiting the celebrated Prambanan Temple complex in central Java, the largest Hindu temple in the archipelago, and the nearby beautiful Borobudur Temple, the world's largest Buddhist temple—both UNESCO World Heritage Sites today. En route he stopped at the town of Mojokerto and nearby Trowulan and Mojoagung, where he admired the remains of one of the great brick gates of the ancient palace of Majapahit.[4] His builder's eye took in the craftmanship—"The extreme perfection and beauty of the brick-work astonished me." There he was gifted a magnificent Hindu bas-relief carved of basalt depicting the famous scene of the goddess Durga slaying the buffalo demon Mahishasura, the culminating event of the ancient *Mahishasuramardini,* and he attended with fascination a Gamelan performance, finding the harmonious clockwork playing of the drums, gongs, zithers, and xylophones of the Gamelan orchestra similar to a gigantic music box: "Very pleasing."[5]

Collecting on his mind, Wallace stayed for a short time at the village of Wonosalem, on the western slope of imposing Arjuno-Welirang volcano, rising to 10,955 feet (3,339 m). This island "garden of the East" was a garden of volcanoes—another borderland of space and time. At Arjuno-Welirang he was now atop the volcanic Sunda Arc that forms the backbone of the island, the product of the Indo-Australian plate diving beneath the Sunda plate. Insect collecting was pretty poor, but he added ninety-eight species of birds to his collection, including peacocks and two species of jungle fowl (*Gallus*), kingfishers, woodpeckers, and the largest and smallest of Javanese birds: massive hornbills (*Buceros*) with a wingspan exceeding five feet, and the tiny yellow-throated hanging parrot (*Loriculus pusillus*), with wings spanning inches. If news of an unfortunate boy tragically falling prey to a tiger nearby gave him pause, he did not say. He didn't say how Ali responded either, but after the brush with the giant python on Buru, the knowledge of human-eating tigers in the vicinity surely put a damper on his young assistant's collecting enthusiasm.

Ready to move on, they returned to Surabaya via riverboat, probably on the Brantas, a tributary of the Mas, and were soon aboard a steamer bound for the western end of Java. They landed at bustling Batavia, then the capital of the Dutch East Indies, now Jakarta, the capital of modern Indonesia and the largest city not only in the country but in all of Southeast Asia, at nearly thirty-four million people. He found the comforts of civilization quite nice, if pricey—the swanky Hôtel des Indes, horses and carriages, and a generous table d'hôte that was a far cry from the diet he had been living on: mostly sago and bird du jour, taking care to pick out the lead shot. Speaking of which, collecting and landscape beckoned, and they soon headed for the hills. First stopping at Bogor, which Wallace knew by its Dutch name of Buitenzorg, he continued on to Megamendung and then up to 4,500-foot (1,400 m) Puncak Pass, where he stayed two weeks on the flanks of the lofty Pangrango and Gunung Gede volcanoes. He was equally delighted with the climate and the collecting at higher elevations, finding fascinating birds that seemed to be endemic to western Java, and among the insect rarities a specimen of the curious caliper butterfly, *Charaxes* (*Polyura*) *dehanii*—so named for the curved pincerlike double tails they sport on each hind wing—deftly caught by a local boy with careful caliper-like fingers while the heedless butterfly was preoccupied. Higher they went, ascending to the volcanic summits of extinct Pangrango and quite active Gede, the highest he had yet climbed in the tropics at just about 10,000 feet (3,048 m), so for the first time he experienced the Humboldtian shift from tropical to temperate flora with elevation.

Beginning at about 8,000 feet (2,400 m) he recognized a decidedly European-looking flora, marked by such familiar groups as honeysuckle,

Saint-John's-wort, viburnum, rhododendron, artemisia, buttercups, violets, and more. A beautiful high-elevation primrose caught his eye—"the rare and beautiful royal cowslip" (*Primula imperialis*, now *P. prolifera*) with a large rosette of coarse basal leaves from which a stout stalk sprouted bearing stacked whorls of golden-yellow flowers. At the very time Wallace was admiring that Javan primrose, incidentally, on the other side of the world his colleague Darwin was putting the finishing touches on a soon-to-be-read paper for the Linnean Society describing a most curious phenomenon of primrose flowers termed *heterostyly* today: individual plants have different flower morphs, some bearing long stamens and short pistils and others, vice versa.[6] The discovery was to rank high on the list of the oddities of nature that only seemed to make sense on his and Wallace's theory of evolution by natural selection.

Here, amid the alpine vegetation of the high, wind-swept volcanic peaks of Java, this primrose and other botanical beauties inspired Wallace to contemplate distribution and dispersal—in particular, Darwin's explanation given in the *Origin*, of north-temperate flora migrating to equatorial regions during glacial periods when the earth's climate was far cooler and migrating north again when climate warmed but also migrating up, seeking refuge in mountain fastnesses where they became marooned on isolated climatic islands. In the fullness of time they diverged and have "become so modified that we now consider them to be distinct species."[7] But then he paired his own historical insights with Darwin's: some might object that Darwin's model fails because Java is an island, with a wide expanse of sea between it and the continent. "This would undoubtedly be a fatal objection," Wallace notes, "were there not abundant evidence to show that Java has been formerly connected with Asia, and that the union must have occurred at about the epoch required"—bolstered by evidence from zoology, as the "great Mammalia of Java," the rhinoceros, tiger, wild ox, and others are found both there and on continental Asia.[8] But not precisely the same species, a key point. Why not? Here is one aspect of species divergence that, in the modern view, neither Darwin nor Wallace could have understood. They attributed such evolutionary change to the "changed conditions" under which the isolated groups now found themselves in, largely because both (but especially Darwin) subscribed to the view that heritable variation is environmentally induced. But in the modern view, mutation arises not so much by environmental induction, generally speaking. Sure, there are environmental mutagens like ultraviolet light and radioactive elements, but these are not driving most mutation. Rather, mutation largely arises by such processes as DNA replication error and unequal crossing over, inversions, deletions, and so on at the chromosomal level. These heritable (genetic) changes accrue regardless of environment, so over time isolated populations become more and more dissimilar even without natural selection pushing them (though that can make it happen faster).

His forays to the Javan alpine zone ended up yielding little by way of biological riches—not least due to the commencement of the wet season—but the biogeographer in him was delighted. From his base camp in Puncak Pass, he related the exhilarating experience to Fanny, his dispatch coming, he wrote evocatively, from "the Mountains of Java."[9]

He departed Java on 1 November 1861, leaving behind a pile of boxes to be shipped to Stevens, listing them on the final page of his specimen notebook:[10]

1. A large case of Buru birds, with a few from Timor and Halmahera thrown in
2. A gin case full of Java birds, marked for his private collection
3. "Box with Java sculpture"
4. A box with peacock and hornbill specimens
5. & 6. Sula birds, collected by Charles Allen
7. A packet of palm leaves from Timor and Manado

A week later they landed at Palembang, south-central Sumatra, a beautiful inland port town on the banks of the Musi River. Immense Sumatra, the world's sixth-largest island, is born of the same tectonic upheaval as its smaller neighbor Java, as evidenced by the many lofty volcanoes along its west coast, aligning with those of Java. But there is a difference: here the Indo-Australian plate subducts obliquely beneath the Sunda plate, not perpendicularly, introducing a shearing motion along the volcanic arc that created the Great Sumatran or Semangko Fault System. The complex plate dynamics make for one of the most seismically active regions on Earth—indeed, Wallace missed by nine months a massive earthquake that shook the island and caused much destruction.[11]

He did not make it to the volcanic district, traveling southwest from Palembang through a low, undulating, geologically young landscape of red clay and stopping about halfway to the west coast at the village of Lubuk Rahman. There he spent a month among the Rembang peoples, collecting amid a forest-and-clearing patchwork dissected by streams. He admired their handsome *rumah adat*, traditional houses built on poles, famous for their high-pitched thatch roofs with dramatically upswept gables,[12] and gloried in being situated "in one of the places unknown to the Royal Geographical Society," as he put it to George Silk—the land of the rhinoceros, the elephant, the tiger, and the

View of eighteenth-century Palembang, Sumatra.

tapir (though all sadly scarce even then), as well as the land of anything-but-scarce monkeys.[13] Although the collecting was not terribly productive, that gave him time to read, ponder, and reflect. He wrote Darwin a chatty letter on how *Origin* was being received and urged him to include ample illustrations in his planned follow-up volume.[14] He also wrote Bates, who had long since sent him a fascinating paper on the butterflies of the Amazon valley. He heaped kudos and accolades on the paper—it was right up his alley, a masterful analysis of the relationships and geographical distribution of butterfly groups throughout the basin, letting them serve as a guide to geological, geographical, and evolutionary history.[15] Wallace pointed out the one crucial point that Bates's data underscored but that was barely touched upon—namely, that the great rivers define the limits of a great many species and varieties—resonant with his own "Monkeys of the Amazon" paper. He trusted, he said, that the paper was but the first of a long series that would establish Bates's fame "and at the same time demonstrate the simplicity & beauty of the Darwinian philosophy"—once again referring to evolution by natural selection as Darwin's idea. There was an indirect Bates connection with another evolutionary insight he had at Lubuk Rahman: it was there that Wallace discovered that the females of the great Mormon swallowtail butterfly *Papilio memnon* occur in two morphs, one of which mimics *Papilio* (*Losaria*) *coon*, the common clubtail.[16]

This discovery was made, in another curious correspondence, at nearly the same time that Bates, back in London, delivered his paper on *Heliconia* butterflies that elucidated the principle of protective resemblance now bearing the name Batesian mimicry.

Among other curious creatures, Wallace glimpsed a forest rhinoceros, acquired a pet siamang (*Symphalangus syndactylus*), the largest of the gibbons, and collected that remarkable forest glider the Sunda "flying lemur," or colugo (*Galeopterus variegatus*)—one of but two species in the genus, distinctive primate relatives placed in their own taxonomic order, Dermoptera.[17] Returning by boat to Palembang, the day before his birthday they made a stop for repairs at Sungairotan, perhaps at the confluence of the Rotan and Musi Rivers, where local hunters acquired a mated pair of great hornbills (*Buceros bicornis*) for him, with a plump pigeon-sized nestling "like a bag of jelly, with head and feet stuck on." The females of these curious birds nest in tree cavities, where they are shut in with their helpless young by a mud wall built by the male, who feeds her through a hole just large enough for her beak to poke through. Marveling, Wallace found this behavior "one of those strange facts in natural history which are 'stranger than fiction.'"[18] It was a fine present for his thirty-ninth birthday.

A week later he departed Sumatra, catching a ferry to outlying Bangka Island and from there a steamer—the *Macassar*, his second ride on that trusty ship—heading to Singapore. Exactly one month after his birthday, he left the East for good aboard the P&O steamer *Emeu*. It was surely bittersweet, bidding farewell to the steadfast and talented Ali, "the faithful companion of almost all my journeyings among the islands of the far East." He gave Ali a parting gift of money, his guns, ammunition, sundry tools, and other items and had a photographic portrait made to remember him by. Ali would return to Ternate, where his wife awaited him. Wallace would never see his friend and helper again, but many years later he received unexpected news of his old companion. In 1907 American herpetologist Thomas Barbour, of Harvard's storied Museum of Comparative Zoology, was on an extended honeymoon with his wife, Rosamond Pierce, traveling the Malay Archipelago in Wallace's footsteps. They were on Ternate, just heading out on a collecting jaunt, when an old man approached them. Barbour later related what happened:

> Here came a real thrill, for I was stopped in the street one day as my wife and I were preparing to climb up to the Crater Lake. With us were Ah Woo with his butterfly net, Indit and Bandoung, our well-trained Javanese

Wallace's right-hand man,
Ali, 1862.

collectors, with shotguns, cloth bags, and a vasculum for carrying the birds.
We were stopped by a wizened old Malay man. I can see him now, with a
faded blue fez on his head. He said, "I am Ali Wallace." I knew at once that
there stood before me Wallace's faithful companion of many years, the boy
who not only helped him collect but nursed him when he was sick. We took
his photograph and sent it to Wallace when we got home. He wrote me a
delightful letter acknowledging it and reminiscing over the time when Ali
had saved his life, nursing him through a terrific attack of malaria. This letter
I have managed to lose, to my eternal chagrin.[19]

It says something about the relationship of the two, one born of mutual
affection, respect, and trust that grows from shared triumphs, trials, and tribulations, that Wallace's steadfast assistant of old would take his name.

While still in Sumatra, Wallace got wind of the possibility of acquiring live
paradise birds in Singapore, and he jumped at the chance. The Zoological Society back in London agreed to pay the asking price of $400 for the pair (about
£100 at the time) and remunerate Wallace himself to the tune of £150 plus free

first-class passage home to care for the birds en route. They were lesser birds of paradise, *Paradisaea papuana* (now *P. minor*), young males just beginning to put on their courtship plumage. What an opportunity if he could manage to get these fabulous birds to London alive! It was no mean feat. He might as well have named them "Trouble" and "Anxiety," two words he used often in describing their journey. But he fortuitously discovered that they would eat cockroaches with relish, which was great, since the bugs were essentially protein, fat, and carb packets on six legs. He was able to catch them onboard and at ports of call along the way; the insect-and-fruit diet kept the birds well fed and healthy. The *Emeu* called at Penang, then Pointe de Galle, Sri Lanka, then north to Mumbai, where he picked up the P&O steamer *Malta* to Suez.

The next leg was overland by train to Alexandria, where two challenges arose: colder weather had set in, and the obstinate train conductors insisted that the birds travel in the baggage car. It was a terrifying prospect, leaving his tropical birds unattended in the potentially fatal cold of a drafty car. He stayed with them in the baggage car all night, and they did fine, to his relief. Now on to Malta via another P&O steamer, the *Ellora*, where he stopped for a week and sent a telegram to Philip Sclater at the Zoological Society via the British and Irish Magnetic Telegraph Company: "The two [paradise birds] have arrived here in perfect health. I wait your instructions." None came, perhaps because what the telegram actually announced was the arrival of two "garadisi bards." Electrical telegraphy, that early text-messaging system, may have been faster than the post but was still rather error-prone.[20] Wallace was soon on a steamer bound for Marseilles, France, where he was able to trade the ship for a local bakery as his cockroach provider. So far, so good. But another nerve-wracking train trip soon followed, three days this time via Paris to the port of Boulogne-sur-Mer on the English Channel. So close and yet so far. . . . On the last day of the month, 31 March 1862, he arrived on English soil, at the Kentish port town of Folkestone, and immediately wrote Sclater: "I have great pleasure in announcing to you the prosperous termination of my journey & the safe arrival in England (I suppose for the first time) of the *Birds of Paradise*."[21] He might have included a few exclamation marks, as much for his achievement in making it home as for this major coup: it was indeed the first time *live* birds of paradise had arrived in England! He would be on the nine o'clock train for London tomorrow, he informed Sclater, arriving at London Bridge at noon to deliver the birds.

It was a triumphant moment made all the sweeter for his memory of bitter loss when, ten years before, his hopes of returning to London with brilliant tropical birds on his arm had gone up in smoke in the mid-Atlantic, along with just about everything else. Now here he was—the First Darwinian was home, his eight-year sojourn crisscrossing the vast Malay Archipelago a triumph in every respect: astonishing collections, paper after remarkable paper, landmark discoveries that included solving one of the greatest scientific questions of the day, and now, as icing on the cake, the rarest of rare birds brought *live* to London. No mere feather in his cap, it was a veritable bird of paradise plume.

———

His heart was surely full as he was reunited with his family and friends and strode into London's scientific salons. More prodigious than prodigal, the son had departed at age thirty-one and was back at thirty-nine, most of his third decade spent abroad achieving great intellectual heights amid dangerous and trying conditions. If Amazonia was Wallace's undergraduate education as a field biologist, the Malay Archipelago was graduate school—an eight-year study of peoples and landscapes; of species richness, variation, distribution; of Earth and life in space and time. Darwin, anxious to meet this wunderkind, immediately invited Wallace to Down House. But Wallace had to politely decline for now, sending Darwin honeycomb from Timor as a gift.[22] He was exhausted and needed to recuperate, take stock of his massive collections, and think about his next moves: where to live and what projects to tackle first. He moved in with his sister and brother-in-law at 5 Westbourne Grove Terrace in central London, a short walk north of Kensington Gardens, where Thomas Sims had moved his struggling photographic business. His mom's cottage lacked enough room, it turned out, and here he had a spacious workroom on the third floor where he could organize and analyze his rich collections. He estimated that they included about three thousand bird skins of one thousand species and perhaps a whopping twenty thousand beetles and butterflies of about seven thousand species—material enough for a lifetime of study![23]

The next few years were truly watershed, both on the science and scientific society front and in his personal life. Wallace returned to an energy-filled, dynamic city, one reflection of which was the Great London Exposition in Kensington, which opened that May to great fanfare—an impressive display of the latest and greatest in Victorian technology, science, music, and art. The intellectual climate was equally dynamic, Darwin's bombshell of a book still very much reverberating. The day after his arrival in London, Wallace learned that he had been elected fellow of the Zoological Society of London, an indication of his growing reputation. He had already been corresponding with Sclater, then secretary of the Zoological Society, and Edward Newman, editor of the society's journal the *Zoologist*, who knew Wallace from his many papers and letter extracts dispatched from the field, both east and west. (He sure had come a long way since his first submission to the *Zoologist* fifteen years before, a list of beetles bagged in Neath, which, as you will recall from chapter 3, Newman dismissed as "scarcely worth publishing," save one.)[24] Now he was introduced and welcomed into the company of naturalists whose names he knew only by reputation and publications, the movers and shakers of British scientific society: the Duke of Argyll; ornithologist John Gould; John Edward Gray, keeper of zoology at the British Museum; up-and-coming Thomas Henry Huxley, already president of the Zoological Society within two years of being elected fellow; Huxley's protege St. George Mivart; and many others. Alfred Newton, co-founder of the British Ornithologists' Union and a fellow of Magdalene College, Cambridge, invited Wallace to the meeting of the British Association for the Advancement of Science (BAAS) held in Cambridge that spring.[25] There he was introduced to the members of the Ornithologists' Union and others, including historian and clergyman Rev. Charles Kingsley.

His first BAAS meeting was memorable for a skirmish he witnessed in the ongoing evolution wars. It involved Huxley, of course, whose reputation as "Darwin's Bulldog" was well established by then—in fact, it was a skirmish at another BAAS meeting, two years earlier in Oxford, that first gave Huxley that pugnacious reputation. Wallace was still in the East then, sailing from Seram to New Guinea, when Huxley squared off with Samuel Wilberforce, the Bishop of Oxford, at the BAAS meeting held in the newly opened Museum of Natural History. Henry Acland and John Ruskin's neo-Gothic design of that cathedralesque museum may have been their way of exalting the study of nature in the natural theology tradition, but the Huxley-Wilberforce debate, as it came to be known, exposed fault lines between science and religion as deep as any in the earthquake-prone Malay Archipelago. The "debate" certainly was a seismic event. When the bishop (coached, it is widely believed, by anti-evolutionist Richard Owen) made a show of derisively asking Huxley if he claimed descent from an ape on his grandmother's or grandfather's side, Huxley threw this back on the bishop: asked to choose between an ape or the likes of Wilberforce, someone who should know better than to inject nothing but ridicule into an important scientific discussion, "I unhesitatingly affirm my preference for the ape!" Or so it is said. The exchange was not recorded, but by all accounts sparks were flying. The jousting with Huxley continued, with

Owen claiming that the brains of humans and apes have fundamental anatomical differences, apes missing in particular a hippocampus minor (calcar avis), warranting the elevation of humans to their own subclass of Mammalia. The implication was that humans were special, divinely created. This came to a head, so to speak, at the BAAS meeting in Cambridge that Wallace attended. When the gauntlet was thrown down by Owen in a paper delivered at the meeting, Huxley was ready—with an assistant he gave a triumphant public dissection of an ape brain, proving Owen wrong with a flourish of his scalpel. At a time when public interest in scientific questions ran high, the "Great Hippocampus Question" and fierce rivalry between Huxley and Owen were widely known. Paleontologist Sir Philip Egerton lampooned the episode in a playful poem titled "Monkeyana" for the satirical magazine *Punch*:

> Says Owen, you can see
> The brain of Chimpanzee
> Is always exceedingly small,
> With the hindermost "horn"
> Of extremity shorn,
> And no "Hippocampus" at all.
> . . .
> Next Huxley replies,
> That Owen he lies,
> And garbles his Latin quotation;
> That his facts are not new,
> His mistakes not a few,
> Detrimental to his reputation.
>
> "To twice slay the slain,"
> By dint of the Brain,
> (Thus Huxley concludes his review)
> Is but labour in vain,
> Unproductive of gain.
> And so I shall bid you "Adieu!"[26]

Kingsley, too, spoofed the event as the "Great Hippopotamus Test" in his famed children's book *The Water Babies*, published the following year, complete with illustrations of Owen and Huxley: If apes have a "hippopotamus major" in their brains just as humans do, why, "What would become of the faith, hope, and charity of immortal millions?" the shocked narrator asks. No, you can depend upon this: "If you have a hippopotamus major in your brain, you are no ape, though you had four hands, no feet, and were more apish than the apes of all aperies."[27]

The Huxley-Owen jousting did not surprise Wallace. While still in the East he had followed reports of *Origin*'s reception in copies of the *Athenaeum* sent to him by Stevens, and it was in the pages of that literary magazine that the two debated the brain anatomy of humans and apes and its implications.[28] Huxley and Owen, he wrote to Darwin, "seem to be at open war," and he appended a P.S. commenting on the shabby mode of argument of his critics, singling out Owen for disingenuously splitting definitional hairs: "First comes Owen with his new interpretation of what naturalists mean by *"creation"* which it turns out is not creation at all, but *'the unknown manner in which species have come into existence'*!!!"[29]

Wallace was not drawn in—not yet. That would come soon enough, and he would prove to be about as devastating as Huxley in his defense of evolution by natural selection. But for now he had work to do, and with his characteristic mix of good cheer and determination, he threw himself into the task of sorting, studying, writing. His first paper, a "Narrative of Search after Birds of Paradise," was delivered at the 27 May 1862 meeting of the Zoological Society, relating his and Charles Allen's adventures (and misadventures) collecting these storied birds—or trying to. This was followed the next month with an exhibition of "Some New and Rare Birds from New Guinea." That was a bit of a puff piece, but over the next three years he had over thirty published papers, notes, and comments on the taxonomy or natural history of Malay Archipelago birds alone. There were entomological papers too—not nearly as many since he farmed out the study of several insect groups to specialists like Francis Polkinghorne Pascoe, but those he wrote would prove incisive.

While Wallace's many taxonomic papers in this period were rewarding and certainly helped bolster his already impressive "trail cred," so to speak—his bonafides as a field naturalist of the first rank—it was his other, more synthetic studies that rocketed his status as a first-rank *philosophical* naturalist. Unsurprisingly, the first of these was on the subject he had been laser focused on since that fateful trip from Bali to Lombok: mapping the distribution of species across the archipelago. Building on his landmark 1859 paper on zoological geography, now Wallace delivered its crucial counterpart: "On the Physical Geography of the Malay Archipelago," read at the Royal Geographical Society meeting of 8 June 1863.[30] The president, Sir Roderick Murchison, was deeply impressed: "He had never heard," he declared afterward, "a paper read of a more luminous character"—one that "so bound together in the most perfect forms all the branches of the science of natural history," fruitfully uniting geography and geology.[31] Wallace acknowledged the insights of British traveler George Windsor Earl, who seventeen years earlier had commented on the Asian and Australian affinities of the western and eastern ends of the archipelago in relation to depth of sea, but he also pointed out that Earl had not fully appreciated the implications of this fact and had indeed muddied the brilliant Indonesian waters by suggesting a former Asia-Australia connection.[32] No, no, no, said Wallace.

Now, fresh from his fourteen thousand-mile-plus journey and armed with an unprecedented distributional database, Wallace's lengthy paper quantified species occurrences, comparing in minute detail similarities and differences in the biota of islands and sets of islands across the archipelago. Significantly, he produced a detailed map with a red line carefully drawn between the islands, from Bali and Lombok to the southwest up to the west of Sulawesi (Celebes) and arcing to the east of the Philippines—the first map of its kind, showing a line of demarcation between what he called the "Indo-Malayan" and "Australo-Malayan" regions. The importance of this act—the simple drawing of a line on a map—cannot be overstated: he at once reinforced the still-new concept of biogeographical regions and provided a powerful visual means of communicating a co-evolutionary conceptual framework for Earth and life. More typical was Sclater's dry schematic approach seen in his important 1858 paper laying out the geographical faunal regions of the world: six square boxes, one for each *regio* with their bird diversity stats, including "Regio Indica" and "Regio Australiana" side by side at the lower right representing Southeast Asia.[33] This was fine but rather academic, Latin and all. There is nothing like the visual impact of a *map* with a *line* to drive the point home. As historian Jane Camerini aptly put it, Wallace's map-as-conceptual-framework served as a means of argumentation and, ultimately, of persuasion, championing Sclater's six regions while driving home the point that, born of Earth history, they are general, not limited to birds.[34]

But just how hard a boundary did Wallace's red line represent? And just how general *are* those regions? Delineating Bali and Lombok was straightforward, but Wallace actually vacillated over the precise placement of the line further north, as did others after him: he could not quite decide if it snaked east or west of puzzling Sulawesi and the Philippines, for example, with their curious mix of Asian- and Australian-affinity animals. Wallace shifted it a couple of times, and Huxley shifted it yet again when, five years later, he coined the term "Wallace's Line."[35] Uncertainties and subsequent analyses based on this or that taxonomic group understandably led later authors to either question whether there was any line at all or draw other lines to refine or complement Wallace's. But ambiguities are surely expected to abound given the ebb and flow (and evolution) of species playing out over eons, crustal plates and sea levels ever in flux in that dynamic tectonic stage now called Wallacea.[36] The wonder of it is that the signal is, in fact, so strong in this sea of islands (especially for mammals and birds), not lost for the noise.

As for the bigger-picture question of Sclater's biogeographical regions, Wallace followed up with another remarkable paper in which he reviewed the state of knowledge on species distribution in several groups to see how far they agreed with the Sclaterian bird-based divisions: mammals (excellent match), reptiles and amphibians (spot on), terrestrial snails (very close), insects (not so great), and plants (pretty badly). There was plenty here to suggest there really was something to Sclater's regions but anomalies enough to suggest additional factors shaping distribution, too, like dispersal ability and prevailing winds. Wallace once again exhorted naturalists to "give more special attention to geographical distribution than has hitherto been done" to provide much needed data to help determine if this sixfold division is the best one to use, or some other. "Some such simple classification of regions is wanted to enable us readily to exhibit broad results, and to show at a glance the external relations of local faunas and floras," he concluded. Here Wallace was laying the groundwork, so to speak, for a rich scientific vein that he was to mine for decades— and indeed is still mined today.[37]

As impressive as these papers were, he followed them with a stunner that took geographical distribution to a whole new level—an evolutionary level.[38] Wallace greatly admired Bates's 1862 paper on Amazon valley "Heliconidae" (members of the modern butterfly subfamilies Danainae, Ithomiinae, and Heliconiinae), in which his friend described remarkable and repeated instances of mimicry between unrelated species coexisting in the same area— correctly intuiting that the basis for this was predation, with palatable *mimics* (termed *Batesian mimics* today) evolving to protectively resemble distasteful *models*. As we have seen, Wallace already knew that certain swallowtails of the Malay Archipelago presented cases of sex-limited mimicry, but now, able to survey virtually all of the species from across the archipelago in his rich collections, he realized not only that this was extremely common but also that these swallowtails provide, like Bates's butterflies, another object lesson in the nature of species and varieties and the action of selection. Here, then, was the perfect case study showing how a diverse group seen through a wide-angle geographical lens could bring profound new insights into focus. In this epic paper, delivered at the Linnean Society in March 1864—with seventy-one pages and eight gorgeous color plates and descriptions of twenty new species!—Wallace built his case methodically. He first classified different forms of geographic and nongeographic variation, from simple variability to geographic races or subspecies—with varieties as "incipient species," to use Darwin's phrase. He then discussed just what *are* species, putting his finger on the modern-sounding criterion of reproductive compatibility.[39] But while acknowledging its importance, he also pointed out the unsuitability of interfertility as the *definition* of species.

His objection was half practical and half logical: the reproductive test "cannot be applied in one case in ten thousand"—such as cases where different forms occur in different areas—"and even if it could be applied, would prove nothing, since it is founded on an assumption of the very question to be decided." He called on morphology for help: species of different areas are marked by definable and more or less constant differences (plural), and if this tends to inflate species numbers some owing to subjectivity in defining these differences, so be it: better to err that way than overlook potentially interesting geographical variation. This all set the stage for the crowning achievement of the paper: the analysis of geographical patterns of variation that are strongly indicative of natural selection at work. In the section "Variation as Specially Influenced by Locality," Wallace detailed curious cases of parallel evolution in such traits as body size, wing shape, and the development of the trademark swallowtail "tails," astutely pointing out that multiple unrelated groups exhibiting the same morphological changes in an area suggest a common selective pressure. In "Mimicry," he chronicled female mimetic forms of *Papilio* species that mimic other unpalatable *Papilio* as well as those that mimic models from different butterfly families altogether. At Huxley's suggestion Wallace wrote an abstract of the paper for the *Reader*, a weekly publication Huxley helped edit.[40] Darwin was deeply impressed: "I have hardly ever in my life been more struck by any paper," he wrote Wallace. "I feel sure that such papers will do more for the spreading of our views on the modification of species than any separate treatises on the single subject itself."[41]

But Darwin was even *more* impressed by an altogether different paper Wallace had lately come out with: yet another closely argued and highly original piece, this one on the hot-button subject of human evolution. It was "grand and most eloquently done," Darwin declared in that same letter. "The great leading idea is quite new to me." This was a subject barely broached in *Origin*, in which Darwin only hinted that "in the distant future" psychology would be based on a new foundation, and "light will be thrown on the origin of man and his history."[42] But all that anyone could think about was what it meant for humans. What were the implications for society's institutions, for scripture, resulting from an organic origin of humanity from some "lower" animal form? Darwin hinted at the implications when, in the privacy of a notebook, he commented that once we allow that species change one into another, "the whole fabric totters & falls."[43] Tied up with the question of human origins was the origin of *race*—a hotter-button topic within this hot-button topic. Recall that in that time of nascent scientific racism it was still hotly debated whether or not each race had an independent divine origin—polygenists, many of whom were slavery apologists who saw nothing wrong with "owning" individuals of different species, akin to what we do to barnyard animals without blinking an eye, versus monogenists, who argued that all races were but variants of one great human species, divinely created, and the enslavement of any people by another morally wrong. (Though some monogenists found a way to justify slavery, too, based on scripture with its Old Testament rhetoric of certain condemned and banished races fit only for subjugation.) Going back to the eighteenth century, the Darwin family had long been passionately abolitionist on moral grounds, but even apart from that, in Darwin's (and Wallace's) evolutionary vision humanity is *one* species—monogenism—albeit one that is geographically variable. The schism between these two schools of thought and belief was manifested in the London scientific scene when in 1863 a splinter polygenist group abandoned the monogenist Ethnological Society to form the Anthropological Society of London. Co-founder James Hunt, sometime speech therapist-turned-anthropologist, claimed that the new society was concerned with scientific facts, natural law, and the study of human physical as well as cultural attributes. Hunt's real stripes were most evident in his almost militant insistence on polygenism, white supremacy, and support for the American Southern states in their bid for secession to preserve the institution of slavery. The split was to last eight years until, soon after Hunt's death, Huxley was able to bring the two organizations back together as the Anthropological Institute of Great Britain and Ireland—which is not to say that the racist and sexist underpinnings of the anthropological science of the day went away.[44]

Yes, in 1859 Darwin preferred to stay in the frying pan and out of the fire, but the gorilla in the room was always the question of human origins and diversification. Others were willing to tread where Darwin dared not go: in 1863 two books appeared on the subject of human origins and antiquity, one by

Huxley and one by Lyell (although Lyell's was frankly something of a disappointment to Darwin, stopping short of explicitly embracing an evolutionary origin for humans).[45] Now Wallace stepped up too, putting his own distinctive stamp on the question and opening a new front in the debate.

Wallace delivered not one but two papers, in fact, one to the Ethnological Society and one to the Anthropological. This was likely a calculated move on Wallace's part, signaling his own thoughts on uniting the two. The Ethnological Society paper came first: "On the Varieties of Man in the Malay Archipelago" was read just a few weeks after his forty-first birthday, in late January 1864.[46] This was the first presentation of his ethnological observations in the context of geography and geology to London's scientific circles.[47] Recall from chapter 9 Wallace's epiphany on Halmahera, where he had "discovered the exact boundary-line between the Malay and Papuan races, and at a spot where no other writer had expected it."[48] This was the thrust of the "Varieties of Man" paper: the geological history of the region shaped the distribution of humans as surely as it shaped the distribution of other organisms. The two lines do not exactly correspond, he noted, with the human line lying further to the east—not surprisingly, he argued, insofar as he expected the enterprising and seafaring Malays to push east into the territory of the Papuans. He discussed the contrasts between these peoples, expressing views that, to our ears, manage to be simultaneously enlightened *and* burdened with the Eurocentric prejudices of the time. For example, he sang the praises of the Malays of northern Sulawesi, in and around Manado—"tractable, industrious, and intelligent." Not so long ago they were but "savages," but once the Dutch government took over, they were readily civilized: "One sees the result now in a beautiful and well cultivated country, in neat and regular villages, and good roads; in a population well fed and well clothed, the greater part of whom are Protestant Christians, who can most of them read and write, and who, if they please, can enjoy a great many of the comforts and luxuries of civilisation." Answering those who denigrate non-Europeans as hopelessly ineducable, worthy only of subjugation, he insisted that here "we have a proof that the absence of civilisation does not necessarily imply the want of capacity to receive it," much to the irritation of James Hunt and his followers. But of course, Wallace was unwittingly giving a backhanded compliment—as if the people of that district *needed* "civilizing" and like children could only respond to a kind of benign paternalistic despotism.[49] Such was the time and the views of even the most enlightened of Westerners.

The "Varieties of Man" paper was well received by the "Ethnologicals," but Wallace decided to deliver his follow-up paper, "The Origin of Human Races and the Antiquity of Man Deduced from the Theory of 'Natural Selection,'" to the breakaway Anthropological Society—in part, he said, because the Ethnological Society meetings were held on the same nights as the Zoological Society's, and he did not want to miss another one of those meetings.[50] His paper was scheduled for 1 March 1864. Huxley, active in the Ethnological Society, could not be persuaded to attend. In a now lost letter to Wallace, he evidently gave his reasons: he had a low opinion of the members of that society, reserving his lowest opinion for its scurrilous president, James Hunt, plus the society was an absurdity, redundant: "There was not the slightest reason for its existence." Wallace agreed about Hunt ("I do not think he is fit to be President"), but he found the papers given to be generally of high quality and pushed back on the suggestion that there was "not the slightest reason" for the society's existence. Wallace's reasons are telling, reflective of the sexism of the times. One of the reasons Hunt gave for forming the renegade Anthropological Society was that the Ethnological had recently become a "ladies society"—a derogatory label meaning that it had begun to allow women to attend meetings. This was taken as a gross affront to Victorian probity—how could gentlemen *possibly* discuss such sensitive topics as genitalia, coming-of-age rituals, or sexual practices of this or that culture in the company of *women*? Papers like that on "Phallic Worship in India," read at the 17 January 1865 meeting of the Anthropological, come to mind.[51] Unheard of, immoral, corrupting—imagine how damaging it would be if women were encouraged to think about, let alone learn about, such subjects, plus the indelicacy of even acknowledging the very existence of a penis!

No, if women were present such topics would just have to be avoided at the meetings, to the detriment of science. In the interests of science, then, and the maintenance of "high standards" of discourse, women should be banned from the meetings. Painfully, to modern readers, Wallace agreed: the establishment of the Anthropological Society, he told Huxley, was a "good protest against the absurdity" of the Ethnologicals admitting women. "Consequently many important & interesting subjects cannot possibly be discussed there—& as the [Royal Geographical Society] is also a ladies Society the Anthropol. is the only place where they can be discussed."[52] He did better by women and the cause of autonomy and equality in later years (see chapter 14), but at that time he was thinking rather narrow-mindedly about the state of the science given the social realities of the time, rather than women's lived realities; advocating for a change in cultural mores probably did not occur to him.

Huxley probably regretted not attending the reading of Wallace's paper—it was characteristically masterful, the first to explicitly discuss human origins and diversification with respect to natural selection (the theory, he said, "promulgated by Mr. Darwin"—deferring *again* to his elder colleague). The most important argument in this famous paper reveals Wallace at his most incisive, seeing a seemingly simple solution that had managed to elude the worthies of London's scientific scene: natural selection was responsible for the *physical* evolution of early humans, but along the way, as the brain was also modified and improved, it eventually reached a tipping point where selection began acting mainly on the mental and moral faculties and stopped acting on the body. In this way, Wallace suggested, our social sympathies and technology eventually *prevented* natural selection from continuing to act on early humans bodily: selection flipped from acting on form to acting on mind. Giving "ascendancy to mind" would in turn trigger a selective cascade and feedback loop, leading to the development of speech (which would further advance the mind), culture, tool use, and beyond. All this change would be going on, yet the animal form of humans would remain stationary. This is what captured Darwin's imagination: "The great leading idea is quite new to me, viz. that during late ages the mind will have been modified more than the body."[53]

His argument also threaded the needle between the monogenists and polygenists: Wallace the monogenist held that all human races had a common origin, constituting one species, but began diverging so long ago that we now had several well-defined "races"—a sort of e pluribus unum argument that gave both monogenists and polygenists food for thought. Hooker immediately wrote Darwin to proclaim that he was "amazed at its excellence—it seems to me a very great move in advance," and Lyell complimented Wallace on its "admirable clearness & fairness," being no small contribution toward resolving the issues between monogenists and polygenists and "clearing the way to a true theory."[54] Interestingly, both Hooker and Lyell commented on Wallace's generosity toward Darwin. Hooker told Darwin that Wallace must be a "very high-minded man"—Darwin agreed—and in his letter to Wallace, Lyell remarked that giving Darwin the whole credit for the theory of natural selection was "very handsome, but if anyone else had done it without allusion to your papers it would have been wrong." As much as he admired Wallace's paper, Darwin was not fully convinced right away, but Wallace's idea grew on him. A few years later, he told Wallace that this was "the best paper that ever appeared in the Anthropological Review!"[55] Indeed, by then Darwin was so impressed by Wallace's idea that he incorporated it into his own model of selection-driven mental and moral evolution in early humans, as presented in his 1871 treatment of human diversification, *The Descent of Man*. The "great leading idea" in that part of the book was Wallace's at the core: if mental and moral faculties are heritable, they must be subject to natural selection, setting the stage for a switch from evolution of form to evolution of mind and thus culture.

For his part, Wallace went on to consider the more immediate implications of his model, and it was not pretty: "It must inevitably follow that the higher— the more intellectual and moral—must displace the lower and more degraded races; and the power of 'natural selection,' still acting on his mental organisation, must ever lead to the more perfect adaptation of man's higher faculties to the conditions of surrounding nature, and to the exigencies of the social state." Here Wallace was in full colonialist mode, taken with his vision for the power of natural selection acting with a terrible inevitability. He built to a crescendo— the "great law" of natural selection would surely see the "higher races" drive the "lower and more degraded" to extinction on the way to some kind of evolutionary utopia, bringing humankind full circle back to but a "single homogeneous race, no individual of which will be inferior to the noblest specimens of existing humanity."[56] Then what? Well, given time, "Mankind will have at length discovered that it was only required of them to develop the capacities of their higher nature, in order to convert this earth, which had so long been the theatre of their unbridled passions, and the scene of unimaginable misery, into as bright a paradise as ever haunted the dreams of seer or poet." It was a vision of a society without inequality, want, government, restrictive laws, or excessive "passions and animal propensities"; a society of the "best of laws," with individual happiness pursued without transgression on others, where everyone had a "perfect sympathy" with one another. He yet retained the lessons of the Owenites. Needless to say, this curious turn toward the utopian raised some eyebrows. He dropped it from later versions of the paper, but it is certainly worthy of our notice, reminiscent as it is of Wallace's earlier idealistic writing—and perhaps foreshadowing his idealistic enthusiasms to come.[57]

———

Wallace was ascendant, his ideas luminescent. In this period he emerged as a devastatingly effective defender of his and Darwin's theory, like the time that Irish clergyman Samuel Haughton published an attack on Darwin's discussion of bee cells in *Origin*, which Haughton argued could only be divinely designed. Wallace's riposte skewered the good reverend's argument with panache, to the great admiration of Darwin, Hooker, and Huxley.[58] Besides his watershed papers were myriad others through the 1860s—species descriptions, anthropological notes, biogeographical observations, and insightful conceptual papers such as a series on the theme of protective coloration in insects and birds as further exemplars of the action of natural selection. These included papers on Australasian pigeons and parrots, relating their unusual coloration to a lack of arboreal predators, and Wallace's remarkable "Philosophy of Birds' Nests" and "Theory of Birds' Nests" papers, in which he disagreed with Darwin over questions of natural versus sexual selection in relation to the sexually dimorphic coloration and nesting behavior of birds. Darwin held that the bright coloration of males was a trait sexually selected by females according to a "taste for the beautiful," while Wallace maintained that natural selection favors drabness in females to better camouflage them on the nest. (They argued to a draw and finally agreed to disagree on the topic.) Then there was "Mimicry, and Other Protective Resemblances among Animals" and "Disguises of Insects," not to mention Wallace's celebrated discovery of warning coloration, later called *aposematism* by Poulton. That was classic Wallacean insight: Darwin, working on sexual selection, could understand bright coloration in adult birds and insects but was puzzled by brightly colored caterpillars—being immatures, sexual selection would seem to be irrelevant, and wouldn't those bright, contrasting colors be detrimental by attracting predators? He asked Bates's opinion, but Bates shrugged and suggested talking to Wallace. Wallace hit the nail on the head. "Bates was quite right, you are the man to apply to in a difficulty," Darwin exclaimed. "I never heard anything more ingenious than your suggestion."[59]

The idea is that caterpillars evolve showy, contrasting colors to advertise their distasteful or toxic properties—a warning to visual predators such as birds. Data were needed to show this, however, so taking a page from Darwin's crowdsourcing playbook, Wallace wrote an open letter for the *Field* magazine asking readers to send in observations on what caterpillars are eaten or avoided by birds. He also petitioned the members of the Entomological Society to undertake experiments or make observations for him, one of whom, John Jenner Weir, rose to the occasion and then some. Weir, a customs officer and talented amateur entomologist and ornithologist (and, incidentally, cat fancier and judge), conducted experiments over the next two years, testing birds in an aviary with a diversity of tempting and not-so-tempting caterpillars— ultimately providing an impressive data set in support of Wallace's theory of warning coloration.[60]

Yes, riding high, Wallace seemed to be crossing another line in those heady days, transcending a boundary not exactly of class but as a field collector admitted to the full and admiring embrace of London's scientific scene, a kind of intellectual rags-to-riches story for the self-made naturalist. But as we shall see, another borderland beckoned, one he found irresistible and that threatened to undo it all.

It was all going so well with family, friends, and science: he did what he could to help Fanny and Thomas's struggling photography business, enjoyed seeing his friends George Silk, Henry Bates, and Richard Spruce (who returned in 1864 after fifteen years in South America, penniless after being defrauded of his life savings and not in the best of health), and enjoyed the company of a *Who's Who* of London's scientific society, dining regularly with the likes of Huxley and Lyell and attending scientific meetings, his opinion sought on matters ornithological, entomological, anthropological, evolutionary. His thoughts turned to domestic bliss, perhaps inspired by seeing his friend Bates, who in 1863 married his childhood sweetheart, Sarah Ann Mason (after they had just had a child). There was his brother John, too, who had five kids by that time (there would be one more within a couple of years). Wallace developed a love interest in Marion Leslie, the daughter of a chess-playing friend that Silk had introduced him to. Unfortunately, that interest was not altogether reciprocated. They were friends, but it was an inauspicious start to their potential romance when the rather shy and awkward Wallace unexpectedly proposed one day—by letter no less—and the young lady was apparently taken unaware. It was a gentle refusal; down but not out he bided his time to get to know her better (he thought) and a year later approached her father. One thing led to another, and suddenly, they were engaged! But it was not to be. Autumn of 1864 rolled around, the wedding date approaching; they had gotten so far as to order invitations and a wedding dress, the arrangements all made, when suddenly without warning "Miss L." broke off the engagement. It was a devastating bolt from the blue—perhaps the greatest shock and letdown of his life. "The blow was very severe," he later wrote in his autobiography. "I have never in my life experienced such intensely painful emotion." Months later he was still despondent, writing plaintively to Darwin that he had "suffered one of those severe disappointments few men have to endure. . . . You may imagine how this has upset me when I tell you that I never in my life before had met with a woman I could love." Now he was paralyzed, incapable of working, he told Darwin, and when Alfred Newton asked him for a contribution to the ornithological journal *Ibis*, he declined: "I have been considerably cut up." He was slowly regaining his old self, he told Newton, but "cannot tell when I shall go at birds again."[61] He needed to shake it off—maybe a change of scenery, routine? This may be why, at least in some measure, he moved out of Fanny and Thomas's home the following month and rented a place for himself and his mother on St. Mark's Crescent. It was a modest but handsome three-story brick house on Regent's Canal, just a stone's throw from the Zoological Gardens at the north end of Regent's Park and not far from Hanover Square, where he often consulted the Zoological Society's library. Of the very little writing he managed in his depths of despair, two pieces worth mentioning were short but aggressive defenses of Darwin and natural selection in letters to the *Reader*, answering a critic who published a severe attack on the Darwin-Wallace theory in the *British Quarterly Review*. Correcting the author's many errors, as he saw it, Wallace concluded with a withering attack of his own: "It is hardly worth while to break such a fly upon the wheel, but it is well to make known as widely as possible to what weak subterfuges those who attempt to stem the flood of modern thought with the worn-out theological mop are at last driven."[62] Ouch! He was surely channeling some of the anguish and anger of his broken engagement.

That heartbreaking episode may have led him to channel other things too. Right about the time of his breakup, perhaps not coincidentally, Wallace's passing interest in an unusual branch of "natural history" ramped up. Recall his early fascination with mesmerism, called hypnotism today. Now his sister Fanny, an ardent spiritualist for some time, urged him to take a look at the extraordinary phenomena of spiritualism—communication with the deceased. The spiritualism movement had started in 1848 with claims of disembodied rapping by three sisters in an obscure corner of upstate New York—unsurprisingly, the same "Burned-Over District" that had produced similar movements such as Millerism and Mormonism earlier that century. By the mid-1860s, spiritualism had become a trans-Atlantic sensation, the sisters now in-demand mediums renowned for their ability to "communicate" with the departed. Despite later confessions that it was all a hoax and revealing their tricks of the trade, the movement had predictably taken on a life of its own. There was no putting that genie, let alone spirits, back in the bottle.[63]

Fanny persuaded her brother to attend a séance with her in July 1865, and from the get-go he was hooked. It is important to note that Wallace's interest in spiritualism was not religious in nature. He was a fierce religious skeptic, and if spiritualism had any trappings of religion, he would have almost certainly dismissed it out of hand. One example of his continued rejection of religious orthodoxy at the time is seen in a letter to Oxford anatomy professor George Rolleston, one of Huxley's protégés. Wallace had severely criticized

Christian missionary work in an essay for the *Reader* titled "How to Civilize Savages." Among other barbs, he pointed out with irony that "the savage may well wonder at our inconsistency in pressing upon him a religion which has so signally failed to improve our own moral character."[64] This provoked a strong response from a friend of Rolleston's, clergyman William Kay at Bishop's College in Kolkata, India. Rolleston forwarded Kay's letter to Wallace, who replied with a point-by-point rebuttal. "The doctrine of future rewards and punishments" as a motive for good behavior or preparation for an afterlife was "radically bad" for everybody, he insisted, and nothing more than bribery and fear. The only way to teach and civilize was by example "through the influence of love and sympathy" and "to have the most absolute respect for the rights of others." Imagine that! Wallace concluded that none of this could be advanced by "the dogmas of any religion."[65]

So, if not religious in nature, then, what was Wallace's interest in spiritualism? He regarded psychological phenomena like mesmerism—something he himself witnessed and took part in—to be real phenomena with a physical basis and seems to have connected this with spiritualism. Certainly the (supposed) trancelike state of mediums seemed much like the "mesmerized" state. "Spirit" was equated with "mind," ineffable and perhaps, to spiritualists, independent of corporeal body. He extrapolated from the individual mesmeric state to the concept of a whole spirit world. Could individuals, through natural ability or training, become conduits between our world and a parallel spirit world? His interest, then, was as a naturalist, with spiritualism an intriguing new branch of natural history in his view.

But however scrupulously objective Wallace tried, or thought he tried, to be, even going to lengths to test and trap the mediums he visited, it is possible that he simply wanted to believe. Wallace may not come off as an especially sentimental, emotional guy, but he was emotionally distraught at the time, and it may not be a coincidence that in those early séances attended with Fanny the "spirits" (acting through the medium) spelled out the names of his dead brothers, Herbert and William. How did they know? Consciously or not, Fanny was the likely source. Whatever was going on, Wallace's personality was such that the fringe nature of spiritualism attracted him all the more. Ever the iconoclast, the more it was dismissed and attacked by mainstream scientists, the more Wallace dug in to defend it—just like transmutation.[66] This led him down more than a slippery slope—he went slip-sliding down into quite the rabbit hole. As he feared, his scientific friends were not simply unimpressed; they were taken aback at what they saw as their colleague's gullibility. But he was unapologetic, trying to get one after another to give the mediums a try and attend séances with him, mostly to no avail. Huxley flatly told him that he really had no interest: "I never cared for gossip in my life and disembodied gossip such as these worthy ghosts supply their friends with, is not more interesting to me than any other." As for the merits of investigating it in the name of science, Huxley pronounced it "too amusing to be fair work and too hard work to be amusing."[67] Considering the potency of Huxley's acid pen, Wallace should have been happy to be let off with this Huxleyan "fuhgedaboudit"—but Huxley was otherwise an admirer of Wallace so likely just shook his head at what he saw as his friend's foibles.[68]

Wallace was just about back to his old self by 1866 and over his rejection by Miss Leslie when another woman came into his life. Actually, she was already in it, thanks to Spruce—soon after Spruce's return in 1864, he invited Wallace along to visit pharmacist and moss specialist William Mitten, who was working on mosses Spruce had collected in South America. Wallace had befriended Mitten's eldest daughter, Annie, then just eighteen years old, enjoying countryside rambles and botanizing with her and her family around the bucolic town of Hurstpierpoint, Sussex, in the south of England. Now, two years later, that friendship blossomed into a romance, simpatico and unfazed by the age difference between them, and they were married on 5 April 1866. Regardless of whether or not it was a case of rebound romance, it was deep and lasting, the start of a life of domestic tranquility together. After an initial honeymoon in Windsor, they spent a month in late summer in North Wales, where the couple botanized and geologized with abandon. And that is not all they did with abandon—their first child was born the following June, in 1867: Herbert Spencer Wallace (nicknamed Bertie), named in honor of both Wallace's late younger brother and the philosopher. Wallace had been deeply impressed with Herbert Spencer's 1860 book *First Principles* ("A truly great work, which goes to the root of everything," he told Darwin), and he and Bates had made a pilgrimage to sit at the guru's feet.[69] (It says something about Wallace and his enthusiasms, and his earnest nature, that he would name his first child after a philosopher he was lately impressed with; one only hopes Annie was a fan too.) Darwin, sending Wallace hearty congratulations on the birth of his son, hoped the lad would "copy his father's style & not his name-sake's"—he never could make heads or tails of Spencer's philosophical circumlocutions.[70]

By now Darwin was becoming increasingly alarmed over Wallace's writings too. Not so much his scientific writings, which continued to be as impressive as ever—comprehensive taxonomic treatments; erudite comments on

Annie Wallace, née Mitten, ca. 1866 (age about twenty-one).

physical geography, anthropology, natural history, and (especially welcome to his colleague) incisive defenses of natural selection. Indeed, in a masterful review of the Duke of Argyll's anti-transmutation *The Reign of Law*, Wallace once again underscored his standing as one of the most eloquent and effective defenders of his and Darwin's evolutionary doctrine.[71] Rather, it was the increasingly frequent papers, comments at meetings, and letters bearing on phrenology, séances, and spiritualism that were concerning to Darwin. Wallace threw down the gauntlet in 1866 with a self-published fifty-seven-page pamphlet titled "The Scientific Aspect of the Supernatural," his first overt writing on spiritualism.[72] The subtitle signals what he was up to: "Indicating the Desirableness of an Experimental Enquiry by Men of Science into the Alleged Powers of Clairvoyants and Mediums." Frustrated by his scientific friends' refusal to engage at all, let alone seriously, with the question of spiritualism, here he was making a case for the *need* for engagement, the necessity of hypothesis testing to determine the veracity of the claims of spiritualism. His carefully chosen epigraphs, one by Sir John Herschel, the astronomer, and the

Alfred Russel Wallace, age
forty-six.

other by none other than Huxley, underscored his point: He quoted Herschel
as noting that naturalists need to have eyes and minds open to occurrences
that "*according to received theories ought not to happen*," as this is what leads to
new discoveries. As for Huxley, Wallace selected a choice comment—"That
the possibilities of nature are infinite is an aphorism with which I am wont to
worry my friends."

Huxley was not put out or put off by all this—again, in his view Wallace's
forays into the "preternatural" may not have been his cup of tea, but they surely
didn't detract from Wallace's remarkable contributions to the scientific understanding of the *natural*. He sang Wallace's praises in his 1863 book *Evidence as
to Man's Place in Nature*: "Once in a generation, a Wallace may be found physically, mentally, and morally qualified to wander unscathed through the tropical
wilds of America and of Asia; to form magnificent collections as he wanders;
and withal to think out sagaciously the conclusions suggested by his collections."[73] Huxley was also behind Wallace's nomination for one of Britain's most
prestigious prizes: the Royal Medal of the Royal Society of London, awarded

for significant contributions "to the advancement of Natural Knowledge." The medal was duly bestowed on 30 November 1868, "in recognition," pronounced Sir Edward Sabine, president of the Royal Society, "of the value of his many contributions to theoretical and practical zoology."[74] Receiving this great honor—the first of many to come—was surely a bittersweet moment for Wallace: his mother had died only two weeks before. How proud she would have been to see her hardworking, passionate, determined son arise amid the pomp and circumstance of the Royal Society, the very pinnacle of British Science, to receive a prestigious medal! Bittersweet, yes, but he probably had a feeling that she was there.

———

The watershed year of 1869 opened with the joyous arrival of Wallace's second child, Violet Isabel, on 25 January—a nice present for his forty-sixth birthday, just two weeks earlier. It was also a year of both intellectual triumph and, to his colleagues, intellectual calamity. The triumph came first: it was a long road, but he finally finished his masterful travelogue *The Malay Archipelago; The Land of the Orang-utan and the Bird of Paradise*, published on 9 March 1869 to much acclaim. Darwin had been encouraging Wallace for years to get his travel book out, hopeful inquiries as to its progress appearing in many of his letters throughout the 1860s. For his part, Wallace's plan after returning from the East had been to first get settled and organize and study his collection—no mean feat itself, on top of which he wanted to get several pressing papers written. But then his affairs of the heart had added further delays. In late 1865, a year after his engagement with Miss Leslie had been broken off, he was still adrift and acknowledged the emotional comfort and aid he longed for in a telling comment to Darwin, who had written with concern: "In reply to your kind inquiries about myself, I can only say that I am ashamed of my laziness. . . . As to my 'Travels,' I cannot bring myself to undertake them yet, and perhaps never shall, unless I should be fortunate enough to get a wife who would incite me thereto and assist me therein—which is not likely."[75] Annie had changed all that, and now he threw himself into writing. He took a highly productive "sabbatical" year in the countryside with his wife and son, living with her family in Hurstpierpoint from summer 1867 to summer 1868 to focus on the project. Now it was out at last, and Wallace graciously dedicated the book to none other than Darwin: "Not only as a token of personal esteem and friendship," it read, "but also to express my deep admiration for his genius and his works."

Darwin received an advance copy in early March. He was as deeply impressed with the book as he was touched by the dedication, "a thing for my children's children to be proud of."[76]

Then came the calamity. That very month came a disconcerting heads-up from Wallace; he had been working on a paper, by the way, a review of the latest editions of Charles Lyell's endlessly revised books *Principles of Geology* and *Elements of Geology*, and he intended to touch on certain "limitations" to natural selection. "I am afraid that Huxley & perhaps yourself will think them weak and unphilosophical." That was an understatement! Darwin waited anxiously as Wallace's paper came out the following month in the *Quarterly Review*. It should have been an unmitigated triumph: Lyell, for the first time in print, explicitly gave up his opposition to evolution by natural selection. But this gave Wallace an opportunity to review progress on the acceptance of the theory and . . . give his own take. Whereas he had earlier held, in his famous 1864 paper, that selection was responsible first for the bodily evolution of humans and at some point switched to acting on cognitive capacities, he now concluded that the brain could not have evolved by natural selection at all and therefore neither could correlated anatomical features such as organs of speech, dexterous hands, and bipedal locomotion. We may have had an initial animal origin and natural evolution, but then at some point, he held, something had guided human evolution along a new path. Religious skeptic Wallace could not bring himself to invoke a Creator, labeling this agency an "Overruling Intelligence" instead."[77] His conclusion was a stunning about-face for the co-discoverer and vigorous defender of the principle of natural selection.

Darwin had been disappointed by Wallace's spiritualism, but he was utterly shocked and dismayed at his disavowal of natural selection: he flatly told Wallace that he saw no need whatsoever to call in some divine agency for humans, and had Wallace not warned him in advance, he would have thought that the comments had been added by someone else. "As you expected I differ grievously from you, & I am very sorry for it."[78] Indeed, Darwin was more than sorry—this was no trivial disagreement; his worst fears that Wallace had irreparably damaged the cause of their joint theory seemingly realized, the First Darwinian turned apostate: "I hope you have not murdered too completely your own and my child."[79] To Darwin and his circle, it was a sad close to an otherwise triumphant decade for Wallace and the cause of science.

CHAPTER 12

A TALE OF TWO
WALLACES?

IT WAS THE BEST of Wallace, it was the worst of Wallace. With apologies to Dickens, that is how his scientific friends likely viewed him in that heady decade of the 1870s. Indeed, to many observers Wallace did seem to reflect Dickensian extremes by the decade's end: wisdom and foolishness, reason and incredulity, hope and despair (for science and humanity). What sense can be made of the seemingly nonsensical: An apparent about-face from natural selection's co-discoverer, rejecting the very thing he had been searching for all those years under the most arduous of conditions? But Wallace's about-face was only apparent in a sense: First, as we have seen he came to believe the spirit world was a part of nature, a newly discovered branch of natural history, and so could and should be investigated by scientific means. He preferred the term "preternatural" (*more than* nature) to "supernatural" (*above* nature). Second, between Wallace's long fascination with mesmerism, equitable view of human capacity, and dedication to natural selection, it's likely that he had always struggled a bit to square his conception of mind with a strict Darwin-Wallace evolutionary framework—rejecting a material origin of mind through natural selection in favor of some overruling intelligence perhaps should not be too surprising, then. As Wallace scholar Charles Smith has pointed out, his positions on evolution by natural selection were more augmented than reversed.[1]

Yes, Wallace was not so much apostate as more Darwinian than Darwin. According to the logic of natural selection, each intermediate step of a trait along an evolutionary pathway must be useful in its time and place to the organism that possesses it. When in *Origin* Darwin argued for the evolution of eyes or honeybees' cells or wings, for example, he made a case for the utility of simpler versions of these complex structures that are useful in their way—proto-wings that make gliding possible as steps toward true flight using lift; cylindrical cells like so many clustered test tubes that morph into hexagons as bees shape each shared wall, economizing on wax; early "eye" versions from pigment spots to pit organs that act as precursors of chambered eyes and are capable, like a pinhole camera, of projecting an image of the world. Early versions of complex organs that arise for one function can even be drafted into playing a very different role, something evolutionary biologists call *co-optation*—perhaps, say, feathers initially arose as scales modified for insulation, setting the stage for getting co-opted into different roles: flight or sexual display.[2]

Such scenarios for eyes or bees' cells or wings all made sense to Wallace. But what about the human brain? Now *that* is one complex organ, perhaps the most complex in the organic world—how it boggles the mind to think about itself! Seriously. Think about the origin of consciousness, abstract thought, and personality, not to mention the astounding capabilities flowing from these: language, music, literature, art, mathematics, invention, abstract reasoning, problem-solving. These abilities are incredible enough in the most "ordinary" of humans; all the more jaw-dropping, then, are the geniuses among us—prodigies of music, mathematics, literature, art, science. Far more than such wonders of nature as eyes or bees' cells or wings, then, the human mind and its organic seat, the brain, seem to stand above all as a pinnacle of sublime complexity. How did this wondrous organ arise? Well, to Darwin the process was no different from any other wondrously complex organ: step-by-step, each early stage useful to its possessor. How best to demonstrate this? In a paleontologist's dream world, every intermediate stage of every structure would be preserved in the fossil record, a boon for researchers to reconstruct evolutionary pathways. But in the real world, fossilization is a hit-or-miss proposition, not to mention that it mainly works for hard parts, not so much soft organs and tissues. Darwin, knowing this, argued that the next best thing is to look around at organisms living today. With complex vertebrate eyes, for example, he pointed to simpler versions of eyes and eye-like structures evident in different species, especially among diverse but divergent groups of invertebrates—not to suggest a direct, linear evolutionary pathway but to give clues to what simpler versions of organs might look like, useful in their time and place.

So too with the brain: in *The Descent of Man*, Darwin made much the same argument, looking around the animal kingdom for simpler expressions of various mental traits of humans. But he also took that a giant step further, applying the idea even to attributes traditionally considered the very hallmarks of humanity: aesthetic sense, morality, compassion, love, justice. With a catalog of case studies gleaned from the literature as well as his own observations, Darwin argued that humans differed in degree, not kind, from the rest of the animal world. Implicit in this argument was the idea that earlier, simpler versions of all such mental attributes, including such abstractions as love, compassion, and a moral compass, were useful to their possessor, representing transitional steps that could lead to an exalted us. But it was a (transitional) bridge too far for Wallace.

On the one hand, Wallace had imbibed such cultural prejudices of the time as the presumed superiority—in its best form—of European-style civilization, ability, and morality and at times expressed stereotypical and often critical views of peoples of different "race" and culture in comparison. On the other hand, recall how he readily called out the degrading influence of "low" elements of European society (slamming the injustices and immorality of unscrupulous traders and sanctimonious exploitative clergy, for example) and took note of the many instances of kindness, morality, loyalty, and generosity of native peoples he met and worked with. Wallace was decidedly nonracist in recognizing common humanity and in his belief that all peoples have essentially the same inherent *capacities* morally as well as mentally, even if not fully realized. Given the opportunity and the right environment, any "savage" could become a mathematician, philosopher, or composer. But how do we account for this ability, lying dormant in the mind of the Dyak, Papuan, or Uaupés native, say, whose ancestors over millennia of untold generations never had any means or need whatsoever to exercise such abilities? So much for transitional stages. "How then," Wallace asked, "was an organ developed so far beyond the needs of its possessor?" No, Wallace concluded—the prodigious capacity of the human mind seemed to represent a break, but not so much *with* nature as *in* nature, mental phenomena to him a higher order of physical reality. It is understandable that others saw him as arguing for discontinuity between the material world and mind—prompting Darwin to score this passage in his copy of Wallace's article and scrawl a triply underlined "No!!!" in the margin.[3]

This thinking went hand in hand with Wallace's growing spiritualism—it was all of a piece, the existence of this parallel spirit world, privy to some larger truths about existence, perhaps all overseen by that overruling intelligence he invoked. He made the link clear in an apologetic letter to Darwin. He understood Darwin's shock, he said, "because a few years back I should myself have

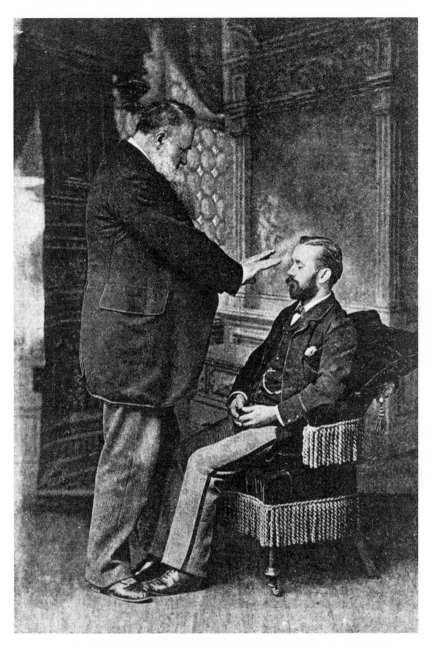

Popularized in the late eighteenth century as a therapeutic technique, mesmerism was all the rage in Wallace's day.

looked at them as equally wild & uncalled for." But there were "remarkable phenomena, physical & mental" that he had been investigating, demonstrating some kind of forces and influences "not yet recognised by science." He knew it sounded crazy, he acknowledged, but pointed to respected intellectuals who held similar views and hoped that Darwin would withhold judgment, as least "till we exhibit some corroborative symptoms of insanity," he half joked.[4] Darwin and most of their circle did not see him as crazy so much as gullible, to their disappointment (and in Darwin's case dismay): How was this possible, for one who had authored such powerfully reasoned papers, masterful in their logic, command of facts, and compelling argument? About the only sympathetic ear was Lyell's, not unexpectedly. He had wrestled mightily with the religious implications of transmutation from the get-go, as we have seen, and at this point, he agreed more with Wallace than Darwin: "I rather hail Wallace's suggestion that there may be a Supreme Will and Power," he told Darwin, one "which may not abdicate its functions of interference, but may guide the forces and laws of nature."[5] Darwin could only shake his head at the two of them.

———

The (apparently . . .) two Wallaces[6] were on display in April of the new year, 1870, when, still riding high on the success of *The Malay Archipelago*, Macmillan brought out *Contributions to the Theory of Natural Selection*—a sort of "greatest hits" of Wallace's papers to date. That book still makes for fascinating one-stop shopping for Wallace's watershed papers—the celebrated Sarawak Law and Ternate papers; his stunning analysis of the Malayan swallowtail butterflies; insightful treatments of mimicry and birds' nests; the most incisive of his vigorous defenses of Darwin and natural selection; the 1864 human evolution paper on how selection switches from body to mind. But of course the collection also included his more heretical views—notably, in the culminating chapter "Limits of Natural Selection as Applied to Man," an elaboration of the argument he had first made in the 1869 review on the inadequacy of natural selection to fully explain human evolution. Here Wallace makes an argument that astounded—and repelled—many of his scientific friends, elaborating on the special features of human anatomy that defied a naturalistic explanation: large brain, speech and language, hands and feet, the distribution of hair on the body, self-consciousness, and moral sensibilities. If selection cannot explain this suite of features that together make humans human, Wallace is led to conclude that "a superior intelligence has guided the development of man in a definite direction, and for a special purpose, just as man guides the development of many animal and vegetable forms."[7]

Think about that: humans as a sort of domesticated animal, literally in this case! Wallace suggests that just as we direct the evolution of certain plants and animals, selecting for such traits as, say, bigger fruits, better milk production, or more meat on the bone, some "superior intelligence" is directing our evolution. And to what end? He only suggests some "special purpose," but perhaps he had in mind a higher level of ultimate self-awareness as a moral being, judging from other writings. Wallace seems to have been merging his spiritualist convictions with his Owenite idealism. As noble as his vision for humanity may have been, it did not go over too well as *science*, needless to say. Bates, for one, was shocked at his friend's "backsliding," as he put it, and wrote Darwin to express concern and the "surprise & bewilderment" he and others were feeling.[8] He exhorted Darwin to rebut Wallace, maybe in a review of the book, but Darwin would have none of it. He was hard at work on *Descent of Man*, due out the next year, and felt that would be answer enough. Besides, Darwin hated the public limelight, and especially getting entangled in any controversy. Wallace's reputation then and today may have fared better if he had taken this page from Darwin's playbook, but it was not to be.

If Wallace's spiritualism was not eye rolling enough to his friends in the scientific scene, a few months before *Contributions* came out he also let loose a pair of letters to the editors of the new journal *Nature* arguing against the public support of science![9] One would have thought that as a newly inducted member of Britain's scientific elite whose landmark insights were made possible by such public institutions as the British Museum (purchaser of many of his collections), he would wholeheartedly support the exciting growth in stature of professional science and the incalculably valuable scientific and educational benefits that flow from scientific exploration and museums. To put this into perspective, bear in mind that the pursuit of science, once called *natural philosophy*, was for centuries the purview of the wealthy and privileged either directly or indirectly—whether self-funded by the wealthy, often aristocrats, or commissioned by such (think of Galileo's House of Medici benefactors). The young yeoman farmer, carpenter, blacksmith, or even merchant lacked the requisite education, skills, resources, and time to pursue any budding taste for science. Such pursuits were undertaken by the privileged, those whose time was their own and had the financial resources for education, books, and equipment. An evolving meritocracy in Britain allowed poor but talented youth to rise to the level of their abilities in such institutions as the Royal Navy, universities, or the Royal Institution, and the practical benefits of science were increasingly recognized in agriculture, medicine, manufacturing, and even natural history—where, for example, the discovery and cultivation of strategically important plants played a major role in British imperial expansion, something that Wallace's friend Richard Spruce played a role in.[10]

Darwin and Wallace represent the growing duality in the pursuit of science as the nineteenth century progressed, independently wealthy Cambridge-educated Darwin emblematic of the talented "country-house" scientist and impecunious and self-educated Wallace of the advancement that mechanics' institutes, libraries, and perseverance made possible.[11] By the mid-nineteenth century, science was still by and large cultivated by the elite, but as Wallace himself demonstrates, it was possible for talented but relatively poor young men (and it was mostly but not exclusively men at that time) to rise to the very center of British science. At this time another largely self-made scientific man, one T. H. Huxley, was just then leading an increasingly vocal group of talented young Turks bent on educational reform and the elevation of science in Victorian Britain, something they accomplished by waging battles in print and in the courts and by assuming leadership positions in the learned societies. In 1864 Huxley founded the X Club, a dining club of nine naturalists and philosophers (of an originally intended ten, hence Roman numeral X) who shared a "devotion to science, pure and free, untrammelled by religious dogmas."[12] One of their aims was to loosen the grip of the Anglican Church on education and other public institutions in Britain, one means of which was by helping provide legal defense funds for individuals who challenged the church and were slapped down—censured theologians and defrocked or deposed clergy accused of heresy for supporting liberal church reform or publishing works of biblical criticism—funds to which Darwin regularly contributed.[13]

Another major agenda item was securing a prominent role for science in education as well as in public life, including through support of scientific institutions for research and the edification of the public. In the year 1864 a competition was opened for design proposals for a new natural history museum in South Kensington, London—seeking to greatly expand the standing of natural history from a mere department of the British Museum to its own glorious place in the sun was one of the few things that Huxley and the X Club circle could agree upon with their anti-transmutation foe Richard Owen at the Royal College of Surgeons. Plans were proceeding apace on Alfred Waterhouse's magnificent mixed Gothic Revival and Romanesque design, but here came Wallace, raining on the parade with a letter to *Nature* in January 1870 in which he decried the expenditure of public money on an institution that benefited, he maintained, but a narrow segment of society—namely, naturalists like himself.[14] He was not opposed to a *popular* museum to "elevate, instruct, and entertain all who visit them," but public funds for an academic museum, expending enormous sums for collections like those he contributed to, was not right: "It will, perhaps, surprise some of your readers to find a naturalist advocating such doctrines as these; but though I love nature much I love justice more," he declared in his letter. The taxpayer should not be burdened to help support an institution "of no interest to the great mass of my countrymen, however interesting to myself." The editors of *Nature* were taken aback, and while they published the letter, they also ran an editorial rebutting it. Was Mr. Wallace saying that "the main result of cultivating science is merely the gratification of those directly engaged in the pursuit"? That those who take no personal interest in the pursuit of science do not benefit from it, and therefore supporting science was an exercise in unjustly taxing the whole community to support the enthusiasms of a few nerds? This they held to be absurd, touting at length the many individual, societal, and ultimately universal benefits direct and indirect that flow from public support of all branches of science. Wallace followed up with another letter, sticking to his guns and maintaining, not entirely convincingly, that achieving physical, social, and intellectual benefits through the public funding of science was doomed to fail and that only curious and intrepid persons self-motivated by a passion for science can contribute great scientific advancements.

It is a curious argument for Wallace to make—a rather libertarian approach to the conduct of science: let those who can and want to do it do it, but do not spend a shilling of public money on their enthusiasms. Did he have his own self-improvement program in mind? Let people fend for themselves; those who have what it takes will make it, like himself and Bates and Spruce. He could have taken a more democratic approach: public funding for science education and institutional support can inspire youth with a passion for the natural world, for one thing, and awaken latent interests and tap talents that would otherwise never come to light. Does pursuing science have to be strictly an up-by-the-bootstraps self-help proposition? In any case, it is worth noting how Wallace's argument was couched: in terms of *equality* and *justice*. It's just another reflection of his utter commitment to social justice: *all* must benefit, equally, if public monies are to be expended. We may or may not agree with his argument about public support for science, but we can at least acknowledge that it came from a noble place.

Wallace's apparent apostasy over human evolution, his embrace of spiritualism, his broadside against public support of science, and a few other choice writings provoked more than a little disappointment, bewilderment, and even consternation among his scientific friends. His achievements in science were undeniable, and in other ways he had indeed become part of the fabric of the vibrant London scientific scene—one indication of which was his recent election as president of the Entomological Society of London, accepting the gavel from his friend Bates. He was also lauded as a devastating defender of Darwin and natural selection—his pen was indeed mightier than any sword, vanquishing their opponents with deft literary thrusts and parries in a stream of reviews, letters, and rebuttals, usually by skillfully showing that the critics did not know what they were talking about. Regardless of whether the critics were coming from a place of genuine or willful ignorance, ignorance it was, and Wallace was just the person to expose the flaws in the anti-evolution diatribes. Darwin rarely joined the fray with him, one conspicuous exception starting with Wallace's "crushing article" (as Darwin put it) to *Nature* rebutting Charles Robert Bree's attack in *An Exposition of Fallacies in the Hypothesis of Mr. Darwin*, which came out in 1872. In response Bree denounced Wallace as "blundering," getting Darwin's ire up. He came to Wallace's defense with his own letter to *Nature*, showing himself capable of brandishing an equally caustic pen when he had to: Wallace's piece was clarity itself, Darwin said, while Bree's letter was "unintelligible." He could not imagine how someone could so completely mistake his (Darwin's) meaning, but then, perhaps nobody "who has read a work formerly published by Dr. Bree on the same subject as his recent one, will be surprised at any amount of misunderstanding on his part."[15] *Touché*!

But however talented and principled a defender he may have been, in other ways Wallace seemed to be exiling himself from the heartland to the borderlands of British science as he flirted with that all-important line drawn in 1859: that line between the up-and-coming evolutionists and their bold vision for the ascendancy of science on one side and the conservative church-and-state establishment on the other. Although he was certainly no supporter of orthodoxy, in other ways his maverick views threatened to undermine the evolutionists' standing—such as the time he wrote approvingly of physician Henry Charlton Bastian's claims to have demonstrated spontaneous generation, something Wallace thought might give evolution by natural selection a head start.[16] Polish German naturalist Anton Dohrn, founder of the venerable marine station at Naples, Italy, that now bears his name, lamented to Darwin how Wallace "completely drifts away" and unfortunately associates himself with such men as Bastian. "His two articles in *Nature* are the worst thing he ever did in his life, and it becomes really difficult for his friends to speak with respect of him."[17] Darwin certainly did not agree with this strong condemnation, but Wallace did worry him.

————

Huxley, too, and his fellow X Clubbers were still shaking their heads, trying to make sense of their wayward comrade-in-evolutionary-arms, when Wallace committed another metaphorical face-plant. It started the very month that he wrote those letters to *Nature*, January 1870, when a note published in the weekly magazine *Scientific Opinion* caught his attention. The editors published it dismissively: "The following choice piece of scientific nonsense has just been sent to us by its author, Mr. J. Hampden, of Swindon." It was a challenge by one John Hampden, a flat-Earth devotee, wagering anywhere from £50 to a whopping £500—tens of thousands in today's money—for anyone to "prove the rotundity and revolution of the world from Scripture, from reason, or from fact"— specifically, by demonstrating curvature in a straight-line railway, river, canal, or the like. Wallace the ace surveyor could not resist—£500?! Yeah, that was like falling off a log easy. He conferred with Lyell, who encouraged him to go for it—he might do some good and "stop these foolish people to have it plainly shown them."[18] He needed the money, after all, and the fruit was so low-hanging it was practically in his pocket. Or so he thought. Wallace wrote at once to accept the challenge, and the two soon settled on the Bedford Canal or Old Bedford River, a flat, slow-flowing waterway in Cambridgeshire that runs nearly six miles in a straight line through the Bedford Level. The site was not accidental—this was where a similar earlier "demonstration" had been conducted by another flat-Earther (and Hampden's inspiration), Samuel Birley Rowbotham, sometime writer, inventor, and Owenite socialist. Rowbotham set up a flag on a boat that was rowed steadily away from a fixed observation point, viewed through a telescope. If there is no curvature to the earth, the flag's level in the field of view should remain unchanged, whereas with a curved surface the flag will eventually drop out of view as it is carried below the horizon with respect to the stationary observer. Atmospheric refraction, however, can appear to "raise" objects below the horizon and render them visible, a well-known form of mirage called *looming*—likely the original inspiration for the expression "castles in the air" as flights of illusory fancy. Such was probably the case with Rowbotham's original observations, but in any event, based on the flag apparently not disappearing, he claimed to have disproved the Earth-as-globe "myth," as he saw it. Hampden, sure that his master was correct, thought the Bedford Canal was thus the perfect site for his challenge.

The agreed-upon arrangement was for Hampden and Wallace to each put up £500, and the two would each name an observer to verify Wallace's observations at the demonstration. Wallace's second was a Mr. Coulcher, a surgeon and amateur astronomer, while Hampden's was a fellow flat-Earther named William Carpenter. John Henry Walsh, editor of the *Field* magazine, was to serve as umpire. Ingeniously, in his demonstration Wallace compensated for refraction by setting a sight line on a bridge as a reference and also installed a pole at the midpoint between the bridge and his observation point to demonstrate the "bump" at that point expected of a curved Earth. Wallace was soon declared the winner and awarded the money, only to have Hampden and his second immediately cry foul, accusing Wallace of cheating and suing for repayment. Wallace should have known that Hampden was not altogether sane—in fact the guy appears to have been something of a paranoid schizophrenic. Wallace should have just given the money back and washed his hands of the whole thing, but once again principle came into play: he felt he had won the bet fair and square and in the process hoped he had taught the delusional flat-Earthers a lesson in science and reason. If only! Long story short, when all was said and done Wallace was embroiled in repeated lawsuits and countersuits dragging on for *fifteen years*, with extensive press coverage. As one example of Wallace's deep frustration and anger as this was all going south, consider this uncharacteristically heated statement concluding a pamphlet he published to defend himself: "Under these circumstances I have issued this reply, every statement in which can be verified by the papers referred to, which are mostly those published by Messrs. Hampden and Carpenter. All who believe Mr. Hampden to be an ignorant but very foul-mouthed libeller, will oblige me by burning unopened and unread any further communications they may receive from him."[19] Strong words from the normally mild-mannered Wallace!

He may have won the lawsuits, but the victory was pyrrhic, costing him dearly in time, energy, and money. After redoing the experiment and reaffirming his results only to have them rejected *again* by Hampden and his partner-in-delusion, writing letter after letter to various periodicals plus self-published pamphlets defending himself, dealing with the stress and aggravation of having himself and his family libeled and threatened with violence (including murder, for which the crazed Hampden was repeatedly jailed), *and* facing constant legal bills—after *all that* a judge ruled that the wager was invalid on a

" Signed by Mr. Carpenter."—Dr. Coulcher's Report.   " Signed ! "

Wallace's demonstration of Earth's curvature: the distant bridge over the Bedford Canal drops below the line of sight of the pole markers as the observer's distance increases.

technicality, and Wallace had to pay back the money after all.[20] Aargh! You can well imagine that none of this did Wallace's reputation any good—in the eyes of the scientific community, it was the height of foolishness, wasting time on a lunatic, if not a charlatan, to "prove" what was well-established science.

Why in the world did he keep going with it? In part it was the principle of the thing, and perhaps, too, he was simply naive—both in thinking that his elegant little demonstration would suddenly convince such absolutely committed flat-Earthers and in believing that people are essentially honest and good. (He was in for a very rude awakening on both counts.) I mentioned, too, that he needed the money, and that alone may have been incentive enough. In the early years following his return from the East, Wallace was quite comfortable thanks to the ever-dependable Stevens not only selling his fine specimens but also making some judicious investments for his friend and client. Unfortunately, however, Wallace was not as shrewd in money matters, to say the least: he practically hemorrhaged money through speculative investments and trying to help support his sister and brother-in-law's ever-precarious photography business. He handed over hundreds upon hundreds of pounds to Thomas and Fanny, all for naught, and he unwisely followed the advice of well-meaning friends who persuaded him to invest in risky ventures: railroads,

Wallace's son Bertie and daughter Violet, ca. 1871.

a slate quarry, a lead mine that promptly went bust. The generous deal he got from Macmillan on *The Malay Archipelago* was a lifeline, as was the fifty-guinea cash award—over £3,000 in today's money—that came with his medal from the Royal Society, but together these were hardly enough to help much. By 1870 he was on somewhat shaky financial grounds with a wife and two children to support. Desperate for income he saw the Hampden wager as money too easy to pass up, to his later regret (the old saying that warns "if something is too good to be true . . ." perhaps not being coined yet).

Financial considerations also played some role in the Wallace family's move to the London hinterlands in the spring of 1870, the first of many such moves the family would make as financial fortunes waxed and (more often) waned. The previous year Wallace had applied for the directorship of a new art and natural history museum soon opening in Bethnal Green, East London. The prospect of a steady income at a job he would enjoy, situated in the suburbs where he and Annie could luxuriate in gardening and rural rambles, was very appealing, and with the encouragement and support of such friends as Lyell, Huxley, and others, he thought he had a good shot at it—good enough to relocate his family. They first rented Holly House on Tanner Street in Barking, about six miles from Bethnal Green, until they could find a more permanent home. Their third child, William Greenell Wallace, was born there the following year: "Bertie has got a brother," he wrote to his sister Fanny, "born last night at 9.30. Annie is pretty well."[21]

A once bucolic suburb several miles from the bustling center of London in a district known for gardens, small farms, and weaving and furniture factories, Bethnal Green was by then a bit frayed around the edges and at the front line of the campaign for beautification and edifying public works in the less prosperous districts of greater London.[22] The lovely 213-acre (86 hectare) Victoria Park, perhaps the oldest purpose-built public park in London, was established there in 1845, and by 1870 plans were well underway to construct a branch of the South Kensington Museum (now the modern Victoria & Albert, on Cromwell Road). The new museum, a hangar-like marble-floored structure of brick and cast iron, was dubbed the Bethnal Green Branch Museum and is now the Young V&A, specializing in objects by and for children. (Wallace would approve: it is certainly far better than the museum's *initial* collections when it opened in 1872: old "Food and Animal Products" displays laying around from the Great Exhibition—stale leftovers.) Unfortunately for Wallace, things did not work out: he was told there was not much funding, and the focus of the new museum would be on the arts and trades rather than science, and anyway the decision was made to administer the new museum from South Kensington after all, so no new director was needed. It was, he confided to Darwin, a "considerable disappointment."[23]

But by then they had moved to Grays, Essex, a more rural idyll where the family was to spend the next four years. Smitten with the place despite being a distant seventeen or so miles from Bethnal Green, he figured he would commute—rail lines were ever-expanding, after all. So much so, incidentally, that railway investment must surely yield dividends, but this was actually one of Wallace's several losing portfolios. His finances were buoyed some by collection proceeds and the continued strong sales of *The Malay Archipelago*, but he still had to supplement his income in various ways. Lyell paid him five shillings an hour to help with editing a new edition of *Geological Evidences of the Antiquity of Man* and other books, for example, and he received payment for some articles and reviews. Beginning in 1869 he also started serving as an examiner in physical geography for different organizations—the Royal Geographical Society, the Indian Civil Engineering College, and the Science and Art Department at the South Kensington museum—bringing in a handsome fifty to sixty pounds or more annually, work that he continued until 1897.[24] Grading over one thousand exam papers in the space of a few weeks each year sounds like a punishment found in one of the lower circles of hell to most of us, but the glass-half-full Wallace made the most of it: the answers given by some of the more clueless pupils provided some comic relief, and he and his fellow examiners would record the most ridiculous ones to share with each another: "We often communicated some of these to our fellow-sufferers, and thus contributed a little hilarity to our otherwise strictly business meetings."[25] Wallace saved these, by the way, and could not resist presenting a selection in his autobiography—nearly ten pages worth! But it was to make a characteristically Wallacean point—namely, to slam the "utter failure" of the educational system in terms that sound quite familiar today. From the inadequacies of a system based on cramming information into brains just long enough to be regurgitated (and promptly forgotten), to the pressure teachers and institutions feel to pass students regardless of their actual knowledge, to degrading the status of teachers by assuming that anyone, with minimal training, is fit to teach, Wallace found the educational system of Victorian Britain severely wanting. But his remedy—to just seek out natural-born teachers and compensate them generously—was perhaps a bit unrealistic too, or was it his naive idealism? "When that is done, no examinations will be advisable or necessary."[26]

In any case Wallace welcomed the extra income as an examiner, as it helped make their dream home possible. In Grays they built a lovely whitewashed, slate-roofed three-story dubbed The Dell, today a building listed as one of the earliest examples of poured concrete construction and the only one of the three houses built by Wallace that survives.[27] Beautiful inside and out, The Dell was spacious and high ceilinged, with a library, a generous drawing room with a bay window, and a dining room opening to a conservatory and greenhouse on the ground floor. Four large bedrooms and a dressing room were found on the second. It had all the latest amenities, including servants' quarters, a well-appointed kitchen, hot and cold running water, coal stoves in each bedroom and bathroom, and beautiful ornamental finishes to boot: stained glass, glazed tile—the architect's drawings even included a bird of paradise weathervane, unfortunately not realized as far as is known. But Alfred and Annie really reveled in the gardens and grounds. Built on a hill overlooking the Thames, Alfred immediately saw its potential. Its four acres included a large chalk pit near the house, plunging a dramatic sixty feet (eighteen meters) deep at its steepest: "A bit of a wilderness that can be made into a splendid imitation of a Welsh valley."[28] Indeed, the vicinity of this picturesque bend of the Thames was dotted

The Dell, the Wallaces' handsome home in Grays, Essex.

with old quarries yielding flints, red-brick earth loess to make bricks, and gravel, sand, and chalk for lime, whitewash, mortar, and cement. Wallace would have known that the chalk had been deposited in a shallow sea during the age of dinosaurs, the Upper Cretaceous period (the name derived from Latin *creta*, chalk) and likely knew, too, of the work of his friend Huxley and geologist Henry Clifton Sorby in identifying the chalk's origin as untold scintillions of *coccoliths*—the calcium carbonate tests of microscopic phytoplankton settling out of the water column over many millions of years. But he would have been fascinated to learn that the sand formation lying atop the chalk of Grays, the Thanet sand, represents a threefold tumultuous ending: of the Cretaceous, the Mesozoic, and the dinosaur's reign, the junction of the sands atop the chalk being a very obvious local marker of that worldwide mass extinction event. The old quarries, now water-filled ponds, are beautiful greenspaces today managed by the Essex Wildlife Trust, their names expressive of the landscape: Chafford Gorges Nature Park, Grays Chalk Quarry Nature Reserve, Lion's Gorge.

At home he and Annie installed a windmill-driven well and laid out a serpentine drive, terraces and winding paths, ponds, a fountain, croquet green, kitchen and vegetable gardens, and an abundance of beautiful flowers, choice shrubs, and stately trees—some originating from Kew, where Wallace hit up

Hooker for seeds and plants. Yes, this was the country idyll he had longed for, leafy and lovely, a spot where he planned to "gather round me all the beauties of the temperate flora which I so much admire" and revel in the garden and greenhouse the rest of his life.[29] As it happened, though, they moved away just four years later, in 1876. It was probably prompted by a combination of factors: he had been unhappy with the climate after all and the challenges it posed for his gardening, for one thing, and also wanted a more direct line into London so he could more easily attend scientific meetings.[30] Perhaps, too, the long shadow of an acutely painful memory cast a pall on The Dell: in what was undoubtedly the greatest blow of their lives, two years after moving to Grays, in April 1874, their eldest child, Bertie, tragically succumbed to scarlet fever. He was just six years old; Wallace was utterly devastated.

————

With the exception of a period following his son's death, Wallace was prolific in the few years he lived at the Dell, publishing some twenty-two book reviews; seventeen scientific letters, reports, addresses, and commentary; ten articles and letters on spiritualism; a dozen letters dealing with the flat-Earther issue; and another seven articles or letters on miscellaneous topics of interest to him—for example, weighing in on the establishment of public parks (in favor, with recommendations), disestablishment of the church (opposed, interestingly enough, with explanations), and more. He kept up a brisk correspondence with Darwin, the two continuing to disagree on certain issues but their friendship steadfast. The two Wallaces are perhaps best reflected in the two books he also published during this period: on the one hand, there was *On Miracles and Modern Spiritualism*, which came out in March 1875: a collection of three of his essays on spiritualism, the centerpiece being his (in)famous apologia "A Defence of Modern Spiritualism," published in the May and June 1874 issues of the *Fortnightly Review*.[31] On the other hand, there was the magisterial tome *The Geographical Distribution of Animals*, issued in two volumes in May 1876.

This amazing book got its start a few years earlier, at the urging of Philip Sclater and Alfred Newton: a comprehensive treatment of the subject was badly needed, and he was just the guy for the job. Wallace later said that had he known just how difficult the task would be he never would have undertaken it—but undertake it he did, throwing himself into the project with aplomb and producing what is fairly hailed as the founding document of modern

evolutionary biogeography. It was no mean feat, poring over stacks of taxonomic and paleontological catalogs of multiple languages covering every region of the world and plagued by uneven treatments, imperfect classifications,
synonymy, and conflicting authorities. For quality control he asked a dozen
or more obliging friends and colleagues to review his material on their area of
specialty, and others were drafted to read the entire manuscript.[32] He built
his treatment around the six Sclaterian biogeographical regions, first giving a
masterful overview of principles of distribution and defining the regions and
then providing an overview of extinct as well as living species, underscoring
the point that the present distribution of animals is necessarily informed by their
past distribution, and thus the distribution of species in the fossil record holds
valuable lessons for us. This constituted the second section of the book—"On
the Distribution of Extinct Animals," the first analysis of its kind. Two more
sections followed, brilliantly distinguishing between what he called *zoological
geography*—characterizing the overall similarities and differences between
regions and subregions—and *geographical zoology*, quantitative distributional
analyses of specific groups, giving insight into patterns of migration. Here Wallace
introduced an innovative device to convey distribution at a glance: for each
family he provided a table dividing each of the six regions into four subregions,
allowing him to essentially check the region/subregion boxes each is found in
for quick distributional reference, all keyed to maps. Speaking of which, the
illustrations for the book were equally innovative. It boasted beautiful colored
maps, including a global map and a series of detailed maps of Sclater's regions
and subregions, each also color-coding ocean depth to indicate a history of
isolation or connectivity. The maps were complemented by a set of twenty
well-executed plates creatively depicting characteristic assemblages of animals
for each region *in their native habitat*—landscape, flora, and fauna brought
together as a harmonious whole, perhaps the earliest pictorial ecological
representations. They were so nicely done and innovative that Wallace hoped
they would give the book a broader reach beyond the confines of zoological
specialists: "My trust is in my pictures & maps to catch the public," he wrote
Darwin.[33]

This was classic Wallace, the undisputed traveler in time and space. His
treatment made Sclater's six realms the textbook system they are today—albeit
much discussed and debated, Sclater's system is not only the starting point but
remains the superstructure for most proposed refinements and alternatives.
While many of Wallace's analyses in *The Geographical Distribution of Animals*
have of course been superseded by later insights into Earth history—notably,

One of Wallace's innovative illustrations: "A Malayan forest, with some of its peculiar birds."

the discovery in the late twentieth century of plate tectonics and the movement and recycling of continental crust—the book remains a landmark work, articulating a grand vision of geographical distribution as a dynamic process of migration, interaction, adaptation, and change—distribution both a *product* of Earth history, ecology, and evolution and a means of gaining *insight* into these. His scientific friends were awed: Huxley complimented him on the "grand volumes," and Darwin expressed his "unbounded admiration." Wallace, he declared, had laid a foundation "for all future work on Distribution," writing a "grand & memorable work which will last for years as the foundation for all future treatises" on the subject.[34] High praise indeed—and he was right on.

That summer the family relocated from Grays to Rose Hill, in the prosperous market town of Dorking, lying about forty miles west in the rolling Surrey Hills just south of London. "Is it not a lovely country?" he commented to Darwin.[35] He certainly had a knack for picking geologically interesting areas, and borderlands no less—his new abode was situated on the edge of the hilly North Downs, the northern rim of a formation called the Wealden anticline, the eroded remains of a great Cretaceous-period dome of buckled strata that stretch across the English Channel to France. It's much like the Usk Inlier in the landscape of Wallace's youth: an eroded arched dome that leaves a telltale pattern of geological strata, those most resistant to erosion—the chalky Downs—standing as hills ringing the more-eroded flat sandy plains within the formation. It makes for a lovely landscape, designated today the "Surrey Hills Area of Outstanding Natural Beauty."

All was well with the family and professionally, too, as he was riding high from the unmitigated success of *Geographical Distribution of Animals*. Bates wrote to invite him to deliver a lecture to the Royal Geographical Society the following spring. He would be happy to, Wallace replied, proposing to speak on "The Comparative Antiquity of Continents, as Indicated by the Distribution of Living and Extinct Animals," and that September he served as the unanimously elected president of the Biology Section of the British Association for the Advancement of Science (BAAS) annual meeting in Glasgow.[36] Now, that was a real mark of esteem, and it should have been a triumph—and indeed it started out that way when Wallace gave a much-acclaimed address at the meeting's opening. It was a characteristically masterful paper given in two parts: "On Some Relations of Living Things to Their Environment" and "Rise and Progress of Modern Views as to the Antiquity and Origin of Man." He set up the first by playfully moving from recent advances in understanding "surface-geology or Earth-sculpture"—alluding to glacially shaped landforms—to what he called "surface-biology"—the coloration of species in relation to locality and the puzzling interrelationships of plants and insects on islands (a subject just then growing in interest to him). In the second part, he kept away from his more heretical notions and focused on making a case that the commonly held idea of linear, progressive human advancement is errone- ous and that human history, if not evolution, is far messier than that, more a two-steps-forward-one-step-back proposition. It went over very well.

So far so good, but alas Wallace had also made a fateful decision to allow a paper on mesmerism and spiritualism to be read in the Anthropology session, a decision colored by his spiritualist convictions and a concern that the scientific community was not giving the subject a fair hearing. A young physicist named William Fletcher Barrett, at the Royal College of Science in Dublin, read his paper titled "On Some Phenomena Associated with Abnormal Conditions of the Mind" to a packed lecture room. Barrett was interested in hypnotism and "thought transference," and although he was not sure he bought into spiritual- ism altogether, he thought there was something to it and encouraged scientific investigation into its validity or invalidity. The discussion following the paper, Wallace presiding, started off amicably enough but predictably devolved into a veritable shouting match despite Wallace's best efforts to keep it on topic, at one point being drawn into the heated argument himself. All this was bad enough, but what followed was worse—he was publicly lambasted by zoolo- gist E. Ray Lankester, a Huxley protégé at University College London, who accused Wallace in a long letter to the *Times* of degrading the BAAS by his "more than questionable" conduct in allowing Barrett's paper to be read. Taken aback, Wallace wrote a rebuttal, but it only added fuel to the fire.[37]

Going from bad to worse, the BAAS meeting had no sooner ended when the hapless Wallace then got sucked into a related matter: the trial of the fa- mous American medium Henry Slade, who had arrived on the London scene the previous summer.[38] Slade was a sensation on both sides of the Atlantic, famous for his séances, at which he would enter a trance, and writing would appear on blank chalk slates held under the table. Wallace had, of course, at- tended several such gatherings with Slade and was convinced that he was the real deal, but unbeknownst to him Lankester and his associates had decided enough was enough and conspired to make an example of Slade by exposing him as a fraud and having him prosecuted. This they did, by going along with the proceedings of a séance until a critical moment when Lankester grabbed the slate from Slade, showing that it already had writing on it before it was placed beneath the table. Slade was charged under the 1824 Vagrancy Act "for the Punishment of idle and disorderly Persons, and Rogues and Vagabonds," a rather broad statute that among many other things outlawed "pretending or Professing to tell Fortunes, or using any subtle Craft, Means, or Device, by Rogues and Palmistry or otherwise, to deceive and impose on any of His Majesty's Subjects."[39] The penalty: up to three months' hard labor. The month-long trial—known as the Great Spiritualist Case at Bow Street—began on 2 October 1876 to a packed courtroom. The trial increasingly took on a circus-like atmosphere—especially when professional magician Nevil Maskelyn took the stand for the prosecution and demonstrated tricks of the trade. Wallace not only attended the proceedings but, toward the end, was even called to the witness stand for the defense. Perhaps the worst part about that, in the eyes of his scientific peers, was that he was prominently identified as "President of the Biological Section of the British Association for the Advancement of Science"—a humiliation for professional science.[40] Slade was duly convicted and given the maximum sentence, but before he could serve time, the conviction was appealed and overturned on a technicality, and he soon fled the country. Appearing as a star witness in support of Slade left yet another stain on Wallace's scientific reputation; he would have been mildly amused, but not surprised, had he known that Darwin helped finance the prosecution.[41] It was to be essentially the last BAAS meeting he would attend, save a brief appearance at the 1881 meeting in York, close to his friend Spruce's home.

It wasn't altogether fair, actually, the severe criticism leveled at Wallace for his, well, enchantment with spiritualism. I mentioned that a great many eminent Victorians were devotees of the movement, or the supernatural generally, and a great many more were dabblers. But more pointedly, some of Wallace's most vocal critics often themselves surreptitiously attended séances—including John Tyndall, William Carpenter, and even Thomas Huxley, despite his early claim to be uninterested in "disembodied gossip." They would claim that their motivation was more in the vein of "know thy enemy" than belief, but it is worth noting that while the three publicly derided Wallace as deluded or a rube, they seemed oddly attracted to spiritualist phenomena and attended séances rather more frequently than would seem warranted for either entertainment value or ghostbusting research.[42] Some critics of Carpenter pointed this out, prompting furious responses from him bordering on the Shakespearean—methinks the guy *"doth protest too much."*[43]

For his part, for better or worse Wallace stuck to his guns and pointed out that many scientific insights celebrated today were initially dismissed as quackery, delusion, or worse. "The whole history of scientific discovery," Wallace declared, "from Galvani and Harvey to Jenner and Franklin, teaches us, that every great advance in science has been rejected by the scientific men of the period"—and, moreover, rejected "with an amount of scepticism and bitterness directly proportioned to the novelty and importance of the new ideas suggested and the extent to which they run counter to received and cherished theories."[44]

———

Restless and fretting over five-year-old Will, who was doing poorly, Wallace acted on a medium's advice and moved the family again just two years after arriving in Dorking. This time it was to bustling Croydon, a market town a bit closer to central London and on a direct rail line—decidedly flatter terrain but a pleasant climate. In March 1878 they moved into a rented house, Waldron Edge, just south of town on Duppas Hill Lane. It was close to Duppas Hill Park, an expansive green that saw jousting matches in medieval times and where the Wallace children undoubtedly frolicked and played. An added bonus, to Wallace's delight, was Croydon's well-established local scientific scene—the Croydon Microscopical and Natural History Club (now the Croydon Natural History & Scientific Society). He was a fairly regular attendee, contributing to paper discussions and treating his fellow members to exhibits of choice specimens from his personal collection—beautiful tropical bird skins on one occasion, butterflies on another, including the fabulous leaf mimic *Kallima paralekta* from Sumatra. He was even part of an (ultimately unsuccessful) effort to change the society's membership rules to allow women to attend meetings.[45]

Just a month after moving to Croydon, Wallace had another book come out: *Tropical Nature, and Other Essays*. Mainly an exposition of Wallace's theory of coloration—including a renewed and vigorous critique of Darwin's sexual selection—the book is perhaps most interesting from a modern perspective for its prescient environmentalism. This is another of the fascinating dots that can be connected through many of Wallace's writings where he veers suddenly into the role of environmental prophet, although his emphasis underwent an evolution of sorts over time. Think back to his eloquent comment on extinction in the famous "Physical Geography of the Malay Archipelago" paper of 1863, warning that if we had the power to preserve species yet failed to act, future generations would charge us with having recklessly allowed them to "perish irrecoverably from the face of the earth, uncared for and unknown"—all the worse for science since every single species represents a piece of the evolutionary puzzle, a strand of the richly woven tapestry of life. Species are like the "individual letters which go to make up one of the volumes of our earth's history," and just as a few missing letters can render a sentence or paragraph unintelligible, "so the extinction of the numerous forms of life which the progress of cultivation invariably entails will necessarily render obscure this invaluable record of the past." But there Wallace was not calling for action to *stop* extinction and indeed seemed to think it was inevitable with the "progress of cultivation." Rather, the preservation he had in mind was establishing comprehensive natural history collections of the kind to which he was contributing. Selectionist that he was, Wallace never wavered from the view that the spread of "civilization" would result in the displacement, if not the wholesale destruction, of native species, but he increasingly came to see that if this spread were rampant—unchecked, unregulated, thoughtless—the destruction visited upon the natural world would ultimately come back to bite us.

Here Wallace was more and more embracing a holistic vision of nature that was surely inspired by his early reading of the great explorer-naturalist Alexander von Humboldt, whose epic *Personal Narrative of a Journey to the Equinoctial Regions of the New Continent* had fired up the Wallace and Darwin generation with a passion for travel and exploration in exotic locales. Humboldt had a grand Enlightenment vision for the integrated study of nature—an ecological vision, yes, but even bigger and broader than that: cosmological, geological, geographical, meteorological, zoological, botanical, and even anthropological since humans are part of nature. It is no coincidence that Humboldt's most profound insights grew out of his studies of the interrelationship of the organic and inorganic worlds—that is, the interrelationship of species and Earth: biogeography! Consider Humboldt's analysis of plant distribution in relation to elevation and latitude in *Essay on the Geography of Plants* (1807). A founding document of biogeography, this was also the first and most famous expression of his holistic "science of the earth."[46] Embarking on his journey to the New World, Humboldt explained his plan: "I shall endeavor," he declared in 1799, "to find out how nature's forces act upon one another, and in what manner the geographic environment exerts its influence on animals and plants. In short, I must find out about the harmony in nature." By "harmony" Humboldt was speaking to the interconnected, integrated, seemingly self-regulating whole of nature in all its glory, and the road to such insights lay in first elucidating patterns in nature in order to infer the processes behind them.

This should sound familiar: Wallace was eminently Humboldtean in his quest to do just that.

Humboldt's philosophy of nature permeated two works that Wallace knew well: *Aspects of Nature* and the first volumes of *Cosmos*, both of which had been translated into English by the late 1840s. Among other things, Humboldt's support for the then heretical notion of *progressive evolution* à la *Vestiges* got Wallace's attention in those early days—he wrote Bates back in 1845 about how he had heard that the "venerable Humboldt," in his celebrated *Cosmos*, "supports in almost every particular its theories," including those relating to the hot-button issue of the transmutation of species.[47] Indeed, when he eventually got his hands on his first *Cosmos* volume, he would have found Humboldt's grand vision front and center in the introduction, a vision of connections between species and thus, by extension, connections "between the laws of the actual distribution of organic beings over the surface of the globe, and the laws of the ideal classification by natural families, analogy of internal organisation, and progressive evolution."[48] It was all of a piece, literally and figuratively.

Humboldt's method and spirit, then, were ever-present in Wallace's own travels and investigations, and when he stepped back and regarded the big biogeographical picture in the 1870s, the German master's holistic vision truly came to the fore. "Every change becomes the centre of an ever-widening circle of effects," Wallace declared in *Geographical Distribution of Animals*. "The different members of the organic world are so bound together by complex relations, that any one change generally involves numerous other changes, often of the most unexpected kind."[49] Recognizing the exquisite web of relations between species and their environment naturally inspires an appreciation for just how easily these are disrupted by misguided human activity and a call to action—as in Humboldt's astute recognition of how forests themselves regulate their own climate, warning of the double danger of clear-cutting: "By felling the trees which cover the tops and sides of mountains, men in all climates seem to bring upon future generations two calamities at once; want of fuel and a scarcity of water." Wallace appreciated this fact too, pointing out that forest destruction likely led to the desertification of much of the Middle East and North Africa: "Much of this extensive area is now bare and arid, and often even of a desert character; a fact no doubt due, in great part, to the destruction of aboriginal forests."[50] Here Wallace was citing the 1864 environmental manifesto of American diplomat and polymathic scholar George Perkins Marsh: *Man and Nature: Or, Physical Geography as Modified by Human Action*. A founding document of the modern environmental movement, *Man and*

Nature was a clarion call to the dangers of rampant environmental degradation that itself drew heavily on Humboldt. Wallace elaborated in *Tropical Nature*, again citing Marsh, where he connected the dots between excessive deforestation, increased temperatures, drought, erosion, and loss of soil fertility. The Wallace who was once admiring of the intensively cultivated districts of Southeast Asia now decried excessive agricultural (and other) clearing that, if unchecked, would eventually lead to the "deterioration of the climate and the permanent impoverishment of the country."[51] He was prescient: twentieth-century ecologists and conservation biologists have well documented the suite of devastating effects of habitat fragmentation and widespread clear-cutting on local and even regional environmental parameters.[52] Darwin's response to *Tropical Nature* was pretty typical of their scientific circle: he gave polite congratulations but did not know quite what to make of it, as favorably disposed as he was toward what we would now call conservation issues. Of course, in his case he was distracted by those parts of the book aimed at undermining sexual selection, an ongoing issue over which the two could not agree.[53] By then Wallace had come to reject Darwin's concept of sexual selection by female choice altogether, or at least that part of Darwin's theory in which mate choice was based on pure aesthetics: a "taste for the beautiful," full stop. Wallace held that bright coloration, melodious song, and curious courtship antics may indeed be attractive to females, but this is because such traits are indicators of health and vigor. Natural selection was ultimately at play here, not Darwin's sexual selection. Again, they agreed to disagree; Darwin was frustrated at times with his wayward friend. (But it is worth pointing out that the modern view aligns with Wallace's far more than Darwin's on this issue.[54])

As we will see, Wallace continued channeling Humboldt (figuratively, in a nonspiritualist sense . . .) as he thought longer and more deeply about such issues. But more immediately, exciting news came to his attention: the City of London was looking to fill the new post of superintendent of Epping Forest, a large six thousand-acre (twenty-four hundred-plus hectare) tract of ancient forest of beech, birch, hornbeam, and oak just northeast of London. Mainly running along a long, broad northeast–southwest trending glacial moraine about twelve miles (nineteen kilometers) long by two and a half or so miles (four kilometers) wide, forming a broad ridge between the meandering Rivers Rode and Lea, Epping Forest was and is a lush mosaic of woodlands, heath, and grasslands crossed by gravely streams and dotted with bogs and ponds, a green island in the ever-flowing Metropolitan Sea. As a Royal Forest since perhaps as early as the twelfth century, hunting there was restricted to the crown, while wealthy freeholders owned plenty of parcels and locals had traditional rights to gather firewood (mainly through pollarding) and graze livestock. Social and technological changes through the nineteenth century increasingly led to another form of usage by the working class: railroads connecting the towns around the forest to greater London meant that the area was accessible to day-trippers on weekends and holidays. An average of fifty thousand people were estimated to visit Epping Forest on Sundays and Mondays each week in 1865, and a whopping two hundred thousand on Easter Monday alone by railway and other conveyances.[55] Despite traditional access privileges for commoners, the forest was nearly lost to development pressures brought on by enclosure—the movement that employed Wallace as a young surveyor and no doubt kindled his socialist streak, benefiting, as it did, the wealthy at the expense of the poor. As was nearly the case with Epping Forest, all was nearly lost by the mid-1860s after the crown sold its rights and enclosures dramatically increased—but happily this prompted a spirited campaign to save the forest, with protests, petitions, campaigns by the newly formed Commons Preservation Society, and fusillades of letters to editors of London newspapers and magazines. Seemingly against all odds, the City of London eventually came to the rescue, purchasing over fifty-five hundred acres of remaining unenclosed land, compensating the landowning freeholders but also setting certain limits on common use. In 1878 the Epping Forest Act was passed by Parliament, and the City of London assumed conservatorship of the forest, the law stipulating that the City shall "at all times keep Epping Forest unenclosed and unbuilt on as an open space for the recreation and enjoyment of the people."[56] Some tracts required restoration, others management—what was needed was a good land manager, a dream job for any socially engaged biogeographer! Plus, he really needed the money. A committee was formed and a job ad circulated. Wallace not only applied for the position but actively campaigned for it with letters and memorials sent by his friends and supporters.[57] He seemed like a natural, but while the movement to save Epping Forest may have been visionary, it is fair to say that the conservators of the City of London Corporation overseeing the place were not. Long story short, Wallace did not get the job—and it may have been his very enthusiasm and vision for what could be done there that did in his application. In November 1878 he published an article in the *Fortnightly Review* sketching out a truly innovative idea for the forest. Sure, the areas that needed restoration could be simply reforested with native trees or even developed into an arboretum of sorts—but how boring is that? How about an arboretum with pizzazz—would it not be amazing to develop a kind

Alfred Russel Wallace,
age fifty-five.

of comparative biogeography garden representing the primary temperate zone forest types of the world?! He envisioned an arboretum of arboreta, so to speak, featuring emblematic trees and shrubs from each forest type in turn: eastern and western North American forests, eastern Europe and West Asia, East Asia and Japan, and the temperate forests of the Southern Hemisphere, including southern South America, Australia, and New Zealand. The idea was, he proudly summed up, "perfectly novel, perfectly practicable, intensely interesting as a great arboricultural experiment"—attractive to all, from the uneducated day-tripper to scientists and students, no more costly than other plans, and consistent with the requirement to maintain the site as forest.[58] The idea was novel, all right—*too* novel for the conservators. The job went to one Alexander McKenzie, a good if very conventional landscape gardener from Scotland who held the post for the next fourteen years, a triumph of the uninspired and unimaginative. Wallace's Epping Forest article may have backfired—ironically, the disappointed Wallace privately complained, he was told that the committee opposed his plan to introduce "foreign trees," saying they wanted a "true English forest," only to busy themselves with planting Lombardy and black poplars, both nonnatives. He chalked that up to typical ignorance: the clueless committee members no doubt thought those trees were British because they are so commonly planted around the city, he confided to his correspondent.[59]

Wallace was again rather ahead of his time: today biogeographical and other educational collections are the highlights of many a botanical garden and arboretum, including Kew Gardens.[60] Being convinced he had visionary ideas was no consolation but was another bitter pill. His third job rejection since returning from the East, this one *really* stressed him out as he was about broke and had been counting on the income. His friend (and fellow spiritualist) Arabella Buckley, who had been personal secretary to Charles Lyell and became close to both the Wallace and Darwin families, tried to help. She wrote Darwin after Wallace's rejection, confidentially alerting him to his colleague's financial straits. With Victorian stoicism Wallace would never reach out for assistance, but he was in real need—might Darwin and his friends help find Wallace some modest post somewhere to ease the financial strain, she wondered? "I will gladly do my best," was Darwin's prompt response. Maybe not a post, but a government pension might be just the thing; he would write Hooker straightaway about it. Which he did—but Hooker shot down the suggestion.[61]

Here Wallace's unscientific enthusiasms and recent debacles came back to haunt him, not the kind of visitation he preferred. Hooker was unsparing: "Wallace has lost caste terribly," he pointedly told Darwin. Not just on account of his nonsensical spiritualism but, even worse, the Barrett disaster at the BAAS, bringing spiritualism into a major scientific meeting and creating a sullying public spectacle that reflected poorly on science. Hooker also claimed that Barrett's paper was approved in an underhanded way by Wallace, "deliberately & against the whole voice of the Committee of his section." That was not quite true, but his anger was real enough. In any case, icing on the Wallace silliness cake for Hooker was "the Lunatic's bet about the Sphericity of the Earth, & pocketing the money." Not the done thing—dishonorable, even, for a scientific man. No, Hooker told Darwin, given all these transgressions he would be hard-pressed to ask fellow scientists to support Wallace in good conscience—not to mention they would have to be up front with the politicians they petitioned about Wallace being a spiritualist, lest that come out later and embarrass some member of Parliament, leaving a black mark on scientists for future consideration for a Civil List pension. Anyway, he concluded, "Wallace's claim is not that he is in need, so much as that he can't find employment"—the unkindest cut of all. Darwin was a bit taken aback—he thought only of Wallace's distress and his service to natural history, he replied to Hooker, not these other issues. Following Hooker's advice, he dropped the matter: "What a mistake & mess I should have made had I not consulted you."[62]

Hooker would later change his mind, but for now as unkind and unfair as his objections were in some ways, on the matter of scientists and pensions he was expressing a very real political concern: Civil List pensions at one time were given out as political favors as well as to aid the accomplished but impecunious. Formalizing the process in the early nineteenth century, the government established eligibility criteria that did include those who made "useful discoveries" in science along with those who were accomplished in literature and the arts. But while some prime ministers were supportive of considering scientists, others were not—between 1838 and 1870, barely more than 10 percent of the awarded pensions went to scientists or, posthumously, their families. Some felt, too, that need should play a role in awarding pensions, while others did not. These questions were much debated over the next several decades, so by the 1870s Hooker's concern about jeopardizing the standing of scientists in consideration for pensions was genuine—this was precisely the time that the X Clubbers were working hard to raise the profile—establish the indispensability, even—of science in public affairs.[63] Thus, in the space of three days the idea of putting Wallace up for a government pension was raised and dropped, unbeknownst to him, of course. Darwin wrote with condolences about the Epping Forest position. "I suppose it was too much to hope that such a body of men should make a good selection. I wish you could obtain some quiet post & thus have leisure for moderate scientific work."[64]

Between their puzzlement over *Tropical Nature* and Wallace's continued involvement in spiritualism and social issues, the word in scientific circles may have been that Wallace's creative potential had been exhausted. Down but not out, though, he continued to plug away. He was just then hard at work on another book, one that would very much get their attention—a book destined to be hailed as another watershed contribution and that would change his financial situation for the better (not so much through sales, decent as they were, but because of the renewed faith it inspired in his scientific acumen). The book was a companion volume of sorts to *Geographical Distribution of Animals*, published in October 1880 to great acclaim: *Island Life, or, The Phenomena and Causes of Insular Faunas and Floras, Including a Revision and Attempted Solution of the Problem of Geological Climates*.

He was back, with bells on: this book was a tour de force, classic Wallace in its breathtaking sweep and originality. Here Wallace treated the special qualities of islands with respect to dispersal and speciation, setting the stage by pointing out biogeographical anomalies: why should such far-flung countries as Britain and Japan be so strikingly similar in flora and fauna, while other areas far closer together—think Australia and New Zealand or, better, Bali and Lombok—are radically different? The great islands of the world present similar anomalies, he pointed out: while some closely resemble the nearest continent in species, others differ dramatically; some are chock-full of endemic species, while others have few. Successfully sleuthing such biogeographical mysteries requires an eye for clues, and Wallace was nothing if not the Sherlock of species distributions past and present. It is evolutionary, my dear reader: Wallace recognized that islands were key to the solution. The time was ripe for a grand synthesis, he declared, as the requisite clues were then available: a theory of evolution by natural selection, plus advances in the understanding of the geographic distribution and systematics of many groups, the fossil and stratigraphic geological records, ocean-floor bathymetry, and the planetary orbital forcing of climatic cycles, which initiates glacial periods. (One area where modern science has greatly superseded Wallace is in his belief in the permanence of continents and ocean basins in geological time—it would be another thirty years before continental drift was proposed by Alfred Wegener and eighty years before it was accepted.)

It is hard to overstate the significance and masterful crafting of *Island Life*—the founding document of modern island biogeography.[65] The book is divided into lengthy parts, the first of which, "The Dispersal of Organisms: Its Phenomena, Laws, and Causes," methodically lays out the *principles* and *processes* that shape geographical distribution in space and time—including evolutionary change itself, a treatment that included a clear discussion of speciation in allopatry, the divergence of subpopulations when spatially separated.[66] This section includes a lengthy treatment of glaciation, including two chapters dedicated to the work of James Croll, the self-taught Scottish mathematician and astronomer. Croll developed a theory of glacial periods stemming from the combined climatic effects of cyclical changes in the tilt of Earth's rotational axis (precession of the equinoxes) and in the circular-to-elliptical shape shifting of earth's orbital path (eccentricity). Wallace closely studied Croll's theory in the 1860s, incorporating it into his 1869 paper "Sir Charles Lyell on Geological Climates and the Origin of Species"—the (in)famous review in which Wallace first announced his reservations over the role of natural selection in human mental evolution. Wallace evidently got a bit carried away extolling Croll's ideas in that paper, later lamenting to Darwin that although he tried to treat the theory fully, "the Editor made me cut out 8 pages!"[67] I feel his pain. Now, in *Island Life*, he gave the Scotsman his due, dedicating two of the longer chapters to the theory of orbital cycles as a cause of glacial cycles—and in so doing laid out remarkably prescient arguments for the dynamics of the periodic advance and retreat of continental glaciers and the attendant effects on the geographical distribution and evolution of species. In the modern view, he may have often been off the mark on the details, but he was surely right on in general principles.[68]

The second section of *Island Life*, "Insular Floras and Faunas," treats a diversity of island systems as case studies, illustrating and extending the principles laid out in part 1. There Wallace opened with a classification of islands, building on Darwin's early geological work on coral atolls defining true oceanic islands (also called hot-spot islands today) as volcanic in origin and never having been connected with a continental area. To this Wallace added continental islands (also called land-bridge islands today) with more complex histories but that have formerly been connected to mainland areas, sometimes cyclically.[69] In the course of his analysis, Wallace came around to Darwin's view about dubious continental extensions and lost continents, torpedoing the idea once and for all and building upon his elder colleague's model of chance dispersal events by wing, wind, and water to get colonists to remote islands (to Darwin's delight).[70]

A great deal more could be said about Wallace's treatment than space here allows.[71] Suffice it to say that the scientific scene was wowed and Darwin, for one, astounded—he declared it "quite excellent" and told Wallace it was the best book he had ever published. To Hooker Darwin admitted that while he "admired it extremely," the book was almost too much for him to take in, so breathtaking in scope were Wallace's analyses and arguments. He didn't agree with everything in there, but on the whole he had to hand it to Wallace. Hooker could only concur that it was an outstanding work. "I am only two thirds through Wallace & it is splendid. What a number of cobwebs he has swept away," he declared, and continued, shaking his head: "That such a man should be a Spiritualist is more wonderful than all the movements of all the plants" (alluding to Darwin's latest work, *The Power of Movement in Plants*, also recently published).[72] But Hooker was more than simply impressed with Wallace's book—he was surprised and deeply touched by the dedication: "To Sir Joseph Dalton Hooker, who, more than any other writer, has advanced our knowledge of the geographical distribution of plants, and especially of insular floras, I dedicate this volume on a kindred subject, as a token of admiration and regard." "There is no one," Hooker wrote Wallace, "from whom I should appreciate the dedication of such a book as yours more happily than from you."[73]

———

That was it—enough was enough. Darwin wasn't going to take no for an answer this time: he rightly saw *Island Life* for the landmark work it was and immediately resolved to get Wallace that pension he so badly needed, spiritualist or not. His contributions to science were far, far bigger than that—just astounding, their depth and breadth! Darwin rallied the troops, writing Buckley, Huxley, Hooker, and others to give them a heads up. They all readily agreed, Hooker's change of heart no doubt inspired by not only the quality of Wallace's book but its kindhearted dedication. Darwin drew up a memorial with the help of Arabella Buckley and Huxley, immediately had it signed by another eleven leading naturalists and politicians, and conveyed the memorial to Prime Minister William Gladstone just after the New Year.[74] The Duke of Argyll, long impressed with Wallace, wrote directly to the prime minister himself. The decision came swiftly: the prime minister was glad to recommend a pension of £200/year for Wallace, generously retroactive to the first of July the previous year! If you think Darwin was elated—"Hurrah–Hurrah!" he wrote Huxley—Wallace was deeply moved, overjoyed, and relieved in equal measure when he heard. He thanked Darwin from the bottom of his heart: "There is no one living to whose kindness in such a matter I could feel myself indebted with so much pleasure & satisfaction."[75]

Darwin's letter bearing the good news arrived at Wallace's home on 8 January 1881—a fine present indeed for his fifty-eighth birthday.

CHAPTER 13

A SOCIALLY ENGAGED
SCIENTIST

"WE ARE ONLY JUST in & are in great confusion," Wallace hurriedly wrote, but encouraged his friend to come by and visit just the same.[1] It was early May 1881, and the family had just moved to Nutwood Cottage, on Frith Hill, Godalming. Another lovely market town south of London, rolling Godalming was not too far from Croyden and Dorking, situated at the edge of the North Downs in the valley of the River Wey. There on a hillside above the river, he cultivated both a garden and friendships with the masters of nearby Charterhouse School—that fine institution, founded in 1611 in London and relocated to Godalming in 1872, was one of the attractions of the place to Wallace and Annie, who wanted an excellent school for their children. "Nutwood" was inspired by the veritable forest of hazels (*Corylus avellana*) and stout oaks (*Quercus robur*) covering their little half acre.[2] Over the course of the next eight years, the couple reveled in their hillside idyll, where they nurtured some thousand plant species between the grounds, garden, and greenhouse—ever-obliging Hooker providing seeds and specimens from Kew.[3]

Nutwood was the second home built by Wallace, thanks to eased financial concerns, but it was his ninth (!) residence since returning from the East, reflecting a peripatetic restlessness matched by his intellectual wanderings. Indeed, by the time Wallace moved in he had already embarked upon another intellectual journey, in characteristically public fashion: two months earlier he had been elected founding president of the Land Nationalisation Society of all things. This appointment stemmed from a remarkable paper Wallace had published in the November 1880 issue of *Contemporary Review*: "How to Nationalize the Land: A Radical Solution of the Irish Land Problem"—a subject that may seem like a new enthusiasm but was in fact part and parcel of

Nutwood Cottage, Godalming, Surrey—Alfred and Annie Wallace's gardening idyll, ca. 1873.

Wallace's long-standing advocacy for social justice in societal advancement.[4] The seeds of his humanitarianism had been planted long ago in his engagement with Owenite socialism, and we have also seen this in his subsequent social commentary—critiques as well as praise of the state of Indigenous societies in Amazonia and the Malay Archipelago, for example—an interest that was turbocharged by his reading of Herbert Spencer's *Social Statics* in the 1860s.

Take Wallace's conclusion in the closing chapter, "Races of Man," in *The Malay Archipelago*: "Before bidding my readers farewell," he began, "I wish to make a few observations on a subject of yet higher interest and deeper importance, which the contemplation of savage life has suggested, and on which I believe that the civilized can learn something from the savage man." We in so-called civilized societies, he continued, may be making progress in attaining, eventually, some ideal social state—"a state of individual freedom and self-government, rendered possible by the equal development and just balance of the intellectual, moral, and physical parts of our nature"—but the egalitarian Wallace observed that many "primitive" societies were practically there already. Near-equality, low crime, the absence of spirit-crushing competition and struggle, little wealth disparity or class or educational distinctions—qualities that many in the West aspired to were very much in evidence already in the forests of Amazonia and Borneo. Sure, Western societies have made greater intellectual or technological strides, but we have not advanced much morally, he maintained. "A deficient morality is the great blot of modern civilization, and the greatest hindrance to true progress," he declared, pointing out that our much-touted technological and commercial progress and the crowded cities and towns it leads to "support and continually renew a mass of human misery and crime *absolutely* greater than has ever existed before." Until this failure is recognized, until we aim to develop the more "sympathetic feelings and moral faculties" of our nature and use these to inform social policies and practices, we will never advance beyond "the better class of savages."[5]

At the time, in 1869, Wallace's eloquent critique got the attention of famed philosopher John Stuart Mill, who had recently co-founded the Land Tenure Reform Association, an organization that aimed to build on the Reform Act of 1867 in advocating for land reform through the elimination of primogeniture and the entails that keep vast properties forever locked up within families and other privileged groups. The existing land laws, a veritable feudal system, originated, by design, at a time when the landowners ran the country, Mill charged, propping up the privileged class. He asked rhetorically if it was any wonder that such laws should require alteration now, when the country belongs (at least in principle) to everyone.[6] Recognizing a simpatico spirit in Wallace, Mill invited him to become secretary of the Land Tenure Reform Association, which met at the venerable Freemason's Tavern in London where the modern De Vere Grand Connaught Rooms stands in Great Queen Street today. Wallace was only too happy to accept. Never very comfortable with public speaking, however, he was less willing to take the podium at the meetings, though at Mill's encouragement he did put forth some of his ideas—such as one adopted at the 9 July 1870 meeting asserting the right of the state to take ownership, with compensation to the owners, of sites of great historic, scientific, or aesthetic value.[7]

Unfortunately, the Land Tenure Reform Association didn't get very far before Mill's death in 1873, and it dissolved not long afterward. Wallace's interest remained strong, however, and through the rest of the decade, he paid close attention to the increasingly vigorous debate over land reform—regarding Ireland in particular, then wholly part of Britain and under its thumb. Ireland had little recovered from the horrific famine of the mid-1840s in which millions died or emigrated, the deadliness of which was greatly exacerbated by the policies and practices of the English absentee landlords who owned most of the land, employing often ruthless agents to extract rents from their impoverished tenants or summarily evict them altogether.[8] The first of the Irish Land Acts, promoting "peasant proprietorship," was passed in 1870 under William Gladstone (with Mill, a member of Parliament at the time, playing a prominent role) but was seen as a half measure that did not go far enough to address the problems (and injustices) inherent in the tenancy system. Conflict erupted again in 1879 with the start of the "Land War" in County Mayo, Ireland, and the subsequent establishment of the Irish Land League advocating for the "Three Fs": Fair Rent, Fixity of Tenure, and Free Sale. Just as that paper by Forbes had provoked Wallace into writing the lucid Sarawak Law paper back in the 1850s, it was one of the Land League's proposals that elicited an immediate and strong response from Wallace, catapulting him into the thick of the land reform movement. Without getting into the details of the Land League's proposal, suffice it to say that Wallace took a dim view, however supportive he was of their goals: "That a scheme so impracticable as this—and even if practicable so unsound and worthless—should be put forth by a body of educated men, who have, presumably, studied the subject, is a noteworthy fact." Ouch!

But not one to simply wag a finger and criticize, Wallace offered solutions. The silver lining to their "unsound and worthless" proposal was that it showed "the importance of a thorough and fearless discussion of all questions relating to the tenure of the land," he said, a discussion that would help clarify fundamental principles and inform legislation. His paper on "How to Nationalize the Land" was intended to be just that—a fearless foray into a hot-button political issue, clarifying fundamental principles and offering a legislatively based "radical solution" to the problem: classic Wallace. While even the likes of Adam Smith, author of *Wealth of Nations*, a founding document of modern capitalism, condemned practices like primogeniture and entail as feudal anachronisms, almost no one questioned the wisdom let alone morality of private land ownership in general. But it was axiomatic to Wallace that this was an evil and more: an evil that begat further evils. He did not advocate for outright state seizure of land, however—that would be unjust, and he agreed that any scheme of government compensation paying market value to those whose land is taken would bankrupt the treasury. His solution was more thoughtful, respectful of individual rights as well as historical precedent, building upon a straightforward principle: "Whatever acts may be done by an individual without injustice or without infringing any rights which others possess or are entitled to claim in law or equity, then acts of a similar nature may be done by the State, also without injustice."[9] Taking a page from the evolutionary and

"What Things Are Coming To; Or, the Boycotted and Land-Leagued Landowner," lampoon of Irish rent boycotts, 1880.

geological playbook—gradualism in all change—he proposed a process for transitioning, over several generations, away from individual property ownership. State ownership did not mean state management, however, avoiding the inefficiencies and possibilities for favoritism that might accompany "state-landlordism." Rather, the state's role would be little more than its present one as tax collector. He separated a *tenant right*, the ownership of all improvements on the land (buildings and other infrastructure), from land ownership per se. A nominal ground rent based on the land's inherent value would be paid to the state for occupancy, while the tenant right becomes the object of ownership that can be bequeathed or sold, even subdivided, but not sublet. What to do about wealthy landlords with extensive holdings and tenants? Their tenants would purchase the tenant right from them at a price determined by the difference between the ground-rent valuation and the average rent paid over the past five years, in payments made over a period of years like a mortgage. If this was financially out of a tenant's reach, Wallace envisioned a co-op or credit union approach to lending the money, to be repaid in installments over a limited period (he did not mention interest). Mill would have been impressed.

There is much more detail, of course, but you get the gist. No radical of the bomb-throwing persuasion, then, what Wallace sought was a solution that balanced the rights of the individual with that of society, with the aim of re-dressing injustices and, in a perhaps naive utopian sort of way, simultaneously helping to cure some of the social ills that stemmed from the current system—rural depopulation and urban overcrowding, for example. Although addressed to the burning issue of the Irish land question, Wallace pointed out that the principles he laid out are "equally applicable to England as to Ireland" and indeed are "of universal application."

That was of course precisely the problem, in the eyes of the privileged classes of England. Wallace closed his essay appealing to "the independent Liberals of Great Britain and to the long-suffering Irish nation," asking only for "a careful perusal, an unprejudiced consideration, and a searching criticism of my proposals." He got searching criticism, all right, and plenty of consideration—his article precipitated a fierce response, both pro and con, and he characteristically met each published criticism with a reply, always up for a debate. He was immediately contacted, too, by a group of like-minded activists who rallied around Wallace to establish the Land Nationalisation Society, by acclaim electing him its first president ("much against my wishes"). He threw himself into the role with his usual enthusiasm, and finding that there was no single, concise overview of the subject, he decided to write one. To reach his intended audience—the "landless classes"—he resolved that the book should be "clear and forcible, moderate in bulk, and issued at a low price."

While working away on the project, Wallace came across a remarkable book by one Henry George, an American journalist-turned-political economist and reformer. Originally from Philadelphia, George had moved to California as a teenager, where he worked first as a printer and then reporter and managing editor of the *San Francisco Times*, taking on political corruption, land specula-tion, wealth monopolization, and railroad and mining interests. In 1879 George published *Progress and Poverty*, seeking to explain the paradox that so bothered Wallace: poverty amid plenty, as the well-known expression goes, and more. Why is there widespread poverty despite overall rising economic tides, and why and how does wealth become concentrated, geographically and demo-graphically? The book's quintessentially nineteenth-century subtitle says it all: *An Inquiry into the Cause of Industrial Depressions and of Increase of Want with Increase of Wealth: The Remedy.* His central point: "Nothing short of making land common property can permanently relieve poverty and check the tendency of wages to the starvation-point. . . . Private property in land always has, and always must, as development proceeds, lead to the enslavement of the laboring class."[10] It was a phenomenon, quickly running to multiple editions, selling millions of copies, and becoming a (maybe *the*) founding document of the Progressive Era.[11]

When Wallace got hold of a copy in early 1881, he was blown away. He re-read it several times over, he told Darwin that summer, something he rarely did (*Origin* being one other prominent case). It's a must-read, he enthused, "the most startlingly novel and original book of the last 20 years," as revolutionary as Adam Smith. Just then Wallace was mainly working in his new garden, out-doors all day "admiring the infinite variety and beauty of vegetable life," he told Darwin. But with the return of short days and cold weather, he planned on throwing himself into his book on the "land question," delighted to have found a potent ally in George. Darwin replied that he would order *Progress and Poverty*, though he admitted it was a subject on which he "utterly" distrusted his own judgment while also much doubting everyone else's. George's book would probably only make his mind even more confounded, he concluded. Such political issues have an allure, but he hoped that Wallace would not "turn renegade to Natural History." I like to think that Darwin may have suppressed an impish impulse to insert a parenthetical "again" into that sentence, but he didn't seem to be feeling his impish self just then. He had recently returned from a family holiday in the Lake District but did not much enjoy it, tiring easily whether walking, socializing, or reading: "Life has become very wea-rysome to me."[12]

He need not have worried about Wallace "turning renegade" to natural history though, not completely: that year Wallace fired off several meaty let-ters regarding ongoing debate over "geological climates" and the cause of ice ages—one of his favorite subjects—and published several reviews of scien-tific books, including a noteworthy review of anthropologist Edward Tylor's *Anthropology: An Introduction to the Study of Man and Civilisation.* There Wal-lace sketched his "mouth-gesture" theory of the origin of language, proposing that language originates in imitative or emotive sounds, where the sound or movements the mouth makes in enunciating words is evocative of the very things being described—soft-flowing sounds conveying words like "smooth-ness," "polish," "oily," for example, contrasting with the harsher-sounding staccato of those conveying the opposite: "rugged," "rough," "gritty." Or con-sider the way pronouncing the word "glue" or "sticky" entails a pulling apart of the tongue and palate; "run" and "fly" are quick to pronounce, while "drag" or "crawl" are slow(er); "in" and "out" are pronounced with inspiration and expiration, and so on. This idea of language originating in onomatopoeia and sound mimicry may not hold much water today, in the view of linguists, but it was a big idea at the time and if nothing else is certainly another example of Wallace's undiminished creativity.[13] Yes, Wallace remained engaged at mul-tiple levels on the natural history front, but there is no question that land reform—and other social causes, as we will see—took up more and more of his time and energy.

In his last-known letter to Darwin, dated 18 October 1881, Wallace sent thanks for the gift of Darwin's latest (and final, as it would happen) book—a treatise on earthworms, giving Wallace a newfound respect for those humble creatures. In return he enclosed copies of two letters on land nationalization he had just penned for the *Mark Lane Express and Agricultural Journal*, a weekly whose editor, William Edwin Bear, applauded Wallace's efforts on behalf of tenant farmers.[14] Darwin probably didn't applaud—as sympathetic as the lib-eral Darwin was on one level, he wished his friend and colleague would focus more on science. It seems resonant, then, that just at the time the sad news of Darwin's death reached Godalming the following April, in 1882, Wallace had no fewer than three papers in the works with the journal *Nature*, two brief scientific book reviews, and a commentary. One review treated the first volume of William Distant's *Rhopalocera Malayana*—an authoritative and beautifully illustrated analysis of the butterflies of peninsular Malaysia and environs. (Wallace would have chuckled at Distant's earnest opening line: "A description of Malay Butterflies needs neither apology nor defence.") His other *Nature* review gave a thumbs-up to the third part of August Weismann's incisive *Stud-ies in the Theory of Descent*, recently translated by Wallace's friend Raphael Meldola, industrial chemist by profession (inventor of the dye Meldola's Blue) and talented amateur entomologist with a special interest in mimicry in butter-flies.[15] In this third collection of papers, Weismann, at the University of Freiburg in Germany, analyzed seasonal dimorphism in butterflies, the coloration of caterpillars, and other forms of variation in light of natural selection. Weis-mann was soon to emerge as a major "next-gen" Darwinian, making seminal evolutionary contributions; Wallace already recognized his talent.

The commentary Wallace published that month was more in-depth, heaping praise on Fritz Müller's recent extension of what is now known as Müllerian mimicry, another triumph of Darwinian insight. Bates's original analysis of mimicry in Amazonian longwing butterflies—Batesian mimicry—brilliantly showed how palatable mimics evolve the color and pattern of unpalatable models through natural selection. He was puzzled, however, by cases where mimic and model, often from distantly related genera or even families, are *both* unpalatable. Wallace was equally puzzled, following Bates's lead in attributing this, unhelpfully, to "unknown local causes." It is surprising that Wallace didn't intuit what was going on, and in 1879 Müller published the first of his papers on the subject, proposing a solution: *mutual* reinforcement of unpalatability by convergence on much the same color and pattern. Now he had a follow-up paper with more evidence, treating the phenomenon in great detail and criticizing Wallace's "unknown local causes" as no explanation at all. Far from put out, this provides another example of Wallace's fair-mindedness: "I may at once say that I admit this criticism to be sound," Wallace wrote in his commentary and proceeded to review (and applaud) Müller's explanation in detail. The science was the thing for Wallace—the ideas came first. He could take criticism when criticism was due. "If these views are correct," he concluded in his piece, "we shall have the satisfaction of knowing that all cases of mimicry are explicable by one general principle; and it seems strange to me now that I should not have seen how readily the principle is applicable to these abnormal cases. The merit of the discovery is however wholly due to Dr. Fritz Müller."[16]

Darwin would have admired Wallace's reviews, especially his commentary on Müller's papers, and would have recognized the same magnanimity that Wallace had so often showed Darwin himself over the discovery of natural selection. But by the time these articles were published, Darwin had been buried with much fanfare in Westminster Abbey. He died at home on 19 April 1882, age seventy-three, from heart failure.[17] Although he had wanted to be buried with his family in the Downe churchyard, his scientific friends and admirers immediately campaigned for interment in Westminster Abbey—honoring the man as well as British science in the eyes of the nation. The family vacillated but acquiesced. The funeral was held one week later, the cathedral packed with mourners and the great and good paying their respects. Pallbearers included two dukes and an earl representing the state—among them Darwin's old nemesis, the Duke of Argyll—Royal Society president William Spottiswoode, American ambassador and poet J. Russell Lowell of

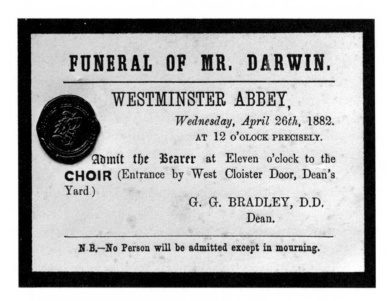

Admission ticket to Darwin's funeral at Westminster Abbey.

Boston, and the four living British scientists of the Darwin-Wallace circle: Hooker, Huxley, Lubbock, and of course Wallace himself—though it may say something about his suspect standing in the eyes of some in the scientific establishment that he was almost overlooked until George Darwin thought of him.[18] Wallace was surely pleased that his esteemed friend should be laid to rest near Newton, considering that he never wavered from his first enthusiastic response to *Origin*: "It is the '*Principia*' of Natural History. It will live as long as the 'Principia' of Newton."[19]

In some ways Darwin's death marked the end of an era, but of course that only meant the beginning of another: increasing recognition of Wallace as the First Darwinian. Coincidentally, just two months after the funeral Wallace was awarded the high honor of a Doctor of Laws (LLD) degree from Trinity College of the University of Dublin—on the recommendation of the Reverend Samuel Haughton no less, he of the scathing "Bee's Cell and the Origin of Species" critique that had roused Wallace to come to Darwin's defense some eighteen years earlier. They had crossed pens, as it were, a couple of times after that too. Now *there's* another example of generosity of spirit, in fine Wallacean tradition. Wallace met Haughton after the ceremony, "enjoying his instructive and witty conversation" with an assembled group of worthies over breakfast. "The brilliant midsummer morning, the cosy room looking over the beautiful gardens, and the highly agreeable and friendly party assembled rendered this one of the many pleasant recollections of my life," Wallace later wrote.[20]

———

In this post-Darwin period, Wallace was to continue to prove the greatest defender of the Darwinian faith even as his extra-scientific interests continued unabated—unlike Darwin, he was quite the intellectual multitasker. In the same month that saw the publication of his three pieces in *Nature*, he also published his small book on land nationalization, a primer of 240 pages dedicated to "The Working Men of England." Its object was to "reveal to them the chief cause of so much poverty in the midst of the ever-increasing wealth, which they create" and explain the reforms necessary to enable them to reap their just rewards and give all who seek it a fair shake in the economic prosperity of the country. Sounds familiar—Wallace would be disappointed to find that some things never seem to change, not least the disparity between rich and poor, which today (near the end of the first quarter of the twenty-first century) has perhaps never been greater in Western societies. Then, as now, moneyed interests tried their best to stifle, if not silence, the Wallaces of the world. Even his longtime publisher, Alexander Macmillan, backed out of an agreement to publish *Land Nationalisation* after receiving threats from powerful quarters, profoundly unhappy with Wallace's treatment of the 1814 Sutherland Evictions in Scotland: "A process of ruin so thoroughly disastrous that it might be deemed scarcely possible to render it more complete," in a few short years turning a "well-conditioned and wholesome" district into "one wide ulcer of wretchedness and woe, as one commentator described it."[21] It indeed was perhaps *the* most shameful episode of the thoroughly shameful Clearances of Scotland, eviction of the local people from their Highland and Lowland homes by absentee landlords, with allegations of atrocities committed in the process.

The London publishing house Trübner & Co. agreed to bring the book out instead, but emotions still ran high over the episode, and Wallace, pursued by the son of one of the accused who had been acquitted of murder for his role in the evictions, threaded a delicate needle by dropping the person's name while upholding the fact of the atrocities in subsequent editions.[22] The piece hit a nerve in others too, and in the following year, Wallace found himself again dueling with his pen, defending his position by writing lengthy letters answering his critics, including the noted economist Alfred Marshall, writing pamphlets for the Land Nationalisation Society, and elaborating on his ideas in a long two-part article in *Macmillan's Magazine* titled "The 'Why' and the 'How' of Land Nationalisation."[23] That one was an invited article, an opportunity to respond to an earlier piece by Cambridge economist and reformer (and postmaster general) Henry Fawcett. Wallace was disappointed and frustrated that the liberal Fawcett, a distinguished champion of the working class and an advocate of women's suffrage, declared land nationalization unjust and a financial impossibility. Wallace's little book was mentioned in the article, but it seemed to Wallace that Fawcett must not have read it—there was zero engagement with his ideas, the economist mainly fixated on Henry George's rather different prescriptions. Wallace's long reply to Fawcett was an opportunity to set the record straight, and in so doing he made a remark worth noting. He (again) acknowledged the huge financial impediment to the state buying out landowners (which is why it was not part of his scheme) but added that even if it *were* possible, doing so would be unacceptable. Why? Because to pay for the land is to tacitly admit that its private ownership is at some level acceptable, *warranting* just compensation. Wallace was all for smoothing the transition to state ownership of land, but he saw private ownership as in principle immoral and thus so, too, was the "class of wealthy idlers, supported out of the produce of land which we have declared rightly to belong to the community."[24] Wallace does not dwell on the point, moving right along to his main argument, but seemingly offhand comments like this one give us insight into Wallace's character: it was always the *principle* of the thing to him.

Perhaps inspired by all this back-and-forth on land nationalization, Macmillan regained his nerve soon enough and in 1885 published Wallace's next missive on a related economic subject: *Bad Times*.[25] Again its wonderfully nineteenth-century subtitle says it all, worth giving in full:

An Essay on the Present Depression of Trade, Tracing It to Its Sources in Enormous Foreign Loans, Excessive War Expenditure, the Increase of Speculation and of Millionaires, and the Depopulation of the Rural Districts; with Suggested Remedies

It's that last bit I love—pure glass-half-full Wallace, ever one to offer solutions! Historians of economics today do not give Wallace much credit—he was more synthesizer and presenter than originator of novel economic ideas, perhaps, but we should still honor his heroic efforts. This particular book, a long essay, actually, arose ironically enough as a result of (surprise, surprise) Wallace's own "economic depression." His investments were by and large a bust, his royalties rather limited. His Civil List pension was certainly a big help, but he was having difficulty making ends meet between school expenses for his two teenagers and a household to support, on top of which he developed a problem with his eyes that cut into his productivity. (His doctor prescribed no reading or writing for an extended period—quite a problem for someone who made his living by the pen!) This helps explain why, when he saw a notice for a contest offering £100 for the best essay on "The Depression of Trade," he jumped at it. Of course, having been "for some time disgusted with the utter nonsense of many of the articles on the subject in the press" (again, surprise surprise), he had clearly been ruminating on the subject and probably would have entered the contest regardless. His essay did not win, but the results were not disastrous like his *last* effort at winning "easy money" (the bet with the crazed flat-Earther, Hampden). The judges were impressed enough to give his essay a sort of honorable mention, asking if they could publish it in part. He turned down that consolation prize, approaching Macmillan instead to publish the piece in its entirety.

His productivity dropped to about half its usual level between 1883 and 1884 owing to his eye affliction, but he couldn't possibly stop working and managed to pen a bevy of essays, addresses, reviews, and letters on subjects scientific and social. On the scientific front, there were reviews of books on bees and elementary mathematics, and he could not resist reporting on the latest botanical beauties blooming at Nutwood Cottage to readers of the *Garden* magazine, like his lovely scarlet-flowered legume *Sutherlandia spectabilis* from South Africa or the white-trumpeted *Datura meteloides* from California. He was happy to share the love too: seeds anyone? "I shall be pleased to send a few to anyone wishing to grow it, who will send me a stamped and directed envelope."[26] One snapshot of the Wallaces and their beloved garden at this time comes from the Irish poet and sometime editor of *Fraser's Magazine*, William Allingham, a kindred spirit whose stirring lines about the infamous eviction of forty-seven Irish tenant farmer families one night in 1861 Wallace would have known and admired: "In early morning twilight, raw and chill | Damp vapours brooding on the barren hill, | Through miles of mire in steady grave array | Threescore well-arm'd police pursue their way | Each tall and bearded man a rifle swings | And under each greatcoat a bayonet clings."[27] Allingham and his wife, Helen, an accomplished artist, lived in nearby Witley and visited the Wallaces periodically. On a warm August day in 1884, they were treated to a tour of the garden's latest productions while their kids Sonny and Evey, eleven and seven, respectively, "raced about" with fifteen-year-old Violet and

Will and Violet Wallace at a favorite family pastime.

thirteen-year-old Willy. "[Wallace] shows us round his garden—rare plants and flowers"—a little "Californian tulip," lovely Canada lily, and eucalypts— "Three kinds, very tender." Later Allingham and Wallace reclined under a tree and discussed spiritualism a good long while.[28]

Allingham related this conversation to another friend, the great poet laureate Alfred, Lord Tennyson, who asked Allingham to bring Wallace around one day. The devout Tennyson's faith had been severely shaken by the sudden death of his dear friend Arthur Hallam back in 1833, especially as he struggled to come to terms with this loss in the context of an uncaring, pitiless nature that seemed to follow from Lyellian deep time, revelations of long-extinct creatures of the former world, and then Darwinian-Wallacean transmutation.[29] It was all too much; he was intrigued by Wallace's spiritualism, though skeptical. That November Wallace took the train from Godalming and was met by the Allinghams and taken to Aldworth House in Haslemere, Tennyson's abode. There the poet praised *Tropical Nature* and peppered Wallace with questions about the tropics, and they discussed the spirit world, mediums, and séances at length. Tennyson was unconvinced: "A great ocean pressing round us on every side, and only leaking in by a few chinks?" he asked Wallace at one point, who apparently shrugged off the question. Somehow politics came up, and Wallace, never one to withhold an unpopular opinion, denounced the worthlessness of the House of Lords and the absurdity of hereditary titles and property. He may or may not have been aware that Tennyson had just had the title Baron Tennyson bestowed, giving him a seat in the House of Lords, that very year. Probably no one would have judged the visit especially successful, but clearly Tennyson kept ruminating over his conversations with Wallace. A month later he commented to Allingham that it was "a very strange thing that, according to Wallace, none of the Spirits that communicate with men ever mention God, or Christ." Wallace, he said, believes in a system he said was far finer than Christianity: "It is Eternal Progress."[30] That was just it.

As readily as Wallace would discuss spiritualism with anyone interested, the subject took a back seat for a time in the mid-1880s to topics such as land nationalization, the "tyranny of capital," Henry George's theories, illegal highway enclosures, and . . . a whole new cause: anti-vaccination. "Oh no!" a modern reader might lament, "not another anti-scientific debacle! What is Wallace thinking?!" But it's important to remember that Wallace was all about the science, as he understood it, and that the understanding of epidemiology, vaccines, and biostatistics were in their infancy in the late Victorian period. We cannot hold those who lived in the past to standards based on modern understanding. We now know that vaccination is a triumph of modern medicine, saving countless millions of lives in the past century, but we have to consider Wallace's critique in light of understanding his time—and on that basis, evidence for the efficacy of vaccination was on shaky ground in the nineteenth century, just when the medical profession was pushing it hard, including through punitive legislation—a red flag for Wallace's sense of justice.

In the Far East, India, and Africa, there had been a centuries-long history of inoculation (or variolation) against smallpox, the horrific viral disease that has killed, disfigured, or blinded untold millions and has about an overall 30 percent mortality rate. The variolation procedure, which was introduced to Britain (not without controversy) in the early eighteenth century, involved purposeful infection using matter taken from the pustules of individuals suffering from a milder form of the disease in hopes that it would induce the same mild disease while conferring immunity to the more deadly form. The technique worked fairly well, but it did have its risks, and a small percentage of those so treated died from the disease. Later in the eighteenth century, several individuals in Europe (including Edward Jenner and Benjamin Jesty in Britain) hit upon the idea of inoculating with material from cowpox pustules instead—a related disease of cows that has mild effects on people but confers immunity to smallpox.[31] This, too, had its detractors, but over time, with analyses of its efficacy and relative risks, *vaccination* against smallpox, as it was dubbed by Jenner (he coined the term from the Latin *vacca*, or "cow"), was not only accepted but codified into law. The 1840 Vaccination Act made variolation illegal and provided for optional vaccination free of charge. But with the rise of professional science in Britain—including, broadly, the medical establishment with its successes and growing authority—the Vaccination Act was amended several times over the next several decades as smallpox epidemics periodically swept through towns and cities, leaving misery and death in their wake.

Vaccination first became compulsory with an 1853 amendment to the Vaccination Act and became increasingly more stringent in the amendments of 1867, 1871, and 1874. Wallace had nothing against vaccination per se early on— he had been vaccinated prior to his travels abroad, and he and Annie had their children vaccinated. But compulsory vaccination now rubbed him the wrong way, and he became increasingly mistrustful of the confident assertions and authority wielded by the medical establishment—his view likely colored by both his commitment to individual liberty as well as his painful experiences with the scientific community over his spiritualist convictions. Once delving into the data, he came to reject claims of the scientific efficacy of vaccination. He had a growing holistic view of health that strongly suggested to him that good nutrition and sanitary living conditions played a major role in reducing disease transmission and susceptibility.

With all this in mind, consider that in the eyes of many the vaccination laws seemed to target the poor, who were often forced to live in squalid conditions that fostered vice and disease. Beginning with the 1867 amendment, "vaccination districts" were created under the authority of the Boards of Guardians, committees that oversaw the enforcement of the Poor Laws, including administration of the hated workhouses, and now were empowered to oversee increasingly punitive fines for parents or guardians not complying with the vaccination law.[32] This seemed wrong to Wallace—bad social policies created unhealthy, disease-ridden conditions, and instead of addressing the cause of this misery, the state was forcing people to undergo an unnecessary, even dangerous, procedure, as he saw it. Consider his reasoning and how he saw the layers of injustice: first, if smallpox vaccination were simply ineffective, neither helping nor harming, compelling people to get vaccinated under penalty of fine or imprisonment was an affront to personal liberty. But if the vaccine were *dangerous*, actually increasing the likelihood of mortality, it was doubly unjust, indeed perverse, that the state should compel it. On top of that, if the poor working class was disproportionately targeted and hurt by the law—social and political policies keeping them in a state of poverty in which they are preyed upon by disease—then penalizing them for it by forcing them to undergo a dangerous treatment was *triply* unjust, beyond perverse, and positively evil. You can appreciate how this issue would push all of his social justice buttons, not least produce outrage at the arrogant and unfeeling establishment running roughshod, as he saw it, over those least able to defend themselves. It's no wonder that the anti-vaccination leagues that sprang up almost immediately in response to the new laws struck a chord with Wallace. He not only sympathized but, in Wallacean fashion, lent the authority of his stature and compelling (and combative) voice to the cause.[33]

The anti-vaccination movement in Britain (and abroad) had a huge following, an alliance of resisters attracted to the cause for various reasons (or a combination), just like today: for some anti-authoritarians it was a matter of personal liberty, for example, while others were fearful that the treatment was dangerous without being particularly efficacious. While many of today's anti-vaccination activists may be steeped in mis- and disinformation, clueless about the science, the activists of Wallace's time had a solid leg to stand on: the state of the science from vaccine manufacture to data analysis was indeed less than rigorous (by modern standards), something that Wallace homed in on in his scathing assessment of the claims of the medical profession of the day.[34] The first indication that he had been thinking about the matter came in the form of a letter written in September 1883 to anti-vaccination campaigner (and fellow spiritualist) William Tebb, who had co-founded the London Society for the Abolition of Compulsory Vaccination in 1880. Wallace later said it was Tebb who first introduced him to anti-vaccinationism sometime in the late 1870s. Encouraged to read up on the subject, he soon cast a critical eye on the prevailing statistical arguments in favor of vaccination and was astonished to come across writings by Spencer, one of his idols, suggesting that the 1840 Vaccination Act actually led to an increase in the incidence of smallpox.[35] The inherent injustice would have been glaringly obvious to him, and he was hooked. Tebb invited Wallace to speak at the 1883 International Anti-Vaccination Congress in Bern, Switzerland, but he had to decline. His letter supporting their cause was soon published, though,[36] and he was also glad to provide an endorsement for Tebb's recent book on the subject: "*Compulsory Vaccination in England* is an admirable little book. The sections on Vaccination in the Army and Navy are themselves absolutely conclusive as to the uselessness of vaccination, and its hurtfulness, if the facts therein stated are correct."[37]

But this was only the beginning: the year 1885 saw the publication of a broadside against smallpox vaccination that Wallace addressed to members of Parliament, which he had distributed as a pamphlet: "To Members of Parliament and Others. Forty-Five Years of Registration Statistics, Proving Vaccination to Be Both Useless and Dangerous." He threw down the gauntlet with the opening sentence: "Having been led to enquire for myself as to the effects of Vaccination in preventing or diminishing Small-pox, I have arrived at results as unexpected as they appear to me to be conclusive." Being a question affecting not merely personal liberty but the health and the very lives of thousands, he said, it was a duty to "make the truth known to all, and especially to those who, on the faith of false or misleading statements, have enforced the practice of vaccination by penal laws." The pamphlet was divided into two parts, the first treating smallpox mortality and vaccines, the second, comparative mortality of the vaccinated and unvaccinated. In the first part, he opened with four "statements of fact" that together undermined pro-vaccination arguments: smallpox mortality has ever so slightly diminished over forty-five years, but, he maintained, there is no evidence to suggest that this decline was due to vaccination, nor has vaccination mitigated the severity of the disease. On the contrary, he charged in his final point, enforced vaccination has led to an *increase* in disease. Wallace then turned to comparative mortality data and skewered blanket claims that vaccination leads to lower mortality by making an important point about the analysis of epidemiological data: rather than lumping into broad "vaccinated" and "unvaccinated" categories, comparative analysis needs to be more fine-grained—for example, broken down demographically. It was a characteristically Wallacean tour de force of lucid argument that concluded with an urgent, all-caps plea:

WE, THEREFORE, SOLEMNLY URGE UPON YOU THE IMMEDIATE REPEAL OF THE INIQUITOUS PENAL LAWS BY WHICH YOU HAVE FORCED UPON US A DANGEROUS AND USELESS OPERATION—AN OPERATION WHICH HAS ADMITTEDLY CAUSED MANY DEATHS, WHICH IS PROBABLY THE CAUSE OF GREATER MORTALITY THAN SMALL-POX ITSELF, BUT WHICH CANNOT BE PROVED TO HAVE EVER SAVED A SINGLE HUMAN LIFE.[38]

Yes, Wallace was a busy guy on the literary front—cheered by some, jeered by others—and at this phase his output mainly took the form of letters, pamphlets, reviews, and articles rather than books (an exception being his

economic policy critique *Bad Times*). Letters and articles streamed from
the study at Nutwood Cottage, many diagnoses and helpful prescriptions for
societal ills among them: "Church Funds: How to Use Them," "How to Cause
Wealth to Be More Equally Distributed," "Illegal Roadside Enclosures," "Three
Acres and a Cow," "State-Tenants versus Freeholders," and on and on. At this
point Wallace's politics were moving decidedly further to the left, but he still
considered himself a member of the Liberal Party. We know that because he
said so, penning "Why I Am a Liberal" for a collection of short testimonials,
"Being Definitions and Personal Confessions of Faith by the Best Minds of the
Liberal Party." The bottom line, for him, was that in light of a world "full of
oppression and wrong," the Liberal Party recognized the need for reform; the
wheels may turn slowly, but turn they do: "Believing that the terrible social
evils which now afflict us can only be remedied by giving to all an equal right
to share in the gifts of nature to man, *I look with confidence to the Liberalism of
the future for a recognition of this fundamental right, and its embodiment in our
constitution and law.*"[39] But as we will see, it was not too long after penning
these words that Wallace lost patience with the Liberal Party. Although he was
already slowly but steadily sliding toward socialism, that process was accelerated by an unlikely opportunity that arose: a journey across North America.

———

What started it all, actually, was an invitation to lecture in Sydney, Australia—
an invite that may have been prompted by his impressive volume *Australasia*
in Stanford's Compendium of Geography and Travel, a hefty and authoritative
six hundred-page treatment of the natural history, geology, and ethnography
of that great region from the South Pacific to the eastern Malay Archipelago.[40]
First published in 1879, this popular work had gone into its fourth edition by
1884. Another journey around the world was the last thing on his mind,
though, and he was not sure he wanted to go. It was an honor, certainly, and
he was always up for an adventure, but that was a *long* journey; it would mean
being away from his beloved family and garden for quite a long time. And there
was his age too; he was in his early sixties now. On the other hand, that was
just it—he was in his sixties . . . if not now, when? He wasn't getting any
younger. But a tempting factor was financial—maybe he could make some
serious money, something that was in chronically short supply in the Wallace
household. "If I had the prospect of clearing £1,000 by a lecturing campaign, I
would go, though it would require a great effort," he wrote to his friend A. C.

Swinton, treasurer of the Land Nationalisation Society.[41] The suggestion was made that he lecture his way across the United States on the way to Australia, an appealing idea in several respects: it could pay well and provide opportunities to see his brother John, in California, whom he had not seen in nearly forty years, as well as the great landscapes, flora, and fauna of North America. A month after writing Swinton, he solicited the advice of Othniel Charles Marsh, the famous (some would say infamous[42]) Yale paleontologist who had met Wallace when visiting London. Did Marsh think there would be interest in Wallace lecturing? More to the point, would it be likely to be a financial success? Marsh sent out inquiries on Wallace's behalf, writing to Daniel Gilman, founding president of Johns Hopkins University, and Augustus Lowell, trustee of the Lowell Institute in Boston. Wallace also reached out to Carl Ernst, editor of the weekly Boston newspaper *The Beacon*, who then also wrote to Gilman and Lowell.[43] Lowell responded immediately, writing Wallace to suggest a course of "6 or 8 lectures & séances" (séances?) at the Lowell Institute the following November.[44] The venerable Lowell Institute had been founded by bequest in 1836 by textile industrialist John Lowell Jr., a scion of one of the first families of Boston. A supporter of the Lyceum movement—U.S. versions of mechanics' institutes—and the Boston Society for the Diffusion of Useful Knowledge, the philanthropically minded Lowell had established his institute in the same self-improvement spirit, providing free public lectures in the arts and sciences, philosophy, and natural history.[45] Lowell, the institute's second trustee, also had a keen interest in science, serving as a member of the Corporation of the Massachusetts Institute of Technology (MIT) and vice president of the American Academy of Arts and Sciences. Following the lead of his predecessor, founding trustee John Amory Lowell, he ensured that scientific luminaries of the United States and Britain were well represented among the Lowell Institute's lecturers.[46] Gilman was less responsive, mainly because he was busy, but when his Johns Hopkins colleagues reached out to Huxley to get a sense of Wallace as lecturer, the response did not inspire confidence: "The substance of what he has to say is sure to be worth listening to, even if it should be about spirit rapping and writing," Huxley allowed, "but I have grave doubts whether his style of speaking is such as to lay hold of a large general audience." But then he lightened the blow a bit with his usual self-deprecating humor: I'm only speculating, he said, and after all, "I hate listening to lectures & have often said I would not hear my own if I could help it."[47] But it was true: Wallace was sure no Huxley when it came to public speaking. Some people—Huxley, certainly, and the late Stephen Jay Gould in more modern times—had the gift of speaking as swashbuckingly as they wrote. Wallace did not: his writing was lucid, penetrating, at times humorous in an understated way, but it's fair to say that it took great effort for him to really engage with an audience.[48] But Gilman and colleagues at Johns Hopkins did eventually invite him and suggested that the nearby Peabody Institute would be the best venue and pay better. That institution was founded in 1857 by philanthropist George Peabody as another kind of Lyceum for public enrichment, renowned then as now for its magnificent five-story library arranged around a central atrium reading room, glass skylights in the soaring ceiling flooding the space with natural light—aptly celebrated as "a cathedral of books."[49]

In the meantime, Wallace still wasn't quite sure he was up for it. Lowell suggested that he confer with poet and art critic Edmund William Gosse, who had lectured at the Lowell Institute in 1884. Wallace quizzed Gosse and others, who encouragingly gave him the lowdown and suggested he get an agent. He soon engaged the Williams Lecture and Musical Bureau ("B. W. Williams, Proprietor") on the recommendation of "parson-naturalist" Rev. John George Wood, author of the popular *Common Objects of* . . . books and illustrated natural histories of field botany, entomology, and more, who had also given lectures at the Lowell Institute in 1883–1884. That winter he finally accepted the Lowell Institute's offer and settled upon a course of eight lectures to be delivered twice weekly in November and December 1886. The idea was to provide an exposition of grand evolutionary and geological themes, from "The Darwinian Theory: What It Is, and How It Has Been Demonstrated" (as ever, he still gave all the credit to his friend Darwin), to coloration and mimicry, island biogeography, and the permanence of continents and ocean basins. *Pre*-pre-Powerpoint, audiovisuals then consisted of one's voice for the "audio" and handmade posters, diagrams, and lantern slides for magic lantern or stereopticon the "visual"—the design and execution of which all took some doing.[50] But scientific lectures were only part of Wallace's agenda. He was keen to connect with American spiritualists and see what he could of the landscape, geology, and natural history of the country—especially the botany. He would ultimately send back loads of plants not only for his and Annie's garden but also for their friend the distinguished horticulturist and garden designer Gertrude Jekyll, who lived just south of town in Busbridge.

Wallace arrived in New York on the afternoon of Saturday 23 October 1886 aboard the steamship *Tower Hill*—a rough passage that vividly reminded him of how seasick he used to get, leading him to immediately drop the idea of proceeding from North America to Australia. It could have been worse—that was hurricane season, after all. But the meals he may have, er, lost on the trip over were more than made up for on his arrival: "As I was pretty hungry I enjoyed my first dinner in America. I had Cod & oyster sauce—most delicious then chicken, sweet potatoes and tomatoes then venison & currant jelly with mashed potatoes, to conclude with apple pie and Italian cream— spiff!! The apple pie being the most delicate & delicious possible. Then Catawba grapes and bananas for dessert."[51] After that logroller of a dinner, he was well fortified for a most remarkable ten months crisscrossing the North American continent.[52]

His itinerary was ambitious. Bookended by Massachusetts and California— the farthest points east and west on his trip—he visited dozens of cities and towns in a total of nineteen states plus two Canadian provinces, counting only stops. Sure, it may not have been quite the fourteen thousand-plus miles he had covered in the Malay Archipelago—and far less dangerous—but it was ambitious nonetheless. He spent his first two months (November and December 1886) based in the Boston area, then the next three, January through March 1887, in Washington, DC. He headed cross-country in early April that year, with stops in no fewer than ten states en route to California, with lectures and excursions in Virginia, West Virginia, Ohio, Indiana, Missouri, Iowa, Kansas, Colorado, Wyoming, and Utah. After nearly two months in California, where he was joyfully reunited with John and met his niece and nephews for the first time, he headed east again, stopping to botanize for a few weeks in Colorado, then on to Chicago, Michigan, Kingston, Ontario, and Montreal, Quebec, where he boarded a steamer, the *Vancouver*, bound for Liverpool.

Consider his first weeks to get a nice sense of his schedule in general—a busy mix of lecturing, social engagements, and sightseeing, the ratio of which varied from place to place. Upon arriving in New York, his first Lowell lecture a week away, Wallace was treated to a whirlwind tour of the city, including a stroll through Central Park and an excursion up the Hudson Valley to West Point, along the way admiring the celebrated Palisades, the great Triassic-age exposure of three-hundred-foot columnar basalt cliffs extending some fifty miles along the west bank of the Hudson River. He likened it to the famous Giant's Causeway formation in Northern Ireland, and indeed geologically they are the same phenomenon. He next took the train to Boston, where his agent met him, and was soon ensconced in the Quincy House Hotel, his home away from home for the next two months; lovely in its day but demolished in 1935, the Quincy stood on the site of the modern City Hall Plaza at Boston City Hall. He had a couple of days to get his diagrams set up in the auditorium

THE FIRST DARWINIAN, Wallace, did not leave a leg for anti-Darwinism to stand on when he had got through his first Lowell lecture last evening. It was a masterpiece of condensed—and as clear and simple as compact—statement, a most beautiful specimen of scientific work. Mr. Wallace, though not an orator, is likely to become a favorite as a lecturer, his manner is so genuinely modest and straightforward.

Newspaper account of Wallace's triumphant opening lecture on his North American tour, *Boston Evening Transcript*, 2 November 1886.

(Huntington Hall, in the Rogers Building on the MIT campus) and then ... his grand debut! Delivered to an audience some eight hundred to nine hundred strong, it was a triumph: "The first Darwinian, Wallace, did not leave a leg for anti-Darwinism to stand on when he got through his first Lowell lecture last evening," declared a reporter for the *Boston Evening Transcript*.[53] Wallace was delighted, enclosing newspaper clippings in a letter to his daughter, Violet, the next day and describing his reception by the attentive and cheering audience as well as the sights he had seen thus far. And the food: "But you ought to see the meals at this hotel!"[54]

The first British Darwinian soon met the first American Darwinian: Asa Gray of Harvard, foremost botanist and foremost champion of Darwin in the United States. Then seventy-six years old, Gray was retired at this point—at least from his professorial duties—but was trying to make progress on his ambitious *Synoptical Flora of North America*. Wallace and Gray surely had an awful lot to talk about given their mutual friendship with Darwin and enthusiasm for biogeography while both being rather unorthodox Darwinians: Wallace a spiritualist and Gray a devout Presbyterian who served as a deacon at the First Church in Cambridge. Gray was a dedicated theistic evolutionist, a position Wallace would migrate toward in the years to come. But little is known of their conversations, though they certainly had a very cordial relationship. They got together several times when Wallace was in Boston, and on one occasion Wallace stayed with the Grays, who put on a dinner in his honor, inviting a *Who's Who* of the scientific faculty of Harvard. Sitting around after dinner, Gray asked Wallace to relate to the assembled group the by then famous story of how he was led to discover natural selection. "This was followed by some interesting conversation"—oh to have been a fly on the wall at the Gray's home that evening![55]

While Wallace's scientific lectures were just that—no spiritualism mixed in—he did not hesitate to underscore his disagreement with Darwin over the evolution of mind and by extension his concept of a purposeful, teleological evolution for humans. There were many in his audiences who fully agreed and others who thought it nonsense, a contrast sometimes reflected in the accounts that duly appeared in the next day's newspapers, lauding Wallace's science and either omitting, politely skirting, or criticizing his little heresy. Among the interesting people introduced by Gray who were more sympathetic than not to his heresy was William James, the philosopher, historian, and early psychologist, and philosopher and mathematician Charles Sanders Peirce—together co-founders of the pragmatism school of philosophy. Both philosophers were dissatisfied with the scientific naturalism then ascendant, with its almost militant rejection of nonmaterialist views. They admired and were fascinated by Wallace, James with his commitment to the philosophy of radical empiricism, as he was to later call it—an empiricism that takes into account diversity of experience, including the psychological, to give insight into causality and meaning—and Peirce as a metaphysical evolutionist, dismissing Darwinian and (early) Wallacean evolution by natural selection as "pseudo-evolutionism," too mechanistic for his tastes.[56] James was rather open-minded about things paranormal too and, though skeptical, attended several séances with Wallace. Gray, James, and Peirce represented an American academic theistic evolutionism that resonated with Wallace's evolutionary teleology, and they were not the only ones: among the distinguished scientists that Wallace met, the geologists James Dwight Dana at Yale and Joseph LeConte at Berkeley were also of this persuasion, and certainly this was true of many of the American spiritualists he got to know too, prominently among them famed lecturer and suffragist Isabella Beecher Hooker (sister of novelist Harriet Beecher Stowe) and Senator Leland Stanford and his wife, Jane, the founders of Stanford University (who, like Wallace, had suffered the anguishing loss of a child, something the balm of spiritualism may have eased a bit). Again, the possibility of a spirit world, a parallel universe of sorts that we could communicate with, was widely entertained in those days in all strata of society. Indeed, it is interesting to note that the single largest and most financially lucrative lecture Wallace gave on his entire tour was on the subject of spiritualism, not science: "If a Man Die, Shall He Live Again" was delivered in San Francisco on 5 June 1887 and was so popular it was quickly printed as a pamphlet and subsequently reprinted more than any of his other works.[57]

———

Wallace held up pretty well on the lecture circuit despite the usual ups and downs. Anyone who has had a projector bulb blow during their presentation can commiserate with the downs he endured: "Lantern failure!" he recorded in his diary when lecturing on "Colours of Animals" at Williams College in Massachusetts, and on another occasion he had a double whammy: "Bad lantern operator and my lamp bad & went out!"[58] Then there were the headaches with his posters and charts. They were rolled up in an unwieldy six-foot-long package, and most railroads insisted they be sent as baggage, which often meant by different train, a recipe for disaster. On a few occasions his visuals failed to show up in time, leaving him scrambling to draw new ones from memory—like the time he arrived early in Bloomington, Indiana, to enjoy some local geologizing with Professor John Branner ahead of his lecture only to find when they got back to campus that his diagrams had failed to show up. Stress! Branner and a student helped Wallace more or less recreate his visuals, and to his relief it worked out: "Went off well not withstanding bad diagrams."[59] But such mishaps were rare, fortunately. All told, including his Lowell Institute presentations Wallace delivered forty-one lectures on his tour; thirty-eight were scientific, two on spiritualism, and one on land and economic reform, the great majority going off without a hitch. (He would have had more lectures, except that his agent dropped the ball and had nothing lined up for Wallace in Washington, DC, following his Peabody Institute engagement.) There were a few other mishaps, though—like getting on the wrong train at Clifton Forge, Virginia, on his way to West Virginia to visit William Edwards, the author of *A Voyage up the River Amazon* who had inspired and encouraged the young Wallace and Bates in their South American ambitions lo those many years ago. He realized his error after a mile or two but had to walk back, missing his correct train. Fortunately, it was no problem—he wired Edwards that he would be a day late and made the best of it, taking the opportunity to explore the local landscape, as he would. Following the Jackson River east from town to where it bends south at Rainbow Gap, he sketched in his diary the "Grand arched strata" visible there on either side of the river—Paleozoic sandstones of the Valley and Ridge Province buckled into a pronounced anti-cline, still very much visible along U.S. Route 220 today.

Geology figures prominently in his American travel diary—just about everywhere he went Wallace made general geological observations, on the lookout especially for evidence of glacial action, a special interest. He spent a day marveling at the spectacular formations of Luray Caverns in central Virginia, discovered only a decade earlier, and admired the impressive fossils on display in the many museums he visited. Museums, in fact, became a special object of study in themselves for Wallace. Recall his belief in "museums for the people"—exhibits and displays should be both instructive and visually appeal-ing, for the edification of visitors. He found his type specimen in the very first American museum he visited: Harvard's Museum of Comparative Zoology (MCZ), founded in 1859 by the famous Swiss American paleontologist and comparative anatomist Louis Agassiz. Despite Agassiz being a creationist thorn in the side of the early Darwinians, under his leadership the MCZ was nonethe-less at the cutting edge of museum pedagogy. His son Alexander, who was first a curator and eventually succeeded his father as director, was glad to meet Wallace and show him around the museum. He had none of his dad's scruples about evolution, though he did cross swords with Darwin over the theory of coral reef and island atoll formation. (Darwin was victorious on that count too.[60]) Alexander agreed with Louis's ideas about museum organization—today the zoological exhibits are still organized on his principles—namely, biogeographical. Rooms of gorgeous wood and glass cases present groupings of animals typical of each biogeographical realm—almost like three-dimensional realizations of those lovely engravings Wallace had executed for his 1876 *Geo-graphical Distribution of Animals*—snapshots of the fauna and flora typical of each region. Wallace was rhapsodic and later wrote an article singing the praises of the MCZ: "We will now pass on to the special feature of the museum, and that which is most to be commended, the presentation to the public of the main facts of the geographical distribution of animals. This is done by means of seven rooms, each one devoted to the characteristic animals of one great division of the earth or ocean." Published in the September 1887 issue of the *Fortnightly Review*, the article not only presented the MCZ as a model for museums back home but, in typical Wallace fashion, offered an even more ambitious vision. Declaring that "a grand opportunity is now afforded for a man of great wealth, who wishes to do something for the intellectual advancement of the masses," he proposed a Museum of Comparative *Paleontology*, designed on the same biogeographical principles, with an added temporal dimension—a museum of species in space and time![61]

He followed up with another museum article in the *Fortnightly*, this one with kudos and accolades for American archaeological and anthropological museums, in particular the Peabody Museum of American Archaeology and Ethnology at Harvard and the Prehistoric Archeology and Anthropology Exhibits of the Smithsonian Institution.[62] In this case it was not so much the arrangement as the richness of the collections that impressed him—something that can be rightly seen as problematic today. The enthusiastic collection—many would now say plundering—of ancient mounds, town sites, and burial sites of Native Americans has yielded great insights into the early Indigenous cultures of the Americas, it is true, but too often this was done with little to no regard for the rights and sensibilities—the humanity, even—of *living* Indigenous peoples, let alone honor for the dead. During his American travels, Wallace had opportunities to visit several ancient Native American mounds. Dotting the landscape of the Southeast and Midwest in the hundreds, the mounds give mute and melancholy testimony to lost cultures. Evocative in their way of the ruins of European antiquity, the ancient mounds were often (mis)interpreted in the nineteenth and twentieth centuries as the vestiges of a fallen civilization toppled by the direct "savage" ancestors of today's native people, like New World versions of the Goths and Visigoths that toppled Rome. His *Fortnightly Review* piece concludes with a mention of the then-current debate over the relationships of current Indigenous groups with those of the distant past, the builders of the mounds and other earthworks the archaeologists were so diligently excavating.[63] As respectful and admiring as Wallace (usually) was of non-Europeans, he was of his time in having no compunction about collecting arrowheads and other artifacts or conducting archaeological investigations that some might consider disrespectful today. But if he was not aware of the injustices of the way Native Americans were treated before, he certainly was after meeting Thomas Bland in Washington, DC. Co-founder of the National Indian Defense Association and editor of the *Council Fire and Arbitrator*, a leading journal advocating for reform in Indian-government relations, Bland asked Wallace to address an assembled party on the subject of land reform in the context of the relationship between rampant land speculation and the theft of lands supposedly guaranteed to the Indians. Wallace obliged and fully agreed about the root of the problem but lamented that given the fever pitch of land speculation in the country at the time most of his listeners just shrugged him off.[64]

If Wallace was not altogether enlightened about the treatment of Native Americans, he was certainly a progressive thinker in related ways. We have seen that in his advocacy for social justice, but in addition he was impressed with American educational trends—especially coeducation of the sexes and higher education for women in the context of women's colleges, trends just gaining momentum in Britain at the time.[65] American women's colleges predated most of their counterparts across the pond and awarded degrees from the get-go—Elmira College was the first, founded in 1855, followed by Vassar in 1861. He lectured on oceanic islands at Vassar in November 1886 to an attentive audience and was impressed with his tour of the campus, noting in his diary the rigor of the curriculum taught by "Lady Professors," listing the courses, admissions requirements, and the facilities, admiring the museum of natural history, art gallery, and observatory—shown around by Maria Mitchell, no less, the famed Nantucket astronomer who joined the Vassar faculty in 1865.[66] Later, in Washington, DC, he was invited to speak at the Woman's Anthropological Institute and developed a lecture on "The Great Problems of Anthropology," focusing on defining "race" and the origins of language.

Even more impressive to him was institutional coeducation—this got his attention at several towns he visited in the Midwest. Approaches varied: sometimes young men and women attended the same institution but took separate classes, while in other cases they actually learned together in the same classroom; Wallace didn't usually elaborate on the details. In Bloomington, Indiana, he noted that "all the public schools have co-education of the sexes. Succeeds well. Boys and girls board out, but meet on perfect equality in the classes. Meet also in debates &c. It succeeds well." In Sioux City, Iowa: "In all schools & colleges there is co-education. Ladies form a considerable proportion of the Professors & Teachers. . . . In these schools & colleges the girls quite hold their own against the boys, often surpassing them in languages." At the University of Kansas, he noted that co-education was the rule and observed that "a lady is professor of Greek, another teaches Latin, French, and German," while over in Manhattan the Kansas Agricultural College (now Kansas State University) offered free education for both sexes, though in typical fashion men's and women's curriculum differed: agricultural theory, chemistry, mathematics, and more for the men, while the women studied "domestic economy," cooking, and horticulture.[67] The fact that Wallace repeatedly noted such American educational trends in his diary is significant—his observations surely resonated with his developing views on women's rights, a natural extension of his sense of social justice that was to come into full expression in the years after his return from the United States.

But in the meantime, there was a grand, expansive continent to see! Even more than geology, the aspect of landscape that Wallace really reveled in was the botany. I mentioned he was constantly sending specimens back to his wife, Annie, and their friend Gertrude Jekyll; he sent quite a few to his father-in-law, William Mitten, too. He was in fact almost manic in his admiration for the wildflowers and could not collect enough of them. He even confessed to dashing off the train whenever it stopped at the smaller stations to quickly dig up any flowers nearby, "keeping a sharp watch for the conductor's cry of 'All aboard!'"[68] He was defeated only once, as far as we know, when he admitted to his father-in-law that he had had it trying to collect the gorgeous sego lily, *Calochortus nuttallii,* in the High Sierra. Botanical elegance itself, this delicate plant bears large flowers singly on a slender stem, each of its three snow-white petals painted with rich yellow and maroon at the base. He found them growing by the hundreds on rocky slopes, and that was just the problem: the bulbs were well protected, growing in and among pesky stones, and the delicate stems broke at the slightest tug. "Absolutely un-dig-up-able!" a frustrated Wallace declared. "I have tried till I am tired, & give it up."[69]

Wallace's botanical interests were an important element of his growing land ethic—everywhere he went he commented on qualities of the landscape, the nature of the vegetation. He had an eye for beauty in any season, as when a population of lovely paper birches and willows struck him on a late November botanizing excursion in western Massachusetts: "I here first noticed the very striking effect of the white-barked birches and yellow-barked willows in the winter landscape."[70] The Christmas ferns (*Polystichum acrostichoides*) he collected there lived for many years in his garden back home. And so did innumerable other plants—in the diary of his American travels he mentions nearly four hundred plants (!), many of which he collected and shipped home.[71] Nutwood Cottage was becoming a veritable botanical garden and nursery. He arrived in the East at the wrong time of year to see many wildflowers in bloom and had to first get through the frigid eastern U.S. winter, spent mainly around DC. But at last: "Our first really good spring botanizing," he remarked in late March. He and paleobotanist and sociologist Lester Ward, an avowed agnostic and monist—believer in a physical basis for mind—took a long walk to High Island, on the Potomac, where Wallace saw such beauties as mayapple (*Podophyllum peltatum*), spring beauty (*Claytonia virginica*), twinleaf (*Jeffersonia diphylla*), and many more wildflowers in bloom for the first time. Wallace later acknowledged that Ward was more advanced than himself in his progressivism (and socialism); they engaged in many a long metaphysical conversation on

Sunday botanical rambles around Washington, but Ward could not quite convince the committed spiritualist Wallace of monism, though the socialism was coming.[72] The most spectacular botanizing was yet to come too, and sooner.

Wallace's stay in DC drew to a close—it had been quite interesting, being shown around and feted about town. He got to be a tourist between the botanizing, lecturing, and séances, visiting the Patent Office, Congress, the Treasury, the Bureau of Engraving and Printing, the White House, and more—but he was no mere tourist. As one indication of Wallace's celebrity, besides professors and naturalists he spent time with a remarkable array of DC personalities: explorer and geologist Wesley Powell, who had led several famed Western expeditions in the late 1860s and was now head of the U.S. Geological Survey (Wallace was a regular guest at Powell's Cosmos Club); Senator Stanford, whom he would visit with again in California; James Brooks, head of the Secret Service; Spencer Fullerton Baird, head of the Smithsonian; and even the president himself, Grover Cleveland, then serving the first of his two nonconsecutive terms in office. Never one for formalities, Wallace found that a meeting to be more endured than savored, the feeling no doubt mutual as Cleveland had little interest in scientific subjects: "I had nothing special to say to him, and he had nothing special to say to me, the result being that we were both rather bored, and glad to get it over as soon as we could."[73] Meeting so eminent a figure as a sitting president might have been far from a highlight of Wallace's Washington visit, but overall he was very pleased: "I met more interesting people there than in any other part of America, and became on terms of intimacy, and even of friendship, with many of them. . . . For many reasons I left Washington with very great regret."[74]

Wallace headed west, for California, on 6 April 1887, taking about a month and a half to get there on a ten-state whistle-stop tour, absorbing the transition from eastern deciduous forest to rolling prairie to plains to high desert to mountains, the local geology becoming more visible as the climate became more arid and the vegetation thinned. His train rolled into Oakland, California, on 23 May, and John was there waiting for him. The brothers had not set eyes upon one another since 1848 when Wallace headed to the Amazon; John, one of the original forty-niners, headed to California the following year. Wallace doesn't say much about the reunion, being rather reserved on family matters as usual, but it must have been a very happy occasion. John had become an engineer, working at turns as engineer and president of the Tuolumne County Water Company, San Joaquin County surveyor, and chief engineer for the San Joaquin & Sierra Nevada Railroad. The tiny town of Wallace in western

Calaveras County, California, was named in his honor, sited at the midpoint of the new rail line he laid in the 1880s across San Joaquin and Calaveras Counties, linking Valley Springs in the Sierras south of Lake Tahoe with Woodbridge and Lodi in the Central Valley to easily connect with markets in San Francisco via steamer. John and Mary's oldest son, John Herbert, became a surveyor in the family tradition and actually surveyed the Wallace townsite.[75] By then John and Mary had five children, the second oldest, William, married with kids of his own.

But Wallace (Alfred, that is) would not meet the rest of the family for another week—first he and John headed to San Francisco, where the beautiful Baldwin Hotel and lecture engagements awaited. Thanks to John's efforts—he had arranged his brother's talks—notices were carried in four newspapers, and Wallace was introduced at his lectures by the distinguished geologist Joseph LeConte of Berkeley, who "alluded to evolution as the greatest idea of modern times, and the lecturer as the greatest living champion of that theory."[76] The lectures were a success, and that weekend the Wallace brothers were treated to an excursion—William Gibbons, doctor, amateur naturalist, and founder of the California Academy of Sciences, arranged for a day trip into the mountains in the company of famed naturalist, author, and environmental philosopher John Muir. The area they visited was known as the "redwood graveyard" to Gibbons, Muir, and their circle—a landscape dotted with enormous stumps of the great California redwoods (*Sequoia sempervirens*), soaring trees that once covered the hills of Alameda and Contra Costa Counties, so lofty they served as guides for ships navigating into San Francisco Bay but now were largely cut to build the burgeoning city.[77] Gibbons showed them the largest stump he had found: a whopping thirty-four feet in diameter. That one was apparently killed by a fire before it was cut, but the redwood graveyard as a whole was a vivid and grim demonstration of the more locust-like side of American society, with its rapacious exploitation of natural resources.

Fifteen years younger than Wallace, the Scottish émigré Muir had not yet gone nova as an icon of wilderness preservation, but he was already a pretty bright star leading a movement advocating for preservation of the Yosemite Valley and environs—just a few years after Wallace's visit, Muir's inspired articles in the *Century Magazine* were to play a key role in establishing Yosemite as the nation's first national park. Wallace visited that glorious locale with his brother and his niece May, their coach driving through the great Wawona Tunnel Tree in the Mariposa Grove (sadly felled in a storm in 1969); beholding El Capitan, Mirror Lake, and Bridal Veil Falls, doubtless dizzy with alternately looking down to admire the wildflowers and up at the Big Trees; taking in the great valleys and their soaring domes; and "pondering over its strange, wild, majestic beauty and the mode of its formation."[78] His jaw had probably barely rearticulated when he finally got to experience magnificent groves of living redwoods and sequoias up close and personal a few days later during a pilgrimage he made to Calaveras Big Trees (now a state park). For three solid days, he clambered and climbed and observed and measured the astounding giant sequoias (*Sequoiadendron giganteum*) in awe: "Of all the natural wonders I saw in America, nothing impressed me so much as these glorious trees."[79]

The *spiritual* Muir and *spiritualist* Wallace certainly met on the common ground of a land ethic, but they had other, related, interests in common too: geology, glacial action, botany, landscape, biogeography, and more. Wallace surely admired Muir's study of the "Post-Glacial History of *Sequoia gigantea*," an effort to understand the geographical distribution of the big trees in light of their past history and what we now call ecology, and he would have nodded in agreement as Muir cautioned against confusing cause and effect in identifying the relationship between these trees and their environment. It is a mistake to assume that the sequoias need wet places to grow, Muir maintained. "On the contrary, the grove is the entire cause of the water being there. Drain off the water and the trees will remain, but cut off the trees, and the streams will vanish. . . . Never was cause more completely mistaken for effect than in the case of these related phenomena of Sequoia woods and perennial streams."[80] The great forests shape their environment, a force of nature—precisely a point Wallace argued in *Tropical Nature*. Muir was an inspiration to Wallace, helping further crystallize the English naturalist's conservation ethic, to use a modern term, as part and parcel of his developing holistic, and idealistic, view of human progress. And Wallace inspired Muir in turn, in more ways than one. Years later Muir undertook an eight-month expedition to South America and Africa, fulfilling a dream inspired by Wallace: "I met Wallace," he later told an interviewer, "and have enjoyed his friendship ever since. I have frequently heard him speak of his travels on the Amazon, and [he] gave me a taste for exploring it myself."[81] But Wallace's influence ran deeper still—Muir's reading of Wallace on tropical nature, his grand vision of global geographical distribution in space and time, and very likely his ideas on the preservation and management of Epping Forest all helped hone Muir's own ideas, informing his "ceaseless search for pattern," as one author put it, very much in the Wallacean tradition and central to his ascendant conservation vision.[82]

Following a visit to Menlo Park, where he toured the estate with the Stanfords and admired the plans for the university they were building in neighboring Palo Alto in memory of their son Leland Jr., Wallace gave one final lecture (on spiritualism) and soon after bid his brother and family farewell, giving his niece May a parting gift of an amber brooch and writing desk. Leaving his package of diagrams to be shipped to Michigan, where his next lectures were scheduled, he headed for Colorado on 7 July 1887. Up and over the Sierras and into the Rockies: Sacramento, Lake Tahoe, Reno, Salt Lake, Cimarron, Gunnison, Salida, Colorado Springs . . . Denver. He knew he was crossing borders of deep time, passing from high mountains to high desert to mountains again, especially given the striking transition from the Front Range of the Rockies to the relatively flat Denver Basin at Colorado Springs and Denver—the abrupt edge of the Great Plains. There he first explored the spectacular Garden of the Gods near Colorado Springs, now a National Natural Landmark, where the convergence of two geological faults dissects the tilted and whimsically eroded sandstones of shallow Paleozoic seas.[83]

In Denver he met back up with the talented botanist Alice Eastwood, who had on their first meeting in California kindly offered to serve as his botanical guide on an extended excursion into the Rockies on his way back east.[84] How appropriate that they botanized on Grays Peak (14,278 ft, 4,352 m), named for the esteemed Harvard botanist, and neighboring Torreys Peak (14,267 ft, 4,348 m), named for Gray's mentor, botanist John Torrey—and it says something about Wallace's vigor in his mid-sixties that he could climb such high-altitude peaks, unacclimated no less. The botany was, appropriately, just spectacular—"The richest & most flowery valley I have seen," he declared with a bold underscore. Mid-July was ideal for the floral gems of the Rocky Mountain meadows, and he listed dozens in his diary, peppered with exclamations—"*Castilleja integra!*" "*Parnassia fimbriata!*" "*Aquilegia coerulea!*" (adding: "Splendid grand!") "*Polemonium humile*, a gem!" And on and on . . . "flowers everywhere! . . . superb!"[85]

———

It was really all downhill after that, literally and figuratively—Wallace headed east on 26 July, descending to the progressively flatter (and far older) landscapes of Illinois, Michigan (two more lectures, at Michigan State University, and all went well), and finally eastern Canada and sea level. His thoughts tending toward home, Annie, and the kids, he was beginning to ponder all he had seen and done over the past ten months. There were landscapes sublime vivid in his mind, though his personal landscape preference was still his native "green & pleasant Land," as Blake so memorably put it; he just couldn't see living without lovely *green* and gardens year-round. Aspects of the natural history, the geology, the wildflowers, and the Big Trees all certainly delighted, but he was disappointed with the zoology—"During more than ten months in America . . . I have not seen one single humming bird, or one rattlesnake, or even any living snake," he lamented. True enough, not a single reptile and maybe one amphibian garner a mention in his diary, birds barely at all, and seemingly he had nary a glimpse of bears, bison, elk, deer, wolves, bobcat, or panthers in the backcountry as far as can be told from his diary. There was the odd pronghorn. Maybe he was so intent on the botany that he was blind to the zoology, but on the other hand, sightings of some mammals merited exclamations suggesting a certain animal deprivation: "Saw a chipmunk!" "Evening saw a skunk!"[86]

But what lessons were to be gleaned from America? There was the good, the bad, and the ugly. In the "good" column was the American spirit of optimism, inquisitiveness (about science especially, he thought), and ingenuity, from wonderful museums to useful inventions: he was taken with trams and snowsheds, and the novelty of square rotating laundry lines glimpsed in people's yards garnered a sketch in his diary. But his favorite invention, and favorite souvenir of his entire trip, was a two-foot folding ruler that he picked up in a shop in Sacramento, "very useful to me ever since," he later wrote. "I have never seen one like it in any English tool-shop, and though it was rather dear [three shillings], it has served as a pleasant and useful memento of my American tour."[87] And the bad? Well, he was decidedly down on chewing gum: "In train a lady chewing gum—saw her at intervals for an hour her jaws going all the time like those of a cow ruminating." And certain American manners of speech annoyed him ("The people here 'enthuse' very much").[88] But the *really* bad was the ugly: ravaged, overexploited landscapes, haphazard cities where slums incongruously abut soaring new edifices, and smoke-belching factories and trains. A passage in an article he was reading in the *Century Magazine* near the end of his journey summed up the ugly for him, and he copied it out in his diary:

A whole huge continent has been so touched by human hands that over a large part of its surface it has been reduced to a state of unkept, sordid ugliness; and it can be brought back into a state of beauty only by further touches of the same hands more intelligently applied.[89]

In this passage lies prescription as well as diagnosis, deeply resonating with Wallace. His great American journey may not have been the financial boon he had hoped for, but he found the experience deeply rewarding in other ways—profound ways, in fact, that very much determined the course he would take for the rest of his life. Darwinism, human progress, social justice, land reform, conservation . . . his ideas were merging, crystallizing. He was sixty-four years old and in some ways was only just beginning.

The steamer *Vancouver* departed Quebec City on the cloudy and windy morning of 12 August 1887, heading down the great Saint Lawrence Seaway bound for Liverpool. He had a mercifully uneventful passage of eight days—merciful considering that exactly thirty-five years before at the conclusion of his first great voyage he was adrift in a lifeboat at this time, uncertain of survival. By 17 August the *Vancouver* had crossed the North Atlantic, pausing to exchange mail and passengers at Portrush on the north coast of Ireland, then passing the grand Giant's Causeway and bearing south toward the Isle of Man. He reached Liverpool late in the evening. Home! Well, almost. Fortified with a good breakfast the next morning, he got through customs in good time and was soon on the express train for London, then another to Godalming, then a cab up Frith Hill, where Annie and the kids eagerly awaited his arrival! All was going like clockwork except for a most curious delay with echoes of his fire-interrupted trip home on the *Helen* thirty-five years earlier. "Suddenly I perceived that the driver's coat was on fire behind—actually in flames!" Wallace shouted to the driver, who looked around and beat the flames with his hands: "All right, sir!" Whew, that was something—the cabbie seems to have put a lighted pipe or some such in his pocket. Then the flames erupted again: "Overcoat, trousers, and cushion were burning," with smoke everywhere. This time the driver stopped, took off his coat, and stomped on it repeatedly. *Sheesh!—really* close that time. Off they went—until he burst into flames a third time. Again Wallace shouted, and now passers-by were stopping, pointing. . . . With their assistance the hapless but unperturbed cabbie finally managed to extinguish the flames once and for all. "A cab-man on fire! No more curious incident occurred during my six thousand miles of travel in America."[90]

CHAPTER 14

ONWARD AND UPWARD

VIOLET WAS ANNOYED with her dad. He was a miserable correspondent, she berated, her letters languishing for weeks unanswered. "My dear Violet," he wrote in his defense,

> If you had letters almost every day about Darwinism, Spiritualism, Vaccination, Socialism, Travelling, Dogs tails, Cats-whiskers, Glaciers, Orchids, &c.—& had books sent you on all these subjects to acknowledge & read, & requests for information on other subjects, & other subjects, and other subjects— and a book to write, and a garden to attend to, & 4 orchid houses, and chess to play, & visitors to see, & calls to make, and plants to name,—and—and, and, and, &c. &c. &c. &c. &c. perhaps you would be a "miserable" letter-writer too! Perhaps also, not![1]

It was true enough—it was late 1896, and he was busier than ever at nearly seventy-four years old, the sage of Dorset, his advice and opinion sought on all manner of subjects, many of which, he said, "I know *nothing*—except general principles, & these go a long way with the *ignorant*." Yes, he was swamped, and he pleaded special dispensation from Violet, signing off: "Your affectionate, much abused Pa."

Did I say sage of Dorset? Shouldn't that be Godalming? No, there's no error: in keeping with Wallace's restless spirit, and with the neighborhood getting a bit too crowded for his taste, in 1889 he and Annie left Nutwood Cottage and were now living in lovely Parkstone, on the east end of the seaside county of Dorset, a good 75 miles (120 km) southwest of Godalming. There they found a "small, very pretty, and uncommon house, with lovely views" on Sandringham Road—called Corfe View, it boasted a fine view of Poole

Alfred Russel Wallace with his wife, Annie, and daughter, Violet, ca. 1905.

Harbor, and (truth-in-advertising) the picturesque ruins of Norman-era Corfe Castle in the Purbeck Hills just beyond. It was quirky, though: "I think Ma is quite pleased with it," Wallace wrote Violet, despite the inconvenience of having no cupboards and a bedroom in the basement, not unusual today but unheard of then. "But that makes it all the more uncommon."[2]

They may have also been attracted by Corfe View's modest size; with Will and Violet on their own, they downsized, with a garden small enough to manage themselves. They immediately saw the potential for gardening: carpets of purple-flowered *Veronica* and wild orchids already adorned the place, and the sandy soil was peaty enough to make his prized rhododendrons and other acid-loving plants happy. As usual, whenever they moved house the plant lovers arranged to ransack the garden left behind, transplanting as much as possible— the botanical beauties they grew from around the world were irreplaceable. They soon built a small pond to grow waterlilies and other aquatics and a rather elaborate orchid house to boot, with three sections to maintain different temperatures.[3] It seemed the ideal location: mild climate, wooded and private yet just a stone's throw from the train station, a lovely park (now Ashley Cross Green) a short stroll away, great for gardening. Idyllic, yes, while it lasted ... by the mid-1890s he was *again* house-hunting: "We think of leaving here," he reported to his friend Raphael Meldola, "as we are getting entirely built round, & the place does not agree with any of us."[4] (Maybe the place was a bit *too* close to that train station: What was he thinking? That's one neighborhood that was bound to grow, after all.)

An awful lot had happened in the decade since his return from America— not least, a string of losses in the early 1890s as family and friends succumbed to the ravages of time one after another: first his old traveling companion Bates died in 1892, followed by his sister Fanny and dear friend Spruce in 1893, then both his brother John and T. H. Huxley, that stalwart of the Darwin-Wallace circle, in 1895. He bore these sad losses with equanimity, certain that the spirit lives on—that was still very much a borderland he frequented, ear to an unseen wall seeking messages from beyond. His garden and his work surely sustained him too, as he remained as creative, and productive, as ever on the writing front. The success of his lectures in the United States made him realize that a general exposition of evolutionary thought was lacking, and there was no better person to write one than himself. Besides, it had now been some thirty years since Darwin's and Wallace's papers on transmutation by natural selection had been read, followed a year later by *On the Origin of Species*, and evidence, data, and new insights and perspectives on the whole process had steadily accumulated in that time. Not in slow and steady Lyellian fashion either—indeed, it was all rather catastrophic for anti-evolutionists.

He threw himself into the writing, turning the innovative charts and diagrams he developed for the lecture tour into figures for the book. The final product, *Darwinism*, was brought out by Macmillan in May 1889. It was another Wallacean tour de force (how many of those had he written by then?), though the title is perhaps lamentable, serving to further eclipse Wallace's independent discovery of natural selection, something Herbert Spencer pointed out at the time: "I regret that you have used the title 'Darwinism' for notwithstanding your qualification of its meaning you will, by using it, tend greatly to confirm the erroneous conception almost universally current."[5] The book's fifteen chapters opened with a sequence methodically laying out how natural selection works. In this he emulated the opening chapters of *Origin*, with a key difference that, to Wallace, made more pedagogical sense.[6] Darwin opened with a chapter on domestication, which to him represented a compelling analogy for descent with modification in nature, then moved to variation in natural species, followed by the struggle for existence—these first three chapters the legs of the deductive stool upholding the fourth and crucial chapter, "Natural Selection."[7] Wallace decided to open with first principles and clearly define just what a species is at the outset since one of the old criticisms of Darwin was that he failed to offer a definition of species despite purporting to explain their origin. Where Darwin opened with two chapters on variation, starting with domestication, and then moved on to struggle, Wallace reversed these: "I commence with the Struggle for Existence, which is really the fundamental phenomenon on which natural selection depends, while the particular facts which illustrate it are comparatively familiar and very interesting."[8] He explained struggle in the context of population pressure, concluding the chapter with a point that gives insight into his, well, "evolving" evolutionary vision. One of the impediments for many in accepting evolution turns on "struggle"— recoiling from a conception of nature seemingly dominated by pain and suffering, violent and sudden death (think predator and prey). This is something people understandably find heartrending, Wallace says, because they think in human terms: "The life full of promise cut short, of hopes and expectations unfulfilled, and of the grief of mourning relatives." But this is misguided. The sentiment, he says, that lies at the heart of Tennyson's "Nature red in tooth and claw" is projected onto nature by our imaginations. In an argument not far from that of the old natural theologians, he points out why this view is mistaken. Think about it: a violent and sudden death in the animal world is "in every way the best"—there *is* no misery and pain (or not much) but a maximum of life and its enjoyment with a *minimum* of suffering and pain. In this way struggle and selection are in keeping with his evolutionary teleology: without the necessity of death and reproduction, *progressive development* of the organic world would be impossible; "It is difficult to even imagine a system by which a greater balance of happiness could have been secured."[9]

Discussing struggle first, Wallace further maintained, had the added advantage of treating variation as a better lead-in to natural selection. He reversed the chapters on variation relative to Darwin, treating variation in nature first, then variation under domestication. His take on domestic varieties has long been the source of misunderstanding and criticism, the charge being that unlike Darwin he somehow didn't realize the significance of domestication for understanding evolution by natural selection. This stems from both a misreading of the opening of his 1858 Ternate essay (where he was dismissing Lyell's claim at the time that domestic varieties undermine arguments for evolutionary change) and his comment in *Darwinism* to the effect that he always thought it was a weakness in Darwin's argument that his esteemed friend lay such stock in domestication. Far from indicating that he did not grasp the lessons taught by domestic varieties, however (as evidenced by his Species Notebook, kept during his Malay Archipelago travels; see chapter 9), he simply thought that it was not *necessary* to stress domestication as an analogue for selection in nature when there was such overwhelming evidence for selection in nature to begin with. "I have endeavoured to secure a firm foundation for the theory in the variations of organisms in a state of nature," he explained—basically a solid foundation of *data* on variation, presented in the extensive charts and diagrams that he referred to throughout the book, "just as Darwin was accustomed to appeal to the facts of variation among dogs and pigeons."[10]

But Wallace's *Darwinism* did far more than simply explain his and Darwin's theory of natural selection. There were important new insights as well, foremost among them his lucid elaboration of the process later called the *Wallace Effect* in speciation (better known as *reinforcement*)—selection for behavioral or physiological traits that reduce the likelihood of hybridization when, in incipient speciation, hybrids suffer lower fitness. Actually, Wallace killed two birds with one stone here, articulating the reinforcement process while torpedoing the arguments of one George John Romanes, a Canadian-born evolutionary biologist and physiologist who, to Wallace's annoyance, posed as Darwin's heir and claimed to provide a crucial missing piece of the Darwin-Wallace theory, which he dubbed "physiological selection." Romanes, whose family had moved to London when he was very young, was indeed a friend and protégé of Darwin and made important contributions to the field now called evolutionary psychology. He was also interested in the outstanding problems of Darwin-Wallace evolution, one of which was how divergence and speciation was possible so long as individuals can interbreed, or hybridize. At the heart of the matter was hybrid sterility: Without it, wouldn't the two entities simply mix and not diverge? And if hybrid sterility is important, so, too, is the question of just how it arises. This was an issue that Huxley had first brought up with Darwin long before, referring to groups of individuals showing hybrid sterility as "physiological species."[11]

Wallace had tackled it back then too, in 1868 sending Darwin an outline of his theory of the "Sterility of Hybrids Produced by Natural Selection"—an elaborate argument that Darwin had a difficult time digesting, so he passed it on to his sons to evaluate, especially the mathematically inclined second son, George, the astronomer.[12] George offered a critique, but Wallace did not find his points insurmountable. To him, if there are degrees of sterility between

varieties based on something heritable, "Is it not probable that natural selection can accumulate these variations. . . . If Nat. Select. can not do this how
do species ever arise, except when a variety is isolated?"[13] Darwin could not
persuade himself that hybrid sterility could arise through natural selection—it
entailed the counterintuitive idea that the degree of sterility itself could be
heritable. He preferred to think hybrid sterility came about as a by-product of
other things happening during divergence, unable to get past his analogy with
grafting in arboriculture: grafting ability varies, and very generally the success
depends on the degree of relationship between the species being grafted, but
no one would say that the ability to be grafted is in *itself* an adaptation. It is
secondary to other traits that define the degree of relationship. Similarly, the
ability to interbreed and produce fertile offspring varied and very generally
was related to the degree of relationship of the two species in question. Hybrid
sterility or interfertility, then, was not in itself an adaptation but arose second-
arily.[14] Wallace wasn't so sure but decided to shelve his argument.

Now, nearly twenty years later, shortly before Wallace departed for the
United States, here was Romanes with a paper on physiological selection (the
name a takeoff on Huxley's physiological species), purporting to provide an
"addition" to natural selection and so a more complete explanation for the origin of species.[15] In fact, Romanes was essentially making a higher-level selection argument in which subgroups of individuals in a population somehow
become reproductively isolated from the greater population while remaining
reproductively (or "physiologically") compatible with one another. They are
favored as a group because of their reproductive compatibility. His proposed
process has been likened to chromosomal aberrations arising and rendering
crosses between individuals with and without the aberration sterile, while
intra-aberration crosses, so to speak, are fertile. Once so isolated, over time
there would be a tendency for the two sets to become more and more divergent, in sympatry. It was a kind of "hopeful monster" scenario, almost instant
speciation.[16]

Wallace was far from impressed—Romanes had made a purely verbal argument, with no examples or evidence from nature, and it didn't hold water.
Wallace was convinced that natural selection was the sole mechanism behind
speciation and was appalled not only to find little critical pushback but to see
reviewers singing Romanes's praises. He fired off a lengthy critique to the *Fortnightly Review*, dusting off his own previously developed ideas on how selection could accumulate degrees of sterility. Raphael Meldola wrote a critical
review in *Nature* too, but Wallace did not think it went far enough. "I wrote

my paper," Wallace told his friend, "quite as much to expose the great *presumption* & ignorance of Romanes in declaring that *Natural Selection* is *not* a theory of the origin of Species,—as it is calculated to do much harm. . . . Romanes poses as the successor of Darwin. This should be stopped before the press & the public finally adopt him as such, & for this reason I wrote my paper."[17] Wallace exposed the logical flaws, assumptions, and lack of supporting data for Romanes's argument; Francis Darwin backed Wallace up, and others followed with an equally dim view of the model. Romanes did not take the criticism well, and his response set the tone for an increasingly acrimonious exchange that went back and forth over the next several years, ending only with Romanes's untimely death in 1894 at age forty-six.[18] But by then Wallace had already demolished the physiological selection idea, opaque as it was. His treatment in *Darwinism* was effectively the end of the ill-conceived theory.

Not one to simply tear down and not build back up, in *Darwinism* Wallace also updated his own theory shared with Darwin years before, presenting it under the title "Can Sterility of Hybrids Have Been Produced by Natural Selection?"[19] In essence he envisioned a two-part process whereby selection leads to hybrid sterility, based on the data suggesting that hybrids often have lower fitness (fertility, viability) than parental forms. In this case not only will the parental populations always increase faster but, insofar as hybrid offspring impose a fitness cost on the parentals—energy and resources wasted, in a sense, by the parental forms in producing subpar offspring—selection will favor the evolution of mechanisms to reduce the likelihood of hybridization to begin with. Mating avoidance might be achieved in a number of ways, from shifts in species recognition or courtship signals, visual or pheromonal, to altered behavior resulting in temporal or spatial differences between individuals of the different subgroups. Thus, in a gradual, stepwise fashion selection would *reinforce* any slight tendency to reduce the waste of hybridization until the reproductive isolation of the two groups, whatever that is based upon, is complete. The modern view of reinforcement, then, is essentially Wallace's and should be called by the name first coined in 1966 honoring his insight: the Wallace Effect.[20]

Another advance in *Darwinism* was Wallace's championing of an exciting new insight that bore on heredity: August Weismann's germplasm theory. Weismann, a professor at the University of Freiburg in Germany, had a special research interest in the nature of variation, and he, too, had been in communication with Darwin over the question for some years. Long impressed with his younger German colleague, Darwin was encouraging and even agreed to write the preface to the English translation of Weismann's *Studies in the Theory of Descent*, published in 1882.[21] There Darwin declared that "at the present time, there is hardly any question in biology of more importance than this of the nature and causes of variability"—it was the essence of evolutionary change, after all.[22] This was his final pronouncement on the subject, as 1882 was the year of Darwin's death. To the very end, he held out hope that his "Provisional hypothesis of Pangenesis," first published in *Variation of Animals and Plants under Domestication* (1868), would be borne out. Pangenesis was Darwin's model of heredity, a theory based on the assumption of blending inheritance—a very old and intuitive idea since it sure looks like the hereditary principle of two parents' blend in their offspring. In Darwin's model each part of the body, all cells, give off ineffable particles he dubbed "gemmules," which basically contain genetic information for that cell type, tissue, or organ. He envisaged that these circulate through the body and accumulate in the reproductive organs, after which they mixed with the gemmules of mates in the process of reproduction.

Yes, it was a tidy theory that to all appearances was in keeping with the observable phenomena of heredity—including the Lamarckian inheritance of acquired characters, something Darwin never really got away from. But things were not looking good for pangenesis. His cousin Francis Galton had carried out experiments testing the hypothesis, involving blood transfusions between rabbits differing in coat color. This blood mixing had no effect whatsoever on the color of the rabbits' subsequent offspring.[23] Galton's experiments led him to develop a variant (appropriately enough) of his cousin's theory, postulating two kinds of gemmule, in essence, one of which is associated with somatic (body) cells and the other of which is confined to germ cells. In this he was to anticipate Weismann, whose observations and experiments with heredity after Darwin's death led to his theory of the "continuity of the germ plasm," which postulated that the germ line is *wholly* sequestered from the soma through the generations such that no alterations of somatic cells can be communicated to the germ cells and are therefore not heritable.[24] Darwin would have found it fascinating, even if the idea of germplasm continuity did provide the nails for the coffins of both pangenesis and inheritance of acquired characters. Weismann's ideas were published in a collection of essays translated in 1889, and Wallace, already appreciating the significance of Weismann's work, was able to get advance copies of the chapters as he was working on *Darwinism*.[25] Weismann's ideas of germplasm continuity and the noninheritance of acquired characters, Wallace declared, aligned beautifully with the facts of heredity and development, underscoring the importance of natural selection "as the one invariable and ever-present factor in all organic change"—*the* factor in fact behind species change.[26]

That is what Wallace was angling for—here he was vanquishing a different kind of enemy of Darwinism: the enemy within, as he saw it, who sought to compromise the integrity of the purist vision. Wallace was aware that even some within the close circle of First Darwinians had been more or less unsure that natural selection could accomplish all that Darwin and Wallace thought it could. Even Darwin himself always had a Lamarckian streak, but now some in the new generation were suggesting a secondary role for natural selection, such as nongradual evolutionary change and other modifications of the original theory. Wallace would have none of it. He may have broken with Darwin over the evolution of human and mind, but other than that, he was fully committed to gradual, stepwise evolutionary change through natural selection. Indeed, in *Darwinism*, his last great evolutionary work, Wallace sought to cement the central role of the "great principle" of natural selection (as he put it) by extending and strengthening it. The strands of his many-faceted arguments were woven into a masterful and beautiful narrative tapestry; it was all of a piece, from his unified theory of coloration and mimicry to the principle of *utility* (adaptation) to the theory of hybrid sterility, all explicable under the "considerably extended" range of natural selection: "Hence it is that some of my critics declare that I am more Darwinian than Darwin himself," he acknowledged. "And in this, I admit, they are not far wrong."[27]

———

Yes, *Darwinism* was masterful—Wallace still wielded a formidable pen and was still a formidable scientific reasoner and defender of the Darwinian faith. It also inspired some innovative and ambitious ideas that were a bit ahead of their time. For example, pondering processes of heredity and evolution while working on *Darwinism*, Wallace got to thinking that some good hard data would settle the outstanding questions once and for all and proposed founding a sort of experimental farm, an "Institute for Experimental Enquiry" in the spirit of the biological field stations and marine labs then being established in Britain and abroad. In those days such institutions were mainly dedicated to nuts-and-bolts taxonomy, embryology, anatomy, and physiology. What was needed, he thought, was an institution dedicated to a "combined and systematic effort to carry out experiments for the purpose of deciding the two great fundamental but disputed points in organic evolution"—namely, inheritance of acquired characters and selection for hybrid sterility. He petitioned the one person he knew well who had conducted such investigations himself: Francis Galton, whose transfusion experiments refuting pangenesis were widely admired. In early 1891 Wallace urged Galton to propose, at the next British Association for the Advancement of Science meeting, the establishment of a standing committee dedicated to overseeing such experiments. Perhaps it could be funded by a Royal Society grant, he suggested, or some wealthy donors. They could cooperate with the Zoological Society, working with some of the animals in their menagerie at Regent's Park, perhaps or, better, acquire some land (a small tenantless farm would be just the thing) and establish an experimental farm of their own. Galton and others were enthusiastic about the idea, but the effort stalled, unfortunately.[28]

Around this same time, Wallace also published a remarkable two-part floristic study in the *Fortnightly Review* inspired by his journey botanizing his way across America and back. At once scientific as well as artistic, with evocative images of plants and environments, it was a masterful lesson in what is called today ecological and historical biogeography: the interplay of historical processes like tectonics and glaciation with more immediate ecological conditions shaped by factors like elevation, precipitation, and temperature. Here Wallace's keen sense of dynamic biogeographical change in time and space came to the fore, painting a picture of the relationships of North American and European (especially alpine) flora understood in the context of glacial cycles. In that vision, suites of northern-affinity species now restricted to polar and alpine regions periodically migrate south during glacial periods, following the arctic conditions that are pushed to the south. There they experience more expansive and continuous suitable habitat and spread far and wide—circumboreal, right across the Northern Hemisphere. When the climate warms again, the boreal conditions contract, northward in latitude and upward in mountains, where these plant species hunker down on sky islands—and present the apparent puzzle of how some of the same species of plant could end up high in both the Rockies and the Alps.

His analysis was grand, big picture, but he concluded with a vision grander and bigger-picture still, an expression of his growing environmentalism inspired by the Big Trees of California. He had witnessed exquisite and stunning vistas across North America, but "neither the thundering waters of Niagara, nor the sublime precipices and cascades of Yosemite, nor the vast expanse of the prairies, nor the exquisite delight of the alpine flora of the Rocky

Mountains" impressed him more than those majestic trees and their magnificent and beautiful forests. "Unfortunately," he warned, "these alone are within the power of man totally to destroy." Let us hope that education will lead to such deep love and admiration for nature that these incredible trees "will be looked upon as a trust for all future generations, and that care will be taken, before it is too late, to preserve not only one or two small patches, but some more extensive tracts of forest, in which they may continue to flourish, in their fullest perfection and beauty, for thousands of years to come, as they have flourished in the past, in all probability for millions of years."[29]

It was deeply insightful—no wonder honors were beginning to be heaped upon him . . . a bit belatedly, some would say, but better late than never. Not that he wanted them—in fact, Wallace tried his best to turn down awards and honors, giving in only when wheedled at length by his friends and family. Take the case of his honorary doctorate from Oxford in 1889. Edward Poulton, professor of zoology at that august university, was a great champion of Darwin and Wallace. With a special interest in animal coloration, it was he who coined the term "aposematism" for Wallace's idea of warning colors, as well as the term "sympatric" for the concept of speciation in a common area. It was also Poulton who, editing Weismann for English translation, had given Wallace advance copies of the German savant's papers and was glad to provide helpful editorial comments on *Darwinism*. He saw what a monumental work that was and already well appreciated it as the latest in a succession of monumental Wallace works, instigating Wallace's nomination for the Oxford degree. It was no coincidence that the news of this prestigious award arrived on Wallace's doorstep the very month that *Darwinism* was published. Poulton probably anticipated his response too:

> I have just received . . . the totally unexpected offer of the Honorary Degree of D.C.L. at the coming Commemoration, and you will probably be surprised and *disgusted* to hear that I have declined it. . . . The fact is, I have at all times a profound distaste for *all* public ceremonials, and at this particular time that distaste is stronger than ever.[30]

He thanked Poulton—he appreciated the gesture but was busy, after all . . . grading exam papers, packing up his books (not to mention his garden) to move house, preparing for the upcoming Land Nationalisation Society annual meeting. No, at any time a ceremony would be a trial, but really, under the present circumstances it would be "a positive punishment." Poulton wasn't going to take no for an answer. Bad timing? No problem—he bumped the date of the award to the autumn, and Wallace reluctantly gave in despite protesting that he really felt himself "too much of an amateur in Natural History, and altogether too ignorant" for such an honor. He was probably glad he acquiesced—the newly minted Dr. Wallace enjoyed himself more than he thought he would.[31]

He "endured" receiving the Darwin Medal of the Royal Society the following year too—the first recipient of this honor but not exactly a grateful recipient: "I am to be dragged up to London again because that stupid Royal Society are so idiotic as to give me another medal," he fumed to his daughter Violet. "I expect it will be a *brass* one this time, but I do not know yet."[32] His sarcasm reflects his feeling of being unappreciated by the Royal Society, its fellows among his most severe (and, he thought, unfair) critics over his "extra-scientific" pursuits. But he went through with it. Then there was a double whammy a couple of years later: medals awarded back-to-back by the Royal Geographical Society and the Linnean Society of London. Ugh. "Isn't it awful!" he wrote Violet, only half-joking. "*Two* medals to receive, *two* speeches to make, neatly to return thanks, and tell them in a polite manner, that I am much obliged, but rather bored!" He was to receive the Founder's Medal of the Royal Geographical first, on 23 May ("3.30—Medal—speech—subside gracefully—congratulations from old friends &c. &c.," he told Violet), and the Gold Medal (now Linnean Medal) of the Linnean Society the very next afternoon. But before showing up at the Linnean, a more pleasant diversion—the grand International Horticultural Exhibition at Earl's Court in central London, the first held in twenty-six years. "A veritable series of gardens fitted for the gods," proclaimed one London newspaper, waxing lyrical about the bewitching "concord of sweetness, fragrance, plants, flowers, and foliage."[33] Then it was off to William Bull's famous orchid house in Chelsea to lose himself amid the exquisite rarities: Bull's was described as "a dream of beauty," by one commentator.[34] Specializing in orchids of the South American tropics, Wallace likely saw some familiar plant faces at Bull's and probably purchased a few too—he and Annie had no fewer than four orchid houses at Corfe View! He would surely have stayed all day if he could, but he had a meeting of the exam-grading examiners to get to—a necessary evil—after which, finally, he planned to catch a cab to the Linnean Society's rooms at Burlington House. Or would he? He confided his fantasy escape to Violet: "Arrive just too late! Medal stolen!! Forged letter telling Council I am ill!! Can't come!!! Bearer to receive it for me!!!! Such Excitement!!!! Universal Collapse!!!!!!!!"[35] That one got both a chuckle and an eye roll, no doubt.

He made it through all the pomp and circumstance for degrees and medals but was going to draw the line at actual membership in the Royal Society. Receiving medals from them was bad enough, but again, he had no warm fuzzy feeling for that venerable society, which he felt could have recognized his considerable scientific contributions by electing him a fellow far sooner.[36] Now his friend William Thiselton-Dyer wanted to put forward his name and asked Wallace if he would have any objection to being a fellow. None, said Wallace. Great, Thiselton-Dyer replied, perhaps a bit surprised, I'll get on it. Steady on, said Wallace: "You asked me, I think, whether I had any objection to being a fellow, & of course I said 'none at all,' but that did not imply that I *wished* to become one." It was hardly worthwhile now, he continued. "Therefore, while thanking you very much for your kind interest in the matter, I really think it will be better to take no further steps towards it." It took a bit of gymnastic cajoling, including a trip to Eastbourne to consult with Huxley in his retirement (he was now an avid gardener), where the two came up with a plan of attack: To turn down a fellowship in the Royal Society, they argued, would be seen as a black mark not simply on the organization but on British science, after all. The position of science in Britain was precarious, the Society sniped at and dismissed as so many scientific socialites, not discoverers. It was Wallace's duty to be a part of it, he wheedled, to show that the country's best scientific minds were behind the Royal Society! It worked; Wallace relented and was duly, if reluctantly, elected a fellow on 1 June 1893.[37]

Such honors were indeed long in coming, in large part due to Wallace's less scientific activities, especially where they seemed to run counter to prevailing scientific thinking, such as the brouhaha over vaccines. But spiritualism and, more and more, socialism were also part of the unique, gifted, and at times irascible personality that was Alfred Russel Wallace. Britain's scientific circles had long since learned to ignore the spiritualism, but socialism was another area that brought Wallace in conflict, to a degree, with his scientific colleagues. Long socialistic in outlook, going way back to his Owenite days, Wallace had something of an epiphany in the months after *Darwinism* came out. In the summer of 1889, he picked up the novel *Looking Backward: 2000–1887* by one Edward Bellamy, sometime lawyer and journalist from Massachusetts. The book astounded Wallace—for one already given to utopian yearnings, *Looking Backward* was a revelation. Published in the United States only the previous year, 1888, the book was already a sensation on both sides of the Atlantic. In the United States it was easily the most-read and talked-about book of its time, with over three hundred thousand copies printed within its first two years of publication. It also inspired a heated cultural conversation, with dozens of books pro and con published over the next decade: critiques, satires, broadside attacks, dystopian takeoffs, imitations, and sequels unofficial (many) and official (one: Bellamy's 1897 *Equality*), not to mention hundreds of Bellamyite "nationalist clubs" across the country—socialist associations that pushed for the nationalization of industry.

Yes, *Looking Backward* sure touched a nerve all right. To understand why, it's important to consider the fraught social, political, and economic climate of the last quarter of the nineteenth century—the constant and often violent struggles between labor and capital, the nonexistent protections for workers toiling in twelve-hour-plus workdays, the horrific slums of the quickly industrializing cities, the grinding rural poverty, and the astronomical disparity between the privileged rich and barely surviving poor. All were issues greatly exacerbated by such economic stresses as the depression of 1873–1879, subsequent recessions and other financial crises, and wage stagnation. The most famous metaphor of *Looking Backward*, appearing in its opening pages, well captures the situation: society, Bellamy said, is like a "prodigious coach," a stagecoach of the privileged riding high, to which the "masses of humanity were harnessed . . . and dragged toilsomely along a very hilly and sandy road."[38] Indeed, this was the America of the Gilded Age; gilded for the few, anyway, at the expense of the many, a time and place where laissez-faire might as well have been *laissez exploitons*—"Let's exploit." It was a time of strife and strikes: the Knights of Labor, founded in 1869 and hundreds of thousands strong by the 1880s, organized frequent strikes for an eight-hour workday (something Robert Owen had been pushing for since the 1810s), higher wages, and improved working conditions—some successful and some not, some peaceful, and many devolving into violence, like the tragic 1886 Haymarket Square riot in Chicago.

Such was the state of society in the mid-1880s, but by the year 2000 when, in a Rip van Winkle-ish kind of way, the protagonist of Bellamy's novel awakens after a 113-year slumber, gone are the riots and strikes, the class struggles, the deep division between the haves and have-nots. Best of all, this was accomplished organically, without violence. You will have to read Bellamy's book to learn just how this happened and the nuts and bolts of his ideal society of the future. It is well worth reading not as a great piece of literature or a rollicking page-turner (it is neither) but as a fascinating historical document: a fable that curiously anticipates some aspects of modernity such as "consumer's cooperatives," rather like today's Costcos and other big warehouse membership clubs, "credit" cards rather like today's debit cards, cable-delivered home entertainment, and more), a book of the times—a socialist utopia that captured the imagination of millions of people for a short time, among them Alfred Russel Wallace.

"I am simply enchanted with it!" wrote Wallace to poet, sometime vicar, and socialist Edward Girdlestone. "And what is more, am *converted to Socialism!*" He enthused over the "great work"—it was genius, beautiful, admirable! He had read many a book advocating socialism before, he said, but this one was different: not just possible but practicable, and desirable. "I never seemed to *wish* to live in other writer's Socialistic worlds. There was too little individuality, too little freedom, too little privacy, and too little variety in them. But I *long* for Bellamy's world & feel that I could be happy in it." He couldn't put it down. Following up with another letter two weeks later, after reading *Looking Backward* for the third time, his admiration had only grown: it was marvelously realistic, compelling, seemingly within reach; he could practically *see* and *feel* it. "I therefore am now a thorough Socialist, & care not who knows it," he proclaimed.[39] But in his enthusiasm for the book's vision of a society of justice, equity, health, and happiness for all, Wallace may have misunderstood, or overlooked, the book's implications for the individuality, freedom, privacy, and variety he had found lacking in other socialist schemes. On a closer read, precisely those virtues were in fact in short supply in Bellamy's future society. As one historian aptly put it, "*Looking Backward* can seem less a forecast of utopia than a precursor of Orwell's *Nineteen Eighty-Four.* . . . Bellamy's future world seems to leave no room for individualism, for difference, for dissent."[40] It was more a model of authoritarian socialism than the apparent harmonious communitarianism that meets the eye, perhaps, but for so many to have seen it as liberating, a dream come true, surely says something about those uncertain, fraught, and inequitable times.

This was certainly Wallace's view, inspired as never before over the possibilities for social improvement. His misgivings over the practicability of earlier socialism-inspired social experiments may explain why he never expressed any interest in taking part in them. As much as he admired Robert Owen's vision, he made no move toward joining any of the Owenite communities that sprouted up in Britain and the United States.[41] Some years later, however, he did pitch his own version of a utopian community of sorts, describing it as "a kind of home-colony of congenial persons."[42] The idea was to get a group of kindred spirits to go in together to purchase a large tract not far from London, maybe an old estate. Each would receive a plot from two or three to ten or more acres in size, according to each person's investment. There would be room enough for gardening and buffers between the houses, and the remainder would be maintained as rural common land, woodlands, fields, and meadows open to all. He was most excited about a property called The Grange, near the market town of Amersham northwest of London. Although several prospective partners looked at the property, the scheme fell through in the end, the £15,000 asking price remaining stubbornly out of reach.[43]

———

"Home colony" or no, Wallace threw himself into this newfound socialistic passion with his usual vigor. It certainly harmonized with his spiritualism, but it's the way Wallace's socialism further honed his sense of social justice that is most interesting. In fact, he points to this in his autobiography. Why is he a socialist, it may be asked? "Because I believe that the highest law for mankind is justice," is Wallace's reply. Far from an advocate of authoritarian socialism, let alone of the violent revolutionary persuasion, Wallace took as his motto the Latin legal phrase *Fiat Justicia Ruat Coelum*—"Let there be justice though the heavens may fall"—and he defined socialism in terms of freedom to live up to one's potential and contribute to the common good. "That is absolute social justice; that is ideal socialism. It is, therefore, the guiding star for all true social reform."[44] There is perhaps no better example of Wallace's commitment to individual liberty, independence, and the opportunity to fully exercise one's faculties than his elevation of women's rights to an absolutely central role in his social thought.[45]

In this he was inspired by *Looking Backward*. We already saw the interest Wallace took in women's education on his American tour. Now Edward Bellamy put forth a startling idea beyond mere coequal education: the progressive improvement not just of society but of people themselves, evolutionarily, thanks to "untrammeled sexual selection." Perfect freedom of choice in marriage, together with freedom from the threat of penury that so often forced women into marriages of necessity (and therefore matches often far from ideal) encourages simpatico relationships. If women are permitted to follow the "natural impulse to seek in marriage the best and noblest of the other sex," the result is mental, moral, and potentially even physical improvement of humankind by sexual selection.[46]

Wait a minute! Now, *here* was a form of female choice Wallace could accept, and more. Recall his disagreement with Darwin over sexual selection, female choice in particular: Wallace did not share Darwin's belief that female birds and insects had an aesthetic sense, let alone one capable of discriminating minute differences in tint or tail length. He proposed the alternative view of female mate choice in animals as more shrewd than aesthetic: selection based on obvious traits of coloration and courtship that signal robust health and vigor. When it came to humans, he rejected female choice completely. But Bellamy opened his eyes to another way sexual selection could operate in humans, not far from his "vigor" hypothesis for animals: female choice in marriage as a means of selecting for beneficial traits in husbands, thus improving the species as a whole over time. As much as Wallace advocated improved living conditions, those benefits accrued only to individuals, in keeping with his rejection of Lamarckian inheritance of acquired traits. The health benefits stemming from simply giving someone better food and plenty of fresh air and exercise could not be passed on to offspring. But female choice in marriage could lead to selective reproduction of individuals with superior traits, including mental and moral. (Note, by the way, that "marriage" was not just a euphemism to Wallace—he wasn't so far ahead of his time that sex and childbearing outside of marriage, let alone contraception even within it, were to be condoned.) It was, of course, a rather naive assumption that women's choices in husbands would immediately trend toward the noble and bright were they freed from financial concerns, and today such traits would not be seen as heritable anyway and so immune to selection.

But in Wallace's day, this was a rather innovative solution to what many commentators saw as a serious problem for the evolutionary future of humanity, lamenting the supposed decline (or at least lack of advancement) of the "race" or species because the less fit seem to do most of the reproducing. Even Darwin expressed some pessimism over the prospect for future improvement of humans on these grounds, lamenting in *The Descent of Man* that in Western society—the pinnacle of civilization in the widely shared view of the time— the patently unfit poor and uneducated seemed to breed like rabbits, while the talented, great, and good often do not marry or do so late and end up having relatively few children. Thus, the supposed "inferior" traits of the former spread at the expense of the "superior" traits of the latter. (It seemed to be lost on Darwin and other fans of natural selection who held this view that those who reproduced the most were by definition the fittest . . . a notion too appalling to contemplate, no doubt.) Wallace, building on both Owen's and Bellamy's ideas regarding the equality of the sexes, was convinced that the legal emancipation of women, with full independence, political rights, and financial

WOMEN'S SUFFRAGE.

A

GREAT DEMONSTRATION

OF

WOMEN

IN THE

COLSTON HALL,

ON

THURSDAY, Nov. 4th, 1880,

Wallace was an early advocate for women's suffrage and other rights.

security, was sure to lead to societal improvement. It started with equality of opportunity for all.[47]

Putting pen to paper, he wrote a remarkable essay in typical Wallacean fashion: "Human Selection," published in the September 1890 issue of the *Fortnightly Review*, a paper he later described as "the most important contribution I have made to the science of sociology and the cause of human progress." He elaborated upon this two years later with "Human Progress: Past and Future," published in the January 1892 issue of *Arena* magazine and in a spate of interviews.[48] Wallace first took aim at the proposals being bandied about by others to solve the supposed social stagnation—proposals he found to be no solutions at all but recipes for disaster. On the one hand, there were the "eugenic" schemes inspired by Francis Galton, Darwin's polymath half cousin best known today as a pioneering statistician, psychologist, and eugenicist with a decidedly "social Darwinist" streak.[49] Beginning with an 1865 essay titled "Hereditary Talent and Character" and continuing in his books *Hereditary Genius* (1869) and *Inquiries into Human Faculty and Its Development* (1883)—where he coined the term "eugenics"—Galton was constantly discussing the problem of the (supposedly) "less fit" out-reproducing the "more fit" members of society. His solution was *not* to limit the reproduction of the former but to introduce incentives to increase reproduction among the latter: establish an endowment, say, to encourage and reward early marriage of talented young ladies and gentlemen (with bonus payments for each kid), maybe based on a system of points for qualities of familial health, intellectual accomplishment, and morals ("Marks for family merit").[50] Wallace found Galton's *positive eugenics*, as it is called today, among the "least objectionable" schemes but said it was also among the "least effective." But he was far more concerned about others running with this Galtonian ball who proposed intrusive forms of state-sanctioned artificial selection of sorts, with state agents permitting marriage for those deemed to have desirable traits and preventing it for those who do not—a form of the *negative eugenics* later taken to such horrific extremes in the twentieth century, from forced sterilizations to fascist death camps. "Nothing can possibly be more objectionable," wrote Wallace. Utterly committed to individual liberty and dignity, any coercive let alone punitive or invasive social engineering scheme outraged him. But a proposal almost as bad in its effects, in his view, came from what we might call "free love" advocates—abolish marriage altogether, they ventured, and enable women to not only pick the most highly desirable partners they can but change these up every few years if they wish—it was all about making better babies. "Detestable," declared Wallace— damaging to family life and parental affection and sure to lead to an increase in pure "sensualism."

No, his solution was infinitely preferable to all these proposals, he was sure. Clean the filthy "Augean stable of our existing social organization"—enable full equality, let all people contribute to the best of their physical or mental ability *and* reap the full reward of their work, and our future will be secured by the laws of human development, a rising tide of human and societal improvement. Sure, he acknowledged, in our society we shield the weak and infirm from the "checks" they might be subject to in nature, enabling them to survive and even have children. But this only reflects the "higher attributes" of our nature: sympathy, compassion, and generosity, which are the essence of humanity.[51] In the future this whole problem will be remedied not by diminishing these qualities—as by brutalizing people, preventing them from marrying or having children—but by rendering the problem moot by addressing social injustices to begin with. "When we allow ourselves to be guided by reason, justice, and public spirit," Wallace concluded in his 1890 essay, "and determine to abolish poverty by recognising the equal rights of all the citizens of our common land to an equal share of the wealth which all combine to produce," then, having secured the well-being of all, we can rest assured that we have set the stage for the natural evolutionary improvement of humanity, committing ourselves to the "cultivated minds and pure instincts of the Women of the Future."[52]

Wallace's commitments to evolution, spiritualism, and social justice—from land nationalization to anti-vaccination to women's rights—crystallized into a singular worldview for him through the 1890s and beyond. "Equality of opportunity" became his watchword and guide. He was no deterministic socialist, no ideologue—he rejected the Marxist dialectic, the "end of history" thinking, and eugenics. His philosophy was to do right by people and set the stage for individual dignity and cooperation; socialism may follow organically.[53] The rights of women were central to this thinking about societal advancement—ensure this and it will lead to incalculable societal and indeed human-evolutionary benefits, was his mantra.

Their daughter, Violet, was certainly left to choose for herself, and whether by choice or not, she remained single. Around the time of her father's human selection papers, Violet was in her early twenties, a kindergarten teacher living in Liverpool. She was very much her father's daughter, and his newsy letters to her are peppered with references to critters she delighted in: "On Saturday, Miss Heaton from Natal called & brought you a fine trap-door spider & trap, and asked if you would like another Chameleon. Of course I said yes!" This was a companion or a replacement—the year before he mentioned to her grandfather William Mitten that Violet had just gotten a new chameleon, "a fine fellow."[54] And Will, too, was very much his father's son in various ways. A socialist and sometime séance attendee, by the mid-1890s he was working as an apprentice electrical engineer in Newcastle. He was also an avid cyclist, cruising around the beautiful Northumberland countryside on weekends.

Wallace's scientific work continued unabated even while laboring away on, well, *labor*—and land, and vaccination, and spiritualism—even monetary policy![55] More and more of his scientific writing turned to geology at that time, especially glaciation and climatology, enthralled as he was with the work of Scottish astronomer and climatologist James Croll, whom you recall from chapter 12 argued that regular variations in the earth's orbit and axial tilt create the climate cycles behind glacial periods.[56] Wallace was a fan of Croll and corresponded extensively with the Scotsman in the 1870s and 1880s, some dozen letters of which were included in Croll's posthumous autobiography—true to form, while extolling the virtues of Croll's theory Wallace also offered his own original modifications.[57] The astronomical theory of ice ages was, dare I say, hotly contested, and as Croll's ideas came under attack from various quarters, Wallace never shrank from a good argument when he thought he had a point. That he readily engaged astronomers over this issue says something about Wallace's chutzpah—even in 1896, six years after Croll's death, Wallace weighed in on the ongoing debate, writing to *Nature* in response to a letter by George Darwin, his old friend Charles Darwin's second son, who was now Plumian Professor of Astronomy at Cambridge. Darwin and an earlier correspondent, Edward Culverwell, had laid out an argument suggesting that Croll's theory was unsupported. Wallace begged to differ: "It now seems opportune, therefore, to lay before your readers the general considerations which lead me to the conclusion that the whole argument they rest upon is unsound."[58]

Then there was his one last swipe at George Romanes. Invited by the Linnean Society to give a lecture in the spring of 1896, he chose as his topic "The Problem of Utility: Are Specific Characters Always or Generally Useful?"[59] Here Wallace squarely took on Romanes's arguments against the idea of "utility"—the uber-selectionist Wallace was convinced that all traits are or were once useful to their bearer or selection would not have allowed them to spread, contra Romanes. Romanes was right about one thing at least: in his less than positive review of Wallace's book *Darwinism* in the *Contemporary Review*, he charged that Wallace's insistence that natural selection is the sole, not merely the main, means of evolutionary change was not Darwinism at all, insofar as Darwin did not hold this view. Rather, Romanes said, it "assuredly is, and always has been, pure *Wallaceism*."[60] He meant it as a barb, but Wallace sure didn't mind—he always did say he was more Darwinian than Darwin, and in his dedication to natural selection he was also ever prepared to fight to the bitter end. In this case he had the last word, as Romanes died in 1894. Yes, Wallace was still the greatest defender of the Darwinian faith, but his dedication to Darwin had its limits. Later that year when Poulton asked Wallace to give a speech at the unveiling of a statue of Darwin at Oxford, he politely but pointedly declined: to Poulton, Wallace told Violet, "I have to write, a kind, careful, but *positive* refusal,—requiring much thought, & an additional gray hair or two to my already *totally* gray head!"[61] Darwin would have been aghast, not at Wallace's refusal but at the very idea of a statue.[62]

How Wallace found time for his myriad personal correspondence, essays, letters to periodicals, addresses, interviews, and books is a study in consummate time management and the skillful wielding of pens. On top of all this was his seemingly incessant gardening—or maybe it was the tranquility of the gardening that made everything else possible? He and Annie each had their own

gardens within the garden and their botanical favorites—stunning orchids for
Wallace, bounteous ferns and primroses for Annie. She had become quite an
adept hybridizer of primroses, deftly wielding her artist's paintbrushes to transfer pollen between the curious "pin" and "thrum" morphs—a favorite experimental subject of Darwin, who would have admired her skill. Spreading the
botanical love, they kept Violet well stocked with plants for her garden in Liverpool too, including three rare ferns they collected near Miller's Dale, in what
is now Peak District National Park. "You remember we found them all at the
top of that forsaken valley where tourists go not," Wallace reminded her—
always one to hike off the beaten path and dig up plants at the drop of a hat.[63]
Botany was in fact a standard theme of family holidays, their finances now secure enough to permit annual travels home and abroad. In one trip to the fabulous Lake District, they hiked some twelve miles one day. "The descent was
rather rough, steep, & tiring," he commented to Mitten. Ascending and descending some three thousand feet, "[having] been on our legs over 9 hours we
were pretty tired."[64] They were probably all the more so carrying a cargo of
plants they dug up along the way, not very unlike the collecting in his tropical
travels all those many years ago, with specimens accumulating almost exponen-
tially! Another time they were off to Devonshire: "The ferns exceeded our wildest imaginations & Ma reveled in them!" he wrote Violet. Indeed, Annie so
reveled in the ferns that on one four-mile hike where they found themselves in
a veritable fern-vana her mother "could hardly be got along." Wallace was not
really complaining—they had never seen so lush a natural fernery anywhere in
Wales, Scotland, or the Lakes, he marveled.[65] Can you imagine hiking with the
two of them? Between the botany, geology, entomology, and every other -ology
that meets the eye, it's a wonder they got much beyond any given trailhead.

Switzerland was a favorite destination too, with its winning combination of
stunning tectonically and glacially sculpted landscape and eponymous alpine
botany. Wallace was always attracted to such borderlands in space and time—
he would have reveled in today's geological understanding—and here was the
very type specimen of alpine orogeny, rugged mountains born of the slow and
inexorable collision of tectonic plates, the African moving northward into the
Eurasian. Their very ruggedness reflects their youth; they are a great "accre-
tionary prism" of the once offshore sediments of a sometime Mesozoic sea
now emplaced accordion-like atop the southern edge of the European plate, a
belt of lofty, steep-sided, and sharp-ridged mountains between plunging wet
valleys.[66] He had gone many times over the years, earlier with such friends as
George Silk and now with his family. These days Violet went there annually

with her friends, becoming quite a hill walker like her parents. In August 1896 she and her mom and five friends went ahead to the alpine village of Adelboden, where "Pa" was to join them after a week in Davos after lecturing at a conference organized by Henry Lunn, the prolific advocate of ecumenism and international cooperation and the founder of a travel company specializing in educational tours.[67] Wallace dreaded it: "I fear it will be an awful place," he confided to Annie, packed with hotels and endless villas and *pensions*, a veritable city squeezed into a narrow valley. "I shall be glad to escape as soon as may be to the rural seclusion of Adelboden."[68] (The socialist Wallace would find Davos even more "awful" today, as the venue for the annual World Economic Forum, an organization of multinational corporations and politicians.) He enjoyed the company of Lunn, with whose family Violet had spent the summer the previous year, and his fellow conferees, but was glad to join Annie, Violet, and friends in their newfound alpine idyll. Adelboden was something of a borderland within the greater Alpine borderland—just north of the Rhône-Simplon Line, in modern geological understanding, the long fault zone that follows the River Rhône, Adelboden lay within a great valley bordered by high ridges of Mesozoic-aged Helvetic metasedimentary rocks, opening to Lake Thunersee to the northeast. There they spent the next two weeks hiking, botanizing, and geologizing to their heart's content.

Will couldn't join them, unfortunately—he was working, most recently in Southampton, which had the virtue of being close enough to Parkstone that he went home regularly. Since graduating he had worked at a series of electrical jobs in and around Newcastle, then in Scotland, spending most of 1895 in Inverkeithing, near Edinburgh, and Govan, Glasgow. Now he and his buddy Mac (short for McAlpine) were bitten by something of a travel bug and were planning an extended trip to the United States, where they figured they would work telegraph and electrical lineman jobs as they explored the country, mainly by bike. They sailed for Boston in March 1897, then made their way west to the Adirondacks, north to Niagara, west again to Chicago, then on to Denver. Ever one for fatherly advice, Wallace gave Will instructions on trapping mammals and preparing skins and skeletons to make some extra money ("You might make nearly as much as by climbing telegraph poles!"). Mammalogist Oldfield Thomas at the British Museum thought his department would purchase the specimens and helpfully recommended a London agent.[69] (An equal-opportunity dad, Wallace also encouraged Violet to do the same in her upcoming trip to Germany—it is not clear how she felt about the idea of trapping and processing small mammals during her continental holiday.)

From a photo by London Stereoscopic Co.

Wishing yourself and the Cause of Socialism every success Believe me Yours very truly

Oct: 8th. 1895 Alfred R. Wallace

Wallace

Wallace, the eternal optimist, campaigned tirelessly for social justice.

Will's journey, which lasted several years, was quite an adventure. He very much enjoyed his "rough" companions: "Their language is frightful," he told his dad, "their morals are apparently absent, but they are not really half as bad as they make out," notwithstanding the occasional fistfight and hatchet wielding. It was hard but invigorating work, especially in the Rocky Mountain winters when the thermometer plunged to single digits, his mustache and beard festooned with icicles.[70] Privately, Wallace was glad Will was enjoying his adventure but worried about his financial security. "He has had the best education I could give him in Electrical Engineering—3 years in College and 3 years in the workshops & at various jobs," he confided to Lester Ward of the Smithsonian; yet, he fretted, "So far, in America, he has been able to get nothing but labourer's or lineman's work at moderate wages, but the bosses always keep them at high pressure for nine hours a day, after which they are of course not fit for much but eating and sleeping." It was terrible, Wallace thought, how most workers could not get a fair shake no matter how well educated: all they could expect was a life of toil followed by an old age of poverty or worse. "Surely the coming century must see the end of the existing system of cut-throat competition, and wealth production based on the misery & starvation of the millions!"[71]

More and more his thoughts were turning to that new century soon to dawn; reflecting on how far they had come, societally, the eternal optimist had high hopes for further progress. He pinned his hopes on the intrinsic goodness of people, perhaps naively believing they would rise to the occasion if they could just be shown the way. In June 1898 he gave an address that gives insight into the path toward which he passionately strove: "Spiritualism and Social Duty," delivered in London at the International Congress of Spiritualists on 23 June 1898 and published the following month. In this crucial address, Wallace explicitly tied his spiritualism and socialism to societal advancement, maintaining that the ethical teachings of spiritualism, with its emphasis on "higher law," were lessons in equality and social justice. We err in emphasizing the afterlife, he argued; rather, we must strive to raise up humanity in *this* life. He believed that the teachings of spiritualism pointed to a path forward: "How to raise the bulk of our people out of that terrible slough of destitution, grinding life-long labour for bare subsistence, and shortened lives" devoid of the uplifting "refinements of art or enjoyment of Nature" so essential to the development of our best selves. "Let us Spiritualists take higher ground. Let us demand Social Justice," he declared, nothing less than equality of opportunity. It was our social duty: "This will be a work worthy of our cause, to which it will give dignity and importance. . . . Our faith, founded on knowledge, has a direct influence on our lives; that it teaches us to work strenuously for the elevation and permanent well-being of all our fellow men."[72]

This matter was weighing heavily on Wallace's mind just then. Not coincidentally, in his address Wallace pointed to "a work published a few weeks since" that underscored the paradox of continued injustice and misery amid plenty—namely, his latest book *The Wonderful Century; Its Successes and Its Failures*. It was a retrospective of the nineteenth century with a distinctly Wallacean spin. He selected as one of its epigraphs a poem by Thomas Lake Harris, the Anglo-American poet, preacher, author of proto-science fiction, and self-styled spiritualist prophet:

> If thou would'st make thy thought, O man, the home
> Where other minds may habit, build it large.
> Make its vast roof translucent to the skies,
> And let the upper glory dawn thereon,
> Till morn and evening, circling round, shall drop
> Their jewelled plumes of sun-flame and of stars.

Now, there's a beautiful vision, evocative of the expansive thought of Wallace himself. The premise of *The Wonderful Century* was to review what Wallace saw as the most notable successes and failures of the century, and it says something about his optimism that the section on "successes" consists of fifteen chapters, while that on "failures," only six. The successes ranged from the practical (modes of travel, labor-saving machinery, photography) to the great intellectual strides: evolution, certainly, but also the latest and greatest insights from geology, physics, astronomy—even a curious essay on the importance of dust ("a source of beauty and essential to life"). The societal failures, intellectual, moral, and social, as he saw it, ranged from blind spots—opposition to "psychical research" and the neglect of phrenology—to the social ills of vaccination, the "demon of greed," the "plunder of the earth," and a growing danger then looming: "Militarism—the curse of civilization."[73]

The "Vampire of war" is what Wallace called militarism in a spirited indictment of governments, his own included, "armed to the teeth, and watching stealthily for some occasion to use their vast armaments for their own aggrandizement and for the injury of their neighbors." It all sounds eerily (and depressingly) familiar, as he decried the "mad race between all the Great Powers . . . to increase the death-dealing power of their weapons," harnessing "the resources of modern science . . . in order to add to the destructive power," and the mind-boggling expenditures on ever-more-expensive armaments and *armies* of armies, funds that could otherwise have been put to genuinely good use to advance society and improve people's lives. Wallace comes close, too, to putting his finger on what is now called the military-industrial complex—indeed, a late Victorian version of the military-industrial-*congressional* complex that U.S. president Dwight D. Eisenhower so urgently warned about in January 1961. Wallace pointed to the "ruling classes—kings and kaisers, ministers and generals, nobles and millionaires . . . the true vampires of our civilization," who foment and fund and benefit from war. "The whole world is now but the gambling table" of the Great Powers, he wrote. Some things don't seem to change.

As Wallace was penning those words, he could hear the latest drumbeat for war picking up tempo as the situation in southern Africa heated up. Britain had tried and failed to annex the Transvaal Republic and Orange Free State in the First Boer War of 1880–1881—the object being control of gold and diamond mines, of course. Then more recently, in 1895, Leander Starr Jameson led a raid of Uitlanders—foreigners, mainly from Britain—attempting (and again failing) to seize control of the Transvaal. Fraught negotiations over the status and rights of the Uitlanders ensued, remaining essentially deadlocked at the time Wallace was writing *The Wonderful Century*. Rising nationalistic voices within Britain claimed grievances against their emigree countrymen and urged action to "protect" them—an early version of the playbook used by Russia to invade Ukraine's Crimean and Donbas regions some 120 years later. Predictably, by June 1899, one year to the month after Wallace's book was published, negotiations broke down, leading to the outbreak of the Second Boer War the following October as the Boer Republics launched preemptive strikes against the British. The war raged on until 1902, the republics eventually coming under British control after a long and bloody guerrilla war by the Boers that was met with a scorched-earth response by the British, with actions that would qualify as war crimes today. Wallace saw it all coming, penning a dozen letters and commentaries over the next few years condemning Britain's actions in the strongest terms.[74] Such was the far-less-than-wonderful conclusion to the "wonderful century."

———

Wallace didn't slow down much in the new century, intellectually or otherwise. The wheels of his brain were always turning—he still had another seven books in him! The brave new century opened with a retrospective of sorts: *Studies Scientific and Social*, a collection of essays in two volumes. It was a sort of "greatest hits II" reprinting his key papers from 1865 to 1899, the first volume dedicated to scientific writings and the second to a range of social, political, and educational topics. Another familiar manifestation of his restlessness was moving house once again, Parkstone now getting a bit too crowded for them. The little utopia of his "home colony for congenial persons" may have fallen through, but one benefit of that project was getting out and about, viewing prospective properties. They really needed to stay in the south of England for maximum gardening potential, and they found just the place not far from their current home, a three-acre southeast-facing grassy hollow surrounded by woods on a hillside in neighboring Broadstone. Like Corfe View it was fairly close to a train station, but this time there was no danger of being crowded in with development, as he told Violet, being strategically sited so that their view could not be obstructed, and it was well buffered by woodlands. He was pleased with himself over his successful negotiation for the property with Lord Wimborne's agent: "At last the deed is done! 'I've met the Douglas in his hall,—the Lion in his den!' and have come out safely! His roar was terrible! but he ended as mildly cooing as any sucking dove!"[75] The main charm of the place, he later noted, was an old, abandoned orchard of gnarled apple, pear, and plum trees—resilient, like him. And useful. Another of the site's many virtues was proximity to their Parkstone residence, so they were able to ransack that garden in turn, transplanting all the choicer plants to their new abode. Then, there was a botanical bonanza when a nursery in Poole liquidated its entire stock at bargain basement prices—he and Annie were soon carting away some thousand trees and shrubs. He served as his own contractor for the house, sparing no expense—beautiful dado rail in the drawing room, teak paneling and bookshelves, spacious verandas, the latest Doulton-glazed ceramic fireplace ("It will be I think quite a feature in the house," he told Will). He estimated the house cost the princely sum of £1,500, "but it will be well worth it."[76] Wallace built an adjacent house too, Tulgey Wood, intending it for Annie after he was gone but in the meantime used by Violet as a kind of kindergarten.[77] She had moved back home around then, perhaps to help out her parents. Her Pa turned eighty soon after moving into the beautiful red-brick home they dubbed Old Orchard, and though her Ma was far younger (fifty-six just then), she had periodic health problems.

Old Orchard was more or less completed by the end of 1902, though they had moved in sooner, camping out in the study and making the best of it.[78] While the carpenters completed the detailing inside, Wallace and Annie labored outside, throwing themselves into their garden, always what made home truly *home* for them. A journalist, Ernest Rann, visited a few years later and commented on the garden's charming "order of disorder," a half-wild, half-cultivated Arcadia of "wild firs, bunches of bracken, hosts of evergreens and subtropical plants, and here a pool with broken irregular edge to add a mirror of nature to the rustic scene." He noted, too, the maps and framed orchid engravings adorning the walls of the house and Wallace's study, with its sizable and well-thumbed "working library," adjoining a conservatory filled with tropical beauties. You would think it might have been difficult for Wallace to get much work done in that study, the conservatory and garden always beckoning, but he had a pretty disciplined routine: at least two solid hours of writing every morning, reading and maybe a short nap in the afternoon, tinkering in the garden or conservatory, and then more reading and writing in the evening. I imagine him, too, taking long walks with Annie and Violet some days through the surrounding meadows, fields, and woodlands and along the lovely Dorset coast. He surely admired the Pinnacles and Old Harry Rocks, shining white chalk sea stacks he could see in the distance from Old Orchard. Those stunning formations mark the easternmost point of the fabulous Jurassic Coast made famous by the talented fossil hunter Mary Anning—Wallace would

Wallace's final home: Old Orchard, in Broadstone, Dorset, built 1902.

applaud the coast's designation as a UNESCO World Heritage Site today. I suspect the whimsical flint nodules weathering out of the cliffs lining those beaches never got old, and how could he resist picking up fossil corals, shells, wood, and fine belemnites, the "thunderbolts" of his youth? Long, straight *Belemnitella mucronata* abounds in that chalk formation. He would have found himself in borderlands of deep time there. Wandering south along the beach of Studland Bay toward Old Harry Rocks is to take a geological stroll back in time, crossing from the Eocene sands and clays of what he knew as the Bagshot Beds (now Poole Formation) to the older London Clay, then suddenly the narrow Reading Formation with its gorgeous red-painted sandstone outcrops, and further on (and earlier still) to the zone of the Upper Cretaceous chalk.[79]

It was an inspiring place, his home, garden, and landscape. The biogeographer no longer traveled the world but brought the world home, populating his garden with botanical beauties from every biogeographical realm across the planet and vicariously enjoying the charms of the Alps through Will's and Violet's regular jaunts. It is easy to imagine him there at Old Orchard, surrounded by his worldwide garden on the edge of that great island, looking out from his study beyond the Cretaceous coastline over the wide sparkling waters of the English Channel, and scanning the horizon with his telescope (and perhaps taking some satisfaction in the fact that he could not see further still, to peninsular Normandy reaching toward him from the continent like a beckoning hand only about 150 km away—yet another proof of the earth's curvature, not that Hampden would have admitted it). He enjoyed turning his telescope skyward too, and it seems somehow appropriate that space turned out to be the final frontier of Wallace's far-ranging explorations of mind.

That all started, maybe not surprisingly, with financial woes. The already hefty price of Old Orchard was getting heftier and heftier with cost overruns and expenses piling up, threatening to break the household bank. He was getting a bit alarmed but had a solution: "I have begun writing an article for an American paper to earn some money to prevent our bankruptcy," he announced to Will.[80] As luck would have it, the *New York Independent* had recently offered him twenty pounds for an article about twenty-five hundred to thirty-five hundred words in length. Necessity being the mother of invention, he proposed to write a longer (and even more remunerative) piece, elaborating an insight he had come to of late as he worked on the fifth and extensively revised edition of *The Wonderful Century*. It was a characteristically grand vision, something he had been thinking about for awhile now: nothing less than an argument for the central position of our solar system, and thus humanity, in the universe—and by extension, therefore, a case for humanity as the central object and purpose of the universe! This was a logical extension and culmination of Wallace's evolutionary teleology, but it should be pointed out that the astronomical part of his analysis was based on the latest and greatest scientific thinking of the day. He had been following the exciting developments in astronomy for some time, highlighting the invention of the spectroscope and spectral analysis in the first edition of *Wonderful Century*—the "New Astronomy," a revolutionary breakthrough that gave insight into the physical and chemical properties of stars and planets for the first time. Now, for the fifth edition, he decided to expand his treatment of astronomy into four chapters, treating the latest insights into the solar system, sun, and stars and culminating in a fascinating chapter on the very "Structure of the Heavens." This, he told Will, would be the subject of his article for the *Independent*. He was angling for a whopping sixty pounds for the piece (which he ultimately got), but even better, his agent suggested he develop it into a book, the subject being so compelling. With luck, he concluded hopefully, "I shall perhaps be able to clear off all the debt before Midsummer."[81]

"Man's Place in the Universe" duly appeared in the 26 February 1903 issue of *The Independent*, and while continuing to work away on the new edition of *Wonderful Century* (published that September), he took a deep dive into the latest astronomical thinking. He was helped along in his crash course by several obliging correspondents, including Irish astronomer Agnes Mary Clerke, the author of widely admired books on astronomy and the history of astronomy who kindly explicated a number of astronomical principles for Wallace and offered comments and corrections on some of his chapter drafts.[82] His new book, *Man's Place in the Universe; A Study of the Results of Scientific Research in Relation to the Unity or Plurality of Worlds*, was published that October to both praise and criticism—something he was both used to and relished at this point, though he was surprised by the almost vehement pushback he got from some quarters of the astronomical community. Using the latest astronomical data to argue for what appeared to be a central location of our solar system in the galaxy (a view then widely held but that did not stand the test of time, by the way), Wallace then pivoted to considering what *seemed* a long string of coincidences, if not fortuitous circumstances, making life on Earth possible, from elemental makeup to physical conditions. For example, our distance from the sun is just right for the needed temperature range, and we have an atmosphere and oceans that function in energy circulation (thus temperature equalization) around the planet; a useful satellite, the moon, which facilitates this through tidal action; vast quantities of water vapor in the atmosphere; dust to provide nucleation sites for rainfall; and so on. In short, conditions favorable to life seemed a long shot, yet here they were. Could that be by design, he wondered? This was basically a forerunner of the idea that became known as the *anthropic principle* later in the century.

What does it all mean? It could all be the result of "one in a thousand million chances" in the infinitude of time, Wallace said, or—and here is where he flirts with the mystical—perhaps those who maintain that the universe is a "manifestation of Mind" are correct, the "orderly development of Living Souls" reason enough for the existence of this clockwork universe and our central position within it the only way things *could* be.[83] He concluded his book by arguing that while the remarkable astronomical discoveries he presented have "no bearing upon the special theological dogmas of the Christian, or of any other religion," they do seem to point to the uniqueness of our place in nature and "that the supreme end and purpose of this vast universe was the production and development of the living soul in the perishable body of man."[84] The book was a success, and in the subsequent editions that quickly followed, he took his argument even further, explicitly linking his astronomical argument with his evolutionary vision for humankind. In a nutshell, insofar as humans are, physically, the end product of a long evolutionary series of modifications, each of which occurred under specific circumstances, the chances of a repeat of this on some other planet would seem to be *exceedingly* remote: humanity, intelligent life, is a unique product of not just the earthly evolutionary process but the very universe itself.

Needless to say, he got a lot of applause from the theistically minded and pushback from those committed to a materialistic understanding of the universe (including humans)—but he was never one to shrink from a good argument, even (especially?) with the scientific community. Wallace was glad to admit he was wrong if grounds were found; otherwise, he relished defending his position to the utmost. It was in that spirit that he turned his attention to an ongoing debate then all the rage in America and Europe: the possibility of life on Mars. This was in part a scientific debate but had captured the public imagination too. Martians?! It had all started with observations of Mars made by Italian astronomer Giovanni Schiaparelli during the 1877 opposition of Mars. At astronomical opposition a planet is more or less aligned with the earth and sun, on the same side of the earth and so *opposite* the sun. Akin to a full moon, a planet in opposition is fully illuminated and high in the night sky and so well placed for observation. At such times professional and amateur astronomers alike excitedly take to their telescopes—Wallace, a longtime astronomy buff, was no exception, and neither was his daughter, Violet: "I suppose you were up till midnight last night observing the snow-cap of Mars, which is now very large, & Mars is nearest the Earth," he wrote her during the opposition of 1892. "Saturn also can be seen in the Evening early with a very narrow ring," he added helpfully. "Venus too could be seen splendidly just before sunrise, so you will have plenty of telescope work." And of course he had plenty of occasions to curse that bane of stargazers everywhere: clouds. "Unfortunately it got bad just as I was going to observe *Mars* at opposition!" he lamented to Violet in 1894.[85]

Schiaparelli observed what seemed to be a network of lines on the surface of Mars during the opposition of 1877, calling them *canali* (channels), which was promptly mistranslated into "canals," a word that implied they were not natural features but built by intelligent beings on Mars. American astronomer Percival Lowell, scion of the distinguished Lowell family of Boston, ran with that ball. Long fascinated by astronomy, he now resolved to study this exciting possibility. Putting the family fortune to good use, he built a state-of-the-art observatory in Flagstaff, Arizona, where the dry desert air was especially conducive to astronomy. Sure enough, during the Martian oppositions of 1893 and 1894 he made extensive observations, mapping Schiaparelli's canali and more— the waxing and waning of polar caps of Mars and seasonal changes in coloration that he interpreted as the seasonal greening up of vegetation found in temperate zones on Earth. Most would have been hard pressed to make much sense of these changes, so many shifting smudges, but Lowell, in thrall to the idea of Martian-built canals, jumped—on an astronomical scale—to sensational conclusions, which he published in his 1906 book *Mars and Its Canals*. The civilization of Mars was dying, he argued, becoming a desert planet, the canals a last-ditch effort to channel water seasonally from the polar regions to agricultural areas at lower latitudes. He had quite the imagination; it was something of a utopian vision, actually—the pacifist Lowell saw planetwide cooperation in a desperate attempt to stave off the collapse of their civilization.

As much as Wallace would have applauded that sentiment, he was completely unimpressed with Lowell's arguments and was surprised that while the astronomical community by and large took an equally dim view, no one seemed to be stepping up to say so. The existence of intelligent Martians would of course have torpedoed Wallace's own arguments about the uniqueness of humanity as intelligent life-forms in the universe. Perhaps in part to correct what he saw as poor scientific reasoning and in part to ensure the foundation of his own ideas remained strong, he took on Lowell in typical fashion. In the revised edition of *Wonderful Century*, in 1903, he had given a fair summary of Schiaparelli's and Lowell's observations, generously acknowledging that "optical illusions do not deceive experienced astronomers for months in succession," even though he found their interpretation "almost certainly erroneous." He also pointed out why he thought Mars could not support life: its relatively low mass meant relatively little atmosphere. "The amount of atmosphere is largely dependent on the mass of the planet," he said, "and this one condition almost certainly renders Mars unsuitable, since its mass is less than one-eighth that of the earth."[86] Now he amplified these arguments and marshaled additional evidence to refute Lowell politely but firmly in a book published in late 1907: *Is Mars Habitable? A Critical Examination of Professor Percival Lowell's Book "Mars and Its Canals," with an Alternative Explanation*. Far more could be said about this fascinating episode, but we will put it down as another battle fought and won and leave it at that.[87]

———

Is Mars Habitable? came out shortly before Wallace's eighty-fifth birthday—of course, he showed no signs of slowing down. Ever the multitasker, while he was working away on these projects there was still a prodigious output of

letters, addresses, editorials, essays, and articles on matters scientific and social. Plus, he somehow found the time to research and write his autobiography
in two hefty volumes, mostly, he said, for Will and Violet. It is more chronicle
than traditional autobiography and deliberately so, titled *My Life; A Record of
Events and Opinions*—and there sure were lots of both events and opinions!
Around this time he also turned his attention to his old friend Spruce, who
had died some dozen years before, chronically ill ever since returning from his
fifteen years of exploration in South America. Wallace did right by his friend,
editing his journals and correspondence and publishing the memoir *Notes of
a Botanist on the Amazon and Andes* in 1908, a fine memorial. The year 1908 was
in fact momentous—a year of remembrance and high honors. In the space of
two weeks, he was invited to give the New Year lecture at the Royal Institution,
was awarded the Copley Medal from the Royal Society ("Your very kind letter
came upon me like a thunderbolt . . ."[88]), and—most prestigious of all—got
word that the nation's top civilian honor was being bestowed upon him by
the king: the Order of Merit, an honor limited to no more than twenty-four
living recipients from the entire Commonwealth. He was deeply appreciative,
but Wallace being Wallace, he could not bring himself to attend the ceremony
at Buckingham Palace to receive the medal, especially given the obligatory
"court dress." Fortunately, it was not necessary: one of the king's equerries
soon came down from London and personally delivered the handsome medal,
a striking gold cross inlaid with red-and-blue enamel and topped with a gold
crown, affixed to a rich blue-and-crimson silk ribbon. He felt "duty-bound" to
wear it to his Royal Institution lecture.[89]

   That year he was also duty-bound to attend the fiftieth anniversary of the
reading of his and Darwin's watershed papers at the Linnean Society that fateful July day in 1858; of the august group involved in the readings, he and
Hooker alone remained now. On the occasion there was yet another high
honor: he was the first recipient of the Darwin-Wallace Medal, a special one
of solid gold awarded with much fanfare—and still the only one of gold ever
made, all the others being silver.[90] His acceptance speech was full of the same
magnanimity that always marked his relationship with Darwin, deferring to his
late friend's priority in discovering natural selection. But more importantly, he
said, he wanted to address the interesting question of *why them?* "Why did so
many of the greatest intellects fail, while Darwin and myself hit upon the solution of this problem?" He chalked it up, on careful consideration, to a "curious
series of correspondences, both in mind and in environment," foremost among
them: beetles! He and Darwin had both been passionate beetle collectors

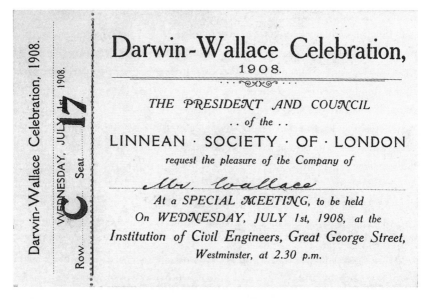

Wallace's invitation to the fiftieth anniversary of the readings of the Darwin and Wallace papers at the Linnean Society of London on 1 July 1858.

when young, he pointed out, a group that in all its bewildering and glorious diversity provides an object lesson in seemingly infinite and minute variation, adaptation, distribution. Add to that a reflective mind, an eye for observation, and a certain stick-to-it-ness and voila! But it all started with beetles. It was typical Wallacean understatement, but he had a point.[91]

It says something that Wallace's final two books, both published at age ninety, reflect his humanitarian interests. He had more and more been turning his energies to social reform, following his own exhortation to his fellow spiritualists to work for the betterment of *this* life. Early in his ninetieth year, he fired his penultimate salvo: *Social Environment and Moral Progress*. ("It is very heretical of course," he wrote his American botanist friend Alice Eastwood.)[92] And later that year, his final one: *The Revolt of Democracy*, a slender and more pointed volume aimed at politicians, addressing the continued labor and wage problems of the day with proposed solutions. The "grand old man" of science worked right up to his final brief illness in early November 1913.[93] He died on 7 November, just two months shy of his ninety-first birthday. Friends suggested burial at Westminster Abbey, but Will and Violet knew that neither their mother nor, especially, their father, would want that.[94] Wallace was buried at Broadstone Cemetery with little ceremony, just as he would have preferred, his monument a fine 7-foot (2.1 m) fossilized tree trunk he had likely collected himself from the fossil forest at nearby Lulworth Cove— *Protocupressinoxylon purbeckensis*, a conifer of the Upper Jurassic first described from the Lower Purbeck Formation of Dorset.[95] His illness and death came so suddenly that everyone—family, friends, his many admirers far and wide— were left thunderstruck, bereft. They were disconsolate, but they likely instinctively knew he was with them still, in one way or another. Had they heard Thomas Jefferson's consoling remarks on the death of American naturalist William Bartram, they would have found them fitting:

> "He is not gone," he urged. "He remains everywhere around you. When you wish to find him, you [need] only to look in his garden, and in his work, and in his green world."

And there we find Alfred Russel Wallace still.

CODA

AUGUST 2010. I was on a bus to London—the Oxford Tube, to be exact, coursing along the M40—when I had an epiphany about Alfred Russel Wallace. It was sort of an evolutionary "road to Damascus" moment as I read a paper published in 1966 by the late historian of science H. Lewis McKinney of the University of Kansas—a paper that opened my eyes to a Wallace of breathtaking depth and breadth, a Wallace whom, truth be told, I had been largely oblivious to. I say *sort of* a road to Damascus moment because there was no conversion, exactly; no, I was already a biologist and knew of and admired Alfred Russel Wallace—one of the great Victorian naturalist-explorers, he of the Wallace Line, astute observer who fortuitously discovered the principle of natural selection, insightful author of the Sarawak Law paper, indefatigable collector who accidentally scooped Darwin and led to the writing of *On the Origin of Species*. Or did he? On the M40 that August morning, I realized not only how little I really knew of Wallace's life and thought but how this common narrative of Wallace as the gung-ho collector who got lucky and stumbled onto evolution by natural selection was really something of a caricature, and a misleading one at that.

Caricatures exaggerate features to amusing effect, and now I realized the collector in Wallace was that exaggerated feature. But it was not comical; it was part of a mythos that had grown up around Wallace—that he was a mere bug collector who by happenstance made some discoveries that mattered, an accidental hero whose real service to evolutionary biology was prying that landmark book out of Darwin as a result of his fluke discovery. Myths and legends take on a life of their own, repeated in textbook after textbook. McKinney's scintillating paper "Alfred Russel Wallace and the Discovery of Natural Selection" gave lie to that view, bringing to my consciousness a remarkable document that he had dubbed the Species Notebook, the most important of Wallace's journals kept during his travels in the Malay Archipelago.[1] A working journal, the Species Notebook was full of jottings and drawings, observations and memoranda, collection tallies, and all sorts of notes-to-self. But it also contained a secret: Wallace had a plan! The notebook revealed a plan to write a book arguing for the reality of transmutation, as evolution was then called.

Not an offhand marginal note about a plan but a fairly clear plan with passages already written out, a structure outlined, arguments sketched—an astonishing array of arguments for transmutation: islands, domestication, the fossil record, morphology, and more. They were built around the anti-transmutation sections of Charles Lyell's landmark *Principles of Geology*. Indeed, facing Lyell head-on—plucky for a lowly, self-educated field collector—was central to the plan for his book. An evolutionary David against a great geological Goliath. McKinney put this remarkable notebook in context—the fortuitous meeting of Wallace and beetle-mad Bates, their correspondence revealing a keen interest in the question of species origins, the hatching of an audacious idea to travel and explore—and collect, yes, but as a means to furnish material for their own study while selling others to finance their explorations. Wallace, I realized, had collected to travel; he did not travel to collect. In other words, Wallace aspired to be a philosophical naturalist, a scientist in modern parlance, and contribute to one of the most profound scientific questions of the day: the nature and origin of species. Collecting was a means to that end in two ways, providing funds and fodder for study and analysis. How had I never heard of this notebook?! What *was* this notebook, anyway, and *where* was it? I hunted for the details in McKinney's paper. A footnote said it resided in the collections of the venerable Linnean Society of London. Really? That was precisely the destination of my London-bound bus at that very moment.

———

I had the privilege of studying and publishing that stunning notebook a few years later, with an awful lot of help along the way, I hasten to add, from the encouragement of my friend Andrew Berry to the tremendous transcription assistance from my wife, Leslie, to the generosity of the Linnean Society itself. It was my homage to Wallace and also to McKinney, who first mused that Wallace's unrealized book on evolution might have been titled *On the Organic Law of Change*.[2] The notebook is a portal not just to another time and place but to a remarkable mind. For me it was also a conduit into a fascinating literature

Wallace reclining in his garden at Old Orchard.

I had little appreciated until then, and I was inspired to fully immerse myself in it. I'm still swimming; what have I learned from my deep dive some dozen or more years later? That Alfred Russel Wallace was one astounding human being.

As I hope this book makes clear, Wallace's life is nothing less than an epic drama—a story of one man's aspiration, inspiration, determination, pluck, luck, generosity, genius—and not a little obstinacy along the way. His is one of the most original minds of the nineteenth century. One manifestation of this is Wallace's voluminous scientific output, a near-constant stream of insightful observations, letters, reports, scientific papers, and books, several hailed then and still regarded today as landmark works. In some areas Wallace made strikingly original contributions; in others he was a great synthesizer, eloquently advancing hypotheses and points of view. His one great blunder in biology, in the modern view, was his insistence that natural selection could not explain the human brain. But consider the range of his scientific interests, the many areas where we think he got it right, the fields in which he made real contributions, even founded or shaped. Wallace was spot-on in his scientific insights more often than not: besides being honored as the co-discoverer of natural selection and the originator of modern evolutionary biogeography (and we could further unpack these into a wealth of individual research threads he pursued, to be sure), there is also the fact that the modern view of how sexual selection works, the evolution of mimicry and protective coloration, and certain aspects of the speciation process (allopatry, reinforcement, and the Wallace Effect) are far closer to Wallace's thinking than to Darwin's. Consider, too, that Wallace made contributions to anthropology, geology, physical geography, climatology, archaeology, taxonomy, systematics, and more. Such prodigious scientific efforts would excite our admiration by themselves—was there *any* Victorian scientist who contributed to a greater range of fields? But in addition there is another manifestation of Wallace's wonderfully original mind: his utter commitment to the improvement of the human condition.

Wallace was arguably the first and greatest of the truly humanitarian scientists. In the last two centuries, several eminent scientists stand out for their efforts to assuage or avert human suffering. Heroic efforts on the medical front were put forth by the likes of Jenner in the eighteenth century, Pasteur in the nineteenth, and others. In the twentieth century, biochemist Linus Pauling, recipient of the 1962 Nobel Peace Prize for his efforts to halt the nuclear arms race, comes to mind, as does 1970 Peace Prize laureate Norman Borlaug for his key role in the green revolution and reducing hunger. There are of course countless other scientists who have undertaken humanitarian work, to their honor. But are there any of Wallace's prominence who passionately threw themselves into advocacy for such a diversity of social causes, doggedly seeking to improve the human condition on multiple fronts, sustained over long years in the face of constant criticism, even ostracism? Wallace tends to be overlooked in this regard because some of his causes were not and in some cases are still not popular with those controlling the levers of power, and it was easy to paint him as some kind of crank owing to his fierce (blind, even) devotion to such eyebrow-raising causes as spiritualism and land nationalization. But Wallace was an eternal optimist who believed in human progress and the potential for continued improvement in body, mind, and spirit, tirelessly campaigning against social and political forces that he believed, rightly or wrongly, kept people down.

There is Wallace's advocacy for land reform, monetary reform, women's rights, and environmental conservation, and his campaigns to rein in the excesses of capitalism, militarism, imperialism, and eugenicists run amok—even his anti-vaccination stance, as much as we now disagree with his criticisms (based as they were on the imperfect scientific knowledge of the day), came from a noble place: a concern for human rights and dignity. *Fiat Justicia Ruat Coelum.* In Wallace's humane view, too, "uncivilized" peoples are neither morally nor intellectually inferior to "civilized" peoples and in fact have more than a few things to teach them, and women are entitled to full rights—in education, marriage, voting, inheritance, and engagement in intellectual society. And let us not forget his gentle soul as husband, father, and friend, or his boundless generosity of spirit—an outgrowth of the guy's utter commitment to fairness, as he saw it. The aftermath of the reading of the Darwin and Wallace papers at the Linnean Society of London could have been ugly indeed had a less generous person scooped Darwin. But no: for Wallace the seeking of scientific truths was the thing, and he doffed his hat to Darwin as the first originator of the theory of natural selection regardless of his own prodigious efforts and independent discovery, even despite his long-held private plans to write his own book arguing for transmutation. Wallace's declared "force of admiration" for Darwin and his accomplishments was as boundless as it was genuine—the very model of magnanimity. Indeed, he never sought glory for himself, even resisting the honors that *were* eventually bestowed practically kicking and screaming.

Was Wallace thus somehow cheated of his rightful place in the sun? No. Despite claims to the contrary from various quarters over the years, there is no evidence *whatsoever* that Darwin pilfered ideas from Wallace or sought to suppress him and seize the limelight, as I have argued based on an analysis of the circumstances surrounding the fateful arrival of Wallace's Ternate essay on Darwin's doorstep and all that followed.[3] Honoring Wallace certainly need not come at the expense of Darwin, who is rightfully lauded for his own prodigious achievements in discovering the principle behind species change and marshaling the eloquent "one long argument" for evolution by natural selection that we know as *On the Origin of Species*—a book for which Wallace had nothing but praise. But the two naturalists were remarkably congruent in the lines of evidence each saw and sought in developing his evolutionary arguments, and they were *together* our first guides to the grand process by which the Tree of Life thrives. Why, then, should Darwin be a household name and not Wallace? Well, as I have argued elsewhere, most likely owing to a combination of reasons, but perhaps foremost among them, I can name three: First, the Darwin name was *already* famous by the mid-nineteenth century thanks initially to grandfather Erasmus, the famed physician, inventor, and best-selling poet, but Charles himself, no slouch, was a rising star in the public eye with the publication of his popular *Voyage of the Beagle*. Second, Darwin soon came out with *Origin*, after which both he *and* Wallace tended to refer to the theory as more the elder naturalist's baby. And third, Wallace's yearslong refusal to take part of the credit, even going so far as to title his major overview of the subject *Darwinism*, reinforced the misperception that the credit was mainly Darwin's. Nonetheless, that Darwin's name should be in lights while Wallace's is largely relegated to obscurity is an injustice that must be remedied—Wallace deserves better.

———

The Wallace Line is an apt metaphor for the many explorations of borderland and boundary that we have seen crisscrossing Wallace's remarkable life— biogeographical and geological, certainly, but there also exists the Wallace

Wallace's final resting place, fittingly marked with a fossilized tree: both travelers in time and space.

Line of science and society, humans and nature, physical and spirit worlds (as he saw it), rich and poor, civilized and savage, privileged and powerless, Wallace's time and ours. The subtitle to historian Ted Benton's 2013 treatment of Wallace's life rhetorically asks: "A thinker for our own times?"[4] Indeed, Alfred Russel Wallace is a thinker and role model for our times and beyond, for the ages, his life of curiosity, adventure, genius, discovery, and advocacy inspired and inspiring in equal measure. Like Walt Whitman he contained multitudes, with a largeness of spirit and diversity of perspectives and experiences that embraced even self-contradiction at times. And that's okay. Like Whitman, too, Wallace beckons to us still from beyond some borderland of being, metamorphosed in a way Ovid would appreciate:

I bequeath myself to the dirt to grow from the grass I love,
If you want me again look for me under your boot-soles.

You will hardly know who I am or what I mean,
But I shall be good health to you nevertheless,
And filter and fibre your blood.

Failing to fetch me at first keep encouraged,
Missing me one place search another,
I stop somewhere waiting for you.[5]

We would do well to keep searching for Alfred Russel Wallace—in the wide world, certainly, but especially within ourselves.

———

ACKNOWLEDGMENTS

I OFFER this book as a modest tribute to the incomparable Alfred Russel Wallace on the occasion of the bicentennial of his birth, keenly aware that it would not have been possible without the kind assistance and encouragement of a remarkable circle of family, friends, and all manner of supporters near and far. First and foremost I thank Leslie Callaham Costa, my multitalented wife of some thirty years, constant collaborator, illustrator, and sometime transcriber and researcher with the Wallace Correspondence Project. Leslie did triple duty, not merely by making it possible for me to lose myself for days at a time in Wallacean Nirvana during the writing of the book but also through her efforts as an incisive (literally . . .) editor and keen-eyed artistic director helping chase down and edit many of the illustrations. This book is as much hers as it is mine.

I am deeply grateful to my editor Eric Crahan at Princeton University Press for so enthusiastically supporting this project from day one and for encouragement, advice, support, and patience throughout the process. Sincere thanks, too, to PUP editorial assistants Barbara Shi and James Collier, Illustration Manager Dimitri Karetnikov, and Editorial Production Manager Elizabeth Byrd for their ready assistance and advice; Senior Designer Heather Hansen for her fine cover design; Production Supervisor Melody Negron of Westchester Publishing Services for ably overseeing the book's progression from copyediting to publication; and copy editor Wendy Lawrence for her thorough and careful work.

Several friends and colleagues kindly read early drafts of the manuscript with a critical eye: many thanks to Jonathan Hodge, David Collard, an anonymous reviewer, and especially George Beccaloni, Charles Smith, and Andrew Berry, who all offered helpful comments, corrections, and suggestions that improved the manuscript greatly. Special thanks to E. J. Tarbox for reading the manuscript and for many an illuminating conversation on the history of ideas. Any errors or oversights, not to mention my admittedly at-times-idiosyncratic writing style and reluctance to rein in certain enthusiasms, are of course my own responsibility. Wallace would understand.

I have benefited tremendously from the kindness and generosity of fellow Wallaceophiles, starting with friends Andrew Berry and George Beccaloni, from whom I have learned a great deal—I am deeply appreciative of our many discussions as well as their encouragement and support over the years. I am grateful to George Beccaloni, Clay Bolt, Andrea Deneau, and Isabelle Charmantier of the Linnean Society of London, and to Tracy Murphy of the Bureau of Land Management Canyons of the Ancients Visitor Center and Museum, for kindly providing images for the book, and to Victor Rafael Limeira da Silva and George Beccaloni for carefully reviewing the maps of Amazonia and the Malay Archipelago, respectively.

Heartfelt thanks to David and Stella Collard for their warm hospitality during a memorable visit to Usk—it was a special pleasure walking in young Wallace's footsteps with such gracious and knowledgeable hosts, and where Clive and Theresa Jones so kindly opened Kensington Cottage, Wallace's birthplace— and their home—to us, and where Clive and Ken Wann graciously gave us a sneak peek of the Alfred Russel Wallace guesthouse under development—I look forward to staying one day! Kudos and accolades are due Charles Smith for his indispensable Alfred Russel Wallace Page website—a treasure trove of writings by and about Wallace and other resources—as well as to the Linnean Society of London for making Wallace's notebooks and other resources readily available, and George Beccaloni for ever-cheerfully responding to my incessant requests for information and for putting the wonderful resources of the Wallace Correspondence Project at my disposal. And speaking of wonderful resources, the Biodiversity Heritage Library, Hathi Trust, and Google Books are a scholar's dream. I am deeply grateful to these organizations and their many contributing institutions for the unparalleled service they provide to scholars worldwide.

I would be remiss, too, if I did not also acknowledge the giants whose shoulders I stand on—I am grateful to the many Wallace and Darwin scholars and Wallace biographers whose works I have eagerly devoured, analyzed, and at times debated, most notably (for this book) the work of George Beccaloni, Barbara Beddall, Ted Benton, Andrew Berry, Janet Browne, Jane Camerini, David Collard, Martin Fichman, Wilma George, Jonathan Hodge, Sandy Knapp, Malcolm Kottler, H. Lewis McKinney, Jim Moore, Penny van Oosterzee, Peter Raby, Michael Shermer, Ross Slotten, Charles Smith, and John van Wyhe. There are many others, too numerous to list here (but are cited in this and my other books), whose insightful papers have helped illuminate my understanding of Wallace, Darwin, and their lives and times.

Closer to home, I owe a great deal to my mentors and friends of Harvard University's Museum of Comparative Zoology (MCZ), where I am privileged to hold a research associate appointment in the Department of Entomology; special thank you to Naomi Pierce, Andrew Berry, Kathy Horton, and the late Ed Wilson. I would like to give a special shout-out to the librarians of Hunter Library of Western Carolina University, the Ernst Mayr Library of the MCZ, and the LuEsther T. Mertz Library of the New York Botanical Garden. My Wallace and Darwin studies have been greatly facilitated by the talented and ever-helpful librarians of these institutions. Last but not least, I thank my colleagues at my home institutions, Western Carolina University and Highlands Biological Station, deeply place-based yet outward-looking institutions whose teaching, training, and research missions are all about supporting exploration and inspiration—lofty and laudable goals that Wallace would surely appreciate.

NOTES

Abbreviations

Correspondence

DCP (Darwin Correspondence Project). Darwin letters are cited in the endnotes by their DCP letter (DCP-LETT) numbers. These may be accessed through Epsilon or the website of the Darwin Correspondence Project at the University of Cambridge Libraries (https://www.darwinproject.ac.uk/).

WCP (Wallace Correspondence Project). Wallace letters are cited in the endnotes by their WCP numbers. These may be accessed through the searchable Epsilon database of Wallace and other correspondence collections at the University of Cambridge Libraries (https://epsilon.ac.uk/).

Manuscripts

Wallace notebooks and journals located at the Linnean Society of London are cited by their manuscript number as follows:

JOURNALS

Wallace's Malay Journals consist of four notebooks with sequentially numbered entries:

MS-178a: first Malay Journal, June 1856–March 1857; entries 1–68.
MS-178b: second Malay Journal, March 1857–March 1858; entries 69–128.
MS-178c: third Malay Journal, March 1858–August 1859; entries 129–192.
MS-178d: fourth Malay Journal, October 1859–May 1861; entries 193–245.

FIELD NOTEBOOKS

MS-177: North American Journal, 1886–1887. Also transcribed and annotated by Charles H. Smith and Megan Derr (2013), *Alfred Russel Wallace's 1886–1887 Travel Diary: The North American Lecture Tour* (Manchester, UK: Siri Scientific Press).
MS-179: Natural History Notebook, 1854

MS-180: Species Notebook, 1855–1859. Also reproduced in facsimile, transcribed, and annotated by J. T. Costa (2013), *On the Organic Law of Change* (Cambridge, MA: Harvard University Press).

MS-182: Palms of the Amazon Notebook, c.1848–c.1852

Darwin's manuscripts and other papers, held by Cambridge University Library, are identified by their CUL-DAR numbers; these documents can be accessed as follows:

Cambridge University Library:
https://cudl.lib.cam.ac.uk/collections/darwin_mss/1;
https://www.lib.cam.ac.uk/collections/departments/manuscripts-university-archives/significant-archival-collections/darwin.

Darwin Online:
http://darwin-online.org.uk/contents.html.

Published Works *(page numbers cited are keyed to the following editions frequently cited in the present text)*

CONTRIBUTIONS

A. R. Wallace (1870), *Contributions to the Theory of Natural Selection. A Series of Essays* (London: Macmillan).

DARWINISM

A. R. Wallace (1889), *Darwinism: An Exposition of the Theory of Natural Selection, with Some of Its Applications* (London: Macmillan).

GEOGRAPHICAL DISTRIBUTION OF ANIMALS

A. R. Wallace (1876), *The Geographical Distribution of Animals; With a Study of the Relations of Living and Extinct Faunas as Elucidating the Past Changes of the Earth's Surface.* 2 vols. (London: Macmillan).

ISLAND LIFE

A. R. Wallace (1880, 2013), *Island Life, or, the Phenomena and Causes of Insular Faunas and Floras, including a Revision and Attempted Solution of the Problem of Geological Climates* (Chicago: University of Chicago Press).

MALAY ARCHIPELAGO

A. R. Wallace (1869), *The Malay Archipelago; The Land of the Orang-utan and the Bird of Paradise* (New York: Harper and Brothers).

MY LIFE

A. R. Wallace (1905), *My Life: A Record of Events and Opinions.* 2 vols. (London: Chapman and Hall).

ORIGIN

C. R. Darwin (1859), *On the Origin of Species by Means of Natural Selection* (London: John Murray).

TRAVEL DIARY

C. H. Smith and M. Derr (2013), *Alfred Russel Wallace's 1886–1887 Travel Diary: The North American Lecture Tour* (Manchester, UK: Siri Scientific Press).

TRAVELS ON THE AMAZON

A. R. Wallace (1853), *A Narrative of Travels on the Amazon and Rio Negro, with an Account of the Native Tribes, and Observations on the Climate, Geology, and Natural History of the Amazon Valley* (London: Reeve).

TROPICAL NATURE

A. R. Wallace (1878), *Tropical Nature and Other Essays* (London: Macmillan).

Preface

1. The "revolutionary" Wallace of the subtitle of the present book is Wallace the iconoclast, the outside-the-box thinker who could see and connect dots no one had thought to connect before, let alone even notice, and had the wherewithal to push back against change-resistant establishments, scientific and social. The curious personality and other factors behind Wallace's "revolutionary" or heretical tendencies are discussed at length in Michael Shermer's compelling 2002 book *In Darwin's Shadow: The Life and Science of Alfred Russel Wallace* (Oxford: Oxford University Press). See also R. Elwyn Hughes (1991), "Alfred Russel Wallace (1823–1913): The making of a scientific non-conformist," *Proceedings of the Royal Institution* 63:175–83.

Chapter 1: A Happy but Downwardly Mobile Family

1. V. G. Walmsley (1959), "The geology of the Usk Inlier (Monmouthshire)," *Quarterly Journal of the Geological Society of London* 114:483–521.

2. *My Life*, 1:23.

3. *My Life*, 1:20–21.

4. *My Life*, 1:24.

5. "The Alfred Russel Wallace Page," Western Kentucky University, http://people.wku.edu /charles.smith/wallace/FAQ.htm#Welsh.

6. Daniel Defoe (1734), *Curious and Diverting Journies, thro' the Whole Island of Great-Britain* (G. Parker), p. 246.

7. A replacement oak was planted on the site by Queen Elizabeth II in 1985.

8. *My Life*, 1:87.

9. A. R. Wallace to G. Silk, 7 October 1903, WCP6588.

10. Lewis Turnor (1830), *History of the Ancient Town and Borough of Hertford* (Hertford: St. Austin and Sons), p. 286.

11. An egregious example was the uninhabited town of Old Sarum in Wiltshire, which nonetheless had parliamentary representation by its land owners (members of the Pitt family).

12. Later in life, Wallace noted his father's conservatism in connection with the passage of the Reform Bill of 1832: "My earliest memories of any political event are those connected with the first Reform Bill of 1832. I recollect my father—a genuine Tory—shaking his head over it as a sad giving-way to the ignorant clamour of the mob, and as being likely to result in some vague but terrible disaster. Then followed the public rejoicings when the Bill was passed, which in our town of Hertford, as I suppose in many others, took the form of a public, open-air free dinner, the broad Fore Street being filled with rows of tables at which the whole of the workers and their families who chose to come feasted to their hearts' content." Alfred R. Wallace (1907), "Personal suffrage: A rational system of representation and election," *Fortnightly Review*, 1 January.

13. William Clarke's *The Boy's Own Book: A Complete Encyclopedia of All the Diversions, Athletic, Scientific, and Recreative, of Boyhood and Youth* was first published in 1828 in London by Vizetelly, Branston and Co. The following year it was brought out in Boston by Munroe and Francis. An instant hit, *The Boy's Own Book* became the go-to manual for generations of boys. American novelist, abolitionist, and women's rights advocate Lydia Maria Child, rightly seeing a need for a similar book for young women, authored *The Girl's Own Book* in 1833.

14. M. A. Wallace to T. Wilson, 5 July 1835, WCP1654.

15. M. A. Wallace to L. Draper, 12 August 1835, WCP1655.

16. Darwin moved into no. 36 Great Marlborough Street on Tuesday, 15 March 1837, just up the street from his brother Erasmus, who lived at no. 43. "It is very pleasant our being so near neighbours," he wrote his cousin (C. Darwin to W. D. Darwin Fox, 12 March 1837, DCP-LETT-348).

Chapter 2: Taking Measure in the Borderlands

1. Launched on 16 October 1821, the School of Arts of Edinburgh (known today as Heriot-Watt University) was the first of these new organizations that became known as mechanics' institutes, dedicated to the Scottish Enlightenment ideal of self-improvement through education: "For the instruction of mechanics in such branches of physical science as are of practical application in their several trades."

2. It has been debated whether mechanics' institutes were initially motivated by Utilitarian and Enlightenment principles or the belief that a scientific education would aid the establishment by preparing workers for industrial society, a form of social control (see, e.g., Steven Shapin and Barry Barnes [1977], "Science, nature and control: Interpreting mechanics' institutes," *Social Studies of Science* 7[1]:31–74). Regardless of their foundational motivation, however, the movement was largely taken over by a Utilitarian and even socialist-utopian agenda by

the mid-1830s, which aided the rapid spread of mechanics' institutes throughout Britain and even abroad in the United States, Ireland, Canada, and Australia.

3. *My Life*, 1:104.

4. Robert Owen (1825), *Two Discourses on a New System of Society; as Delivered in the Hall of Representatives at Washington* (Philadelphia: Atkinson and Alexander).

5. Daniel Feller (1998), "'The spirit of improvement': The America of William Maclure and Robert Owen," *Indiana Magazine of History* 94:89–98. See also Arthur Bestor (1971), *Backwoods Utopias: The Sectarian Origins and the Owenite Phase of Communitarian Socialism in America, 1663–1829* (Philadelphia: University of Pennsylvania Press), chap. 5.

6. Arriving in New Harmony via the Ohio River in the winter of 1825–1826 aboard the keel boat *Philanthropist* (the "Boatload of Knowledge"), the accomplished group included American entomologist and conchologist Thomas Say, French artist and naturalist Charles-Alexandre Lesueur, French educators Marie Fretageot and William S. Phiquepal and several of their students, Swiss artist Balthazar Abernasser, artist and musician Virginia DuPalais, physician William Price and family, and a host of other notables. See Donald E. Pitzer (1989), "The original boatload of knowledge down the Ohio River: William Maclure's and Robert Owen's transfer of science and education to the midwest, 1825–1826," *Ohio Journal of Science* 89(5):128–42. The town is a National Historic Landmark District today.

7. D. Thomson (1955), "Queenwood College, Hampshire: A mid-19th century experiment in science teaching," *Annals of Science* 11(3):246–54.

8. Edward Royle (1974), *Victorian Infidels: The Origins of the British Secularist Movement, 1791–1866* (Manchester: Manchester University Press), app. 1 and 2, pp. 294–301; *My Life* 1:87.

9. The quoted passage comes from the opening section of pt. 1, "The author's profession of faith," in Thomas Paine (1794), *The Age of Reason: Being an Investigation of True and Fabulous Theology* (Paris: Barras). Paine's *Age of Reason* has been much reprinted; two readily available editions can be found in *Thomas Paine: Collected Writings*, collected and edited by Eric Foner (1995, Library of America) and *The Life and Major Writings of Thomas Paine*, collected and edited by Philip S. Foner, uncle of Eric Foner (1993, Citadel Press).

10. Paine, *Rights of Man*, pt. 2.

11. *My Life*, 1:89.

12. V. Robinson to A. R. Wallace, January 1907, WCP1407; A. R. Wallace to V. Robinson, 14 January 1907, WCP4293.

13. Greta Jones (2002), "Alfred Russel Wallace, Robert Owen and the theory of natural selection," *British Journal for the History of Science* 35:73–96.

14. *My Life*, 1:104.

15. Acts of the Parliament of Great Britain, Inclosure Act 1773, sect. 1, How arable lands shall be fenced, https://www.legislation.gov.uk/apgb/Geo3/13/81/contents.

16. Gordon E. Mingay (2014), *Parliamentary Enclosure in England: An Introduction to Its Causes, Incidence and Impact, 1750–1850* (London: Routledge).

17. Acts of the Parliament of Great Britain, Tithe Act 1836, sect. 12, Meaning of the words "person," "lands," "tithes," "parish," "parochial," "landowner," "tithe owner," as used in this Act, https://www.legislation.gov.uk/ukpga/Will4/6-7/71/section/12; H. C. Prince (1959), "The tithe surveys of the mid-nineteenth century," *Agricultural History Review* 7(1):14–26; Roger J.

P. Kain and Hugh C. Prince (2006), *The Tithe Surveys of England and Wales* (Cambridge: Cambridge University Press), chaps. 1 and 2.

18. *My Life*, 1:152.

19. *My Life*, 1:106, 115.

20. See pp. 15–17 in George Beccaloni (2008), "Homes sweet homes: A biographical tour of Wallace's many places of residence," pp. 7–43, in Charles H. Smith and George Beccaloni, eds., *Natural Selection and Beyond: The Intellectual Legacy of Alfred Russel Wallace* (Oxford: Oxford University Press).

21. *My Life*, 1:106–7.

22. See James A. Secord (2000), *Visions of Science: Books and Readers at the Dawn of the Victorian Age* (Chicago: University of Chicago Press), chap. 2.

23. The Soulbury Stone, as it is now known, was accused of being a menace to motorists in 2016. In the words of one commentator: "The Soulbury Stone has been sitting contentedly, keeping itself to itself for around 11,000 [years] when it was deposited in its resting place from Derbyshire during the last Ice-Age. There isn't anything very unusual about that of course except for the fact that this large stone is [set] right in the middle of a modern road. . . . Everything was just going splendidly for 11,000 years until a dozy motorist drove into it earlier this year and tried to sue the local authority for £1,800 damages." There were some calls to move the rock, provoking an outcry. Fortunately, the town council decided to paint stripes on the pavement around the rock rather than relocate it—the lesser of the evils. See Stephen Liddell (website), "The Soulbury Stone," https://stephenliddell.co.uk/2016/04/18/the-soulbury-stone/.

24. In the years before Swiss naturalists Louis Agassiz and Jean de Charpentier promulgated the theory of ice ages and vast extensions of glacial ice, the prevailing model was one of an earth marked by slow oscillation of both land/sea level and climate; erratic boulders—some as big as a house—were postulated to have been carried on icebergs at a distant time, a colder epoch when much of Europe was under the sea (as evidenced by marine fossils and formations inland), and icebergs could float farther south than they do today, depositing their burden as they slowly melt. One year after Wallace observed the Soulbury Stone, Charles Darwin published a short paper titled "Note on a rock seen on an iceberg in 61° south latitude" in the *Journal of the Royal Geographical Society* (9 March 1839, pp. 528–29), describing such a phenomenon deep in the Southern Hemisphere while on the *Beagle* voyage. "Every fact on the transportation of fragments of rock by ice is of importance, as throwing light on the problem of 'erratic boulders,' which has so long perplexed geologists," he noted. Incidentally, Darwin's hometown of Shrewsbury is home to its own erratic, a smallish smooth boulder known as the Bellstone.

25. A. R. Wallace to J. Wallace, 11 January 1840, WCP337; A. R. Wallace to G. Silk, 12 January 1840, WCP338; A. R. Wallace to G. Silk, 15 January 1840, WCP336.

26. *My Life*, 1:170.

27. M. S. Rosenbaum (2007), "The building stones of Ludlow: A walk through the town," *Proceedings of the Shropshire Geological Society* 12:5–38.

28. English geologist Roderick Impey Murchison first recognized the distinctiveness of certain sedimentary strata of South Wales in the early 1830s, proposing the name "Silurian" in honor of the ancient Silures tribe in 1835 ("On the Silurian system of rocks," *Philosophical Magazine*, series 3, 7:46–52), inspired by fellow geologist Adam Sedgwick, who had named an earlier series of Welsh rocks the Cambrian, for Cambria, the Roman Latin name for Wales. In Neath Wallace

would have had access to Murchison's three-volume 1839 treatise *The Silurian System* (London: John Murray).

29. *My Life*, 1:167.

30. A. R. Wallace to H. E. Wallace, March 1842, WCP339. This is part of a long rhyming letter from Alfred to his younger brother Herbert ["Edward"], then boarding at a school in Essex. The letter opens: "Dear Herbert, As I plainly see | By your last letter unto me | That you have 'taen [taken] to Poesy | I'll try to get my pen to go | And write away a page or so (see also *My Life* 1:178).

31. *My Life*, 1:196.

32. *My Life*, 1:192.

33. Wallace's copies of Lindley's *Elements* (4th ed., 1841) and Swainson's *Treatise* (1835) reside in the Linnean Society of London. The annotation quoted from his copy of Swainson appears on p. 5. Wallace slightly altered the quote from Darwin (1839), *Journal of Researches into the Geology and Natural History of the Various Countries Visited by H.M.S.* Beagle (London: Colburn), p. 604 (emphasis in original).

34. A. R. Wallace to F. Wallace, probably written in 1842–1843, WCP6671.

35. Transcript by Thomas Vere Wallace on pp. 98–99, WCP5531. Probably written May 1836–April 1843.

36. Charles William Sutton (1810–1875), s.v. "Jones, Thomas," *Dictionary of National Biography, 1885–1900* (London: Smith, Elder), 30:170.

37. R. E. Hughes (1989), "Alfred Russel Wallace; some notes on the Welsh connection," *British Journal for the History of Science* 22:401–18.

38. *My Life*, 1:199.

39. A similar circumstance—dissatisfaction with another naturalist—later prompted Wallace to write what became known as the Sarawak Law paper in 1855, in that case prompted by Edward Forbes and his "polarity theory." See p. 192 in Alfred R. Wallace (1855), "On the law which has regulated the introduction of new species," *Annals and Magazine of Natural History*, 2nd series, 16:184–96.

40. "On a probable means of procuring plane and curved specula of great size, with a few remarks on fixed telescopes," A. R. Wallace to W. H. Fox Talbot, 12 April 1843, WCP1792.1680. Wallace's paper was not read at the subsequent meeting of the British Association, nor is any further correspondence with Talbot known. In the 1850s both chemical deposition and electroplating processes for depositing an ultrathin layer of silver on a ground glass mirror surface were introduced by Karl August von Steinheil in Germany and Léon Foucault in France, abandoning speculum mirrors and revolutionizing the production of large telescope mirrors. A vacuumdeposition process on glass was later developed using silver and, more commonly, aluminum. Spinning liquid-mirror telescopes have mirrors made with a reflective liquid rotated at a constant speed around a vertical axis, causing the liquid surface to assume a parabolical shape. The concept was first recognized by Isaac Newton and later developed in an 1850 paper by Ernesto Capocci of the Naples Observatory (perhaps with indirect input from Talbot, via his Italian colleague the optician Giovanne Battista Amici; see Charles H. Smith [2006], "Reflections on Wallace," *Nature* 443:33–34). The first working laboratory liquid-mirror telescope was built by New Zealand inventor Henry Skey in 1872; there are several in operation today.

41. "The South-Wales Farmer," in *My Life* 1:206–22.

42. David Williams (1955), *The Rebecca Riots: A Study in Agrarian Discontent* (Cardiff: University of Wales Press); David Howell (1988), "The Rebecca riots," pp. 113–38, in Trevor Herbert and Gareth Elwyn Jones, eds., *People and Protest: Wales, 1815–1880* (Cardiff: University of Wales Press); David J. V. Jones (1989), *Rebecca's Children: A Study of Rural Society, Crime and Protest* (Oxford: Oxford University Press).

43. Wallace scholar Charles H. Smith quotes the following notice printed in the *Hereford Times* issue of 19 October 1844: "Kington. Mechanics' Institution. Some time ago a prize was awarded to Mr. A. R. Wallace, one of its members, for an essay on 'the best method of conducting the Kington Mechanics' Institution.' We are informed that it possesses great merit." He further notes that "The 2 November issue of the same paper contains an anonymous letter to the editor agreeing as to the quality of the essay but noting that it had been the only essay submitted for the prize! The *Hereford Journal* issue of 22 October printed the same first letter, though it referred to Wallace as 'Mr. A. Wallace.'" See the Alfred Russell Wallace Page, http://people.wku.edu/charles.smith/wallace/misc.htm, item 4.

44. "An essay, on the best method of conducting the Kington Mechanic's Institution," pp. 66–70, in Richard Parry, ed. (1845), *The History of Kington* (Kington, UK).

45. Wallace used very similar language in a different context some twenty years later, in a passage decrying the hypocrisy of upholding species as evidence of the Creator's handiwork yet allowing their senseless destruction despite being easily able to do otherwise: "Yet, with a strange inconsistency, seeing many of them perish irrecoverably from the face of the earth, uncared for and unknown." From Wallace's 1863 paper "On the physical geography of the Malay Archipelago," *Journal of the Royal Geographical Society* 33:217–34. See chapter 12.

46. Alfred R. Wallace (1843/1905), "The advantages of varied knowledge," lecture; see *My Life* 1:201–5. It is worth noting that Wallace's later arguments against an evolutionary origin for the human mind are based on some of the same concerns as those expressed here—namely, that our "great store of mental wealth" should lie unused. See chapters 11 and 12.

Chapter 3: Beetling and Big Questions

1. Alfred R. Wallace (1845), [Letter to the editor on ARW's early experiments with mesmerism in Leicester; "Journal of Mesmerism" column], *Critic* (London) 2(19):45 (10 May).

2. *My Life*, 1:236.

3. *My Life*, 1:237.

4. Bates had three short notes on insects published in the *Zoologist* in 1843, the journal's inaugural year, including "Notes on Coleopterous insects frequenting damp places" (pp. 114–15), "Notes on the seasons of appearance of *Polyommatus argiolus*" (p. 199), and "Note on the occurrence of *Colias edusa* in Leicestershire" (pp. 330–31). He also published a fascinating fourth note (p. 156) asking if the editor would "admit communications of a general critical nature, on arrangements, natural systems, &c., which your correspondent thinks would materially advance the increasing importance of your periodical." The editor noted in response that he would be "obliged for the opinions of our readers on this subject." Bates's inquiry is noteworthy in showing that the then eighteen-year-old was no mere collector but was interested in principles of classification and relationship—perhaps an added inspiration to Wallace.

5. It is no wonder that Leicestershire has a long and venerable "beetling" tradition. The county has produced an impressive number of professional and amateur coleopterists over the past two

centuries; see Derek Lott (2009), *The Leicestershire Coleopterists: 200 Years of Beetle-Hunting* (Loughborough, UK: Loughborough Naturalists' Club).

6. Information on the Leicester Mechanics' Institute and Literary and Philosophical Society is derived from the fascinating accounts of Gerald T. Rimmington (1975), "Education, politics and society in Leicester, 1833–1940" (PhD thesis, University of Nottingham, chap. 1), and Patrick Boylan, ed. (2010), *Exchanging Ideas Dispassionately and without Animosity: The Leicester Literary and Philosophical Society. 1835–2010* (Leicester, UK: Leicester Literary and Philosophical Society).

7. Boylan, *Exchanging Ideas Dispassionately*, pp. 3, 8.

8. Rimmington, "Education, Politics and Society in Leicester," pp. 13, 14.

9. *My Life*, 1:232.

10. *My Life*, 1:232.

11. See James Moore (1997), "Wallace's Malthusian moment: The common context revisited," pp. 290–311, in B. Lightman, ed., *Victorian Science in Context* (Chicago: University of Chicago Press) for a compelling treatment of the geographic, economic, and social factors that compose the "common context," as Moore puts it, for Wallace's later discovery of the principle of natural selection.

12. The county of Neath Port Talbot has produced a series of fine walking tours to follow in the footsteps of its most famous sons, including, among others, actor Richard Burton, Romantic artist Joseph Mallord William Turner, and of course Alfred Russel Wallace. The Wallace Trail has five-to-six-mile and ten-to-eleven-mile versions: "Start off at Neath Castle, a Norman stronghold which still displays an impressive twin-towered gatehouse. You will visit the Mechanic's Institute and Neath Town Hall where Wallace attended lectures before heading to Neath Abbey. The Abbey, founded in 1130, is one of South Wales' most impressive monastic remains. A further exploration of Neath Abbey Iron Works will see you head towards the Neath and Tenant Canals before ending the walk at the Church of St Illtyd on the bank of the River Neath." See "In Their Footsteps," https://dramaticheart.wales/home-2/plan-your-visit/in-their-footsteps/.

13. F. Wallace to M. A. Greenell Wallace, 11 September 1844, WCP1257; 19 October 1844, WCP1260; and 14 May 1845, WCP1267; F. Wallace to W. Wallace, 26 September 1844, WCP1259.

14. See Fanny Wallace's letters of 14 May 1845, WCP1267; 22 May 1845, WCP1268; 1 June 1845, WCP1269; 7 July 1845, WCP1270; Fanny's account of enslaved individuals being driven to market is given in her letter of 29 November 1845 to her mother. See WCP1263.

15. F. Wallace to M. A. Greenell Wallace, 10 October 1845, WCP1273.

16. Mark Casson (2009), *The World's First Railway System: Enterprise, Competition, and Regulation on the Railway Network in Victorian Britain* (Oxford: Oxford University Press), pp. 29, 289, 298, 320; *My Life*, 1:243.

17. See, for example, tithe commutation apportionment hearings advertised in the 13 March and 5 June 1846 issues of the *Cambrian* at Charles Smith's Alfred Russel Wallace page, http://people.wku.edu/charles.smith/wallace/bib1.htm, papers S1bb and S1bc.

18. Gordon Roderick (1993), "Technical instruction committees in South Wales, United Kingdom, 1889–1903 (part 1)," *Vocational Aspect of Education* 45(1):59–70, 62.

19. History Points, "Former Mechanics' Institute, Neath," https://historypoints.org/index.php?page=former-mechanics-institute-neath.

20. M. Jones to A. R. Wallace, 5 July 1895, WCP3159.

21. Louise Miskell (2006), *Intelligent Town: An Urban History of Swansea, 1780–1855* (Cardiff: University of Wales Press).

22. Excerpts from the following letters: A. R. Wallace to H. W. Bates, 26 June 1845, WCP342; 3 October 1845, WCP343; 13 October 1845, WCP344; 11 April 1846, WCP340; 3 May 1846, WCP341.

23. A. R. Wallace to H. W. Bates, August 1846, WCP347.

24. *My Life*, 1:248.

25. A. R. Wallace to H. W. Bates, 26 June 1845, WCP342.

26. Alfred R. Wallace (1847), "Capture of *Trichius fasciatus* near Neath [excerpt from a letter sent from Neath, Wales]," *Zoologist* 5:1676.

27. Tony Ramsay (2017), "Fforest Fawr Geopark—a UNESCO Global Geopark distinguished by its geological, industrial and cultural heritage," *Proceedings of the Geologists' Association* 128:500–509.

28. Charles Davidson (1906), "The earthquake in South Wales," *Nature* 74:225–26.

29. A. R. Wallace to H. W. Bates, 26 June 1845, WCP342.

30. A. R. Wallace to H. W. Bates, 9 November 1845, WCP345.

31. Charles Darwin, newly converted to the heretical idea of transmutation in 1837, immediately grasped these implications: admit species change, he confided in a private notebook, and "the whole fabric totters and falls . . . the fabric falls!" Notebook C, pp. 76–77, http://darwin-online.org.uk/content/frameset?itemID=CUL-DAR122.-&viewtype=text&pageseq=1.

32. See Adrian Desmond (1989), *The Politics of Evolution: Morphology, Medicine, and Reform in Radical London* (Chicago: University of Chicago Press); James A. Secord (2000), *Victorian Sensation: The Extraordinary Publication, Reception, and Secret Authorship of* Vestiges of the Natural History of Creation (Chicago: University of Chicago Press).

33. See Richard Holmes (2010), *The Age of Wonder: The Romantic Generation and the Discovery of the Beauty and Terror of Science* (New York: Vintage).

34. Astronomer Nicolas-Louis de Lacaille, who observed at the Cape of Good Hope between 1750 and 1754, introduced fourteen new constellations, all but one (Mensa, honoring Table Mountain) symbolizing instruments of Enlightenment science and art: Antlia (pneumatic pump), Caelum (chisel), Circinus (drafting compass), Fornax (furnace), Horologium (clock), Microscopium (microscope), Norma (framing or carpenter's square), Octans (octant), Pictor (painter's easel), Pyxis (mariner's compass), Reticulum (telescope reticle), Sculptor (sculptor's tools), and Telescopium (telescope).

35. Robert Chambers (1844), *Vestiges of the Natural History of Creation* (London: W. and R. Chambers), pp. 154, 156. It is worth noting, too, that when pitching his own transmutational theory in *On the Origin of Species*, Darwin selected as epigraphs quotes bearing on this question from two philosophers: William Whewell on how the Divine works through natural law and Francis Bacon on how it is equally as important to become proficient in studying the book of God's works (nature) as the book of God's word (scripture). See frontmatter commentary in James T. Costa (2009), *The Annotated Origin* (Cambridge: Harvard University Press).

36. Chambers, *Vestiges of the Natural History of Creation*, pp. 235, 296.

37. Secord, *Victorian Sensation*, pp. 168–69.

38. Darwin's eventual geology professor, Rev. Adam Sedgwick, vehemently denounced *Vestiges* in a lengthy review published in the July 1845 *Edinburgh Review*. In April of that year, he shared his contempt for the "foul book" and his suspicion that it was authored by a woman in a letter to his geological colleague Charles Lyell:

[*Vestiges*] filled me with such inexpressible disgust that I threw [it] down. . . . I cannot but think the work is from a woman's pen, it is so well dressed and so graceful in its externals. I do not think the "beast man" could have done this part so well. Again, the reading, though extensive, is very shallow; and the author perpetually shoots ahead of his facts, and leaps to a conclusion, as if the toilsome way up the hill of Truth were to be passed over with the light skip of an opera-dancer. This mistake was woman's from the first. A. Sedgwick to C. Lyell, 9 April 1845, in John Willis Clark (1890), *The Life and Letters of the Rev. Adam Sedgwick*, 2 vols. (Cambridge: Cambridge University Press), 2:84–85.

39. A. R. Wallace to H. W. Bates, 28 December 1845, WCP346.

40. A. R. Wallace to H. W. Bates, 28 December 1845, WCP346.

41. A. R. Wallace to H. W. Bates, 11 April 1846, WCP340.

42. A. R. Wallace to H. W. Bates, 11 October 1847, WCP348.

Chapter 4: Paradise Gained . . .

1. William H. Edwards (1847), *A Voyage up the River Amazon: Including a Residence at Pará* (London: John Murray), p. iii.

2. Ross A. Slotten (2004), *The Heretic in Darwin's Court: The Life of Alfred Russel Wallace* (New York: Columbia University Press), pp. 42–43.

3. A. R. Wallace to W. J. Hooker, 30 March 1848, WCP3802.

4. There is an extensive literature on the colonial context of eighteenth- and nineteenth-century natural history collecting, with a focus on Wallace and Bates. See, for example, Janet Browne (1992), "A science of empire: British biogeography before Darwin," *Revue d'Histoire des Sciences* 45:453–75; Jane Camerini (1996), "Wallace in the field," *Osiris*, 2nd ser., 11:44–65; Martin Fichman (2004), *An Elusive Victorian: The Evolution of Alfred Russel Wallace* (Chicago: University of Chicago Press), pp. 22–24; and Melinda B. Fagan (2008), "Theory and practice in the field: Wallace's work in natural history (1844–1858)," pp. 66–90, in Charles H. Smith and G. Beccaloni, eds., *Natural History and Beyond: The Intellectual Legacy of Alfred Russel Wallace* (Oxford: Oxford University Press). See also the excellent introductory discussion in Victor Rafael Limeira-DaSilva (2022), "The itinerary of Alfred Russel Wallace's Amazonian journey (1848–1852): A source for researchers and readers," *Notes and Records of the Royal Society* 76:633–52.

5. J. Wallace to F. Wallace, 5 May 1849, WCP5572; H. E. Wallace to F. Wallace, 7 May 1849, WCP392; September 1849, WCP393.

6. A. R. Wallace to G. Silk, 16 June 1848, WCP406.

7. *Travels on the Amazon*, p. 3.

8. *Travels on the Amazon*, p. 4.

9. *Travels on the Amazon*, p. 16.

10. Slotten, *Heretic in Darwin's Court*, p. 52, gives the breakdown of insects in Wallace and Bates's first shipment: 553 species of butterflies and moths, 450 species of beetle, and some 400 species from other insect orders. See also A. R. Wallace to W. J. Hooker, 20 August 1848, WCP3798.

11. J. Augusto Correio to J. Antonio Correio Seixus, September 1848, WCP7080.

12. Henry Walter Bates (1863), *The Naturalist on the River Amazons* (London: John Murray), 1:139.

13. *Travels on the Amazon*, pp. 83–84.

14. R. Spruce to W. J. Hooker, 3 August 1849, WCP4899.

15. Samuel Stevens (1849), "Journey to explore the province of Pará," *Annals and Magazine of Natural History*, 2nd ser., 3(13):74–75; A. R. Wallace and H. W. Bates to S. Stevens, 23 October 1848, WCP3744.

16. *Travels on the Amazon*, p. 103.

17. A. R. Wallace to Neath Mechanics Institution, February 1849, WCP829.

18. H. E. Wallace to F. Sims, 7 May 1849, WCP392.

19. Joanna Klein, "This week surfers will ride a wave in the Amazon," *New York Times*, 14 March 2016.

20. *Travels on the Amazon*, p. 116.

21. *Travels on the Amazon*, pp. 130–32.

22. *Travels on the Amazon*, pp. 121–22.

23. R. Spruce to W. J. Hooker, 3 August 1849, WCP4899.

24. Bates, *Naturalist on the River Amazons*, 1:223.

25. *Travels on the Amazon*, p. 136.

26. H. E. Wallace to F. Sims, September 1849, WCP393.

27. In fact the letter was only being penned back in London about the time Wallace arrived in Santarém, on 21 August 1849. It was sent "with compliments" to Stevens, eventually reaching Wallace in Manaus in late December 1849; see WCP5492.

28. *Travels on the Amazon*, p. 143.

29. A. C. Roosevelt et al. (1996), "Paleoindian cave dwellers in the Amazon: The peopling of the Americas," *Science* 272:373–84.

30. H. E. Wallace to M. A. Wallace, 12 November 1849, WCP3534.

31. *My Life*, 1:279.

32. Walter F. Cannon (1961), "The impact of uniformitarianism: Two letters from John Herschel to Charles Lyell, 1836–1837," *Proceedings of the American Philosophical Society* 105(3):301–14.

33. Samuel Stevens (1850), "Journey to explore the natural history of the Amazon River," *Annals and Magazine of Natural History*, 2nd ser. 6(36):494–95; see also WCP4268. Emphasis in the printed original may have been added by Stevens.

34. Richard Spruce (1908), *Notes of a Botanist on the Amazon and Andes*, ed. Alfred R. Wallace (London: Macmillan), 1:202, 291.

35. *Travels on the Amazon*, p. 167.

36. J. Podos and M. Cohn-Haft (2019), "Extremely loud mating songs at close range in white bellbirds," *Current Biology* 29(20):R1068–69.

37. It has been suggested that Wallace and Bates's friendship remained severely strained. That may be so, but in his book *The Naturalist on the River Amazons* (1863), Bates writes in fond terms of the brief time they spent together at Manaus as they waited for the rains to let up. Bates refers to the "pleasant society" of Wallace and their other expatriate companions and of forest rambles and plans with Wallace: "The miseries of our long river voyages were soon forgotten, and in two or three weeks we began to talk of further explorations. Mean-time we had almost daily rambles in the neighbouring forest" (Bates, *Naturalist on the River Amazons*, 1:341). They ultimately agreed to divide and conquer, Bates proceeding up the Rio Solimões and Wallace up the Rio Negro (1:347), as Bates related to their agent Samuel Stevens. John Hemming takes a similar view in his 2015 book *Naturalists in Paradise: Wallace, Bates, and Spruce in the Amazon* (London:

Thames and Hudson): "So it is wrong to imagine that the two had quarrelled seriously. They simply had different approaches to travel and collecting, and preferred to do it alone" (p. 117).

38. Stevens, "Journey to explore the natural history," pp. 495–96; see also WCP4269.

39. Alfred R. Wallace (1850), "On the umbrella bird (*Cephalopterus ornatus*), 'Ueramimbé,' L. G. [From a letter dated 10 March 1850, Barra do Rio Negro (Manaus); communicated by Samuel Stevens to the Zoological Society of London meeting of 23 July 1850], *Proceedings of the Zoological Society of London* 18:206–7.

40. *Travels on the Amazon*, pp. 176–78, 419–20.

41. See, for example, M. H. Horn, S. B. Correa, P. Parolin, et al. (2011), "Seed dispersal by fishes in tropical and temperate fresh waters: The growing evidence," *Acta Oecologica* 37:561–77; S. B. Correa, R. Costa-Pereira, T. Fleming, et al. (2015), "Neotropical fish-fruit interactions: Eco-evolutionary dynamics and conservation," *Biological Reviews* 90:1263–78; Mauricio Camargo Zorro (2018), "The fishes and the Igapó forest 30 years after Goulding," pp. 209–27, in Randall W. Myster, ed., *Igapó (Black-Water Flooded Forests) of the Amazon Basin* (Berlin: Springer).

42. *Travels on the Amazon*, pp. 178–79.

43. *Travels on the Amazon*, p. 180.

44. J. Wallace to M. A. Wallace, 22 June 1851, WCP1629.

45. There is no evidence that Wallace ever hooked up with any local ladies, Indigenous or otherwise, unlike, say, Spruce. In one suggestive letter, written from San Carlos de Rio Negro in Venezuela, Spruce wrote of being besieged by the "naughty" *moças* (girls) of the town, including two prostitutes: "My chastity was sorely assailed, and you no doubt tremble at the narrowness of my escape (if indeed I escaped at all, which I leave entirely to your judgment to decide)." But he went on to describe frankly how he invited a "buxom widow" he hired as a housekeeper to move in and sleep with him (she turned him down) and hinted that her younger sisters "provided other 'delicacies' which a *bachelor* can with difficulty obtain, and the *muchachas* [girls] make themselves useful in various ways." R. Spruce to A. R. Wallace, 2 July 1853, WCP351.

46. H. E. Wallace to F. Sims and M. A. Wallace, 30 August 1850, WCP394.

47. *My Life*, 1:278.

Chapter 5: . . . and Paradise Lost

1. *Travels on the Amazon*, p. 194.

2. *Travels on the Amazon*, p. 195.

3. Raoni Valle (2009), "Petroglyphs in the Lower Negro River Basin, NW Brazilian Amazon—a Preliminary View," Congresso Internacional da IFRAO 2009, Piauí, Brazil.

4. *My Life*, 1:316; Alfred R. Wallace (1853), "On the Rio Negro" [A paper read at the RGS meeting of 13 June 1853], *Journal of the Royal Geographical Society* 23:212–17.

5. *Travels on the Amazon*, p. 207.

6. *Travels on the Amazon*, p. 213.

7. See, for example, A. A. de Oliveira and S. A. Mori (1999), "A central Amazonian terra firme forest. I. High tree species richness on poor soils," *Biodiversity and Conservation* 8:1219–44.

8. *Travels on the Amazon*, p. 227.

9. H. E. Wallace to R. Spruce, 29 December 1850, WCP1656.

10. *Travels on the Amazon*, p. 229; emphasis in original.

11. A. R. Wallace to T. Sims, 20 January 1851, WCP390.

12. For an excellent account of the history and mapping of the Casiquiare Canal, see A. Hamilton Rice (1921), "The Rio Negro, the Casiquiare Canal, and the upper Orinoco, September 1919–April 1920," *Geographical Journal* 58(5):321–43.

13. Alexander von Humboldt (Helen Maria Williams, trans.) (1819–1829), *Personal Narrative of Travels to the Equinoctial Regions of the New Continent, during the Years 1799–1804*, pt. 1, 7 vols. (London: Longman, Hurst, Rees, Orme and Brown), 5:195–96, 288.

14. *Travels on the Amazon*, p. 241.

15. *Travels on the Amazon*, p. 264.

16. *Travels on the Amazon*, p. 253.

17. In his palms book, p. 21, Wallace noted that "the greater part, if not all of the Piassaba now imported, comes, however, from the Rio Negro, where several hundred tons are cut annually and sent to Para, from which place scarcely a vessel sails for England without its forming a part of her cargo."

18. *Travels on the Amazon*, p. 255.

19. A helpful introduction to Rousseau's life and thought can be found at the online *Stanford Encyclopedia of Philosophy* (https://plato.stanford.edu/entries/rousseau/). Sections 2 and 3, "Conjectural history and moral psychology" and "Political philosophy," provide an overview of Rousseau's thinking on the origins of inequality.

20. *Travels on the Amazon*, pp. 257–61.

21. R. Spruce to G. Bentham, 1 April 1851, WCP6719; see also Richard Spruce (1908), *Notes of a Botanist on the Amazon and Andes*, ed. Alfred R. Wallace, vol. 1 (London: Macmillan), pp. 208–11.

22. *Travels on the Amazon*, p. 275.

23. *Travels on the Amazon*, p. 277.

24. *Travels on the Amazon*, p. 279.

25. H. W. Bates to M. A. Wallace, 13 June 1851, WCP1658. In *Travels on the Amazon* (p. 323), Wallace mentions that he eventually received the letter Miller wrote informing him of his brother's illness and in the same batch of mail learned of Miller's own death. He had no idea of his brother's fate until he returned to Manaus the following April.

26. H. W. Bates to M. A. Wallace, 18 October 1851, WCP1659; if Bates wrote this letter, Wallace does not appear to have received it.

27. Caapí, or ayahuasca, still used ritually by Indigenous peoples throughout the Amazon Basin, is prepared from the woody Caapí vine (*Banisteriopsis caapi*, Malpighiaceae) together with *Psychotria viridis* (Rubiaceae) and other plants. See J. C. Callaway et al. (2005), "Phytochemical analyses of *Banisteriopsis caapi* and *Psychotria viridis*," *Journal of Psychoactive Drugs* 37(2):145–50. *B. caapi* was described by specimens collected by Spruce in 1854.

28. *Travels on the Amazon*, p. 299; John Heming (2015), *Naturalists in Paradise: Wallace, Bates, and Spruce in the Amazon* (London: Thames and Hudson), p. 165, clarifies that Wallace was the first "non-Brazilian" foreigner to witness and describe such a ceremony because the Portuguese government had excluded other Europeans from Brazil during the colonial period. Spix, Martius, and Natterer were among the very first to be allowed to travel up the Amazon after Brazilian independence. Only Natterer voyaged up the Uaupés, but as all his papers were later lost in a fire, there is no record of the customs he may have witnessed there.

29. *Travels on the Amazon*, p. 308.

30. *Travels on the Amazon*, pp. 300–1, 312.

31. *Travels on the Amazon*, p. 306.

32. *Travels on the Amazon*, pp. 307–8.

33. *Travels on the Amazon*, p. 318.

34. Spruce, *Notes of a Botanist*, p. 267; R. Spruce to J. Smith, 28 December 1851, WCP6721.

35. Word of H. Edward Wallace's death had evidently reached Spruce after Alfred Wallace's departure from Manaus, as Spruce mentions it in his 28 December 1851 letter to Smith (WCP6721): "Wallace's younger brother, who came out from Liverpool along with me, died last May. He had gone there, poor fellow, to embark for England, took the yellow fever, and died in a few days." He next saw Alfred at São Joaquim in February 1852, visiting when he was still ill but recovering. Edward's last words were communicated by Bates to Mary Ann Wallace (18 October 1851, WCP1659); see also *My Life*, 1:282. Wallace dedicated chapter 19 of volume 1 of *My Life* to memorializing his brother.

36. *Travels on the Amazon*, p. 346.

37. *Travels on the Amazon*, p. 352. Legendary twentieth-century Harvard botanist-explorer Richard E. Shultes recommended the ocoki tree as an excellent candidate for domestication, crediting Wallace for providing the earliest account of the valued fruit for Western science; see Richard E. Shultes (1989), "*Pouteria ucuqui* (Sapotaceae), a little-known Amazonian fruit tree worthy of domestication," *Economic Botany* 43(1):125–27.

38. *Travels on the Amazon*, p. 356.

39. Wallace, "On the Rio Negro," *Journal of the Royal Geographical Society* 23:212–17.

40. *Travels on the Amazon*, p. 382.

41. *Travels on the Amazon*, p. 375.

42. R. Spruce to A. R. Wallace, 21 November 1863, WCP380.

43. Spruce wrote Wallace in October 1852 from São Jeronimo (WCP350), describing his eighteen-day journey there from São Gabriel and plans to travel to the upper Uaupés.

Chapter 6: Down but Not Out

1. *Travels on the Amazon*, p. 396; *My Life*, 1:305; Dante Alighieri, *The Divine Comedy: Inferno*, cantos 12, 14, and 21 (Translated by Grant White and illustrated by Gustave Doré; New York: Pantheon, 1965).

2. Wallace, *Travels on the Amazon*, p. 398.

3. The Atlantis Fracture Zone was named for the celebrated Woods Hole Oceanographic Institution research vessel *Atlantis*, which provided the earliest detailed bathymetric data of the North Atlantic. These data were used by Marie Tharp and Bruce Heezen at Columbia in their pioneering mapping and analysis of the mid-Atlantic ridge and central rift valley in the 1950s, paving the way to an understanding of seafloor spreading and plate tectonics. See Marie Tharp (1999), "Connect the dots: Mapping the seafloor and discovering the mid-ocean ridge," in L. Lippensett, ed., *Lamont-Doherty Earth Observatory of Columbia: Twelve Perspectives on the First Fifty Years, 1949–1999* (New York: Lamont-Doherty Earth Observatory); Betsy Mason (2021), "Marie Tharp's groundbreaking maps brought the seafloor to the world," *Science News*, https://www.sciencenews.org/article/marie-tharp-maps-plate-tectonics-seafloor-cartography. Wallace later became interested in the contours of the ocean floor in the context of sea level rise and fall and its effect on the geographical distribution of species. Like nearly all naturalists of his day,

he held to a model of mainly static continents and ocean basins, with continental movement restricted to elevation and subsidence that, with other factors, could create land bridges and other corridors of migration; see C. H. Smith, J. T. Costa, and M. Glaubrecht (2019), "Alfred Russel Wallace's 'Die Permanenz der Continente und Oceane,'" *Archives of Natural History* 46(2):265–82.

4. *Travels on the Amazon*, p. 401.

5. *My Life*, 1:310.

6. *My Life*, 1:310.

7. A. R. Wallace to R. Spruce, 19 September 1852, WCP349; reprinted in *My Life*, 1:302–9.

8. *Travels on the Amazon*, p. 403; A. R. Wallace to R. Spruce, 19 September 1852, WCP349.

9. Stevens published three sets of letter extracts in the *Annals and Magazine of Natural History*: "Journey to explore the province of Pará," from a letter from Wallace and Bates dated 23 October 1848, Belém (Pará), published in the January 1849 issue (2nd ser., 3:74–75); "Journey to explore the natural history of South America," from a letter from Wallace dated 12 September 1849, Santarém, published in the February 1850 issue (2nd ser., 5:156–57); and "Journey to explore the natural history of the Amazon River," from letters dated 15 November 1849, Santarém, and 20 March 1850, Manaus (Barra), published in the December 1850 issue (2nd ser., 6:494–96).

10. *Transactions of the Entomological Society of London*, meeting of 4 October 1852, n.s., 2:29.

11. Alfred R. Wallace (1852), Letter to the editor in the "Proceedings of Natural-History Collectors in Foreign Countries" section of the *Zoologist* 10:3641–43.

12. Several years later, perhaps with his visits to these disparate and sometimes widely separated libraries of the learned societies in mind, Wallace set down in a notebook one of his practical proposals: "Formation of a Complete Library of Natural History," suggesting that the libraries of the Linnean, Entomological, Zoological, and other societies might be housed under one roof, where members of any of the societies would be free to use all the volumes of the joint collection. One benefit, he suggested, was that expensive natural history works would be more likely donated to a single shared library than multiple copies to several separate libraries. It is not clear, however, that he ever advanced the plan. See J. T. Costa (2013), *On the Organic Law of Change: A Facsimile Edition and Annotated Transcription of Alfred Russel Wallace's Species Notebook of 1855–1859* (Cambridge, MA: Harvard University Press), p. 168.

13. J. Wallace to M. A. Wallace, 10 January 1853, WCP1637.

14. Alfred R. Wallace (1852), "On the monkeys of the Amazon," *Proceedings of the Zoological Society of London* 20:107–10, reprinted in the December 1854 issue of *Annals and Magazine of Natural History*, 2nd ser., 14:451–54.

15. See brief overview in J. T. Costa (2019), "Historical and ecological biogeography," pp. 305–7, in C. Smith, J. T. Costa, and D. Collard, eds. (2019), *An Alfred Russel Wallace Companion* (Chicago: University of Chicago Press).

16. Wallace, "On the monkeys of the Amazon," pp. 109–10.

17. A. R. Wallace to H. W. Bates, 11 October 1847, WCP348.

18. A. R. Wallace to T. Sims, 20 January 1851, WCP390.

19. Alfred R. Wallace (1854), "On the habits of the butterflies of the Amazon Valley" [A paper read at the Entomological Society of London meetings of 7 Nov. and 5 Dec. 1853], *Transactions of the Entomological Society of London*, n.s., 2, pt. 8:253–64.

20. See Heliconiini at the Tree of Life, http://tolweb.org/Heliconiini/70208, and associated resources and C. D. Jiggins (2016), *The Ecology and Evolution of Heliconius Butterflies*

(Oxford: Oxford University Press). Pollen feeding in adult heliconiines was first described by L. E. Gilbert (1972), "Pollen feeding and reproductive biology of *Heliconius* butterflies," *Proceedings of the National Academy of Sciences USA* 69:1403–7; see also Fletcher J. Young and Stephen H. Montgomery (2020), "Pollen feeding in *Heliconius* butterflies: The singular evolution of an adaptive suite," *Proceedings of the Royal Society of London B* 2020:28720201304.

21. Wallace, "Butterflies of the Amazon Valley," p. 258.

22. In the most comprehensive analysis to date, lepidopterist Marianne Espeland and a team of fourteen colleagues from labs worldwide showed that the family Nymphalidae likely arose in the Cretaceous about ninety-two million years ago, the heliconiine subfamily about fifty million years ago (mid-Eocene), and tribe Heliconiini about thirty-six million years ago (Eocene/Oligocene). See Marianne Espeland et al. (2018), "A comprehensive and dated phylogenomic analysis of butterflies," *Current Biology* 28:770–78.

23. C. Hoorn, F. P. Wesselingh, H. ter Steege, et al. (2010), "Amazonia through time: Andean uplift, climate change, landscape evolution, and biodiversity," *Science* 330:927–31.

24. This phenomenon was famously treated by Bates in a now-classic 1862 paper, "Contributions to an insect fauna of the Amazon Valley. Lepidoptera: Heliconidae," *Transactions of the Linnean Society of London* 23(3):495–566. This was the first elucidation of the phenomenon later dubbed Batesian mimicry, in which a palatable species (mimic) evolves to closely resemble an unpalatable or dangerous one (model). Bates recognized natural selection at work here, and this tour de force of a paper, coming just three years after *On the Origin of Species*, was one of the earliest applications of Darwin's (and Wallace's) ideas. A related form of mimicry was later described by German-Brazilian naturalist Fritz Müller, where two (or more) unpalatable species converge in color and pattern, reinforcing the warning coloration signal. The heliconiine mimicry complexes are thought to be largely Müllerian in nature today.

25. Henry Walter Bates (1864), *The Naturalist on the River Amazons* (London: John Murray), 2:346.

26. A. R. Wallace to H. Bates, 10 December 1861, WCP377.

27. Heliconiine butterflies have since become an important model system in ecology and evolution, the focus of a diverse and vibrant research program including investigations of selection, mimicry, hybridization, speciation, plant-insect interactions, and more. For an overview, see Jiggins, *Ecology and Evolution of* Heliconius *Butterflies*; the dated but still useful review by K. S. Brown Jr. (1981), "The biology of *Heliconius* and related genera," *Annual Review of Entomology* 26:427–56; and R. M. Merrill, K. K. Dasmahapatra, J. W. Davey, et al. (2015), "The diversification of *Heliconius* butterflies: What have we learned in 150 years?," *Journal of Evolutionary Biology* 28(8):1417–38. For recent genetic analyses, see K. K. Dasmahapatra, J. R. Walters, A. D. Briscoe, et al. [*Heliconius* Genome Consortium] (2012), "Butterfly genome reveals promiscuous exchange of mimicry adaptations among species," *Nature* 487:94–98; I. J. Garzón-Orduña and A.V.Z. Brower (2018), "Quantified reproductive isolation in *Heliconius* butterflies: Implications for introgression and hybrid speciation," *Ecology and Evolution* 8:1186–95; W. O. McMillan, L. Livraghi, C. Concha, and J. J. Hanly (2020), "From patterning genes to process: Unraveling the gene regulatory networks that pattern *Heliconius* wings," *Frontiers in Ecology and Evolution* 8:221, doi:10.3389/fevo.2020.00221.

28. C.F.P. von Martius (1823–1853), *Historia Naturalis Palmarum*, 3 vols., Munich.

29. Alfred R. Wallace (1853), *Palm Trees of the Amazon and Their Uses* (London: John Van Voorst), p. vi.

30. R. Spruce to W. J. Hooker, 5 January 1855, WCP5527.

31. In their 2002 overview of Wallace's work on palms, botanists Sandra Knapp, Lynn Sanders, and William Baker conclude that Wallace's contributions can be deemed substantial in two main respects: he identified and named several palm species new to science, and his book constitutes the first field guide to tropical palms. The Museum of Economic Botany at Kew holds nine palm specimens from Wallace and Bates. Of the fourteen palms in his book that Wallace thought were new to science, four species are still known by the names he gave them: *Leopoldina major*, *L. piassaba*, *Euterpe catinga*, and *Mauritia carana*. See S. Knapp, L. Sanders, and W. Baker (2002), "Alfred Russel Wallace and the palms of the Amazon," *Palms* 46(3):109–19.

32. C. R. Darwin to H. W. Bates, 3 December 1861, DCP-LETT-3338; C. R. Darwin to J. D. Hooker, 15 and 22 May 1863, DCP-LETT-4167; J. D. Hooker to C. R. Darwin, 24 May 1863, DCP-LETT-4169.

33. Historian and explorer John Heming, former director of the Royal Geographical Society and author of the fine 2015 book *Naturalists in Paradise*, summed up Wallace's *Travels on the Amazon* nicely: despite its errors, it is "a lively travel book, a wonderful read and a fine achievement . . . totally truthful, with no exaggeration, sometimes humorous, and largely devoid of the prejudices one would expect of a young Englishman of that period." John Heming (2015), *Naturalists in Paradise: Wallace, Spruce, and Bates on the Amazon* (London: Thames and Hudson), p. 298.

34. Bates returned from Amazonia in 1859, and from 1864 until his death in 1892 he served as assistant secretary of the Royal Geographical Society. His best-known works include his watershed paper "Contributions to an insect fauna of the Amazon Valley" and his travel memoir *The Naturalist on the River Amazons* (1863). Later in life Bates also contributed several volumes on beetles to the *Biologia Centrali-Americana* series. Wallace wrote an obituary for Bates, praising his friend's contributions but lamenting that the "confinement and constant strain" and "mere drudgery" of his position at the Royal Geographical Society had kept him from more congenial and valuable scientific pursuits and with "little doubt . . . weakened his constitution and shortened a valuable life." Alfred R. Wallace (1892), "H. W. Bates, the naturalist of the Amazons," *Nature* 45:398–99.

35. H. W. Bates to S. Stevens, 23 and 31 December 1850 [erroneously dated 1851 in *Zoologist* 9:3142–44].

36. A. R. Wallace to R. Spruce, 19 September 1852, WCP349; *My Life*, 1:309.

37. Latham, a founder of the field of ethnology, held that all human races and ethnicities, the "varieties of man," as he put it, constituted a single diverse species (monogenism). The presentation of human cultural groups in a natural history context, his "museum of man," was groundbreaking in some ways but also reflected the prevailing view of cultural hierarchy. The biogeographical arrangement of the peoples of the world was novel, presented west to east like a map with representations of the environments in which people lived, underscoring Latham's view of ethnology as a branch of zoology and his belief that such an arrangement would be "both instructive and amusing, and afford a clearer conception than can be obtained elsewhere of the manner in which the varieties of man, animals, and plants, are distributed over the globe." Samuel Philips and F.K.J. Shenton (1859), *Guide to the Crystal Palace and Its Park and Gardens*, revised ed. (Sydenham: Crystal Palace Library), p. 91. Another innovation was the omission of European peoples from the exhibit, a seeming oddity but perhaps a subtle way to encourage visitors to look around and consider themselves and their fellow visitors in relation to those on display, as suggested by the

satirical magazine *Punch*: the curators had evidently not thought it necessary to include a collection of Europeans, perhaps because "there will always be found among the visitors themselves a collection of living curiosities of the various populations of Europe." *Punch's Handbooks to the Crystal Palace* 27(678):8 [1854]. The presentation was also rather predicable in that the exhibits were intended to represent the march of progress, from the dinosaur models in the garden to the upper terrace lined with twenty-four statues personifying countries and trade centers important to the British empire. See Sadiah Qureshi (2011), "Robert Gordon Latham, displayed peoples, and the natural history of race, 1854–1866," *Historical Journal* 54(1):143–66; Jeffrey Auerbach (2015), "Empire under glass: The British Empire and the Crystal Palace, 1851–1911," pp. 111–41, in John McAleer and John M. MacKenzie, eds., *Exhibiting the Empire: Cultures of Display and the British Empire* (Manchester, UK: Manchester University Press); and Tulse Hill Terry (2007), "Statues and Fountains in Crystal Palace Park," Syndenham Town Forum, 12 December 2007, https://sydenham.org.uk/forum/viewtopic.php?t=1538.

38. In the published guide to the ethnological exhibit, Latham quoted at length from Wallace's *Travels on the Amazon* account of Amazonian Indigenous peoples, praising it as "one of the best we have"; see pp. 64–71 in Robert Latham (1854), *The Natural History Department of the Crystal Palace Described*, pt. 1, Ethnology, Crystal Palace Library, Sydenham. In *My Life* (1:322–23), Wallace recounted the problems with the Amerindian exhibit in Latham's "museum of man" and characteristically offered a remedy:

> To be successful and life-like, such groups should be each completely isolated in a deep recess, with three sides representing houses or huts, or the forest, or river-bank, while the open front should be enclosed by a single sheet of plate-glass, and the group should be seen at a distance of at least ten or fifteen feet. In this way, with a carefully arranged illumination from above and an artistic colouring of the figures and accessories, each group might be made to appear as life-like as some of the best figures at Madame Tussaud's, or as the grand interiors of cathedrals, which were then exhibited at the Diorama.

He thought quite a lot about effective, didactic museum exhibits, writing several papers and reviews on the subject; see pp. 90–91 in James T. Costa (2019), "Field study, collecting, and systematic representation," pp. 67–95, in Smith, Costa, and Collard, *An Alfred Russel Wallace Companion*; pp. 351–56 in Andrew Berry (2002), *Infinite Tropics: An Alfred Russel Wallace Anthology* (London: Verso).

39. A. R. Wallace to R. I. Murchison, June 1853, WCP4308; see also J. Brooke to A. R. Wallace, 1 April 1853, WCP3072.

40. H. U. Addington to H. N. Shaw, 19 August 1853, WCP3640; A. R. Wallace to H. N. Shaw, 27 August 1853, WCP3559; H. U. Addington to H. N. Shaw, 6 September 1853, WCP3639.

41. In fact, Fremantle captained the *Juno* until 1857 when, accused of "overstrict discipline," he was recalled and relieved of his command; see Royal Museums Greenwich, https://collections.rmg.co.uk/archive/objects/491758.html.

42. Edward Newman, presidential address, Entomological Society of London, 23 January 1854; see *Transactions of the Entomological Society of London*, n.s., 2(1852–1853):147.

43. A. R. Wallace to H. N. Shaw, 8 February 1854, WCP 3558; J. Wodehouse to H. N. Shaw, 9 February 1854, WCP3644; J. Wodehouse to H. N. Shaw, 14 February 1854, WCP3647.

44. H. N. Shaw to J. Wodehouse, 16 February 1854, WCP4310; J. Wodehouse to A. R. Wallace, 24 February 1854, WCP3648; J. Wodehouse to A. R. Wallace, 1 March 1854, WCP3643.

45. *My Life*, 1:340; Kees Rookmaaker and John van Wyhe (2012), "In Alfred Russel Wallace's shadow: His forgotten assistant, Charles Allen (1839–1892)," *Journal of the Malaysian Branch of the Royal Asiatic Society* 85(2):17–54.

46. See George Beccaloni (2020), "Portraits of Alfred Russel Wallace," vol. 3, doi: 10.13140/RG.2.2.16414.69447.

47. A. R. Wallace to G. Silk, 19 and 26 March 1854, WCP352.

48. Also known as Pompey's Pillar (a misnomer), Diocletian's Column once supported a large statue of the Roman emperor Diocletian. The surviving inscription on the base reads: "Publius, governor of Egypt, [set this up to] the most revered emperor, the guardian-god of Alexandria, Diocletian the invincible . . ." See LSA Database, Oxford University, http://laststatues.classics.ox.ac.uk/database/discussion.php?id=1246.

49. A. R. Wallace to G. Silk, 19 and 26 March 1854, WCP352; *My Life*, 1:334.

50. *My Life*, 1:336.

Chapter 7: Sarawak and the Law

1. *Malay Archipelago*, p. 32.

2. Peter G. Rowe and Limin Hee (2019), *A City in Blue and Green: The Singapore Story* (Singapore: Springer Nature Singapore), chap. 2, "Early Days."

3. The growth and prosperity of Singapore has been attributed to Raffles's vision of freedom and free enterprise, as reflected in the inscription on his memorial in Westminster Abbey: "He founded an emporium at Singapore, where in establishing freedom of person as the right of the soil, and freedom of trade as the right of the port, he secured to the British flag the maritime superiority of the eastern seas" (see Westminster Abbey, "Stamford Raffles," https://www.westminster-abbey.org/abbey-commemorations/commemorations/stamford-raffles). An accomplished naturalist as well as statesman, the memorial also reminds us that Raffles served as the first president of the Zoological Society of London. As it happened, the first mammal Wallace collected on the island of Borneo was a macaque, *Macaca fascicularis*, first described by Raffles in 1821.

4. See, for example, Lucille H. Brockway (2002), *Science and Colonial Expansion: The Role of the British Royal Botanic Gardens* (New Haven, CT: Yale University Press); Patricia Fara (2003), *Sex, Botany, and Empire: The Story of Carl Linnaeus and Joseph Banks* (London: Icon Books).

5. Nearly a decade after Wallace departed South America, his friend Richard Spruce was commissioned by geographer Clements Markham to procure *Cinchona* for cultivation in British India. Spruce studied the cultivation of red cinchona, *Cinchona pubescens*, in Ecuador for several years under arduous conditions and exported a quantity of seeds and young plants to British India, where it was successfully cultivated for quinine. Spruce was later awarded a pension in retirement largely in recognition of these efforts, as Wallace mentioned in the obituary he wrote for his friend. Alfred R. Wallace [1894], "Richard Spruce, Ph.D., F.R.G.S," *Nature* 49:317–19.

6. After Raffles's departure from Singapore and death in 1826, support for his botanical and experimental garden waned. It closed in 1829 and was briefly revived in the mid-1830s but closed permanently after a decade. Ironically, its lack of support was largely due to the steadily declining value of nutmeg, which the British themselves were ultimately responsible for beginning with those plants pirated from Pulau Run. In 1859 a new botanical garden was established in the Tanglin district, just a couple of miles from Fort Canning Hill Park. Originally founded

more on the model of landscaped pleasure garden than experimental garden, the Singapore Botanic Garden later engaged in important scientific and economic botanical research in addition to ornamental horticulture. Named a UNESCO World Heritage Site, the Singapore Botanic Garden's many features now include an evolution garden—a botanical walk through time that Wallace would surely have enjoyed. See Tin Seng Lim (2005), "Singapore Botanic Gardens," in *Singapore Infopedia*, https://eresources.nlb.gov.sg/infopedia/articles/SIP_545 _2005-01-24.html.

7. Alfred R. Wallace (1854), "Letters from the eastern archipelago," *Literary Gazette and Journal of the Belles Lettres, Science, and Art* 1961:739.

8. A. R. Wallace to S. Stevens, 9 May 1854, WCP4259; Alfred R. Wallace (1854), [Letter from Alfred R. Wallace dated 9 May 1854, Singapore], *Zoologist* 12(142):4395–97.

9. The impressive Wallace Education Centre, which includes the Wallace Environmental Learning Laboratory (WELL), a "one-stop learning centre for scientific research, national education, community service, teacher training and international exchange," can also be found on the Wallace Trail, occupying a renovated cow shed from the old dairy farm.

10. Alfred R. Wallace (1854), [Letter from Alfred R. Wallace dated 9 May 1854, Singapore], *Zoologist* 12(142):4395–97.

11. Although early accounts of tiger attack fatalities in Singapore were often exaggerated, with only about 159 deaths officially attributed to tigers between 1831 and 1890 (disproportionately, Chinese laborers), historians believe this is a significant undercount. The last recorded wild tiger in Singapore was killed in October 1930. See Miles A. Powell (2016), "People in peril, environments at risk: Coolies, tigers, and colonial Singapore's ecology of poverty," *Environment and History* 22(3):455–82.

12. A. R. Wallace to M. A. Wallace, 28 May 1854, WCP354; *My Life*, 1:338.

13. Charles Smith (2008), "Alfred Russel Wallace, journalist," *Archives of Natural History* 35(2):203–8. Wallace ultimately had four pieces published in the *Literary Gazette*, two in 1854 ("Letters from the Eastern Archipelago," 1961:739, 19 August; "Letter from Singapore," 1978:1077–78, 16 December) and two in 1855 ("Letter from Sarawak," 2003:366, 9 June; "Borneo," 2023:683–84, 27 October).

14. Grahame J. H. Oliver and Avijit Gupta (2019), *A Field Guide to the Geology of Singapore*, 2nd ed. (Singapore: Lee Kong Chian Natural History Museum), pp. 6–7.

15. *Malay Archipelago*, pp. 39–40.

16. W. C. Hewitson (1862–1866), *Illustrations of New Species of Exotic Butterflies Selected Chiefly from the Collections of W. Wilson Saunders and William C. Hewitson* [Agrias and Nymphalis], vol. 3 (London: John van Voorst).

17. *Malay Archipelago*, p. 41.

18. *Malay Archipelago*, p. 43.

19. Wallace's "magnificent" specimen was most likely *Papilio palinurus*, the banded peacock. Alfred R. Wallace (1855), "The entomology of Malacca," *Zoologist* 13(149):4636–39; sent to Samuel Stevens, 25 November 1854, WCP4260.

20. The intricate eyespots, or ocelli, of the great Argus pheasant feature a stunning trompe l'oeil ball-and-socket pattern that, with subtle shading, gives the impression of three-dimensionality. They were cited by the anti-transmutationist Duke of Argyll, who declared the eyespots "the one instance in Nature (and, as far as I know, only one) in which ornament takes the form of pictorial representation." Darwin later argued for the gradual, stepwise evolution of this and other

complex and beautiful ornaments by female choice, a form of sexual selection. In *Descent of Man* he devoted considerable space to the case of the peacock and argus pheasant; see Charles Darwin (1871), *The Descent of Man, and Selection in Relation to Sex*, vol. 1 (London: John Murray), chap. 14. While he agreed that such coloration patterns as these eyespots arose by selection, Wallace famously disagreed with Darwin over the mechanism, favoring natural selection over female choice. For an excellent treatment of Darwin and Wallace's disagreement over sexual selection, see Helena Cronin (1991), *The Ant and the Peacock: Altruism and Sexual Selection from Darwin to Today* (Cambridge: Cambridge University Press), especially chapters 5–9.

21. It is unclear when rhinos disappeared from Gunung Ledang, but Indian elephants (*Elaphas maximus indicus*) were known to persist in the park until the late twentieth century. Recent mammal surveys using camera trapping and other techniques documented thirty-seven mammal species, including the rare Sumatran serow (*Capricornis sumatraensis*), a kind of montane forest goat relative. See the review by Ain Ahmad Bakri Faiznur et al. (2020), "The first record of Sumatran serow, *Capricornis sumatraensis* (Bovidae, Cetartiodactyla), in Gunung Ledang Johor National Park, a tropical forest remnant on the southern Malay Peninsula," *Mammal Study* 45(3):259–64.

22. In contrast to Wallace's time, just two extant rhinoceros species are recognized today; the one formerly found on the Malay peninsula is the Javan rhinoceros, *R. sondaicus*, now critically endangered and restricted to fewer than one hundred individuals living in a tiny portion of Java. The elephants of Malaysia have fared a bit better but are still endangered. Populations of the Indian elephant (*Elephas maximus indicus*) in peninsular Malaysia today number around twelve hundred or so.

23. A. R. Wallace to M. A. Wallace, July 1854, WCP355; 30 September 1854, WCP357.

24. Bob Reece (1982), *The Name of Brooke: The End of White Rajah Rule in Sarawak* (New York: Oxford University Press); (2004), *The White Rajahs of Sarawak: A Borneo Dynasty* (Singapore: Archipelago Press).

25. A. R. Wallace to M. A. Wallace, July 1854, WCP355; 30 September 1854, WCP357.

26. Spenser St. John (1879), *The Life of Sir James Brooke: Rajah of Sarawak: From His Personal Papers and Correspondence* (Edinburgh: William Blackwood and Sons), p. 274.

27. A Wallace Center will be one of several sites in an archaeological park under development on the Santubong peninsula. See Hafizuddin Tajuddin, Faridatul Akma Abd Latif, and Salina Mohamed Ali (2018), "Preserving and enhancing the cultural landscape of Kampung Santubong, through eco-village approach," *Built Environment Journal* 15(1):33–40; Sarawak Heritage Society, *Misc. Heritage News*, March 2016, "Sarawak Minister of Tourism Restates Plans for Wallace Point, Santubong," https://sarawakheritagesociety.com/misc-heritage-news-march-2016/; and "Exploring Santubong Archaeological Sites," https://sites.google.com/view/ssf2193/theme-3-sites/exploring-santubong-archaeological-sites.

28. Robert Hall and H. Tim Breitfield (2017), "Nature and demise of the Proto-South China Sea," *Bulletin of the Geological Society of Malaysia* 63:61–76; H. Tim Breitfield et al. (2018), "Unravelling the stratigraphy and sedimentation history of the uppermost Cretaceous to Eocene sediments of the Kuching Zone in west Sarawak (Malaysia), Borneo," *Journal of Asian Earth Sciences* 160:200–223. See also Ramlah Zainudin et al. (2010), "Genetic structure of *Hylarana erythraea* (Amphibia: Anura: Ranidae) from Malaysia," *Zoological Studies* 49(5):688–702; K. M. Wong and L. Neo (2019), "Species richness, lineages, geography, and the forest matrix: Borneo's 'Middle Sarawak' phenomenon," *Gardens' Bulletin Singapore* 71(suppl. 2):463–96.

29. It is possible that Wallace visited the Singapore Library, which opened in January 1845, but as this library did not carry much by way of natural history and scientific works at the time, it would have been of very limited interest to Wallace. See Porscha Fermanis (2017), "The Singapore Library, 1845–1874," https://southhem.org/2017/04/19/the-singapore-library-1845-1873/; Lara Atkin, Sarah Comyn, Porscha Fermanis, and Nathan Garvey (2019), *Early Public Libraries and Colonial Citizenship in the British Southern Hemisphere* (Cham, Switzerland: Palgrave Macmillan).

30. Even Lyell's geological friends and colleagues found the idea of suites of the same species cycling around to be a bit much, contrary to the strongly directional evidence of the fossil record; this idea of Lyell's was caricatured by Henry de la Beche in a famous cartoon titled "Awful Changes. Man found only in a fossil state—Reappearance of Ichthyosaur!" where Professor Ichthyosaur (Lyell himself, in barrister's robes) holds forth to the attentive young ichthyosaurs on the skull of a long-extinct creature—a human—observing that its insignificant teeth and puny jaw made one wonder how the creature could have procured food. See Martin J. S. Rudwick (1975), "Caricature as a source for the history of science: De la Beche's anti-Lyellian sketches of 1831," *Isis* 66:534–60; Dov Ospovat (1977), "Lyell's theory of climate," *Journal of the History of Biology* 10(2):317–39.

31. Pietro Corsi (1978), "The importance of French transformist ideas for the second volume of Lyell's *Principles of Geology*," *British Journal for the History of Science* 11(3):221–44.

32. See J. T. Costa (2013), "Engaging with Lyell: Alfred Russel Wallace's Sarawak law and Ternate papers as reactions to Charles Lyell's *Principles of Geology*," *Theory in Biosciences* 132(4):225–37.

33. See Frances Darwin, ed. (1909), *The Foundations of the Origin of Species* (Cambridge: Cambridge University Press) [1969; New York: Kraus Reprint]; James T. Costa (2017), *Darwin's Backyard: How Small Experiments Led to a Big Theory* (New York: W. W. Norton).

34. Natural History Notebook (MS-179), http://linnean-online.org/wallace_notes.html.

35. William Whewell (1837), *History of the Inductive Sciences from the Earliest to the Present* (London: John W. Parker), 3:579–80.

36. C. R. Darwin to C.J.F. Bunbury, 21 April 1856, DCP-LETT-1856.

37. R. C. Tytler (1854), "Miscellaneous notes on the fauna of Dacca, including remarks made on the line of march from Barrackpore to that station," *Annals and Magazine of Natural History* 14 (ser. 2): 168–77; F. J. Pictet, (1853–1854). *Traité de Paléontologie*. 4 vols. (Paris: J.-B. Bailliére).

38. *My Life*, 1:354.

39. Historian H. Lewis McKinney first pointed out the significance of Wallace's notes on Pictet in the formulation of the Sarawak Law paper, showing that Wallace incorporated several of Pictet's points into his analysis; see H. Lewis McKinney (1972), *Wallace and Natural Selection* (New Haven, CT: Yale University Press), pp. 46–49.

40. Edward Forbes (1854), "On the manifestation of polarity in the distribution of organized beings in time," *Notices of the Proceedings of the Meetings of the Members of the Royal Institution* 1:428–33.

41. The Species Notebook (MS-180) resides at the Linnean Society of London. For a facsimile and commentary on this remarkable notebook, see J. T. Costa (2013), *On the Organic Law of Change* (Cambridge, MA: Harvard University Press) and (2014), *Wallace, Darwin, and the Origin of Species* (Cambridge, MA: Harvard University Press).

42. It has been suggested that because Wallace did not explicitly state that his "law" entailed the material derivation of one species from another, it cannot be claimed that he was talking about evolution. Perhaps, the argument goes, he was talking about special creation of one species modeled on a preexisting closely related one. Fair enough, but to suggest that Wallace did not have transmutation in mind here is to ignore his consistent rejection of arguments for creation or design and his interest in amassing evidence for the reality of transmutation, as seen in his Species Notebook; see Costa, *On the Organic Law of Change.*

43. Wallace's discussion of the difficulties of achieving a natural classification and his branching tree metaphor are found on p. 187 in Alfred R. Wallace (1855), "On the law which has regulated the introduction of new species," *Annals and Magazine of Natural History,* 2nd series, 16:184–96. Darwin independently came to the same insights; for example, "If used in simple earnestness the natural system ought to be a genealogical [one]" (1842), and "Finally, then, we see that all the leading facts in the affinities and classification of organic beings can be explained on the theory of the natural system being simply a genealogical one" (1844), pp. 36 and 212 in F. Darwin, *Foundations of the Origin of Species.* He also expressed the concept in *On the Origin of Species* (p. 420): "All true classification is genealogical . . . community of descent is the hidden bond which naturalists have been unconsciously seeking."

44. For a detailed analysis of the Sarawak Law paper, including an annotated facsimile of the paper, see pp. 144–73 in Costa, *Wallace, Darwin, and the Origin of Species.*

45. Wallace, "On the law which has regulated the introduction of new species," pp. 184–96.

46. See Alex Shoumatoff (2017), *The Wasting of Borneo: Dispatches from a Vanishing World* (Boston: Beacon Press).

47. Reminders of the mining days, rusting steam engines in the forest, concrete wells, and still-accessible (albeit dangerous, so do not enter!) tunnels with stonework, are present at the northwestern end of the bukit, not far up the modern Guning Ngeli trail head along the overgrown coal road that runs more or less parallel to the modern road.

48. Species Notebook (MS-180); see Costa, *On the Organic Law of Change* for an annotated edition of this notebook.

49. A. R. Wallace to F. Sims, 25 June 1855, WCP359.

50. A selection of Wallace's striking beetles from Simunjan are illustrated in chapter 4 of *The Malay Archipelago,* featuring "*Neocerambyx aeneas, Cladognathus tarandus, Diurus furcellatus, Ectatorhinus wallacei, Megacriodes saundersii,* and *Cyriopalpus wallacei.*"

51. Mel Sunquist and Fiona Sunquist (2017), *Wild Cats of the World* (Chicago: University of Chicago Press), pp. 48–51.

52. Alexander Nater, Maja P. Mattle-Greminger, Anton Nurcahyo, et al. (2017), "Morphometric, behavioral, and genomic evidence for a new orangutan species," *Current Biology* 27:3487–98; April Reece (2017), "New orangutan species identified," *Nature* 551:151. This species was inducted into the ranks of the critically endangered immediately upon being formally recognized. It is the rarest great ape in the world, with fewer than eight hundred individuals estimated to remain in its homeland of the Batang Toru forest of western Sumatra, a mere 540 square miles (1,400 square km). In 2018 Lithuanian artist Ernest Zacharevic, whose playful orangutan murals adorn the walls of Kuching, worked with the Orangutan Information Centre and the Sumatran Orangutan Society to draw attention to the catastrophic scale of forest destruction in the region by carving a giant "SOS" distress call into a twenty-hectare abandoned oil palm plantation at Bukit Mas, Sumatra. See Splash and Burn, "Art for Change," https://www.splashandburn.org/.

53. Wallace's accounts of hunting orangs as well as his observations on their biology are found in chapter 4 of the American edition of *Malay Archipelago* and chapter 4 of volume 1 of the United Kingdom edition (both published in 1869).

54. Wallace published three papers on orangs in 1856: "Some account of an infant Orang-utan," *Annals and Magazine of Natural History*, 2nd ser., 17:386–90; "On the Orang-utan or Mias of Borneo," *Annals and Magazine of Natural History*, 2nd ser., 17:471–76; and "On the habits of the Orang-utan of Borneo," *Annals and Magazine of Natural History*, 2nd ser., 18:26–32. See also the letter from A. R. Wallace to G. Waterhouse, 8 May 1855, WCP781.

55. Transmutation Notebook C, p. C79; P. H. Barrett, P. J. Gautrey, S. Herbert, D. Kohn, and S. Smith, eds. (1987), *Charles Darwin's Notebooks, 1836–1844* (Ithaca, NY: Cornell University Press), p. 264; see additional orang entries on pp. 545, 551, and 554.

56. For a comparison of Wallace's and Darwin's views on the human-primate relationship generally and their studies of orangs in particular, see pp. 125–29 in Costa, *Wallace, Darwin, and the Origin of Species*. Cambridge, MA: Harvard University Press; and John Van Wyhe and Peter C. Kjaergaard (2015), "Going the whole orang: Darwin, Wallace and the natural history of orangutans," *Studies in History and Philosophy of Biological and Biomedical Sciences* 51:53–63. For analyses of the orang accounts of Wallace and other naturalists in the context of nineteenth-century scientific and social debates about the nature of humans, see Ted Benton (1997), "Where to draw the line? Alfred Russel Wallace in Borneo," *Studies in Travel Writing* 1:96–116; and Tiffany Tsao (2013), "The multiplicity of humanity in the orangutan adoption accounts of Alfred Russel Wallace and William Temple Hornaday," *Clio* 43:1–31.

57. A. R. Wallace to F. Sims, 25 June 1855, WCP359. See also Alfred R. Wallace (1856), "Some account of an infant 'Orang-utan,'" *Annals and Magazine of Natural History*, 2nd ser., 17:386–90, and "A new kind of baby," *Chambers's Journal*, 3rd ser., 6:325–27.

58. An interesting analysis of the tension inherent in the dichotomy between seeing orangs as human-like or pets as well as scientific specimens can be found in Shira Shmuely (2020), "Alfred Wallace's baby Orangutan: Game, pet, specimen," *Journal of the History of Biology* 53(3):321–43.

59. A. R. Wallace to J. Wallace, 20 April 1855, WCP1829; A. R. Wallace to F. Sims, 25 June 1855, WCP359; A. R. Wallace to F. Sims, 28 September and 17 October 1855, WCP360.

60. Alfred R. Wallace (1855), "Proceedings of natural-history collectors in foreign countries," *Zoologist* 13:4803–7; WCP4261.

61. Alfred R. Wallace (1855), "Borneo," *Literary Gazette and Journal of the Belles Lettres, Science, and Art* 2023:683–84; WCP612. See also A. R. Wallace to M. A. Wallace, 25 December 1855, WCP361 and *My Life*, 1:345–47.

62. For an insightful treatment, see Jeremy Vetter (2015), "Politics, paternalism, and progressive social evolution: Observations on colonial policy in the scientific travels of Alfred Russel Wallace," *Victorian Review* 41(2):113–31.

63. H. Tim Breitfeld and Robert Hall (2018), "The eastern Sundaland margin in the latest Cretaceous to late Eocene: Sediment provenance and depositional setting of the Kuching and Sibu zones of Borneo," *Gondwana Research* 63:34–64; Hans Hazebroek, "Geology and Geomorphology," chap. 2 in Jayasilan Mohd-Azlan et al., eds. (2016), *Life from Headwaters to the Coast: Gunung Penrissen Roof of Western Borneo* (Sarawak: Universiti Malaysia Sarawak).

64. Here is Wallace's wonderful paean to the durian, in full:

The durian grows on a large and lofty forest-tree, somewhat resembling an elm in its general character, but with a more smooth and scaly bark. The fruit is round or slightly oval, about the size of a large cocoa-nut, of a green color, and covered all over with short stout spines the bases of which touch each other, and are consequently somewhat hexagonal, while the points are very strong and sharp. It is so completely armed, that if the stalk is broken off it is a difficult matter to lift one from the ground. The outer rind is so thick and tough, that from whatever height it may fall it is never broken. From the base to the apex five very faint lines may be traced, over which the spines arch a little; these are the sutures of the carpels, and show where the fruit may be divided with a heavy knife and a strong hand. The five cells are satiny-white within, and are each filled with an oval mass of cream-colored pulp, imbedded in which are two or three seeds about the size of chestnuts. This pulp is the eatable part, and its consistence and flavor are indescribable. A rich butter-like custard highly flavored with almonds gives the best general idea of it, but intermingled with it come wafts of flavor that call to mind cream-cheese, onion-sauce, brown-sherry, and other incongruities. Then there is a rich glutinous smoothness in the pulp which nothing else possesses, but which adds to its delicacy. It is neither acid, nor sweet, nor juicy, yet one feels the want of none of these qualities, for it is perfect as it is. It produces no nausea or other bad effect, and the more you eat of it the less you feel inclined to stop. In fact, to eat durians, is a new sensation worth a voyage to the East to experience. (Alfred R. Wallace [1856], "On the bamboo and durian of Borneo" [In a letter to Sir W. J. Hooker], *Hooker's Journal of Botany and Kew Garden Miscellany* 8:225–30; see also *Malay Archipelago*, pp. 85–86)

65. Wallace, *Malay Archipelago*, pp. 86–87.

66. Today adventurous hikers with a science or history bent can do the "Wallace Trail" up Serumbu, to the site of Rajah Brooke's modest mountain cottage, now reconstructed thanks to a campaign spearheaded by the Brooke Heritage Committee and the Brooke Trust (https://www.brooketrust.org/). See also "Retracing Brooke's and Wallace's Footsteps," YouTube video, https://youtu.be/7PgwGseoeCA.

67. Wallace, *Malay Archipelago*, pp. 95–97.

68. Alfred R. Wallace (1856), "Notes of a journey up the Sadong River, in north-west Borneo" [Communicated to the RGS meeting of 10 Nov. 1856], *Proceedings of the Royal Geographical Society of London* 1(6):193–205. Incorporated into *Malay Archipelago*, chaps. 5 and 6: "Journey in the interior" and "Borneo—the Dyaks."

69. Alfred R. Wallace (1856), "Observations on the zoology of Borneo" [dated 10 March 1856, Singapore], *Zoologist* 14:5113–17.

70. James Brooke wrote to his nephew John Brooke Johnson Brooke that he was sorry to see Wallace, his "pleasing and intellectual companion," go; 27 January–7 February 1856, WCP3791.

71. A. R. Wallace to F. Sims, 20 February 1856, WCP362.

72. Rajah Brooke reported to Wallace in one letter that Charles, who started going by his middle name Martin, was unhappy: "They say he is not clever at book[s] and when here he appeared damped & disheartened." In another he reported that Charles was "miserable at the mission" and moved on to other employment; J. Brooke to A. R. Wallace, 4 July 1856, WCP3073

and 5 November 1856, WCP3074. See also Kees Rookmaaker and John van Wyhe (2012), "In Alfred Russel Wallace's shadow: His forgotten assistant, Charles Allen (1839–1892)," *Journal of the Malaysian Branch of the Royal Asiatic Society* 85(2):17–54.

73. John van Wyhe and Gerrell Drawhorn (2015), "'I am Ali Wallace': The Malay assistant of Alfred Russel Wallace," *Journal of the Malaysian Branch of the Royal Asiatic Society* 88(1):3–31.

74. A. R. Wallace to F. Sims, 20 February 1856, WCP362; 21 April 1856, WCP363.

75. Alfred R. Wallace (1856), "Attempts at a natural arrangement of birds," *Annals and Magazine of Natural History*, 2nd ser., 18:193–216.

76. J. Brooke to A. R. Wallace, 4 July 1856, WCP3073.

77. Leonard G. Wilson, ed. (1970), *Sir Charles Lyell's Scientific Journals on the Species Question* (New Haven, CT: Yale University Press), pp. 6, 66.

78. C. Lyell to C. R. Darwin, 1–2 May 1856, DCP-LETT-1862; C. R. Darwin to J. D. Hooker, 9 May 1856, DCP-LETT-1870.

79. C. R. Darwin, 1839–1881 Journal (CUL-DAR158.1-76), p. 34v; see http://darwin-online .org.uk/content/frameset?itemID=CUL-DAR158.1-76&viewtype=text&pageseq=1.

80. A. R. Wallace to S. Stevens, 12 May 1856, WCP1702.

Chapter 8: Crossing the Line(s)

1. First–fourth Malay Journal (MS-178a–d), entries 1–245: Wallace's best-selling travel memoir *The Malay Archipelago* (1869) was derived from these four surviving journals chronicling his journey.

2. The VOC—the first truly public company, issuing the earliest known initial public offering in 1602 on the world's first full-scale stock exchange, in Amsterdam—laid the foundation for the modern multinational corporations and capital markets that dominate the global economic system today. There is some irony in Wallace benefiting from and supporting the Dutch colonial apparatus that the VOC established while deploring the capitalist system it is based on; see Femme S. Gaastra (2003), *The Dutch East India Company: Expansion and Decline* (Zutphen, the Netherlands: Walburg Pers).

3. Three banteng subspecies are recognized today: *B. javanicus javanicus* (Java and Bali), *B. j. lowi* (Borneo), and *B. j. birmanicus* (peninsular Southeast Asia); see, for example, H. Matsubayashi et al. (2015), "First molecular data on Bornean banteng *Bos javanicus lowi* (Cetartiodactyla, Bovidae) from Sabah, Malaysian Borneo," *Mammalia* 78(4):523–31; M. Qiptiyah et al. (2019), "Phylogenetic position of Javan banteng (*Bos javanicus javanicus*) from conservation area in Java base on mtDNA analysism," *Biodiversitas* 20:3352–57.

4. Charles Lyell, *Principles of Geology* (1835, 4th ed.), 2:226.

5. Such measurements are made possible by the remarkable network of forty-two geostationary GPS markers across the region, monitored by satellite in the Geodynamics of South and South-East Asia (GEODYSSEA) project covering an area of about four thousand square kilometers; see W. Simons et al. (1999), "Observing plate motions in S.E. Asia: Geodetic results of the GEODYSSEA project," *Geophysical Research Letters* 26(15):2081–84; W. Simons et al. (2007), "A decade of GPS in Southeast Asia: Resolving Sundaland motion and boundaries," *Journal of Geophysical Research* 112:B06420, doi:10.1029/2005JB003868.

6. Simon Winchester (2003), *Krakatoa: The Day the World Exploded: August 27, 1883* (New York: HarperCollins). See also the Smithsonian Institution's Global Volcanism observatory entry for Krakatau at https://volcano.si.edu/volcano.cfm?vn=262000.

7. International Federation of Red Cross and Red Crescent Societies, *Situation Report*: https://reliefweb.int/report/indonesia/indonesia-earthquakes-and-tsunamis-sunda-straits -tsunami-emergency-plan-action-0.

8. NASA Earth Observatory, "Tectonic Uplift near Sumatra," https://earthobservatory.nasa .gov/images/5449/tectonic-uplift-near-sumatra; see also Suyarso (2008), "Topographic changes after 2004 and 2005 earthquakes at Simeulue and Nias Islands identified using uplifted reefs," *Journal of Coastal Development* 12(1):20–29.

9. Darwin later wrote in his *Journal of Researches* (1839, p. 379): "The most remarkable effect (or perhaps speaking more correctly, cause) of this earthquake was the permanent elevation of the land. Captain FitzRoy having twice visited the island of Santa Maria, for the purpose of examining every circumstance with extreme accuracy, has brought a mass of evidence in proof of such elevation, far more conclusive than that on which geologists on most other occasions place implicit faith." Later printings of the *Journal of Researches* became known as *The Voyage of the Beagle*.

10. *Malay Archipelago*, pp. 17, 18.

11. *Malay Archipelago*, p. 178.

12. By the time of Wallace's visit in 1856, the Sasaks and the Balinese had had a long and more or less stable, if uneasy, coexistence under Balinese rule, but periodic rebellions by the Sasaks (including one just a year earlier, in 1855) underscored the perennial tensions that came to a head some thirty years after Wallace's departure from the East, when the Sasak chiefs persuaded the Dutch East India Company, which knew an opportunity when it saw one, to intervene and overthrow the Balinese ruler. This culminated in a series of bloody battles in 1894. See J. Stephen Lansing (2009), *Priests and Programmers: Technologies of Power in the Engineered Landscape of Bali* (Princeton, NJ: Princeton University Press), pp. 19–22; Robert Pringle (2004), *A Short History of Bali: Indonesia's Hindu Realm* (Sydney: Allen & Unwin), chap. 5.

13. UNESCO, "Rinjani-Lombok UNESCO Global Geopark," https://en.unesco.org/global -geoparks/rinjani-lombok.

14. *Malay Archipelago*, p. 163.

15. First Malay Journal (MS-178a), entry 3.

16. First Malay Journal (MS-178a), entries 15, 16.

17. Species Notebook (MS-180), inside front cover; see J. T. Costa (2013), *On the Organic Law of Change* (Cambridge, MA: Harvard University Press), p. 18. See also first Malay Journal (MS-178a), entry 13 and *Malay Archipelago*, p. 171.

18. First Malay Journal (MS-178a), entry 7.

19. Wallace first described this phenomenon in an 1863 paper ("List of birds collected in the island of Bouru (one of the Moluccas), with descriptions of the new species," *Proceedings of the Zoological Society of London* 1863:18–28, 26–28) and later in *Malay Archipelago* (pp. 403–5). See also Jared M. Diamond (1982), "Mimicry of friarbirds by orioles," *Auk* 99(2):187–96.

20. Species Notebook (MS-180), p. 55; see Costa, *On the Organic Law of Change*, p. 138.

21. First Malay Journal (MS-178a), p. 9, entry 8a.

22. This jungle-fowl specimen, along with a specimen of the upright Bali duck, went to Charles Darwin, who was just then studying the variations of domesticated species. Wallace

had evidently received word, presumably through Stevens, of Darwin's request for domestic specimens; see Darwin's memorandum dated December 1855, DCP-LETT-1812 and WCP4758; and A. R. Wallace to S. Stevens, 21 August 1856, WCP1703.

23. George Robert Gray of the British Museum described *Megapodius gouldii* from a specimen collected by Wallace on Lombok: G. R. Gray (1861), "List of species composing the family Megapodiidae, with descriptions of new species and some account of the habits of the species," *Proceedings of the Zoological Society of London* (25 June):288–96, plates 32–34.

24. *Malay Archipelago*, p. 165.

25. See C. Barry Cox and Peter D. Moore (2010), *Biogeography: An Ecological and Evolutionary Approach* (New York: John Wiley and Sons), pp. 6–8; and Peter J. Bowler (2003), *Evolution: The History of an Idea*, 3rd ed. (Berkeley: University of California Press), pp. 76–79.

26. Gray published his findings in a long article titled "Statistics of the flora of the Northern United States," appearing in two parts in the *American Journal of Science and Arts* for 1856 (2nd ser., 22:204–32) and 1857 (2nd ser., 23:62–84, 369–403). See also C. R. Darwin to A. Gray, 12 October 1856, DCP-LETT-1973.

27. Mr. Ross was most likely a child of Scotsman John Clunies-Ross, who co-settled the tiny Cocos or Keeling Archipelago in the 1820s with his family along with Englishman Alexander Hare and a group variously described as a harem or escaped slaves. The HMS *Beagle* stopped there in April 1836, and Darwin observed that they lived "nominally in a state of freedom, and certainly are so, as far as regards their personal treatment; but in most other points they are considered as slaves." See C. R. Darwin (1839), *Narrative of the Surveying Voyages of His Majesty's Ships* Adventure *and* Beagle *between the Years 1826 and 1836*. Vol. 3, *Journal and Remarks, 1832–1836* (London: Henry Colburn), p. 540.

28. This consignment was shipped to Singapore and then London on the *City of Bristol*; Natural History Notebook (MS-179), verso, p. 113. See also A. R. Wallace to S. Stevens, 21 August 1856, WCP1703.

29. Wallace's eventual detailed treatment of the faunal discontinuity, "On the zoological geography of the Malay Archipelago," was not read until 1859 and was published the following year in the *Journal of the Proceedings of the Linnean Society* (4:172–84). Stevens published part of this letter in the January 1857 issue of *Zoologist* in the meantime, which included Wallace's announcement of the discovery: Alfred R. Wallace (1857), "Proceedings of natural-history collectors in foreign countries, from a letter dated 21 Aug. 1856, Ampanam, Lombok; communicated by Samuel Stevens," *Zoologist* 15:5414–16. See also A. R. Wallace to S. Stevens, 21 August 1856, WCP1703.

30. Wallace stayed in this district upon his arrival, at the Sociëteit De Harmonie on what is now busy Jalan Riburane, just up the road from the fort. It is now home to a performing arts center, Gedung Kesenian Sulsel (South Sulawesi Art Center), and perhaps the only place in the archipelago where Wallace stayed that is still extant.

31. A. R. Wallace to S. Stevens, 27 September 1856, WCP1704.

32. *Malay Archipelago*, p. 225.

33. Before internationally agreed-upon rules of naming species, *synonymy* of taxonomic names—in which different naturalists knowingly or unknowingly bestowed different names on the same species—created rampant confusion that we still live with. Correcting duplicate names over the years is one reason that many currently recognized genus and species names

differ from those given in Wallace's day, as we see throughout this book. One of the earliest rules adopted to remedy the synonymy problem is that of *priority*, in which the first published description and name of a species has precedence over later ones. Some naturalists, however, found this inconvenient for whatever reason (sometimes very personal ones), preferring a name that did not have priority. Such was the case with this birdwing butterfly collected by Wallace when he found that British entomologist Edward Doubleday had changed the long-accepted name *Ornithoptera remus*, which had been given erroneously by Linnaeus. Like many taxonomists, Linnaeus had been fooled by the very different appearance of males and females of this species and, thinking them different species, bestowed different names. Wallace called out Doubleday's "most erroneous as well as inconvenient interpretation of the law of priority" in a note read at the Entomological Society of London meeting of 3 May 1858, pointing out that Cramer's name should have priority as the first to correctly diagnose the two sexes. Wallace's note was published later that year: Alfred R. Wallace (1858), "A disputed case of priority in nomenclature," *Proceedings of the Entomological Society of London* (1858–1859):23–24. Cramer's name is the one recognized today, with subspecies *T. hypolitus cellularis* described from Sulawesi in 1895.

34. *Malay Archipelago*, pp. 231–32; first Malay Journal (MS-178a), entry 41.

35. *Malay Archipelago*, p. 230; first Malay Journal (MS-178a), entry 39.

36. See Hans Hägerdal (2010), "The slaves of Timor: Life and death on the fringes of early colonial society," *Itinerario* 34(2):19–44; (2020), "Slaves and slave trade in the Timor area: Between indigenous structures and external impact," *Journal of Social History*, Special Issue: Slave Trade and Slavery in Asia—New Perspectives. 54(1):15–33.

37. M. A. Wallace to J. and M. Wallace, 16 September 1856, WCP5579; A. R. Wallace to J. and M. Wallace, 6 December 1856, WCP5580; A. R. Wallace to F. Sims, 10 December 1856, WCP365.

38. Alfred R. Wallace (1857), [Letter concerning collecting, dated 1 Dec. 1856, Macassar; communicated by Samuel Stevens], *Zoologist* 15:5652–57.

39. This consignment was sent to Batavia (Jakarta) and then shipped to London on the *Margaret West*; Natural History Notebook (MS-179), pp. 109–12. See also A. R. Wallace to S. Stevens, 1 December 1856, WCP4262.

40. First Malay Journal (MS-178a), p. 9, entry 49.

41. *Malay Archipelago*, pp. 420–21; first Malay Journal (MS-178a), entry 41.

42. The distinction here is a bit simplified; some monogenists were nonetheless slavery supporters, citing the biblically sanctioned condemnation of certain "cursed" peoples. One prominent example was American clergyman and naturalist John Bachman of Charleston, South Carolina, a slaveowner who attacked polygenist arguments in such publications as *The Doctrine of the Unity of the Human Race Examined on the Principles of Science* (1850, Charleston, SC: C. Canning); see Lester D. Stephens (2000), *Science, Race, and Religion in the American South: John Bachman and the Charleston Circle of Naturalists, 1815–1895* (Chapel Hill: University of North Carolina Press).

43. Indeed, a fresh methodological approach as well as perspective, eroding the unspoken but accepted practice of a "division of labor" between the data-generating worker in the field and the higher-status armchair intellectual of the salons of London drawing inferences from these data. That he dared presume as a mere field naturalist to draw his own inferences and generate new knowledge based on his own firsthand observations is another example of Wallace's independent streak and perhaps contrarian nature. See Henrika Kuklick (1991), *The Savage*

Within: The Social History of British Anthropology, 1885–1914 (New York: Cambridge University Press) and Jeremy Vetter (2006), "Wallace's *other* line: Human biogeography and field practice in the eastern colonial tropics," *Journal of the History of Biology* 39:89–123.

44. For excellent treatments of Wallace's investigation of "human biogeography" and its broader context, see George W. Stocking (1987), *Victorian Anthropology* (New York: Free Press), and Vetter (2006), "Wallace's *other* line," pp. 89–123.

45. Holger Warnk (2020), "From trading post to town: Some notes on the history of urbanisation in far eastern Indonesia c. 1800–1940," pp. 273–88, in Sandra Kurfürst and Stefanie Wehner, eds., *Southeast Asian Transformations: Urban and Rural Developments in the 21st Century* (Bielefeld, Germany: Transcript Verlag).

46. Wallace, *Malay Archipelago*, p. 444; first Malay Journal (MS-178a), entry 66.

47. Plumes of birds of paradise, and sometimes even whole birds, were in high demand by the millinery industry (see *Fashioning Feathers: Dead Birds, Millinery Crafts and the Plumage Trade*, exhibition co-curated by Drs. Merle Patchett and Liz Gomez in association with the Material Culture Institute of the University of Alberta, FAB Gallery, University of Alberta, https://fashioningfeathers.info/about/), with costs peaking at as much as twenty-four dollars per plume in the London wholesale market in the early twentieth century during which time an estimated eighty thousand bird of paradise skins were auctioned in London, Paris, and Amsterdam. See also Pamela Swadling (2019), *Plumes from Paradise: Trade Cycles in Outer Southeast Asia and Their Impact on New Guinea and Nearby Islands until 1920* (Sydney, Australia: Sydney University Press).

48. There was often a synergy, if not tension, between Wallace's interests as a philosophical naturalist and the economic realities of how his travels were funded. He had to focus largely on the most salable bird and insect groups but hoped that extensive collections of these would also provide ample material for his biogeographical and transmutational studies. See Melinda Bonnie Fagan (2008), "Theory and practice in the field: Wallace's work in natural history, 1844–1858," pp. 66–90, in Charles H. Smith and George Beccaloni, eds. *Natural Selection and Beyond: The Intellectual Legacy of Alfred Russel Wallace* (Oxford: Oxford University Press).

49. *Malay Archipelago*, p. 440.

50. Note that the names of two of these rarities were given by the distinguished French lepidopterist Jean Baptiste Alphonse Déchauffour Boisduval in honor of his countryman the explorer Jules Dumont d'Urville, commander of the famed eighteenth-century voyages of *La Coquille* and *Astrolabe*, which yielded extensive zoological and botanical collections.

51. *Malay Archipelago*, p. 434; first Malay Journal (MS-178a), entry 59.

52. Wallace included an illustration of the Aru bird of paradise hunters in their blinds as the frontispiece to volume 2 of the London (Macmillan) edition of *The Malay Archipelago*.

53. *Malay Archipelago*, p. 449. The full quote reads: "This consideration must surely tell us that all living things were *not* made for man. Many of them have no relation to him. The cycle of their existence has gone on independently of his, and is disturbed or broken by every advance in man's intellectual development; and their happiness and enjoyments, their loves and hates, their struggles for existence, their vigorous life and early death, would seem to be immediately related to their own well-being and perpetuation alone, limited only by the equal well-being and perpetuation of the numberless other organisms with which each is more or less intimately connected."

54. Much later, Wallace and others speculated that these birds of paradise and their raucous calls provided an interesting lesson in cultural biogeography, inspiring the Islands of Wák-Wák in *The Arabian Nights*. Wallace discussed this in a long two-part article published in 1904: Alfred R. Wallace (1904), "The birds of paradise in the Arabian Nights," pt. 1 and 2, *Independent Review* 2(7):379–91 (April) and 2(8):561–71 (May). See also chapter 5, " On the translation of myths between the Old and New World," in Charles Gould (1886), *Mythical Monsters* (London: W. H. Allen).

55. Wallace made extensive notes on the behavior and morphology of the great and king birds of paradise in his Species Notebook, including sketches; see Costa, *On the Organic Law of Change*, pp. 170–177, 198–207. These notebook entries served as a draft for Wallace's subsequent paper on the great bird of paradise published in the December 1857 issue of the *Annals and Magazine of Natural History*: Alfred R. Wallace (1857), "On the Great Bird of Paradise, *Paradisea apoda*, Linn.; 'Burong mati' (dead bird) of the Malays; 'Fanéhan' of the natives of Aru," *Annals and Magazine of Natural History*, 2nd ser., 20:411–16.

56. "The stings and bites and ceaseless irritation caused by these pests of the tropical forests would be borne uncomplainingly," wrote Wallace, "but to be kept prisoner by them in so rich and unexplored a country, where rare and beautiful creatures are to be met with in every forest ramble—a country reached by such a long and tedious voyage, and which might not in the present century be again visited for the same purpose—is a punishment too severe for a naturalist to pass over in silence." *Malay Archipelago*, p. 466.

57. Second Malay Journal (MS-178b), entries 71, 81, 84.

58. First Malay Journal (MS-178a), entry 63.

59. Second Malay Journal (MS-178b), entry 71.

60. Second Malay Journal (MS-178b), entry 71.

61. Second Malay Journal (MS-178b), entry 83.

62. Second Malay Journal (MS-178b), entry 93; Alfred R. Wallace (1858), "On the Arru Islands" [Communicated to the RGS meeting of 22 February 1858], *Proceedings of the Royal Geographical Society of London* 2(3):163–70.

63. A. R. Wallace to S. Stevens, 10 March and 15 May 1857, WCP4746; published as Alfred R. Wallace (1857), [Letter and postscript concerning collecting dated 10 March 1857, Dobbo, Aru Islands, and 15 May, Dobbo; to Samuel Stevens and communicated by him to the ESL meeting of 5 Oct. 1857], *Proceedings of the Entomological Society of London* 1856–1857:91–93.

64. There are thirty-nine bird of paradise species (family Paradisaeidae) currently recognized, including paradise crows (genus *Lycocorax*), manucodes (*Maucodia* and *Phonygammus*), paradigallas (*Paradigalla*), astrapias (*Astrapia*), parotias (*Parotia*), sicklebills (*Epimachus* and *Drepanornis*), riflebirds (*Ptiloris*), and a host of species called birds of paradise proper in the genera *Paradisaea*, *Cicinnurus*, *Semioptera*, *Paradisornis*, *Seleucidis*, and *Lophorina*. Most species occur in New Guinea, with some on outlying islands and south to the East Coast of Australia. For an authoritative and visually stunning treatment, see Tim Laman and Edwin Scholes (2013), *Birds of Paradise: Revealing the World's Most Extraordinary Birds* (Washington, DC: National Geographic and Cornell Laboratory of Ornithology).

65. J. Brooke to A. R. Wallace, 5 November 1856, WCP3074. Note that while some historians have argued that Wallace did not go to the East with a view toward the subject of transmutation, this comment from James Brooke shows that Wallace was indeed actively discussing the subject at the time.

66. H. W. Bates to A. R. Wallace, 19 and 23 November 1856, WCP824.

67. C. R. Darwin to A. R. Wallace, 1 May 1857, WCP1839; A. R. Wallace to C. R. Darwin, 27 September 1857, WCP4080.

68. A. R. Wallace to H. W. Bates, 4 January 1858, WCP366.

69. Species Notebook (MS-180), pp. 34–53; see Costa, *On the Organic Law of Change*, pp. 96–134. Wallace's planned book is discussed extensively in J. T. Costa (2013), "Engaging with Lyell: Alfred Russel Wallace's Sarawak Law and Ternate papers as reactions to Charles Lyell's *Principles of Geology*," *Theory in Biosciences* 132(4):225–37, and (2014), *Wallace, Darwin, and the Origin of Species* (Cambridge, MA: Harvard University Press).

70. Species Notebook (MS-180), p. 51; Costa, *On the Organic Law of Change*, p. 130.

71. Historian H. Lewis McKinney first suggested that this book by Wallace might have been titled *On the Organic Law of Change*, based on Wallace's heading for the lengthy section of the Species Notebook (MS-180), in which he rebuts Lyell's anti-transmutation arguments: "Note for organic law of change." In homage to McKinney, I adopted this as the title to my annotated transcription of the Species Notebook (MS-180): Costa, *On the Organic Law of Change*. See p. 98 for the section "Note for organic law of change."

72. E. Blyth to C. R. Darwin, 8 December 1855, DCP-LETT-1792.

73. C. R. Darwin to A. R. Wallace, 22 December 1857, WCP1840. By the time Wallace received this letter, he had already discovered the principle that was to become known as natural selection and penned his Ternate essay.

Chapter 9: Eureka

1. A. R. Wallace to H. W. Bates, 4 January 1858, WCP366.

2. C. R. Darwin to A. R. Wallace, 1 May 1857, WCP1839.

3. See Species Notebook (MS-180) and J. T. Costa (2013), *On the Organic Law of Change* (Cambridge, MA: Harvard University Press) for his original field observations and sketches of the great bird of paradise (pp. 170–77, 198–207) and the butterflies *Ornithoptera priamus poseidon* and *Papilio euchenor* (pp. 458–65).

4. Alfred R. Wallace (1857), "On the natural history of the Aru Islands," *Annals and Magazine of Natural History*, 2nd ser., 20(suppl.):473–85; (1858) "On the Arru Islands," *Proceedings of the Royal Geographical Society of London* 2(3):163–70; (1858), "Note on the theory of permanent and geographical varieties," *Zoologist* 16:5887–88.

5. The key passage from *Principles of Geology* (1835, 4th ed., 3:152–56) is quoted on p. 480 of Wallace, "On the natural history of the Aru Islands," pp. 473–85, and on p. 50 of the Species Notebook (MS-180); Costa, *On the Organic Law of Change*, pp. 128–29.

6. Species Notebook (MS-180) pp. 50–51; Costa, *On the Organic Law of Change*, pp. 128, 130.

7. A. R. Wallace to S. Stevens, 10 March 1857, WCP4746; read at the 5 October 1857 meeting of the Entomological Society of London and later published in the *Proceedings* for that year: Alfred R. Wallace (1857), [Letter and postscript concerning collecting dated 10 March 1857, Dobbo, Aru Islands, and 15 May, Dobbo; to Samuel Stevens], *Proceedings of the Entomological Society of London* 1856–1857:91–93.

8. Natural History Notebook (MS-179), pp. 109–10, verso.

9. Quotes in this paragraph are taken from the second Malay Journal (MS-178b), entries 107–11.

10. Adam Brumm, Adhi Agus Oktaviana, Basran Burhan, et al. (2021), "Oldest cave art found in Sulawesi," *Science Advances* 7:eabd4648.

11. Natural History Notebook (MS-179), p. 108, verso.

12. Second Malay Journal (MS-178b), entry 112.

13. Second Malay Journal (MS-178b), entry 116.

14. Second Malay Journal (MS-178b), entry 116.

15. The snake incident was memorialized in *The Malay Archipelago*, including an illustration titled "Ejecting an Intruder." The skin of the hapless snake now resides in the collection of the Linnean Society of London, donated in 1958 by Wallace's grandson A.J.R. Wallace; see p. 87 in Leonie Berwick and Isabelle Charmantier, eds. (2020), *L: 50 Objects, Stories and Discoveries from the Linnean Society of London* (London: Linnean Society of London).

16. Eli A. Silver and Casey J. Moore (1978), "The Molucca Sea Collision Zone, Indonesia," *Journal of Geophysical Research: Solid Earth* 83(B4):1681–91; R. Hall, M. G. Audley-Charles, F. T. Banner et al. (1988), "Late Palaeogene-Quaternary geology of Halmahera, eastern Indonesia: Initiation of a volcanic island arc," *Journal of the Geological Society of London* 145:577–90; R. Hall (2000), "Neogene history of collision in the Halmahera region, Indonesia," *Proceedings of the Indonesian Petroleum Association 27th Annual Convention*, pp. 487–93.

17. The location of Wallace's Ternate house was previously identified as the Santiong House, until entomologist and Wallace scholar George Beccaloni pointed out that the locality did not fully match Wallace's description of the site. For an account of the sleuthing to reveal the actual (most likely) location, see George Beccaloni and Paul Whincup (2019), "The location of Alfred Russel Wallace's legendary house on Ternate Island, Indonesia," doi:10.13140/RG.2.2.11813.86242; Paul Whincup (2020), "The quest for Alfred Russel Wallace's house on Ternate, Maluku Islands, Indonesia," *Journal of the Royal Society of Western Australia* 103:530–40.

18. The narrow north–south alley that runs between Jalal Pipit and Jl. Juma Puasa, just to the west of the mosque abutting the Wallace house site, was renamed Lorong Alfred Russel Wallace in 2010.

19. Species Notebook (MS-180), pp. 108–9; Costa, *On the Organic Law of Change*, pp. 244–47.

20. Alfred R. Wallace (1891), *Natural Selection and Tropical Nature* (London: Macmillan).

21. See, for example, James P. Huzel (1969), "Malthus, the Poor Law, and population in early nineteenth-century England," *Economic History Review* 22(3):430–52; Russell Dean (1995), "Owenism and the Malthusian population question, 1815–1835," *History of Political Economy* 27(3):579–97; E. A. Wrigley and Richard Smith (2020), "Malthus and the poor law," *Historical Journal* 63 (Special Issue 1: Malthusian Moments):33–62. A comprehensive treatment of the Poor Law debates is given by George R. Boyer (1990), *An Economic History of the English Poor Law, 1750–1850* (Cambridge: Cambridge University Press).

22. For example, Wallace later dedicated several pages to discussing "checks to population" of the Hill Dyaks of Borneo and the people of Java, Celebes, and other areas; e.g., *Malay Archipelago*, pp. 100–102, 108–9, 264–65, 597.

23. Costa, *On the Organic Law of Change*, p. 430.

24. C. Darwin to A. R. Wallace, 6 April 1859, WCP1842.

25. Quoted from "Dispersion of Seeds," one of Thoreau's last essays, penned about 1860–1861 before his untimely death from tuberculosis in 1862. Thoreau's journals show that his study of seed dispersal and plant migration was influenced by Darwin's and Wallace's ideas, having read

On the Origin of Species. It is resonant that Wallace's insight into transmutation, inspired in part by the dispersal of seeds, should in turn play a role in inspiring Thoreau's insight into seed dispersal and what we now call the ecological process of succession. See Henry David Thoreau and Bradley P. Dean, eds. (1993), *Faith in a Seed: The Dispersion of Seeds and Other Late Natural History Writings* (Washington, DC: Island Press); and chapter 3 in Michael Benjamin Berger (2000), *Thoreau's Late Career and* The Dispersion of Seeds: *The Saunterer's Synoptic Vision* (Rochester, NY: Camden House).

26. See J. T. Costa (2013), "Engaging with Lyell: Alfred Russel Wallace's Sarawak Law and Ternate papers as reactions to Charles Lyell's *Principles of Geology*," *Theory in Biosciences* 132(4):225–37. For a detailed treatment of the essay, including an annotated facsimile, see pp. 195–213 in J. T. Costa (2014), *Wallace, Darwin, and the Origin of Species* (Cambridge, MA: Harvard University Press).

27. And thus was born the common misunderstanding that Wallace thought domestic varieties teach us nothing about transmutation (unlike Darwin, who made domestication/artificial selection a key analogy in making his case for natural selection). It is not true that Wallace somehow did not understand or accept lessons from domestication: his opening point in the Ternate essay was rhetorical, a device to undermine Lyell's argument. Domestic varieties are not natural, Wallace pointed out, so we cannot use them to make pronouncements on species in nature per se. But he well understood that the very existence of domesticated varieties, the profound changes produced by selective breeding, is evidence of transmutation. This is clear from his thought experiment on progressive divergence of dog breeds in the Species Notebook, where the development of canine varieties is central to one of his protransmutation arguments. These breeds represent continued divergence from the parental type: "Is not the change of one original animal into two such different animals as the Greyhound & the bulldog a transmutation?" he asks in the Species Notebook (MS-180) p. 41. See Costa, *On the Organic Law of Change*, pp. 106–11, and *Wallace, Darwin, and the Origin of Species*, pp. 95–96, 136–37 for a discussion of Wallace's canine varieties argument for transmutation. See also the treatment by historian Jonathan Hodge (2023), "On revisiting Wallace's 1858 theory of natural selection," in Pierre-Olivier Méthot, ed., *Philosophy, History and Biology: Essays in Honor of Jean Gayon*. History, Philosophy and Theory of the Life Sciences Series (Cham, Switzerland: Springer Nature).

28. Quotes taken from pp. 58, 59, and 62 in Alfred R. Wallace (1858), "On the tendency of varieties to depart indefinitely from the original type" [Dated Feb. 1858, Ternate; third part of "On the tendency of species to form varieties"; and "On the perpetuation of varieties and species by natural means of selection" by Charles Darwin and Alfred Wallace; communicated by Sir Charles Lyell and Joseph D. Hooker to the LSL meeting of 1 July 1858], *Journal of the Proceedings of the Linnean Society* (Zoology) 3(9):53–62.

29. Wallace signed the essay "Ternate, February, 1858," but as mentioned it was most likely written in Dodinga on Halmahera. Writing out a clearer "fair copy" with cover letter in Ternate may be why he signed it thus. The precise timing of the mailing of the essay has become a subject of great interest to Darwin and Wallace scholars and enthusiasts, some of whom have accused Darwin of lying about the date and circumstances around his receipt of the essay. While it is possible that Darwin received Wallace's essay somewhat earlier than he implied, he ultimately did the right thing and forwarded it to Lyell, as Wallace had asked, and there is no evidence whatsoever that he somehow stole ideas from Wallace's essay—there was nothing in the essay

that was not already part of Darwin's own formulation of the theory. See the following, in chronological order of publication: John van Wyhe and Kees Rookmaaker (2012), "A new theory to explain the receipt of Wallace's Ternate essay by Darwin in 1858," *Biological Journal of the Linnean Society of London* 105:249–52; R. Davies (2012), "How Charles Darwin received Wallace's Ternate paper 15 days earlier than he claimed: A comment on van Wyhe and Rookmaaker (2012)," *Biological Journal of the Linnean Society of London* 105:472–77; Charles H. Smith (2013), "A further look at the 1858 Wallace-Darwin mail delivery question," *Biological Journal of the Linnean Society of London* 108(3):715–18; Charles H. Smith (2014), "Wallace, Darwin and Ternate 1858," *Notes and Records* 68:165–70. See also the detailed treatment of this subject in Costa, *Darwin, Wallace, and the Origin of Species.*

30. C. Darwin to C. Lyell, 18 June 1858, WCP5647.

31. C. Darwin to C. Lyell, 25 June 1858, WCP5648; 26 June 1858, WCP5649.

32. J. D. Hooker and C. Lyell to the Linnean Society of London, 30 June 1858, DCP-LETT-2299; Charles Darwin and Alfred R. Wallace (1858), "On the tendency of species to form varieties; and on the perpetuation of varieties and species by natural means of selection," *Journal of the Proceedings of the Linnean Society of London* (Zoology) 3:45–62.

33. The carved wooden float is now in the collection of the British Museum (artifact no. Oc1935,1014.9), one of a dozen Wallace artifacts in the collection, most from New Guinea, donated in 1935 by Wallace's children. See https://www.britishmuseum.org/collection/term/BIOG129344. See also https://wallacefund.myspecies.info/wallace-artifacts-institutions.

34. Third Malay Journal (MS-178c), entry 134.

35. Third Malay Journal (MS-178c), entry 130.

36. Third Malay Journal (MS-178c), entry 135.

37. Alfred R. Wallace (1860), "Notes of a voyage to New Guinea" [communicated to the RGS meeting of 27 June 1859], *Journal of the Royal Geographical Society* 30:172–77.

38. After the prince's ship had departed, Wallace wrote, "We were then a little quiet again, and got something to eat; for while the vessels were here every bit of fish or vegetable was taken on board, and I had often to make a small parroquet serve for two meals." *Malay Archipelago*, p. 509.

39. Third Malay Journal (MS-178c), entry 139.

40. Third Malay Journal (MS-178c), entry 146.

41. In *The Malay Archipelago* (pp. 505–6), Wallace declared these flies "the most curious and novel" of all the insects he collected in New Guinea. British entomologist W. W. Saunders dubbed Wallace's novel flies *Elaphomia*, "Deer flies" (from the ancient Greek *élaphos*, "deer") in a paper read at the Entomological Society of London on 2 May 1859. Owing, however, to a long publishing delay, this description did not appear in print until late 1861, by which time German entomologist Adolf Gerstaecker had already published a description naming them *Phytalmia*, the genus used today according to the rules of publication priority for scientific names. For a fascinating discussion of these flies in the context of nineteenth-century collecting, transport, and scientific publishing, see Matthias Glaubrecht and Marion Kotrba (2004), "Alfred Russel Wallace's discovery of 'curious horned flies' and the aftermath," *Archives of Natural History* 31(2):275–99.

42. Today the term "Alfur" is often used generically for Indigenous peoples of the Maluku Islands (Moluccas).

43. Robert Chambers (1844), *Vestiges of the Natural History of Creation* (London: John Churchill), p. 296.

44. Chambers, *Vestiges of the Natural History of Creation*, p. 222.

45. The typological (and problematic!) concepts of "Malay" and "Papuan" emerged in the nineteenth-century science of race; for a fascinating history of these terms, see Chris Ballard (2008), "'Oceanic Negroes': British anthropology of Papuans, 1820–1869," pp. 157–201, in Bronwen Douglas and Chris Ballard, eds., *Foreign Bodies: Oceania and the Science of Race, 1750–1940* (Canberra: ANU E Press). Wallace recorded "racial" observations of Malayan, Papuan, "Alfuro," and other peoples in his Species Notebook (MS-180), pp. 65–66, 104–6, 134; see Costa, *On the Organic Law of Change*, pp. 158–61, 236–41, 296–97.

46. In his journal, Wallace wrote at the time: "I was much interested in the indigenes or alfuros of this part of Gilolo, of which a large population are settled in the neighbouring interior & numbers are daily seen in the village, either bringing their produce for sale or engaged by the Chinese or Ternate traders. A careful examination has strengthened my previous idea that they are a mixed race"; third Malay Journal (MS-178c), entry 154.

47. A. R. Wallace to G. Silk, 30 November 1858, WCP370; *Malay Archipelago*, p. 323.

48. Wallace, together with his contemporaries the Scottish physician, colonial administrator, and philologist John Crawfurd and English navigator and writer George Windsor Earl (who coined the term "Indu-nesia," forerunner of the modern name Indonesia) contributed significantly to a racial categorization of the peoples of the Malay Archipelago region, especially Papuans, that continues to have sociopolitical reverberations today. He also elevated the stature of field work over armchair theorizing in the nascent field of anthropology. See Jeremy Vetter (2006), "Wallace's *other* line: Human biogeography and field practice in the eastern colonial tropics," *Journal of the History of Biology* 39:89–123; and Ballard, "'Oceanic Negroes,'" pp. 157–201.

49. C. R. Darwin to A. Gray, 4 July 1858, WCP5650.

50. C. R. Darwin to J. D. Hooker, 5 July 1858, WCP5299.

51. C. R. Darwin to J. D. Hooker, 13 July 1858, WCP5298.

52. C. R. Darwin to C. Lyell, 18 July 1858, WCP5651.

53. Francis Darwin, ed. (1887–1888), *Life and Letters of Charles Darwin, including an Autobiographical Chapter*, 3 vols. (London: John Murray), 2:126.

54. Darwin and Wallace, "On the tendency of species to form varieties," pp. 45–62.

55. A. R. Wallace to J. D. Hooker, 6 October 1858, WCP1454.

56. A. R. Wallace to M. A. Wallace, 6 October 1858, WCP369.

57. A. R. Wallace to G. Silk, 30 November 1858, WCP370.

58. C. R. Darwin to J. D. Hooker, 23 January 1859, WCP5329.

Chapter 10: Island Hopper

1. A. R. Wallace to S. Smith, 29 October 1858, WCP1705; extracts published in the *Proceedings of the Zoological Society of London* 27 (1859):129–30.

2. See the species account and photos of *S. wallacii* in e-bird, https://ebird.org/species /walsta2, and Wallace's illustration in *Malay Archipelago*, chap. 24.

3. Gray initially placed the new species in the genus *Paradisaea*, subgenus *Semioptera*, and ornithologist John Gould soon after elevated the subgenus, making *Semioptera* a monotypic

genus. See George R. Gray (1859), [Notes of Mr. G. R. Gray on the sketch of a new form of Paradise-Bird], *Proceedings of the Zoological Society of London* 27(2):130; and Alfred R. Wallace (1860), "Notes on *Semioptera wallacii*, Gray" [Extract from a letter to John Gould dated 30 September 1859, Amboyna; communicated to the Zoological Society of London meeting of 24 January 1860], *Proceedings of the Zoological Society of London* 28:61. Several elements of the description Wallace gave Gray in this letter are also found in his Species Notebook—perhaps a draft; see Species Notebook (MS-180), p. 135; J. T. Costa (2013), *On the Organic Law of Change* (Cambridge, MA: Harvard University Press), pp. 298–99.

4. The Wallace and Ali statue was unveiled on 30 August 2019; see National University of Singapore Lee Kong Chian Natural History Museum, https://lkcnhm.nus.edu.sg/wallace-ali -statue-launch/.

5. Alfred R. Wallace (1859), [Extracts from a letter from Alfred R. Wallace to Samuel Stevens dated 28 January 1859, Batchian; communicated by Stevens to the ESL meeting of 6 June 1859], *Transactions of the Entomological Society of London*, n.s., 5:70–71. Wallace's suggested name for his new birdwing butterfly, *O. croesus*, was adopted by Gray in his later description: G. R. Gray (1859), "On a new species of the Family Papilionidae from Batchian," *Proceedings of the Zoological Society of London* 27:424–25; Natural History Notebook (MS-179), p. 105, verso.

6. Today this butterfly is recognized as a subspecies of the widely distributed blue mountain butterfly: *Papilio ulysses telemachus*.

7. C. R. Darwin to A. R. Wallace, 25 January 1859, WCP1841.

8. Species Notebook (MS-180), p. 14, verso; Costa, *On the Organic Law of Change*, pp. 430–31.

9. I provide a detailed comparative analysis of Darwin's and Wallace's respective lines of argument in James T. Costa (2014), *Wallace, Darwin, and the Origin of Species* (Cambridge, MA: Harvard University Press). Wallace's book *Darwinism* (1889) is in some ways Wallace's complement to the *Origin*; although published much later, it is a lucid treatment of the subject and structured in a way that Wallace felt was more pedagogically useful for explaining the principles of evolution by natural selection. See chapter 14 of this volume for a further discussion of *Darwinism*.

10. Wallace described his collecting on Bacan in *Malay Archipelago*, chapter 24.

11. Albeit likely introduced by people, according to modern thinking.

12. Natural History Notebook (MS-179), pp. 104–5, verso.

13. Frederick Smith (1860), "Catalogue of hymenopterous insects collected by Mr. A. R. Wallace in the islands of Bachian, Kaisaa, Amboyna, Gilolo, and at Dory in New Guinea," *Journal of the Proceedings of the Linnean Society of London* 5:93–143. The world's largest bee species, Wallace's giant bee had not been observed for over a century and was thought to be extinct until 1981 when the bees were found living inside arboreal termite nests by American entomologist Adam Messer. Another thirty-seven years went by, and then a couple of specimens popped up on Ebay. A year later, in 2019, American conservationist and photographer Clay Bolt and colleagues rediscovered the bees once again on Bacan, obtaining the first video and photographs of this species in the field. It appears that Wallace's giant bee, a resin bee, is an obligate inquiline of nests of the arboreal termite *Microcerotermes amboinensis*, harvesting resin from trees such as the lofty dipterocarp *Anisoptera thurifera* to line its galleries and presumably wall its nest off from the termites whose home it invades. See A. C. Messer (1984), "*Chalicodoma pluto*: The world's

largest bee rediscovered living communally in termite nests (Hymenoptera: Megachilidae)," *Journal of the Kansas Entomological Society* 57(1):165–68; Clay Bolt (2019), "Rediscovering Wallace's giant bee," Re:wild, https://www.rewild.org/news/rediscovering-wallaces-giant-bee -in-search-of-raja-ofu-the-king-of-bees.

14. *Malay Archipelago*, p. 349.

15. A. R. Wallace to T. Sims, 25 April 1859, WCP371.

16. A. R. Wallace to P. L. Sclater, March 1859, WCP4275; Alfred R. Wallace (1859), "Letter from Mr. Wallace concerning the geographical distribution of birds" [Dated March 1859, Batchian], *Ibis* 1:449–54.

17. A. R. Wallace to F. P. Pascoe, 20 July 1859, WCP1463; emphasis in original.

18. Darwin was dismayed that Hooker invoked continental extensionism to explain the flora of the circum-Antarctic islands and that Charles Lyell seemed to encourage the idea on the flimsiest evidence. He objected to Lyell:

> My blood gets hot with passion and turns cold alternately at the geological strides, which many of your disciples are taking. Here, poor Forbes made a continent to North America and another (or the same) to the Gulf weed; Hooker makes one from New Zealand to South America and round the World to Kerguelen Land, Here is Wollaston speaking of Madeira and P. Santo "as the sure and certain witnesses of a former continent." Here is Woodward writes to me, if you grant a continent over 200 or 300 miles of ocean depths (as if that was nothing), why not extend a continent to every island in the Pacific and Atlantic Oceans? And all this within the existence of recent species! If you do not stop this, if there be a lower region for the punishment of geologists, I believe, my great master, you will go there! (C. R. Darwin to C. Lyell, 16 June 1856, DCP-LETT-1902)

He followed this letter with another detailing his rebuttal of continental extensionism; see C. R. Darwin to C. Lyell, 25 June 1856, DCP-LETT-1910. Darwin's favored hypothesis of chance colonization of oceanic islands, and how he defended his view in the face of widespread opposition, is discussed in chapter 5 of James T. Costa (2017), *Darwin's Backyard: How Small Experiments Led to a Big Theory* (New York: W. W. Norton).

19. C. R. Darwin (1842), *The Structure and Distribution of Coral Reefs* (London: Smith, Elder); see also Alistair Sponsel's excellent 2018 book *Darwin's Evolving Identity: Adventure, Ambition, and the Sin of Speculation* (Chicago: University of Chicago Press).

20. As we will see in chapter 12 of this volume, Wallace eventually wrote a foundational text on the nature of oceanic islands and the chance migrants that populate them: Alfred R. Wallace (1880), *Island Life: Or, the Phenomena and Causes of Insular Faunas and Floras, including a Revision and Attempted Solution of the Problem of Geological Climates* (London: Macmillan).

21. C. R. Darwin to A. R. Wallace, 9 August 1859, DCP-LETT-2480 and WCP1843; see Alfred R. Wallace (1860), "On the zoological geography of the Malay Archipelago," *Journal of the Proceedings of the Linnean Society: Zoology* 4:172–84.

22. Third Malay Journal (MS-178c), entry 180.

23. The vivid coloration of this prized butterfly stems from unique micro- and nanoscale structures that combine the optical effects of multilayer interference, diffraction gratings, and photonic crystals; see Mathias Kolle, Pedro M. Salgard-Cunha, Maik R. J. Scherer, et al. (2010), "Mimicking the colourful wing scale structure of the *Papilio blumei* butterfly," *Nature Nanotechnology* 5:511–15.

24. Two anoa species are currently recognized: *Bubalus depressicornis*, now called the lowland anoa, and the mountain anoa *B. quarlesi*, described in 1910; see J. A. Burton, S. Hedges, and A. H. Mustari (2005), "The taxonomic status, distribution and conservation of the lowland anoa *Bubalus depressicornis* and mountain anoa *Bubalus quarlesi*," *Mammal Review* 35(1):25–50. Northeast Sulawesi is the site of the Tangkoko Nature Reserve today, home to what is currently the world's largest monument to Wallace: a 5-foot (1.5 m) bust of Wallace sitting atop an 8.5-foot (2.6 m) plinth. This site was first proposed for preservation by botanist Sijfert Koorders of Java and the Dutch Indies Society for the Protection of Nature (which Koorders chaired) in 1913, not coincidentally the year of Wallace's death. A great admirer of Wallace, Koorders later named the monotypic tree genus *Wallaceodendron* in his honor, the sole species of which is the Sulawesi endemic *W. celebicum*. Wallace's association with north Sulawesi was also recognized by traveler and naturalist Francis Henry Hill Guillemard, who, exploring the area in 1883, dubbed the gulf on which Tangkoko is located Wallace Bay. See the account by Wallace scholar George Beccaloni on the Alfred Russell Wallace Memorial Fund website: https://wallacefund.myspecies .info/content/impressive-new-monument-alfred-rusel-wallace-sulawesi-indonesia.

25. Third Malay Journal (MS-178c), entry 191.

26. C. R. Darwin to A. R. Wallace, 6 April 1859, DCP-LETT-2449 and WCP1842.

27. The paragraph reads:

My work is now nearly finished; but as it will take me two or three more years to complete it, and as my health is far from strong, I have been urged to publish this Abstract. I have more especially been induced to do this, as Mr. Wallace, who is now studying the natural history of the Malay archipelago, has arrived at almost exactly the same general conclusions that I have on the origin of species. Last year he sent to me a memoir on this subject, with a request that I would forward it to Sir Charles Lyell, who sent it to the Linnean Society, and it is published in the third volume of the Journal of that Society. Sir C. Lyell and Dr. Hooker, who both knew of my work—the latter having read my sketch of 1844—honoured me by thinking it advisable to publish, with Mr. Wallace's excellent memoir, some brief extracts from my manuscripts. (*Origin*, pp. 1–2)

28. Natural History Notebook (MS-179), p. 103, verso.

29. A. R. Wallace to S. Stevens, 22 October 1859, WCP4276; Alfred R. Wallace (1860), [Extracts from letter concerning collecting dated 22 Oct. 1859, Amboyna], *Ibis* 2:197–99.

30. Alfred R. Wallace (1860), "The ornithology of Northern Celebes," *Ibis* 2:140–47.

31. Alfred R. Wallace (1861), "On the ornithology of Ceram and Waigiou," *Ibis* 3:283–91.

32. A. R. Wallace to S. Stevens, 26 November 1859, WCP4277; Alfred R. Wallace (1860), [Extracts from letters to Samuel Stevens concerning collecting dated 26 November 1859, Awaiya, Ceram, 31 December 1859, Passo, Amboyna, and 14 February 1860, Passo], *Ibis* 2:305–6.

33. A. R. Wallace to S. Stevens, 31 December 1859, WCP4278; Alfred R. Wallace (1860), [Extracts from letters to Samuel Stevens concerning collecting dated 26 November 1859, Awaiya, Ceram, 31 December 1859, Passo, Amboyna, and 14 February 1860, Passo], *Ibis* 2:305–6.

34. C. R. Darwin to: J. S. Henslow, 11 November 1859, DCP-LETT-2522; T. C. Eyton, 24 November 1859, DCP-LETT-2546; T. H. Huxley, 15 October 1859, DCP-LETT-2505; H. Falconer, 11 November 1859, DCP-LETT-2524.

35. C. R. Darwin to A. R. Wallace, 13 November 1859, DCP-LETT-2529 and WCP1844.

36. C. R. Darwin to A. R. Wallace, 18 May 1860, DCP-LETT-2807 and WCP1846.

37. When Wallace departed Sarawak in February 1856, Charles stayed behind to train to become a teacher at the mission, but, unhappy there, he soon moved on to the Borneo Company Ltd., a mining company, and was probably in their employ when Wallace invited him to come collect for him; see Kees Rookmaaker and John van Wyhe (2012), "In Alfred Russel Wallace's shadow: His forgotten assistant, Charles Allen (1839–1892)," *Journal of the Malaysian Branch of the Royal Asiatic Society* 85(2):17–54. Planning for his return to New Guinea, Wallace wrote Stevens from Paso, Ambon, via Marseilles asking him to send three inexpensive guns by the expedited overland route as soon as possible, "to be ready for my next year's campaign to New Guinea" since he now had Charles working for him and needed another set of collecting equipment. A. R. Wallace to S. Stevens, 14 February 1860, WCP4279.

38. *Malay Archipelago*, p. 367; fourth Malay Journal (MS-178d), entry 204.

39. A brief note in the January 1861 issue of the *Ibis*, p. 118, sent June 1860 from Seram, mentions the few notable specimens collected by Wallace there and his disappointment with the site:

> Mr. Wallace's collections from Amboyna and Ceram have arrived in England. The greatest novelty in them is a beautiful new *Basilornis* with an erect crest, making the second of the genus. Other species of interest are *Lorius domicella, Eos rubra, Trichophorus flavicaudus*, and *Tanysiptera dea* (?). Mr. Wallace's latest letters, dated from Ceram, in June last, speak of the probability of his return to England being not long delayed. He had been much disappointed with the results of an expedition to the northern part of the island, and was then intending to go to Mysol, which was expected to prove a good locality.

40. *Malay Archipelago*, p. 382; fourth Malay Journal (MS-178d), entry 215.

41. The ordeal of the voyage from Seram to Waigeo is recounted in chapter 35 of *Malay Archipelago* and the fourth Malay Journal (MS-178d), entries 214–29.

42. *Malay Archipelago*, p. 520.

43. *Malay Archipelago*, p. 523.

44. See Species Notebook (MS-180); Costa, *On the Organic Law of Change*, pp. 350–58 for Wallace's original field observations and sketches of the red bird of paradise, which served as the basis for his treatment in *Malay Archipelago* (pp. 529–31, 558–60). See also Tim Laman and Edwin Scholes (2013), *Birds of Paradise: Revealing the World's Most Extraordinary Birds* (Washington, DC: National Geographic and Cornell Laboratory of Ornithology), pp. 12–13, 217.

45. See, for example, Russell D. Gray, Quentin D. Atkinson, and Simon J. Greenhill (2011), "Language evolution and human history: What a difference a date makes," *Philosophical Transactions of the Royal Society B* 366:1090–100; Pavel Duda and Jan Zrzavý (2016), "Human population history revealed by a supertree approach," *Nature Scientific Reports* 6:29890.

46. *Malay Archipelago*, p. 538.

47. The Malay anchor is illustrated in chapter 37 of *Malay Archipelago*, p. 546.

48. Wallace elaborated a bit:

> My first crew ran away; two men were lost for a month on a desert island; we were ten times aground on coral reefs; we lost four anchors; the sails were devoured by rats; the small boat was lost astern; we were thirty-eight days on the voyage home, which should not have taken twelve; we were many times short of food and water; we had no compass-lamp, owing to there not being a drop of oil in Waigiou when we left; and to crown all, during the whole of our voyages from Goram by Ceram to Waigiou, and from Waigiou

to Ternate, occupying in all seventy-eight days, or only twelve days short of three months (all in what was supposed to be the favorable season), we had not one single day of fair wind! We were always close braced up, always struggling against wind, tide, and leeway, and in a vessel that would scarcely sail nearer than eight points from the wind. Every seaman will admit that my first voyage in my own boat was a most unlucky one. (*Malay Archipelago*, p. 550)

49. A. R. Wallace to G. Silk, 1 September 1860 and 2 January 1861, WCP373.

50. A. R. Wallace to H. W. Bates, 24 December 1860, WCP374.

51. A. R. Wallace to G. Silk, 1 September 1860 and 2 January 1861, WCP373.

52. A. R. Wallace to S. Stevens, 7 December 1860, WCP4751; Alfred R. Wallace (1861), [Extracts from letter to Samuel Stevens concerning collecting dated 7 December 1860, Ternate], *Ibis* 3:211–12.

53. Wallace, "On the ornithology of Ceram and Waigiou," pp. 283–91.

54. See chapters 3–5 in Katharine G. Davidson (1994), "The Portuguese colonisation of Timor: The final stage, 1850–1912" (PhD diss., University of New South Wales, Australia); fourth Malay Journal (MS-178d), entries 239–40.

55. Alfred R. Wallace (1861), "Notes on the ornithology of Timor" [Dated 20 April 1861, Delli, Timor], *Ibis* 3:347–51. Wallace described several of his Timor bird species in a paper delivered to the Zoological Society of London: Alfred R. Wallace (1863), "A list of the birds inhabiting the islands of Timor, Flores, and Lombock, with descriptions of the new species" [A paper read at the ZSL meeting of 24 November 1863], *Proceedings of the Zoological Society of London* 1863:480–97.

56. Alfred R. Wallace, Notebook 5, Species Register (1858–1861), Natural History Museum, London, cited in Rookmaaker and Wyhe, "In Alfred Russel Wallace's shadow," pp. 17–54.

57. Wallace described these and other bird species in a paper delivered to the Zoological Society of London: Alfred R. Wallace (1863), "List of birds collected in the island of Bouru (one of the Moluccas), with descriptions of the new species" [A paper read at the ZSL meeting of 13 January 1863], *Proceedings of the Zoological Society of London* 1863:18–36. See also Kees Rookmaaker and John van Wyhe (2018), "A price list of birds collected by Alfred Russel Wallace inserted in *The Ibis* of 1863," *Bulletin of the British Ornithologists' Club* 138(4):335–45.

58. This species was described by Wallace in 1862: Alfred R. Wallace (1862), "Descriptions of three new species of *Pitta* from the Moluccas" [A paper read at the ZSL meeting of 24 June 1862], *Proceedings of the Zoological Society of London* 1862:187–88. In this paper Wallace dubbed it *Pitta* (now *Erythropitta*) *rubrinucha*. He opted for a descriptive specific epithet, referring to the red patch on the nape of the neck, but it would have made a fine tribute to the indefatigable Ali, who persevered in collecting this species, if he had named it for him: *Erythropitta aliae*?

59. *Malay Archipelago*, p. 395.

60. *Origin*, p. 490.

61. Walt Whitman (1855), *Song of Myself*, cantos 3, 44 (self-pub).

Chapter 11: First Darwinian

1. "I am as you will see now commencing my retreat westwards, I have left the wild & savage Moluccas & New Guinea for Java the garden of the East." A. R. Wallace to M. A. Wallace, 20 July 1861, WCP375.

2. "Will you, next time you visit my mother, make me a little plan of her cottage showing the rooms & their dimensions, so that I may see if there will be room enough for me on my return. I shall want a good large room for my collections, & as when I can decide exactly on my return it would be as well to get a little larger house beforehand if necessary." A. R. Wallace to T. Sims, 15 March 1861, WCP3351.

3. A. R. Wallace to M. A. Wallace, 20 July 1861, WCP375.

4. Mojokerto is the site of discovery of the Mojokerto Child, an early human fossil skullcap found in February 1936 by a team led by German Dutch paleontologist Ralph von Koenigswald. Originally named *Pithecanthropus modjokertensis*, the fossil is now thought to represent a young *Homo erectus* individual, dated to a maximum age of about 1.5 million years. See Michael J. Morwood et al. (2003), "Revised age for Mojokerto, an early *Homo erectus* cranium from East Java, Indonesia," *Australian Archaeology* 57:1–4. Forty-five years before this discovery, the first early human fossils to be found in Asia were also found on Java, some fifty miles west of Mojokerto: Java Man, also recognized as *Homo erectus* today, was discovered by Dutch paleoanthropologist Eugène Dubois at Trinil, in a broad loop of the Solo River. It is interesting to note that Dubois undertook his search for early humans in the Malay Archipelago because Lyell and Wallace argued that Southeast Asia was the cradle of humanity, disagreeing with Darwin, who pointed to Africa as the continent of human origins. See Carl C. Swisher, Garniss H. Curtis, and Roger Lewin (2000), *Java Man: How Two Geologists Changed Our Understanding of Human Evolution* (Chicago: University of Chicago Press), pp. 58–59.

5. Wallace later donated the bas-relief, which dates to the Majapahit era, to the Charterhouse School Museum in Godalming, Surrey, close to where he lived from 1881 to 1889. In 2002, however, the school auctioned off many of its prized antiquities, including Wallace's bas-relief, much to the outrage of archaeologists (*Guardian*, "Charterhouse treasures go to auction as academics rail," https://www.theguardian.com/uk/2002/jul/27/arts.schools). It fetched £2,629, the buyer perhaps unaware of its connection to the cofounder of evolutionary biology. See George Beccaloni's article "Who bought Wallace's Javanese carving?" at Alfred Russel Wallace Memorial Fund, https://wallacefund.myspecies.info/who-bought-wallaces-javanese-carving; see also *Malay Archipelago*, pp. 111–13, for Wallace's account of the carving.

6. Darwin's paper "On the two forms, or dimorphic condition, in the species of *Primula*, and on their remarkable sexual relations" was read on 21 November 1861 and published the following year (*Journal of the Proceedings of the Linnean Society of London (Botany)* 6:77–96). Darwin's discovery, and its significance, is discussed in chapter 6 in James T. Costa (2017), *Darwin's Backyard: How Small Experiments Led to a Big Theory* (New York: W. W. Norton). Interestingly, at the very meeting where Darwin's paper on *Primula* was read, Henry Walter Bates also presented his paper on mimicry in Amazonian butterflies—yet another fascinating phenomenon only explicable in light of the Darwin and Wallace theory.

7. *Origin*, pp. 365–82, "Dispersal during the glacial period."

8. *Malay Archipelago*, p. 130.

9. A. R. Wallace to F. Sims, 10 October 1861, WCP376.

10. Natural History Notebook (MS-179), p. 99, verso.

11. The massive Sumatran earthquake of 16 February 1861 struck just off the west coast and resulted in several thousand fatalities. Estimated at magnitude 8.5, this earthquake has recently been identified as the culmination of a surprisingly long-lived (thirty-two year) slow-slip event,

refining the understanding of earthquakes at tectonic plate boundaries. See Rishav Mallick et al. (2021), "Long-lived shallow slow-slip events on the Sunda megathrust," *Nature Geoscience* 14:327–33; and Maya Wei-Haas (2021), "An earthquake lasted 32 years. Scientists want to know how," *National Geographic*, https://www.nationalgeographic.co.uk/science-and-technology /2021/06/an-earthquake-lasted-32-years-and-scientists-want-to-know-how.

12. Wallace stayed in a *passangrahan*, Dutch government guest houses built about every dozen miles or so along the main highways (*Malay Archipelago*, pp. 134–35). Anthropologist Gerrell Drawhorn has identified a good candidate for Wallace's passangrahan in Lubuk Rahman: https://wallacefund.myspecies.info/visit-wallace-s-sumatran-collection-site-lobo-raman-june -2012. Wallace described the village in *Malay Archipelago*, pp. 135–36. While he admired the fine native houses, he was surprised by the unsanitary practice of eliminating waste into smelly open-pit cesspools directly underneath. He speculated that this unfortunate practice was a vestige of a maritime past when such houses were built right over the water, conveniently flushing out waste with the tides. In the transition from siting houses on the coast to rivers and streams and then to the uplands, Wallace thought these Sumatrans retained a waste removal practice now rendered ineffective. Although he did not say so, it may have occurred to him that this could be a curious cultural example of mismatched "structure" (morphology) and "habit" (behavior or instinct) indicative of evolutionary transitions with changed environment—in this case house design retained despite altered living conditions, resulting in a maladaptive (at least malodorous) mismatch that may eventually lead to some adaptive change in one or the other. In his Species Notebook (MS-180, p. 53), Wallace discussed changed habits (behavior) and constant structure as an argument against design, suggesting, first, that animal structure is not divinely designed for some optimal function and, second, that behavior changes adaptively more quickly than structure. In *Origin* (p. 185) Darwin took a more overtly evolutionary view on cases where "habits and structure [are] not at all in agreement," citing them as evidence of transition where habits are altered first, with form perhaps to follow. See James T. Costa (2013), *On the Organic Law of Change* (Cambridge, MA: Harvard University Press), pp. 134–35.

13. A. R. Wallace to G. Silk, 22 December 1861, WCP378.

14. A. R. Wallace to C. R. Darwin, 30 November 1861, WCP4109 and DCP-LETT-3334.

15. A. R. Wallace to H. W. Bates, 10 December 1861, WCP377; Henry W. Bates (1860), "Contributions to an insect fauna of the Amazon valley. Diurnal Lepidoptera," *Transactions of the Entomological Society of London* 5:223–28, 335–61.

16. *Malay Archipelago*, pp. 138–40. To underscore just how remarkable this phenomenon is, Wallace gave a rather memorable analogy: Imagine an Englishman with two wives, one, say, Malay and the other Papuan. Instead of children with the respective wives reflecting a mix of the characteristics of their parents, he said, imagine that all the boys were just like their father and the girls, their mother. That would be thought odd enough, but the case of the butterflies is odder still: "Each mother is capable not only of producing male offspring like the father, and female like herself, but also other females like her fellow-wife, and altogether differing from herself!"

17. Misidentified as a lemur in Wallace's day, *Galeopterus* helped fuel speculations of a former African–Southeast Asian land bridge, with almost comical consequences.

18. *Malay Archipelago*, pp. 146–47.

19. *My Life*, 1:382–83; Thomas Barbour (1943), *Naturalist at Large* (Boston: Little, Brown), p. 42. See also John van Wyhe and Gerrell M. Drawhorn (2015), "'I am Ali Wallace': The Malay

assistant of Alfred Russel Wallace," *Journal of the Malaysian Branch of the Royal Asiatic Society* 88(1):3–31.

20. A. R. Wallace to P. L. Sclater, 18 March 1862, WCP1722.

21. A. R. Wallace to P. L. Sclater, 31 March 1862, WCP1723.

22. A. R. Wallace to C. R. Darwin, 7 April 1862, WCP1847 and DCP-LETT-3496.

23. *My Life*, 1:385.

24. Alfred R. Wallace (1847), "Capture of *Trichius fasciatus* near Neath" [excerpt from a letter sent from Neath, Wales], *Zoologist* 5:1676.

25. A. R. Wallace to A. Newton, 12 August 1862, WCP4000.

26. Philip Edgerton, "Monkeyana," *Punch*, 18 May 1861.

27. Charles Kingsley (1863), *The Water-Babies: A Fairy Tale for a Land-Baby* (London: Macmillan), p. 156. For detailed treatments of the Huxley-Owen hippocampus debate and its cultural context, see Charles G. Gross (1993), "Huxley versus Owen: The hippocampus minor and evolution," *Trends in Neuroscience* 16(12):493–98; C. M. Owen, A. Howard, and D. K. Binder (2009), "Hippocampus minor, *calcar avis*, and the Huxley-Owen debate," *Neurosurgery* 65(6):1098–104; Piers J. Hale (2013), "Monkeys into men and men into monkeys: Chance and contingency in the evolution of man, mind and morals in Charles Kingsley's *Water Babies*," *Journal of the History of Biology* 46(4):551–97.

28. Huxley's and Owen's letters debating the brain anatomy of humans and apes appeared in the *Athenaeum*, March and April 1861; see L. G. Wilson (1996), "The gorilla and the question of human origins: The brain controversy," *Journal of the History of Medicine and Allied Sciences* 51:184–207.

29. A. R. Wallace to C. R. Darwin, 30 November 1861, WCP4109; emphases in original.

30. Alfred R. Wallace (1863), "On the physical geography of the Malay Archipelago" [a paper read at the RGS meeting of 8 June 1863], *Journal of the Royal Geographical Society* 33:217–34.

31. Roderick I. Murchison (1863), [Remarks in "Discussion on Mr. Wallace's Paper"], *Proceedings of the Royal Geographical Society* 7:210–12.

32. Earl (also spelled Earle) described the general contours of what are now called the Sahul and Sunda Shelves, noting that "it will be found that all of the countries [islands] lying upon these banks partake of the character of the continents to which they are attached." See p. 359 in G. Windsor Earle (1845), "On the physical structure and arrangement of the islands of the Indian Archipelago," *Journal of the Royal Geographical Society* 15:358–65.

33. Philip L. Sclater (1858), "On the general geographical distribution of the members of the class Aves," *Journal of the Proceedings of the Linnean Society of London: Zoology* 2(7):130–45.

34. Jane R. Camerini (1993), "Evolution, biogeography, and maps: An early history of Wallace's Line," *Isis* 84:700–727.

35. Huxley coined the term "Wallace's Line" in a paper on gallinaceous birds read at the 14 May 1868 meeting of the Zoological Society of London: p. 313 in Thomas H. Huxley (1868), "On the classification and distribution of the Alectoromorphae and Heteromorphae," *Proceedings of the Zoological Society of London* 1868:294–319.

36. For early reviews of Wallace's Line and efforts at refining it, see Ernst Mayr (1944), "Wallace's Line in the light of recent zoogeographic studies," *Quarterly Review of Biology* 19(1):1–14; and George Gaylord Simpson (1977), "Too many lines: The limits of the Oriental and Australian zoogeographic regions," *Proceedings of the American Philosophical Society* 121(2):107–20. For a

brief overview of "The problem of Wallacea," see chap. 10, pp. 323–25, in Charles H. Smith, James T. Costa, and David Collard, eds. (2019), *An Alfred Russel Wallace Companion* (Chicago: University of Chicago Press). For an excellent treatment of Wallacea in the context of tectonic and environmental history, see Penny van Oosterzee (1997), *Where Worlds Collide: The Wallace Line* (Ithaca, NY: Cornell University Press).

37. Wallace's paper "On the geographical distribution of animal life" was read at the 31 August 1863 meeting of Section D, Zoology and Botany, of the BAAS meeting in Newcastle-upon-Tyne and later published as "On some anomalies in zoological and botanical geography" in the *Edinburgh New Philosophical Journal* (19:1–15) for 7 January 1864 and the *Natural History Review* (4:111–23) for January 1864. While Sclater's six original "ontological divisions of the earth's surface" continue to be used today, many alternatives have been proposed over the years, some taxon-specific (plants, freshwater fish), some environmental (marine vs. terrestrial), and some based on new quantitative approaches to assessing the geographical distribution of diversity. See, for example: Miklos D. F. Udvardy (1975), "A classification of the biogeographical provinces of the world" [prepared as a contribution to UNESCO's Man and the Biosphere Programme, project no. 8], International Union for Conservation of Nature Occasional paper no. 18; C. Barry Cox (2001), "The biogeographic regions reconsidered," *Journal of Biogeography* 28:511–23; R. Abell et al. (2008), "Freshwater ecoregions of the world: A new map of biogeographic units for freshwater biodiversity conservation," *BioScience* 58:403–14; Holger Kreft and Walter Jetz (2010), "A framework for delineating biogeographical regions based on species distributions," *Journal of Biogeography* 37:2029–2053; D. M. Olson et al. (2001), "Terrestrial ecoregions of the world: A new map of life on Earth," *Bioscience* 51(11):933–38; Ben G. Holt et al. (2013), "An update of Wallace's zoogeographic regions of the world," *Science* 339:74–78; J. J. Morrone (2015), "Biogeographical regionalisation of the world: A reappraisal," *Australian Systematic Botany* 28:81–90.

38. Alfred R. Wallace (1865), "On the phenomena of variation and geographical distribution as illustrated by the Papilionidae of the Malayan region" [A paper read at the Linnean Society of London meeting of 17 March 1864], *Transactions of the Linnean Society of London* 25(pt. 1):1–71. Space constraints preclude a full exposition of this remarkable paper. My brief discussion benefited greatly from the lucid analyses of evolutionary biologist James Mallet: see, especially, Mallet's papers of 2004, "Poulton, Wallace and Jordan: How discoveries in *Papilio* butterflies led to a new species concept 100 years ago," *Systematics and Biodiversity* 1(4):441–52; 2008, "Wallace and the species concept of the early Darwinians," pp. 102–13, in C. H. Smith and G. Beccaloni, eds., *Natural Selection and Beyond: The Intellectual Legacy of Alfred Russell Wallace* (Oxford: Oxford University Press); and 2009, "Alfred Russel Wallace and the Darwinian species concept: His paper on the swallowtail butterflies (Papilionidae) of 1865," *Gayana* 73(suppl.):37–43.

39. It is worth pointing out that although the modern *biological species concept* with its reproductive compatibility criterion for diagnosing species is often attributed to twentieth-century Harvard zoologist Ernst Mayr, in fact this concept has its origin in the nineteenth century and is attributable mainly to Wallace, as evolutionary biologist James Mallet has pointed out. In the early twentieth century, Oxford biologist Edward B. Poulton clearly articulated the modern species concept (based on what he called *syngamy*) in his presidential address to the Entomological Society of London in January 1904, "What is a species?" (*Proceedings of the Entomological Society of London* 1903:77–116). In the process Poulton coined, among others, the terms *sympatry*

and *asympatry*, the former still in use and the latter now changed to *allopatry*. Mallet has pointed out that Poulton's address was likely inspired by a gift from Wallace given shortly before the address: a bound collection of three seminal papers on mimicry, Bates (1862), Wallace (1865), and an 1869 paper by British-born South African entomologist Roland Trimen (1869). Remarkably, Wallace's gift included his *personal* copy of Bates's 1862 paper, bearing the inscription "Mr. A. R. Wallace from his old traveling companion—the author." Mayr and others later built upon Poulton's concept, Mayr masterfully reviewing the issue and cementing the concept of species as reproductive communities in his important 1942 book *Systematics and Origin of Species* (New York: Columbia University Press). See Mallet, "Poulton, Wallace and Jordan," pp. 441–52, and Mallet, "Wallace and the species concept," pp. 102–13.

40. Alfred R. Wallace (1864), "Mr. Wallace on the phenomena of variation and geographical distribution as illustrated by the Malayan Papilionidae," *Reader* 3:491b–493b. First published in 1863, the *Reader* was a short-lived weekly newsmagazine that covered art, religion, history, and science. Huxley and physicist John Tyndall served as editors of the science section, and Wallace was a regular contributor. It ceased publication in 1867, however, largely a victim of the increasingly antagonistic relationship between religion and science in those post-*Origin* years in Britain, much of that exacerbated by Huxley's inflammatory style. After it ceased publication, Huxley and other scientific contributors decided to set up their own scientific journal; thus was the esteemed journal *Nature* born in 1869.

41. C. R. Darwin to A. R. Wallace, 28 May 1864, WCP1858 and DCP-LETT-4510.

42. *Origin*, p. 488.

43. Transmutation Notebook C (1838), pp. 76–77; Paul H. Barrett et al., eds. (1987), *Charles Darwin's Notebooks, 1836–1844* (Ithaca, NY: Cornell University Press).

44. This was evident in the very first paper delivered before the new society, by Hunt himself: "On the Negro's place in nature," *Memoirs of the Anthropological Society of London* 1(1863–1864): 1–64, pp. 54–55. Ironically, Hunt uses language echoing Darwin's in his notebook in insisting that the arguments of abolitionists "shall fall, and shall be replaced by a new fabric built upon a more solid foundation" (p. 60). For a brief overview of the emergence of the Anthropological Society in opposition to the Ethnological Society in the context of the social and political climate of Britain at that time, and the response of the Darwin-Wallace school, see George W. Stocking (1987), *Victorian Anthropology* (New York: Free Press), pp. 245–62. See also Efram Sera-Shriar's insightful analysis of James Hunt and Thomas Henry Huxley in the context of the Ethnological and Anthropological Society schism: Efram Sera-Shriar (2013), "Observing human difference: James Hunt, Thomas Huxley and competing disciplinary strategies in the 1860s," *Annals of Science* 70(4):461–91.

45. Thomas H. Huxley (1863), *Evidence as to Man's Place in Nature* (London: Williams and Norgate); Charles Lyell (1863), *Geological Evidences of the Antiquity of Man* (London: John Murray).

46. "On the varieties of man in the Malay Archipelago" was originally delivered at the BAAS meeting of 1 September 1863 at Newcastle-upon-Tyne, an abstract of which was printed in the "Notices and abstracts of miscellaneous communications to the sections" portion of the *Report of the British Association for the Advancement of Science* 33(1863):147–48 (reprinted in the *Reader* 2(43):483a–c [24 October 1863] and *Anthropological Review* 1(3):441–44 [November 1863]). The paper was later presented as "On the varieties of man in the Malay Archipelago" at the

Ethnological Society of London meeting of 26 January 1864 and printed in the *Transactions of the Ethnological Society of London*, n.s., 3:196–215 (1865). This paper was largely incorporated into chapter 40 of *The Malay Archipelago*.

47. Wallace delivered this paper several months earlier at the 1863 British Association meeting in Newcastle.

48. *Malay Archipelago*, p. 323.

49. Jeremy Vetter (2015), "Politics, paternalism, and progressive social evolution: Observations on colonial policy in the scientific travels of Alfred Russel Wallace," *Victorian Review* 41(2):113–31, especially pp. 121–22 in regard to Menado.

50. Wallace's paper "The origin of human races and the antiquity of man deduced from the theory of natural selection" was read at the Anthropological Society of London meeting of 1 March 1864 and published in the *Journal of the Anthropological Society of London* (2:clviii–clxx) (1864), followed by an account of related discussion on pp. clxx–clxxxvii . The paper was reprinted as "'Natural selection' applied to man" in the *Natural History Review*, n.s., no. 15:328–36 (July 1864). Considered by Wallace to be one of his most important papers, he included a revised version titled "The development of human races under the law of natural selection" in his 1870 collection *Contributions to the Theory of Natural Selection*, pp. 303–31.

51. Wallace contributed comments to a discussion of "Linga puja, or phallic worship in India," a paper by E. Sellon read at the Anthropological Society of London meeting of 17 January 1865. *Journal of the Anthropological Society of London* 3(9):cxviii–cxix (cxiv–cxxi).

52. A. R. Wallace to T. H. Huxley, 26 February 1864, WCP3751. Historian Efram Sera-Shriar has shown that for all of Huxley's dislike of Hunt they were not so different in key aspects of their thinking, including the matter of admitting women to scientific meetings (perhaps prompted by Wallace on the issue). When in 1868–1869 Huxley became president of the Ethnological Society, he reversed the decision to admit women over the vehement protests of female members. He then tried to thread the needle by creating two types of meetings, "ordinary meetings" from which women would be barred and "special meetings" on popular subjects that they would be permitted to attend; see Sera-Shriar, "Observing human difference," pp. 461–91, 471–73.

53. In his 1864 *Journal of the Anthropological Society of London* paper (p. clxiii), Wallace wrote:

> Thus man, by the mere capacity of clothing himself, and making weapons and tools, has taken away from nature that power of changing the external form and structure which she exercises over all other animals. As the competing races by which they are surrounded, the climate, the vegetation, or the animals which serve them for food, are slowly changing, they must undergo a corresponding change in their structure, habits, and constitution, to keep them in harmony with the new conditions—to enable them to live and maintain their numbers. But man does this by means of his intellect alone; which enables him with an unchanged body still to keep in harmony with the changing universe. (Wallace [1864], *Journal of the Anthropological Society of London* 2:clviii–clxx)

54. C. R. Darwin to A. R. Wallace, 28 May 1864, WCP1858 and DCP-LETT-4510; J. D. Hooker to C. R. Darwin, 14 May 1864, WCP 5296; C. Lyell to A. R. Wallace, 22 May 1864, WCP2074.

55. C. R. Darwin to A. R. Wallace, 26 January 1870, WCP1931 and DCP-LETT-7086.

56. Polygenist and virulent racist James Hunt was appalled at the idea of human races becoming homogenized into one, attacking Wallace on this point in the pages of the *Anthropological Review*. Wallace defended his view: see Alfred R. Wallace (1867), [Mr. Wallace on natural selection applied to anthropology], *Anthropological Review* 5:103–5.

57. See sociologist Ted Benton's interesting discussion of Wallace's "origin of human races" paper: pp. 29–32 in Ted Benton (2009), "Race, sex and the 'earthly paradise': Wallace versus Darwin on human evolution and prospects," *Sociological Review* 57(2 suppl.):23–46.

58. Alfred R. Wallace (1863), "Remarks on the Rev. S. Haughton's paper on the bee's cell, and on the 'Origin of Species,'" *Annals and Magazine of Natural History*, 3rd ser., 12:303–9.

59. C. R. Darwin to A. R. Wallace, 23 February 1867, WCP609; C. R. Darwin to A. R. Wallace, 26 February 1867, WCP1875; A. R. Wallace to C. R. Darwin, 2 March 1867, WCP4081; see also Alfred R. Wallace (1867), [Wallace's explanation of brilliant colors in caterpillar larvae, and others' comments thereon, presented at the Entomological Society of London meeting of 4 March 1867], *Proceedings of the Entomological Society of London* 1867:80–81, reprinted in *Zoologist*, 2nd ser., 2:717–18 (April 1867).

60. Wallace, [Wallace's explanation of brilliant colors in caterpillar larvae], pp. 717–18 and "Caterpillars and Birds," *Field* 29:206a–b; John Jenner Weir (1869), "On insects and insectivorous birds; and especially on the relation between the colour and the edibility of Lepidoptera and their larvae," *Transactions of the Entomological Society of London* 1869:21–26; (1870), "Further observations on the relation between the colour and the edibility of Lepidoptera and their larvae," *Transactions of the Entomological Society of London* 1870:337–39. See also A. R. Wallace to C. R. Darwin, 10 March 1869, WCP6651.

61. *My Life*, 1:410; A. R. Wallace to C. R. Darwin, 20 January 1865, WCP4101; A. R. Wallace to A. Newton, 19 February 1865, WCP4006.

62. Alfred R. Wallace (1865), "The *British Quarterly* and Darwin," *Reader* 5:77c–78a; and (1865), "The *British Quarterly* Reviewer and Darwin," *Reader* 5:173a–b.

63. Barbara Weisberg (2004), *Talking to the Dead: Kate and Maggie Fox and the Rise of Spiritualism* (New York: Harper Collins). For more on the psychology behind the tendency to believe fraudsters even in the face of evidence, see Carl Sagan's illuminating discussion of "baloney detection" in his 1996 book *The Demon Haunted World: Science as a Candle in the Dark* (New York: Random House). "Spirit-rapping" is discussed on pp. 242–43.

64. Alfred R. Wallace (1865), "How to civilize savages," *Reader* 5:671–72.

65. A. R. Wallace to G. Rolleston, 23 September 1865, WCP6656.

66. See Michael Shermer's excellent treatment of Wallace as "heretical scientist" in his 2002 book *In Darwin's Shadow: The Life and Science of Alfred Russel Wallace* (Oxford: Oxford University Press).

67. T. H. Huxley to A. R. Wallace, 27 November 1866, WCP2544.

68. Himself a self-avowed agnostic (a term he coined), Huxley put spiritualism on a par with other "superstitions," commenting to the prime minister that Wallace's beliefs were "not worse than the prevailing superstitions of the country." C. R. Darwin to T. H. Huxley, 27 November 1880, WCP6655.

69. A. R. Wallace to C. R. Darwin, 30 September 1862, WCP1852.

70. C. R. Darwin to A. R. Wallace, 12 and 13 October 1867, WCP1883.

71. Alfred R. Wallace (1867), "Creation by law" [Essay-review of *The Reign of Law* by the Duke of Argyll, 1867], *Quarterly Journal of Science* 4:471–88.

72. See Charles H. Smith's Alfred Russel Wallace Page, http://people.wku.edu/charles.smith /wallace/S118A.htm.

73. Huxley, *Evidence as to Man's Place*, pp. 24–25.

74. Edward Sabine (1869), "Anniversary Meeting," *Proceedings of the Royal Society of London* 17:133–54, 148.

75. A. R. Wallace to C. R. Darwin, 2 October 1865, WCP4906 and DCP-LETT-4906.

76. C. R. Darwin to A. R. Wallace, 5 March 1869, WCP6642 and DCP-LETT-6642.

77. The full concluding statement reads as follows:

> Such, we believe, is the direction in which we shall find the true reconciliation of Science with Theology on this most momentous problem. Let us fearlessly admit that the mind of man (itself the living proof of a supreme mind) is able to trace, and to a considerable extent has traced, the laws by means of which the organic no less than the inorganic world has been developed. But let us not shut our eyes to the evidence that an Overruling Intelligence has watched over the action of those laws, so directing variations and so determining their accumulation, as finally to produce an organization sufficiently perfect to admit of, and even to aid in, the indefinite advancement of our mental and moral nature. (Alfred R. Wallace [1869], "Sir Charles Lyell on geological climates and the origin of species" [A. R. Wallace, Review of *Principles of Geology* (10th ed.), 1867–1868, and *Elements of Geology* (6th ed.), 1865 (both by Sir Charles Lyell)], *Quarterly Review* 126:359–94).

78. C. R. Darwin to A. R. Wallace, 14 April 1869, WCP1920 and DCP-LETT-6706.

79. C. R. Darwin to A. R. Wallace, 27 March 1869, WCP6684 and DCP-LETT-6684.

Chapter 12: A Tale of Two Wallaces?

1. See Charles Smith's illuminating discussions of this subject in Charles H. Smith (1992, 1999), "Alfred Russel Wallace on Spiritualism, Man, and Evolution: An Analytical Essay," https:// people.wku.edu/charles.smith/essays/ARWPAMPH.htm; Charles H. Smith (2008), "Wallace, spiritualism, and beyond: 'Change' or 'no change'?," pp. 391–423, in Charles H. Smith and George Beccaloni, eds., *Natural Selection and Beyond: The Intellectual Legacy of Alfred Russel Wallace* (Oxford: Oxford University Press; Charles H. Smith (2019), "Wallace and the 'preternormal,'" pp. 41–66, in Charles H. Smith, James T. Costa, and David Collard, eds., *An Alfred Russel Wallace Companion* (Chicago: University of Chicago Press).

2. Darwin discussed the origin of complex organs (including of vision and flight), evolutionary transitions, and the absence of transitional forms in the fossil record in chapter 6 of *Origin*, "Difficulties on theory." The evolution of bees' cells is treated in chapter 7, "Instinct." For explanatory notes on the key points made by Darwin in these chapters, see James T. Costa (2009), *The Annotated Origin: A Facsimile of the First Edition of "On the Origin of Species"* (Cambridge, MA: Harvard University Press).

3. The quote appears on p. 392 of Wallace's paper in the April 1869 issue of the *Quarterly Review*, published anonymously but referred to in *My Life*, 1:406: Alfred R. Wallace (1869), "Sir Charles Lyell on geological climates and the 'Origin of Species'" [book review], *Quarterly*

Review 126:359–94. Regarding Darwin scoring the passage in his copy of the paper, see James Marchant, ed. (1916), *Alfred Russel Wallace: Letters and Reminiscences* (London: Cassell), 1:240.

4. A. R. Wallace to C. R. Darwin, 18 April 1869, WCP1921 and DCP-LETT-6703. To Wallace's point about respected figures giving credence to spiritualism, historian Malcolm Jay Kottler expressed this best in the opening sentence of his fine 1974 paper "Alfred Russel Wallace, the origin of man, and spiritualism:" "It has long been forgotten, ignored, or perhaps never known by historians of science that in the second half of the nineteenth century a considerable number of renowned scientists were favorably disposed toward such psychical phenomena as telepathy, clairvoyance, precognition, levitation, slate writing, spirit communication, spirit materialization, and spirit photography," *Isis* 65(2):145. See also the excellent overview by Richard J. Noakes (2004), "Spiritualism, science, and the supernatural in mid-Victorian Britain," pp. 23–43, in Nicola Bown, Carolyn Burdett, and Pamela Thurschwell, eds., *The Victorian Supernatural* (Cambridge: Cambridge University Press).

5. C. Lyell to C. R. Darwin, 5 May 1869, DCP-LETT-6728.

6. I am far from the first to portray these seemingly contradictory sides of Wallace as "two Wallaces"—indeed, the first may have been the editors of *The Lancet*, in the 23 September 1876 issue (no. 2769, pp. 431–33): "Side by side with acumen we find obtusity; a vigorous and exacting judgment compounded with impressibility amounting to a credulous abnegation of intelligence." Zoologist George Romanes also expressed this Wallacean duality in a swipe at Wallace's 1889 book *Darwinism.* See Gareth Nelson (2008), "The two Wallaces then and now," *Linnean* special issue no. 9 (Survival of the Fittest):25–34. But as Wallace scholar Charles Smith has pointed out, there never were "two Wallaces," just two sets of understandings of him.

7. Alfred R. Wallace (1870), *Contributions to the Theory of Natural Selection. A Series of Essays* (London: Macmillan), p. 359.

8. H. W. Bates to C. R. Darwin, 20 May 1870, DCP-LETT-7197.

9. Alfred R. Wallace (1870), "Government aid to science" [Letters], *Nature* 1:288–89 (13 January 1870) and 1:315 (20 January 1870).

10. See, for example, Richard H. Grove (1996), *Green Imperialism: Colonial Expansion, Tropical Island Edens and the Origins of Environmentalism, 1600–1860* (Cambridge: Cambridge University Press); Lucile H. Brockway (2002), *Science and Colonial Expansion: The Role of the British Royal Botanic Gardens* (New Haven, CT: Yale University Press).

11. For an illuminating case study on the professionalization of science in the nineteenth century, see Soraya de Chadarevian (1996), "Laboratory science versus country-house experiments: The controversy between Julius Sachs and Charles Darwin," *British Journal for the History of Science* 29(1):17–41.

12. The X Club membership included, besides Huxley, surgeon George Busk; chemist Sir Edward Frankland; mathematician Thomas Archer Hirst; botanist and Kew Gardens director Joseph Dalton Hooker; naturalist, archaeologist, and future member of Parliament John Lubbock; philosopher Herbert Spencer; mathematician and linguist William Spottiswoode; and the Irish physicist John Tyndall. The definitive history of the X Club is provided by Ruth Barton (1998), *The X Club: Power and Authority in Victorian Science* (Chicago: University of Chicago Press).

13. Prominent examples include the cases of John William Colenso, Bishop of Natal, who was deposed in 1863 for his works of biblical criticism contrary to church doctrine, and liberal Anglican theologians Henry Bristow Wilson and Rowland Williams, tried for heresy for

contributing to the 1860 *Essays and Reviews*, a collection of essays challenging church authority and the conservative interpretation of biblical history. Being published just four months after *Origin* was too close for comfort, the volume not to be tolerated by church authorities. Bishop Samuel Wilberforce rose to the occasion with a scathing review, appearing anonymously in 1861 ([Review of] *Essays and Reviews*," *Quarterly Review* 109:248–301). An overview of the *Essays and Reviews* controversy and the involvement of the X Club in its defense is given in vol. 9, app. 6 of the *Correspondence of Charles Darwin* (F. Burkhardt, J. A. Secord, et al., eds. [Cambridge: Cambridge University Press]). See also W. H. Brock and R. M. Macleod (1976), "The scientists' declaration: Reflexions on science and belief in the wake of *Essays and Reviews*, 1864–5," *British Journal for the History of Science* 9:39–66; Ieuan Ellis (1980), *Seven against Christ: A Study of "Essays and Reviews,"* Studies in the History of Christian Thought, no 23 (Leiden, Netherlands: E. J. Brill); Jeff Guy (1983), *The Heretic: A Study of the Life of John William Colenso, 1814–1883* (Pietermaritzberg, SA: University of Natal Press; Johannesburg, SA: Ravan Press).

14. Alfred R. Wallace (1870), [Letters on government aid to science], *Nature* 1(11, 13):288–89 (January) and 12(20):315 (January); see also the editorial "Government aid to science," *Nature* 1(11, 13):279–80 (January).

15. Alfred R. Wallace (1872), "The last attack on Darwinism," *Nature* 6:237–39 (25 July 1872). Bree's response was published in the 1 August 1872 issue of *Nature*: Charles R. Bree (1872), "Bree on Darwinism," *Nature* 6:260, prompting Darwin's letter: Charles R. Darwin (1872), "Bree on Darwinism," *Nature* 6:279 (8 August). See also C. R. Darwin to A. R. Wallace, 27 July 1872, WCP1952 and 3 August 1872, WCP4645; A. R. Wallace to C. R. Darwin, 4 August 1872, WCP1953 and DCP-LETT-8450.

16. As he was preparing his review of Bastian's book, Wallace wrote Darwin: "I am now reviewing a much more important book and one that, if I mistake not, will really compel you sooner or later to modify some of your views, though it will not at all affect the main doctrine of Natural Selection as applied to the higher animals. I allude, of course, to Bastian's 'Beginnings of Life'. . . . My first notice of it will I think appear in 'Nature' next week" (A. R. Wallace to C. R. Darwin, 4 August 1872, WCP1953). Wallace's long review was published in two parts: Alfred R. Wallace (1872), [Review of *The Beginnings of Life: Being Some Account of the Nature, Modes of Origin, and Transformations of Lower Organisms*, by H. Charlton Bastian, 1872], pt. 1, *Nature* 6:284–87 (8 August); pt. 2, *Nature* 6:299–303 (15 August).

17. A. Dohrn to C. R. Darwin, 21 August 1872, DCP-LETT-8481.

18. *My Life*, 2:365.

19. Alfred R. Wallace (1871), "Reply to Mr. Hampden's charges against Mr. Wallace" [Privately printed pamphlet responding to John Hampden], Alfred Russel Wallace Page, https://people.wku.edu/charles.smith/wallace/S202.htm.

20. Wallace's account of "Hampden and the flat-earth" debacle is given in *My Life*, 2:365–76. For a comprehensive treatment of the flat Earth movement and the Hampden affair, see Christine Garwood (2007), *Flat Earth: The History of an Infamous Idea* (London: Macmillan). A brief overview is found in Christine Garwood (2001), "Alfred Russel Wallace and the flat earth controversy," *Endeavour* 25(4):139–43. See also Wallace's initial letter to Hampden: A. R. Wallace to J. Hampden, 15 January 1870, WCP4989. This and other letters from Wallace were published in a pamphlet by Hampden titled *Is Water Level or Convex After All? The Bedford Canal Swindle Detected and Exposed, Etc.* (Swindon: Alfred Bull, 1870). Additional letters from Wallace

regarding the Bedford Canal experiment and Hampden can also be found at Charles Smith's Alfred Russel Wallace Page, http://people.wku.edu/charles.smith/index1.htm, papers S162, S163, S163a, S179aa, S220a, S220b, S228a, S248a, S248b, S248ab, S248ad, and S252c.

21. Wallace continued: "The baby weighs more than Bertie did, & seems to be about an inch taller, and decidedly better looking than Bertie was at the same age." A. R. Wallace to F. Sims, 31 December 1871, WCP397.

22. In her excellent history *On Exhibit: Victorians and Their Museums* (2000, Charlottesville: University of Virginia Press), historian Barbara J. Black writes:

> At no point did museology more closely resemble philanthropy than when the South Kensington's continuous renovations literally sent its discarded parts to be reassembled in London's poorest district. . . . On 24 June 1872 the South Kensington opened in East London, rechristened the Bethnal Green Branch. In its new working-class milieu, the museum competed directly with the public house by offering evening hours and specially targeted exhibits that provided "an excellent antidote" to the "peculiar temptations" of the bank holiday. Curious as to the institution's effect, one newspaper sent [an observer] . . . to study the poor in their museum and offer reports as the following: "There would be hope for the British workman if he took to collecting." As the museum's guidebooks indicate, the focus was on both the relevant and the improved. Thus exhibits featured the principal local trades of weaving and furniture making, and the purpose of the Food Collection, the museum's most popular exhibit, was "to show the nature and sources of the food which we daily use" and to answer the question "what are the substances or elements which, together, constitute my body?" (p. 33)

23. A. R. Wallace to C. R. Darwin, 31 August 1872, WCP1955 and DCP-LETT-8498; see also H. Cole to C. Lyell, 3 July 1872, WCP2286.

24. A. R. Wallace to F. Galton, 15 December 1868, WCP4664; see also Alfred R. Wallace (1870), "Report to the council, by the examiner in physical geography for 1870," *Proceedings of the Royal Geographical Society of London* 14(3):255–56.

25. *My Life*, 2:407.

26. See *My Life*, 2:407–16, for Wallace's examples of absurd student exam answers and 2:417 for his critique of the educational system.

27. George Beccaloni has rightly pointed out the irony that this building should be listed for its architectural merits with no regard whatsoever for its illustrious builder and owner. In 2002 the Wallace Fund, in cooperation with the Thurrock Local History Society, the Heritage Forum, and the Thurrock Council, unveiled a plaque on the house commemorating Wallace. For more information about The Dell, see Beccaloni's blog post at https://wallacefund.myspecies.info /node/1292/revisions/3912/view, Wallace's account in *My Life* (2:91–93), and pp. 30–33 in George Beccaloni (2008), "Homes sweet homes: A biographical tour of Wallace's many places of residence," pp. 7–43, in Smith and Beccaloni, *Natural Selection and Beyond.*

28. A. R. Wallace to C. R. Darwin, 24 November 1870, WCP1937 and DCP-LETT-7382.

29. A. R. Wallace to C. R. Darwin, 24 November 1870, WCP1937 and DCP-LETT-7382.

30. In 1873 he complained to Darwin that "the last three months I have been living in a perpetual hurricane, for my house is fully exposed to the South west & the wind howls around it at night terrifically." Postscript to A. R. Wallace to C. R. Darwin, 14 January 1873, WCP4108 and DCP-LETT-8736.

31. Alfred R. Wallace (1874), "A defence of modern spiritualism," pts. 1 and 2, *Fortnightly Review* 15 (n.s.):630–57 (1 May 1874) and 785–807 (1 June 1874).

32. For example, Wallace described his plan for the book's structure in a letter to Newton, sending a copy of his manuscript for review. Newton was happy to oblige and go over Wallace's "grand work." A. R. Wallace to A. Newton, 14 February 1875, WCP4036; A. Newton to A. R. Wallace, 15 March 1875, WCP2312.

33. A. R. Wallace to C. R. Darwin, 7 November 1875, WCP1965, and DCP-LETT-10247.

34. C. R. Darwin to A. R. Wallace, 5 June 1876, WCP1966 and DCP-LETT-10531; T. H. Huxley to A. R. Wallace, 4 October 1876, WCP2342.

35. A. R. Wallace to C. R. Darwin, 7 June 1876, WCP1967 and DCP-LETT-10535.

36. A. R. Wallace to H. W. Bates, 10 November 1876, WCP3565; this important paper was read at the 25 June 1877 meeting of the Royal Geographical Society, the last meeting before the summer recess, and published later that year: Alfred R. Wallace (1877), "The comparative antiquity of continents, as indicated by the distribution of living and extinct animals," *Proceedings of the Royal Geographical Society* 21(6):505–34.

37. Alfred R. Wallace (1876), [Letter dated 18 September 1876, Glasgow; one of several printed as "A Spirit Medium"], *Times* (London), no. 28738:4f (19 September).

38. For a detailed treatment of the Slade trial and Wallace's involvement in it, see Ross Slotten (2004), *The Heretic in Darwin's Court: The Life of Alfred Russel Wallace* (New York: Columbia University Press), pp. 337–47. See also pp. 98–105 in Richard Milner (2008), "Charles Darwin: Ghostbuster, muse and magistrate," *Linnean*, special issue no. 9 (Survival of the Fittest):97–117.

39. Legislation.gov.uk, "Vagrancy Act 1824," https://www.legislation.gov.uk/ukpga/Geo4/5/83/made.

40. A transcription of Wallace's testimony was published in the *Spiritualist* for 3 November 1876: "Evidence of Mr. A. R. Wallace, President of the Biological Section of the British Association for the Advancement of Science," part of an article titled "Evidence in defence of Dr. Slade," *Spiritualist* 9(14):161, 164 (160–161, 164–165).

41. Lankester revealed Darwin's role in an 1896 recollection: E. R. Lankester (1896), "Charles Robert Darwin," pp. 4835–93, in C. D. Warner, ed., *Library of the World's Best Literature Ancient and Modern* (New York: R. S. Peale and J. A. Hill), 2:4391.

42. Huxley's first séance was at Darwin's prompting. Darwin's brother Erasmus hosted a séance in London in January 1874. Charles excused himself early on, but those who stayed included his wife, Emma, and son George Darwin, his brother-in-law Hensleigh Wedgwood and his wife, novelist George Eliot (Mary Anne Evans), George Henry Lewes, and Huxley—who later provided Charles with a detailed report ("Report of Séance," 27 January 1874, DCP-LETT-9256). Hensleigh later became a committed spiritualist and authored several articles for the *Journal of the Society for Psychical Research*. For accounts of the January 1874 séance and the interest of Hensleigh Wedgwood and other family members in spiritualism, see 2:216–17 in Henrietta E. Litchfield, ed. (1915), *Emma Darwin: A Century of Family Letters, 1792–1896*, 2 vols. (London: John Murray); pp. 404–6 in Janet Browne (2002), *Charles Darwin: The Power of Place* (London: Pimlico); and pp. 305, 324–25 in Barbara Wedgwood and Hensleigh Wedgwood (1980), *The Wedgwood Circle, 1730–1897: Four Generations of a Family and Their Friends* (London: Studio Vista).

43. Ross Slotten first suggested this in his excellent 2004 Wallace biography, *Alfred Russel Wallace: The Heretic in Darwin's Court*, pp. 348–49, citing letters from William Barrett, *Spectator*, 28 October 1876, pp. 1343–44 and William Crookes, *Nature*, 1 November 1877, pp. 7–8.

44. Alfred R. Wallace (1877), [Letter to the editor: "Mr. Wallace and Reichenbach's Odyle"], *Nature* 17:8 (1 November 1877).

45. Paul Sowan and Jean Byatt, in their 1974 paper on Wallace's membership in the Croydon Microscopical and Natural History Club, related how Wallace teamed up with Rev. E. M. Geldart and Dr. Alfred Carpenter to amend the membership rules to allow the attendance of ladies. An initial motion by Geldart having failed, at the 18 February 1880 meeting Wallace proposed a compromise addition to the rules: "That the reader of a paper be allowed the privilege of having lady visitors introduced on the occasion when his paper is read, on announcing his wish at the previous meeting." Carpenter seconded. The motion was much debated but then roundly voted down. "Thus Wallace the Socialist, Carpenter the Liberal, and Geldart, failed in their early attempts to introduce ladies into the Society, even as visitors" (p. 90). The first women were granted membership seventeen years later, in 1897, and the first woman was elected a member of the council in 1913, the year of Wallace's death. See P. W. Sowan and J. I. Byatt (1974), "Alfred Russel Wallace [1823–1913]: His residence in Croydon [1878–1881] and his membership of the Croyden Microscopical and Natural History Club," *Proceedings of the Croydon Natural History and Scientific Society* 15:83–97.

46. See the definitive translation and illuminating analysis of this watershed document provided by Stephen T. Jackson and Sylvie Romanowski (2009), *Essay on the Geography of Plants* (Chicago: University of Chicago Press).

47. A. R. Wallace to H. W. Bates, 28 December 1845, WCP346.

48. Alexander von Humboldt (1849), *Cosmos: Sketch of a Physical Description of the Universe*, 6th ed. (London: Longman, Brown, Green, and Longmans; John Murray), 1:48. For treatments of Humboldt's influence on Wallace, see Charles H. Smith (2013), "Early Humboldtian influences on Alfred Russel Wallace's scheme of nature," presented at "Alfred Russel Wallace and His Legacy," Royal Society of London meeting of 21 October 2013 (DLPS Faculty Publications, Paper 73, https://digitalcommons.wku.edu/dlps_fac_pub/73/) and "The early evolution of Wallace as a thinker," pp. 11–40, in Smith, Costa, and Collard, *An Alfred Russel Wallace Companion*. Treatments of Humboldt's holistic environmental vision can be found in Anne Buttimer (2004), "Poetics, aesthetics and Humboldtean science," pp. 63–78, in W. Gamerith, P. Messerli, P. Mausberger, and H. Wanner, eds., *Alpenwelt—Geburgswelten, Inseln, Brucken, Grenzen: Tagungsbericht und Wissenchaftsliche Abhandlungen, 54* (Bern, Switzerland: Geographisches Institut, University of Bern); Aaron Sachs (2006), *The Humboldt Current: Nineteenth-Century Exploration and the Roots of American Environmentalism* (New York: Viking); Laura Dassow Walls (2009), *The Passage to Cosmos: Alexander von Humboldt and the Shaping of America* (Chicago: University of Chicago Press); Andrea Wulf (2015), *The Invention of Nature: Alexander von Humboldt's New World* (New York: Alfred A. Knopf).

49. *Geographical Distribution of Animals*, 1:44.

50. Quotes are taken from Alexander von Humboldt and Aimé Bonpland (1825), *Personal Narrative of Travels to the Equinoctial Regions of the New Continent, During the Years 1799–1804*, 2nd ed. (London: Longman, Hurst, Rees, Orme, Brown, & Green), 4:143; Wallace, *Geographical Distribution of Animals*, 1:200. Humboldt also discussed the "threefold manner" in which forests

moderate temperature (cooling shade, evaporation, and radiation) in *Aspects of Nature* (1849, 1:126–28).

51. *Tropical Nature*, pp. 19–21.

52. See Mark Lomolino's incisive discussion of Wallace as a foundational thinker for modern conservation biology: Mark Lomolino (2019), "Wallace at the foundations of biogeography and the frontiers of conservation biology," pp. 341–55, in Smith, Costa, and Collard, *An Alfred Russel Wallace Companion*.

53. Technologically driven human expansion and its attendant environmental impacts have been occurring since time immemorial, but the scope and scale of these impacts, and their trajectory, were glaringly obvious by the mid-nineteenth century, prompting some like Marsh and Wallace to raise an alarm. Unlike Wallace, Darwin was not one to get involved in social issues, but lending his support to campaigns for certain conservation measures was one of the few exceptions. One prominent example is Darwin's signing on to a petition to the governor of Mauritius to protect the last remaining giant tortoises of the Indian Ocean. Once found on several islands of the Seychelles, these tortoises were by the mid-1870s restricted to Aldabra Atoll, near Mauritius, and threatened by continued habitat destruction. The petition had its intended effect, and the Aldabra giant tortoise (*Aldabrachelys gigantea*) was rescued from the brink of extinction. See C. R. Darwin et al. (1875), [Memorial to A. H. Gordon, Governor of Mauritius, requesting the protection of the giant tortoise on Aldabra], *Transactions of the Royal Society of Arts and Sciences of Mauritius*, n.s., 8:106–9.

54. See the excellent treatment of Wallace and Darwin's yearslong debate over sexual selection by Malcolm J. Kottler (1980), "Darwin, Wallace, and the origin of sexual dimorphism," *Proceedings of the American Philosophical Society* 124(3):203–26.

55. Table 1 in R. L. Layton (1985), "Recreation, management and landscape in Epping Forest: c. 1800–1984," *Field Studies* 6:269–90.

56. Layton, "Recreation, management and landscape," pp. 269–90.

57. For example: A. R. Wallace to J. D. Hooker, 27 August 1878, WCP3812; A. R. Wallace to C. R. Darwin, 14 September 1878, WCP1977 and DCP-LETT-11693; C. R. Darwin to A. R. Wallace, 16 September 1878, WCP1978 and DCP-LETT-11695; A. R. Wallace to A. Macmillan, 27 September 1878, WCP3374; W. Caruthers to A. R. Wallace, 18 October 1878, WCP2382.

58. Alfred R. Wallace (1878), "Epping Forest," *Fortnightly Review*, n.s., 24, o.s., 30:628–45.

59. A. R. Wallace to W. Caruthers, 9 December 1879, WCP644.

60. The Arboretum at Kew Gardens features biogeographical collections including temperate East Asia (central and western China, South Korea, Japan, and Taiwan), Europe (including the Mediterranean), North America, Vietnam, and the Caucasus. See https://www.kew.org/kew-gardens/plants/arboretum. Many botanical gardens today have embraced the educational potential of their collections, often featuring didactic gardens or plant groupings with approaches from the taxonomic or biogeographical to climatic or thematic (e.g., pollination, evolution).

61. For a thorough account of the matter of Wallace's pension and Darwin's role in securing it, see Ralph Colp Jr. (1992), "'I will gladly do my best': How Charles Darwin obtained a Civil List pension for Alfred Russel Wallace," *Isis* 83(1):2–26.

62. A. B. Buckley to C. R. Darwin, 16 December 1879, WCP6777 and DCP-LETT-12358; C. R. Darwin to A. B. Buckley, 17 December 1879, DCP-LETT-12361; C. R. Darwin to J. D. Hooker, 17

December 1879, DCP-LETT-12360; J. D. Hooker to C. R. Darwin, 18 December 1879, WCP5307 and DCP-LETT-12362; C. R. Darwin to J. D. Hooker, 19 December 1879, DCP-LETT-12363.

63. Roy MacLeod (1970), "Science and the Civil List, 1824–1914," *Technology and Society* 6:47–55; see also "Civil List pensions" in the January and April 1871 *Quarterly Review* 130:407–31.

64. C. R. Darwin to A. R. Wallace, 5 January 1880, WCP1980 and DCP-LETT-12401.

65. See mammalogist Lawrence R. Heaney's fine overview in his introduction to the University of Chicago Press facsimile reprint: Lawrence R. Heaney (2013), "Introduction and Commentary," pp. xi–lxxi, in Alfred R. Wallace (1880, 2013), *Island Life, or, the Phenomena and Causes of Insular Faunas and Floras, including a Revision and Attempted Solution of the Problem of Geological Climates* (Chicago: University of Chicago Press).

66. See the section "How new species arise from a variable species," pp. 59–60, in Wallace, *Island Life.*

67. A. R. Wallace to C. R. Darwin, 18 April 1869, WCP1921.

68. As Heaney (p. xxix) aptly notes: "The example that Wallace set in Part I of *Island Life* of broad, rigorous synthesis came to some incorrect conclusions, but the questions he raised and the framework he established remain a large part of the foundation of evolutionary biogeography today."

69. As Heaney points out in his introduction and commentary (see note 64), there is a third island class that Wallace and Darwin were unaware of: island arcs that develop in connection with plate subduction and multiple volcanic plumes, with far more complex histories than either oceanic or land-bridge islands. Without realizing it Wallace was intimately familiar with this type of island, as they are exceedingly common in the tectonically dynamic Malay Archipelago.

70. For the definitive treatment of Wallace's evolving view of land bridges and dispersal, see Martin Fichman (1977), "Wallace, zoogeography, and the problem of land bridges," *Journal of the History of Biology* 10(1):45–63.

71. Wallace himself summed up the grand sweep of his treatment (*Island Life*, pp. 511–12):

> I trust that the reader who has followed me throughout will be imbued with the conviction that ever presses upon myself, of the complete interdependence of organic and inorganic nature. Not only does the marvellous structure of each organized being involve the whole past history of the earth, but such apparently unimportant facts as the presence of certain types of plants or animals in one island rather than in another are now shown to be dependent on the long series of past geological changes; on those marvellous astronomical revolutions which cause a periodic variation of terrestrial climates; on the apparently fortuitous action of storms and currents in the conveyance of germs; and on the endlessly varied actions and reactions of organized beings on each other. And although these various causes are far too complex in their combined action to enable us to follow them out in the case of any one species, yet their broad results are clearly recognizable.

72. C. R. Darwin to A. R. Wallace, 3 November 1880, WCP1651 and DCP-LETT-12791; J. D. Hooker to C. R. Darwin, 22 November 1880, WCP5297 and DCP-LETT-12838; C. R. Darwin to J. D. Hooker, 23 November 1880, WCP5294 and DCP-LETT-12841.

73. J. D. Hooker to A. R. Wallace, 24 August 1880, WCP1511.

74. C. R. Darwin (1880), [Memorial of A. R. Wallace for a Civil List Pension], CUL-DAR91.95-98, Darwin Online, http://darwin-online.org.uk/manuscripts.html. In addition to Darwin, the memorial was signed by George James Allman, Henry Walter Bates, Henry Austin Bruce (home secretary), William Henry Flower, Albert Günther, Joseph Dalton Hooker, Thomas Henry Huxley, John Lubbock (a naturalist as well as a member of Parliament), Andrew Crombie Ramsay, Philip Lutley Sclater, and William Spottiswoode.

75. See C. R. Darwin to A. B. Buckley, 31 October 1880, WCP7141 and DCP-LETT-12785; C. R. Darwin to W. E. Gladstone, 4 January 1881, DCP-LETT-12975 (cover letter with memorial); C R. Darwin to A. B. Buckley, 4 January 1881, WCP7155 and DCP-LETT-12977; C. R. Darwin to A. R. Wallace, 7 January 1881, WCP1988 and DCP-LETT-12985; W. Gladstone to C. R. Darwin, 6 January 1881, WCP7156 and DCP-LETT-12981; C. R. Darwin to T. H. Huxley, 7 January 1881, WCP3769 and DCP-LETT-12986; A. R. Wallace to C. R. Darwin, 8 January 1881, WCP1989 and 29 January 1881, WCP1991 and DCP-LETT-13033.

Chapter 13: A Socially Engaged Scientist

1. A. R. Wallace to R. Meldola, 5 May 1881, WCP4605.

2. Nutwood Cottage was demolished in 1970, soon after which a rather uninspired terrace of attached brick units was built—also called Nutwood despite nearly all the nut-bearing trees having been removed in the process.

3. *My Life*, 2:103.

4. Alfred R. Wallace (1880), "How to nationalize the land: A radical solution of the Irish land problem," *Contemporary Review* 38:716–36.

5. *Malay Archipelago*, pp. 596–98.

6. John Stuart Mill (1871), *Programme of the Land Tenure Reform Association, with an Explanatory Statement* (London: Longmans, Green, Reader, and Dyer). For an overview of Mill's involvement in land reform, see David E. Martin (1981), *John Stuart Mill and the Land Question*, Occasional Papers in Economic and Social History, no. 9 (Hull, UK: University of Hull).

7. J. S. Mill to A. R. Wallace: July 1870, WCP6311; April 1871, WCP 6314; 30 April 1871, WCP6807; Wallace's proposal was adopted as item X of the platform of the Land Tenure Reform Association: "To obtain for the State the power to take possession (with a view to their preservation) of all Natural Objects, or Artificial Constructions attached to the soil, which are of historical, scientific, or artistic interest, together with so much of the surrounding land as may be thought necessary; the owners being compensated for the value of the land so taken." In this Wallace anticipated by twenty-four years the establishment of the National Trust, founded in January 1895 "to promote the permanent preservation for the benefit of the Nation of lands and tenements (including buildings) of beauty or historic interest."

8. Only a small percentage of Irish farmers owned their own land, while most were tenants. In 1870, for example, these figures were 3 percent and 97 percent, respectively; Paul Bew (2007), *Ireland: The Politics of Enmity, 1789–2006* (Oxford: Oxford University Press,) p. 568.

9. Wallace, "How to nationalize the land," p. 718.

10. Henry George (1886), *Progress and Poverty*, 4th ed. (Garden City, NY: Doubleday, Page), pp. ix–x.

11. *Progress and Poverty* inspired the single-tax movement later called Georgism or Geoism, a central tenet of which is the idea of a tax based on the value of land rather than wages, targeting unearned income (ground rent) monopolized by landowners. The influence of George and *Progress and Poverty* in the late nineteenth century cannot be overstated, inspiring myriad single-tax advocacy groups, political parties, and even experimental communities based on its principles, such as Fairhope, Alabama, and Arden, Delaware. For an especially illuminating treatment of George and his life and thought, see Christopher England (2015), "Land and liberty: Henry George, the single tax movement, and the origins of 20th century liberalism" (PhD diss., Georgetown University).

12. A. R. Wallace to C. R. Darwin, 9 July 1881, WCP1992 and DCP-LETT-13238; C. R. Darwin to A. R. Wallace, 12 July 1881, WCP1993 and DCP-LETT-13243.

13. These examples are Wallace's—see Alfred R. Wallace (1881), [Review of *Anthropology: An Introduction to the Study of Man and Civilisation* by Edward B. Tylor], *Nature* 24:242–45. Wallace later elaborated on the theory in an article in the *Fortnightly Review* (1895): "The expressiveness of Speech, Or Mouth-gesture as a Factor in the Origin of Language" (n.s., 58, o.s., 64:528–43). Wallace considered the concept an important extension of the onomatopoeic and mimetic theory of vocal-language origin. Most modern linguists would agree that many *words* arise from onomatopoeia and sound mimicry but not language per se. Incidentally, knowing of William Gladstone's scholarly interest in the language of the Homerian epics, Wallace sent him a copy of his 1895 article. Gladstone replied with thanks and examples of his own connecting sound and sense in language; see W. E. Gladstone to A. R. Wallace, 18 October 1895, WCP5630; A. R. Wallace to W. E. Gladstone, 22 October 1895, WCP1435.

14. A. R. Wallace to C. R. Darwin, 18 October 1881, WCP1994. Darwin's book *The Formation of Vegetable Mould, through the Action of Worms, with Observations on Their Habits*, was published by John Murray in October 1881.

15. An active member of the British scientific scene, the energetic and multi-talented Meldola was a friend and correspondent of both Wallace and Darwin. As Wallace was the recipient of numerous awards, it is fitting that one is annually bestowed in his honor: the Meldola Medal (since 2008 the Harrison-Meldola Memorial Prizes) of the Royal Society of Chemistry. See Anthony Travis's excellent treatment of Meldola's life and many scientific contributions: Anthony S. Travis (2010), "Raphael Meldola and the nineteenth-century neo-Darwinians," *Journal for General Philosophy of Science* 41(1):143–72.

16. Alfred R. Wallace (1882), "Dr. Fritz Müller on some difficult cases of mimicry," *Nature* 26:86–87 (25 May). Wallace's book reviews also published in May 1882 issues of *Nature* include [Review of *Rhopalocera Malayana: A Description of the Butterflies of the Malay Peninsula* by William L. Distant], *Nature* 26:6–7 (4 May) and [Review of *Studies in the Theory of Descent* pt. III by August Weismann], *Nature* 26:52–53 (18 May).

17. Distinguished historian and Darwin biographer Janet Browne gives a moving account of Darwin's final days and the aftermath in her fine 2002 biography *Charles Darwin: The Power of Place* (New York: Alfred A. Knopf), pp. 491–97.

18. G. Darwin to T. H. Huxley, [?]22 April 1882, WCP3757.

19. A. R. Wallace to G. Silk, 1 September 1860, WCP373.

20. *My Life*, 2:89.

21. Alfred R. Wallace (1882), *Land Nationalisation; Its Necessity and Its Aims; Being a Comparison of the System of Landlord and Tenant with That of Occupying Ownership in Their Influence*

on the Well-Being of the People (London: Trübner); Wallace quotes Hugh Miller on the Sutherland evictions on pp. 60–61 of *Land Nationalization*. See Sir Thomas Martin Devine's authoritative works on this chapter in Scottish history: *Clanship to Crofter's War: The Social Transformation of the Scottish Highlands* (2013, Manchester: Manchester University Press) and *The Scottish Clearances: A History of the Dispossessed, 1600–1900* (2018, London: Allen Lane/Penguin).

22. See app. 7, "Correspondence with Mr. A. Russell Wallace," pp. 77–92, in Thomas Sellar (1883), *The Sutherland Evictions of 1814: Former and Recent Statements Respecting Them Examined* (London: Longmans, Green). Wallace's part of the correspondence is also given by Charles Smith, S368b, "Correspondence with Thomas Sellars," Alfred Russel Wallace Page, http://people.wku.edu/charles.smith/wallace/S368B.htm.

23. Alfred R. Wallace (1883), "The 'why' and the 'how' of land nationalisation," parts 1 and 2, *Macmillan's Magazine* 48:357–68 (September 1883: no. 287) and 48:485–93 (October 1883: no. 288).

24. Wallace (1883), "'The why' and the 'how' of land nationalisation," p. 490. Fawcett's article "State socialism and the nationalisation of the land" was written as a chapter for a new edition of his *Manual of Political Economy* (Macmillan, 1883) and published in pamphlet form by Macmillan that same year.

25. Alfred R. Wallace (1885), *Bad Times: An Essay on the Present Depression of Trade, Tracing It to Its Sources in Enormous Foreign Loans, Excessive War Expenditure, the Increase of Speculation and of Millionaires, and the Depopulation of the Rural Districts; with Suggested Remedies* (New York: Macmillan).

26. Alfred R. Wallace (1884), "*Sutherlandia spectabilis*," *Garden* 25:441b (24 May); "*Datura meteloides*," *Garden* 26:352a (25 October).

27. Best known today as "The Eviction," Allingham's poem tells the story of the overnight eviction of forty-seven families (244 men, women, and children) by John George "Black Jack" Adair in Derryveagh, County Donegal, Ireland, in 1861. It was originally part of "Tenants at Will," a poem constituting chapter 7 (pp. 137–52) of Allingham's *Laurence Bloomfield in Ireland: A Modern Poem* (1864, Macmillan).

28. This account is given in H. Allingham and D. Radford, eds. (1907), *William Allingham: A Diary* (London: Macmillan), pp. 329–35.

29. Tennyson's famous 1850 poem *In Memoriam*, an elegy for Arthur Hallam, includes the immortal lines "Tis better to have loved and lost | Than never to have loved at all" (canto 27), and these despairing lines raging against nature in canto 55: "So careful of the type?' but no. | From scarped cliff and quarried stone | She cries, 'A thousand types are gone: | I care for nothing, all shall go.'" He then asks about the fate of humanity—would we "Who trusted God was love indeed | And love Creation's final law— | Tho' Nature, red in tooth and claw | With ravine, shriek'd against his creed— | Who loved, who suffer'd countless ills, | Who battled for the True, the Just, | Be blown about the desert dust, | Or seal'd within the iron hills?"—ending up but dust or fossil.

30. Allingham and Radford, *William Allingham*, p. 339.

31. Donald R. Hopkins (2002), *The Greatest Killer: Smallpox in History* (Chicago: University of Chicago Press); see also the US Centers for Disease Control and Prevention, https://www.cdc.gov/smallpox/history/history.html.

32. Under the 1867 Vaccination Act, evidence of vaccination was to be presented within a week of the birth of a child. If vaccination was not administered within three months, the

parents or guardians were subject to cumulative fines starting at twenty shillings (1 pound, which in 1870 was equivalent to about 109 pounds today).

33. For authoritative treatments of Wallace's involvement in the anti-vaccination movement, see chapter 4 in Charles H. Smith (1991), *Alfred Russel Wallace: An Anthology of His Shorter Writings* (Oxford: Oxford University Press); Martin Fichman and Jennifer E. Keelan (2007), "Resister's logic: The anti-vaccination arguments of Alfred Russel Wallace and their role in the debates over compulsory vaccination in England, 1870–1907," *Studies in History and Philosophy of Biological and Biomedical Sciences* 38:585–607; Martin Fichman (2008), "Alfred Russel Wallace and anti-vaccinationism in the late Victorian cultural context, 1870–1907," pp. 305–319, in Charles H. Smith and George Beccaloni, eds. (2010), *Natural Selection and Beyond: The Intellectual Legacy of Alfred Russel Wallace* (Oxford: Oxford University Press); and pp. 215–27 in Martin Fichman (2019), "Wallace as social critic, sociologist, and societal 'prophet,'" pp. 191–233, in Charles H. Smith, James T. Costa, and David Collard, eds., *An Alfred Russel Wallace Companion* (Chicago: University of Chicago Press). For a brief overview, see Thomas P. Weber (2010), "Alfred Russel Wallace and the antivaccination movement in Victorian England," *Emerging Infectious Diseases* 16(4):664–68.

34. Anti-vaccination leagues gained momentum despite the periodic smallpox epidemics that continued to kill or maim thousands in Britain. Matters came to a head in 1885 in Leicester, the scene of a large joint demonstration of anti-vaccination organizations from many towns and cities across the country. This demonstration led to a Royal Commission on the vaccination mandate, which ultimately recommended dropping the compulsory requirement in the statute and allowing exemption for conscientious objectors. For an excellent comprehensive history of the subject, see Nadja Durbach (2004), *Bodily Matters: The Anti-vaccination Movement in England, 1853–1907* (Durham, NC: Duke University Press). More concise treatments are provided by Stanley Williamson (1984), "Anti-vaccination leagues," *Archives of Disease in Childhood* 59:1195–96; Dorothy Porter and Roy Porter (1988), "The politics of prevention: Anti-vaccinationism and public health in nineteenth-century England," *Medical History* 32:231–52; J. D. Swales (1992), "The Leicester anti-vaccination movement," *Lancet* 340:1019–21; and Robert M. Wolfe and Lisa K. Sharp (2002), "Anti-vaccinationists past and present," *British Medical Journal* 325(7361):430–32.

35. *My Life*, 2:351.

36. Alfred R. Wallace (1883), [Letter of support to the Berne International Anti-vaccination Congress held on 28 September 1883 and following days], *Vaccination Inquirer and Health Review* 5:160.

37. Alfred R. Wallace (1884), [Endorsement of William Tebb's *Compulsory Vaccination in England*], *Vaccination Inquirer and Health Review* 5:235.

38. Alfred R. Wallace (1885), [*To Members of Parliament and Others*] *Forty-Five Years of Registration Statistics, Proving Vaccination to Be Both Useless and Dangerous* (London: E. W. Allen); quote taken from pp. 3 and 36. See also Wallace's "Vaccination judged by its results," *Pall Mall Gazette*, no. 6250:1–2 (24 March 1885).

39. This piece appears as "Alfred Russel Wallace, LL.D." on pp. 103–4 in Andrew Reid, ed. (1885), *Why I Am a Liberal: Being Definitions and Personal Confessions of Faith by the Best Minds of the Liberal Party* (London: Cassell); emphasis in original.

40. In his autobiography (*My Life*, 2:105), Wallace states that in late 1885 he was invited to deliver a series of lectures at the Lowell Institute in Boston. It is clear from his correspondence,

however, that the late 1885 invitation to lecture came from Australia. See A. R. Wallace to A. C. Swinton, 23 December 1885, WCP5714; and O. C. Marsh, 23 January 1886, WCP5360; Alfred R. Wallace (1879), *Australasia*, vol. 1, Stanford's Compendium of Geography and Travel (London: Edward Stanford (2nd ed. 1880, 3rd ed. 1883, 4th ed. 1884, 5th ed. 1888).

41. A. R. Wallace to A. C. Swinton, 23 December 1885, WCP5714.

42. Marsh was embroiled in a yearslong bitter and public rivalry with paleontologist Edward Drinker Cope of the Academy of Natural Sciences of Philadelphia, each of whom went to great lengths to humiliate and undermine the other and, worse, resorted to bribery, theft, and even the spiteful destruction of fossils in the field lest the other find them. This shameful episode in the history of paleontology is known as the Bone Wars. See David Rains Wallace (1999), *The Bonehunters' Revenge: Dinosaurs, Greed, and the Greatest Scientific Feud of the Gilded Age* (New York: Houghton Mifflin); and Mark Jaffe (2000), *The Gilded Dinosaur: The Fossil War between E. D. Cope and O. C. Marsh and the Rise of American Science* (New York: Crown).

43. A. R. Wallace to O. C. Marsh, 23 January 1886, WCP5360; C. W. Ernst to D. C. Gilman, 2 February 1886, WCP4853; O. C. Marsh to D. C. Gilman, 12 February 1886, WCP4854.

44. Mentioned in A. R. Wallace to E. W. Gosse, 4 March 1886, WCP4855; although evidently not a spiritualist himself, Lowell knew many people who were and likely thought taking part in séances would be an added attraction to Wallace.

45. The Lowell Institute continues to provide a vibrant and diverse menu of lectures and programs in the arts and sciences, many in partnership with public and private cultural institutions of Boston and environs; see its website at http://www.lowellinstitute.org/.

46. "In the long line of eminent men who have lectured on their several specialties for the Lowell Institute may be mentioned, in science, the names of Silliman, Lyell, Agassiz, Gray, Levering, Rogers, Cooke, Wyman, Peirce, Tyndall, Whitney, Newcomb, Ball, Proctor, Young, Langley, Gould, Wallace, Geikie, Dawson, Cross, G. H. Darwin, Farlow, and Goodale." Harriet Knight Smith (1898), *The History of The Lowell Institute* (Boston: Lamson, Wolffe), pp. 30–31.

47. T. H. Huxley to H. N. Martin, 4 March 1886, WCP4855.

48. See James Woods's insightful essay on "Wallace as writer," describing Wallace's "cool, all-seeing eye, that instinct for narrative, the frequent dark humour—as crisp and flavourful as a cool Sancerre," *Current Biology* 23(24):R1072–73.

49. Franklin Parker (1997), *George Peabody: A Biography* (Nashville: Vanderbilt University Press), p. 197.

50. "All this took a great deal of time, and the maps and diagrams forming a large package, about six feet long in a waterproof canvas case, caused me much trouble, as some of the railways refused to take it by passenger trains, and I had to send it as goods; and in one case it got delayed nearly a week, and I had to give my lectures with hastily made rough copies from recollection" (*My Life*, 2:106). Note that in his autobiography Wallace omits mention of his initial invitation, from Sydney, Australia, and erroneously states that the invitation to lecture came from the Lowell Institute in late 1885.

51. A. R. Wallace to A. Wallace, 23 October 1886, WCP422.

52. Wallace's North American travels were recorded in a notebook that now resides in the library of the Linnean Society of London. See North American Journal (MS-177), http://linnean-online.org/54016/. This notebook was helpfully transcribed and annotated in *Travel Diary*.

53. *Boston Evening Transcript*, 2 November 1886, p. 4.

54. A. R. Wallace to V. I. Wallace, 2 November 1886, WCP424.

55. *Travel Diary*, p. 22; A. Gray to A. R. Wallace, 13 November 1886, WCP2194.

56. James was skeptical of spiritualism but fascinated by its possibilities, undertaking investigations into mediums, thought transference, and the like. He was a co-founder of the American Society for Psychical Research, inspired by William Barrett, the British researcher whose paper at the Glasgow British Association for the Advancement of Science meeting caused Wallace such headaches back in the 1870s. Finding British academia hostile to the idea of psychical phenomena, Barrett and a group of colleagues relocated to the United States in 1884; Philip P. Wiener (1946), "The evolutionism and pragmaticism of Peirce," *Journal of the History of Ideas* 7(3):321–50, p. 331n26.

57. *Travel Diary*, p. 109; *My Life*, 2:160; see the extensive reprint list for this lecture at Charles Smith's Alfred Russel Wallace Page, paper S398, http://people.wku.edu/charles.smith/wallace/bib2.htm.

58. *Travel Diary*, pp. 23, 85.

59. *Travel Diary*, pp. 87–88.

60. Much more could be said about this fascinating topic, but it is outside our purview here. See, however, the fine treatment by David Dobbs (2005): *Reef Madness: Charles Darwin, Alexander Agassiz, and the Meaning of Coral* (New York: Pantheon).

61. Alfred R. Wallace (1887), "American museums. The Museum of Comparative Zoology, Harvard University," *Fortnightly Review*, n.s., 42, o.s., 48:347–59. In his article Wallace also dinged museums back home with their "interminable series of over-crowded wall-cases" and "old and long-exploded arrangement" of specimens, "often quite at variance with the knowledge of the day as to the affinities of the different groups." As might be expected, some curators took exception to his comments. William Flower, director of the Natural History Museum, took it as an unjust attack on himself and the curators. Wallace apologized for inadvertently hurting anyone's feelings in a pair of letters, one in the 1 November 1887 issue of *Fortnightly Review* and the other in the 6 October 1887 issue of *Nature*.

62. Alfred R. Wallace (1887), "American museums. Museums of American pre-historic archaeology," *Fortnightly Review*, n.s., 42, o.s., 48:665–75.

63. Partly because the native people had been long since dispossessed and displaced, partly because even where they were not displaced they themselves often did not have knowledge of the ancient mound builders (any cultural continuity long since disrupted), and partly due to a form of racism—an Anglocentrism that assumed that the native people now living did not have the capacity or know-how to build the ancient earthworks—the archaeologists blithely excavating away were often knowing or unknowing participants in a great injustice against the native peoples, an injustice that is only in recent years beginning to be addressed.

64. *Travel Diary*, p. 55; in *My Life* (2:129–30) Wallace recounted this episode:

> Another evening I was asked by Dr. T. A. Bland, editor of *The Council Fire*, and friend of the Indians, who had seen the evils of land-speculation in leading to the robbery of land granted as Indian reserves, to give some of his friends a short address, explaining my views on land reform. I note in my journal, "preached on 'Land Nationalization,' talk afterwards." At this time, however, the one subject of private interest everywhere in America was land-speculation, and nobody could see anything bad in it. My ideas, therefore, seemed very wild, and I don't think I made a convert.

65. Cheltenham Ladies' College, opening in 1853, was the first girls' school in Britain to offer a rigorous academic education and University College London the first British institution of higher education to confer degrees on women. Girton College, the first all-women's college at Cambridge, was established in 1869, while in Oxford Somerville College and Lady Margaret Hall were established a decade later, 1879, followed by St. Hugh's in 1886. Others would follow, but it would be decades before young women at Oxford and Cambridge would be able to receive degrees for their efforts (1920 at Oxford, 1948 at Cambridge).

66. *Travel Diary*, pp. 26–27.

67. *Travel Diary*, pp. 87, 93, 95.

68. *My Life*, 2:176.

69. A. R. Wallace to W. Mitten, 10 July 1887, WCP454.

70. *Travel Diary*, p. 24; *My Life*, 2:111.

71. *Travel Diary*, app. 6, "List of plants observed/collected," pp. 255–58.

72. *Travel Diary*, pp. 69–70; *My Life*, 2:117–18.

73. *My Life*, 2:134.

74. *My Life*, 2:135.

75. Salvatore John Manna writes: "The first rail was laid at Brack's Landing on the Moke-lumne River near Woodbridge in April 1882 and headed east, entering Calaveras County in October. John laid out the line and son John Herbert, the Wallace family's newest surveyor, surveyed the town site the railroad created in Calaveras just over the San Joaquin County border. 'A station will be located and the nucleus formed of a town to be named Wallace,' reported the *Lodi Sentinel*, 'in honor of Mr. Wallace, the engineer whose efficient work as surveyor for the company merits this 'signal honor.'" See p. 12n20 in Salvatore John Manna (2008), "A brothers' reunion: Evolution's champion Alfred Russel Wallace and Forty-Niner John Wallace," *California History* 85(4):4–25, 70–71.

76. Notice in the *Daily Alta California*, quoted in *Travel Diary*, p. 105n2.

77. W. P. Gibbons to J. Muir, 23 May 1887; Online Archive of California, Pacific Library Holt-Atherton Special Collections, John Muir Correspondence, 1856–1914, letter muir05_0827-md-1, https://oac.cdlib.org/ark:/13030/kt987038p4/?brand=oac4.

78. *My Life*, 2:162.

79. *My Life*, 2:163.

80. John Muir (1876), "On the post-glacial history of *Sequoia gigantea*," *Proceedings of the American Association for the Advancement of Science* 25:242–52.

81. Quoted on pp. 17 and 21 in Manna, "Brothers' reunion," pp. 4–25, 70–71, citing Michael P. Branch, ed. (2001), *John Muir's Last Journey. South to the Amazon and East to Africa: Unpublished Journals and Selected Correspondence* (Washington, DC: Island Press, 2001), p. 275; *New York Times*, 21 April 1912.

82. Wallace was no isolated influence on Muir among the great naturalists of the time, certainly—Muir spent time in the field with Gray and Hooker, read Darwin as well as Wallace, and was intimately familiar with leading American naturalists of his day besides Gray (e.g., John Torrey, Henry Fairfield Osborn, Jospeh LeConte, and others). See R. M. McDowell (2010), "Biogeography in the life and literature of John Muir: A ceaseless search for pattern," *Journal of Biogeography* 37:1629–36; C. Michael Hall (1993), "John Muir's travels in Australasia, 1903–1904:

Their significance for conservation and environmental thought," chap. 13, in Sally M. Miller, ed., *John Muir: Life and Work* (Albuquerque: University of New Mexico Press).

83. Marcus R. Ross et al. (2010), "Garden of the Gods at Colorado Springs: Paleozoic and Mesozoic sedimentation and tectonics," GSA Field Guide, vol. 18, in Lisa A. Morgan and Steven L. Quane, eds., *Through the Generations: Geological and Anthropogenic Field Excursions in the Rocky Mountains from Modern to Ancient* (Boulder: Geological Society of America).

84. Eastwood later headed to California, eventually (1894) becoming head of the Botany Department at the California Academy of Sciences, a position she held until 1949; *Travel Diary*, p. 233.

85. *Travel Diary*, pp. 140–44.

86. *Travel Diary*, p. 152; 25 April 1887, Greencastle Junction, IN, p. 87; 9 July 1887, Donner Lake, CA, p. 129.

87. *My Life*, 2:171.

88. *Travel Diary*, pp. 83, 100.

89. Quoted from the *Century Magazine* for June 1887; *Travel Diary*, p. 147.

90. *My Life*, 2:200–201.

Chapter 14: Onward and Upward

1. A. R. Wallace to V. I. Wallace, 27 November 1896, WCP278.

2. A. R. Wallace to V. I. Wallace, 16 May 1889, WCP202.

3. Within a few years, in 1899, Wallace was exulting over his orchids and water lilies to Violet: "Ma will have told you all the news and of my superhuman labours (not yet near over) with a huge case—bigger far than your trunk!—full of Indian orchids, *and* another box of blue water lilies! How the potting of the orchids takes 10 hours a day & fetching bushels of sphagnum, how a warm water pond has been made and now it has to be enlarged to hold *more* blue water lilies, so that next holidays you may perhaps sit in the summer house & look down upon the azure flowers of the S. African water lily." A. R. Wallace to V. I. Wallace, 22 May 1899, WCP323.

4. See pp. 35–37 in George Beccaloni (2008), "Homes sweet homes: A biographical tour of Wallace's many places of residence," pp. 7–43, in Charles H. Smith and George Beccaloni, eds., *Natural Selection and Beyond: The Intellectual Legacy of Alfred Russel Wallace* (Oxford: Oxford University Press); pp. 27–28 in Ahren Lester (2014), "Homing In: Alfred Russel Wallace's homes in Britain (1852 to 1913)," *Linnean* 30(2):22–32.

5. H. Spencer to A. R. Wallace, 18 May 1889, WCP2017.

6. Wallace struggled a bit to decide how best to approach his *Popular Sketch of Darwinism*, the original title of *Darwinism*. To Arabella Fisher (formerly Buckley) he wrote: "I really think I shall be able to arrange the whole subject more intelligibly than Darwin did, and simplify it immensely by leaving out the endless discussion of collateral details and difficulties which in the 'Origin of Species' confuse the main issue." A. R. Wallace to A. Fisher, 16 February 1888, WCP5623.

7. For discussions of how and why Darwin structured *On the Origin of Species* as he did, see James T. Costa (2009), *The Annotated Origin: A Facsimile of the First Edition of* On the Origin of Species (Cambridge, MA: Harvard University Press) and (2009), "Darwinian revelation: Tracing the origin and evolution of an idea," *BioScience* 59:886–94.

8. *Darwinism*, p. vii.

9. *Darwinism*, pp. 39–40.

10. *Darwinism*, p. vi.

11. For example, Huxley's 1863 book *Evidence as to Man's Place in Nature* (New York: Appleton), pp. 127–28.

12. A. R. Wallace to C. R. Darwin, 1 March 1868, WCP1889 and DCP-LETT-5966. Actually, Darwin's sons had about as difficult a time as their dad in wrapping their heads around Wallace's argument, as Darwin told Wallace: "I do not feel that I shall grapple with the sterility argument just yet. . . . I have tried once or twice & it has made my stomach feel as if it had been placed in a vice.—Your paper has driven 3 of my children half-mad—One sat up to 12 oclock over it"; C. R. Darwin to A. R. Wallace, 17 March 1868, WCP1891 and DCP-LETT-6018.

13. A. R. Wallace to C. R. Darwin, 24 March 1868, WCP1894 and DCP-LETT-6045. Wallace's comment about isolation is pertinent—in their time Darwin and Wallace downplayed the necessity of isolation, or what is called speciation in *allopatry* today, in favor of what is called *sympatric* speciation (largely competition-driven). Wallace also flirted with "good of the species" thinking in his arguments here, something that Darwin flatly rejected. Both eventually came round to seeing an important role for isolation in speciation, but Wallace also invoked selection acting to increase hybrid sterility where diverging populations do come into contact— the Wallace Effect, or reinforcement.

14. Darwin discussed this at length in chapter 8 of *Origin*; for annotations to Darwin's argument, including his grafting analogy, see Costa, *Annotated Origin*.

15. George J. Romanes (1886), "Physiological selection; an additional suggestion on the origin of species," *Zoological Journal of the Linnean Society* 19:337–411. This is part of a larger problem first pointed out by Scots engineer Fleeming Jenkin, who argued that blending during reproduction would swamp new favorable variations and prevent their spread by natural selection. Thus, interbreeding (hybridization) of potential incipient varieties blends and also prevents continued divergence (speciation). Darwin acknowledged the difficulty, commenting to Hooker that although Jenkin had given him headaches, his critique had been "of more real use to me, than any other Essay or Review." (C. R. Darwin to J. D. Hooker, 16 January 1869, DCP-LETT-6557). Darwin attempted to address Jenkin's critique in the fifth edition of *Origin*. In part his solution was to invoke the isolation of subpopulations.

16. See, for example, Donald R. Fordyke (2020), "Revisiting George Romanes' 'physiological selection' (1886)," *Biological Theory* 15:143–47. Fordyke proposes a "chromosomal" vs. "genic" selection distinction as a modern expression of what Romanes was arguing for—variants presumably arising by chromosomal inversion, or duplication or deletion of some chromosomal region. A challenge for such chromosomal "hopeful monsters" (a term originated by twentieth-century German American geneticist Richard Goldschmidt to describe large mutations) is their rarity, just one of them at a time. In principle new lineages (subspecies, species) may arise in this way at times, but it is considered to be of secondary or tertiary importance relative to other evolutionary processes.

17. A. R. Wallace to R. Meldola, 28 August 1886, WCP4496.

18. The acrimony was exacerbated by Romanes's unseemly potshots at Wallace. Of Wallace's chapter on "Darwinism applied to man" in *Darwinism*, Romanes wrote condescendingly in a review that here "we encounter the Wallace of spiritualism and astrology, the Wallace of

vaccination and the land question, the Wallace of incapacity and absurdity." See p. 831 in George J. Romanes (1890), "Darwin's latest critics," *Nineteenth Century* 27:823–32. Wallace was outraged—not least because he never embraced astrology and also because in any case he viewed the whole attack as hitting below the belt. Unbeknownst to Romanes, while in Canada on his American tour Wallace was shown and allowed to make notes on letters Romanes had written to Darwin and others years before purporting to support spiritualism. This Wallace found to be the very height of hypocrisy after such a vehement personal attack. Wallace informed Romanes that he knew all about his "secret" dalliance with spiritualism, in detail, and encouraged (taunted?) him to come clean with the scientific community. Wallace recounted the episode in *My Life*, 2:309–26.

19. *Darwinism*, pp. 179–80. In fairness to Romanes, he altered his theory of physiological selection over time, emphasizing the role of isolation more in setting the stage for divergence. Although even in this case he perhaps overstated things, suggesting that selection *itself* is isolation, his views on isolation in general are more consonant with modern evolutionary biology; see John E. Lesch (1975), "The role of isolation in evolution: George J. Romanes and John T. Gulick," *Isis* 66(4):483–503.

20. The term "Wallace Effect" was coined by botanist and evolutionary biologist Verne Grant in a 1966 paper, "The selective origin of incompatibility barriers in the genus *Gilia*," *American Naturalist* 100:99–118. The concepts of *reinforcement* and the modern *biological species concept* are often attributed to twentieth-century evolutionary biologists: Russian American geneticist Theodosius Dobzhanzky, in the case of the former, and German American evolutionary biologist Ernst Mayr, in the case of the latter. *Both* concepts were first articulated by Wallace. See Norman A. Johnson (2008), "Direct selection for reproductive isolation: The Wallace Effect and reinforcement," pp. 114–24, and James Mallet (2008), "Wallace and the species concept of the early Darwinians," pp. 102–13, both in Smith and Beccaloni, *Natural Selection and Beyond*.

21. August Weismann (R. Meldola, trans.) (1882), *Studies in the Theory of Descent*, 2 vols. (London: Sampson, Low, Marston, Searle and Rivington).

22. For an overview of the early struggles to make sense of the nature of genetic variation, see James T. Costa (2021), "There is hardly any question in biology of more importance: Charles Darwin and the nature of variation," pp. 25–54, in D. Pfennig, ed., *Phenotypic Plasticity and Evolution: Causes, Consequences, Controversies* (Boca Raton: CRC Press).

23. Darwin was unsure how his postulated gemmules were transmitted, but his language in *Variation in Animals and Plants under Domestication* strongly suggested that they are diffused through the circulatory system. When Galton argued that his blood transfusion experiments refuted blood-borne gemmules, Darwin protested that he never claimed this per se, leading Galton to take him to task over being sent on a "false quest" (see F. Galton [1871], "Experiments in pangenesis, by breeding from rabbits of a pure variety, into whose circulation blood taken from other varieties had previously been largely transfused," *Proceedings of the Royal Society of London* 19:393–410; [1871b], "Pangenesis," *Nature* 4:5–6; C. R. Darwin [1871], "Pangenesis," *Nature* 3:502–3). Galton then experimented with "Siamesed" rats, surgically connecting the rodents to permit a more complete exchange of fluids. This experiment, too, failed to support pangenesis, leading Galton to abandon his cousin's theory. He then developed his own theory of inheritance based on what he termed "stirps" (from Latin *stirpes*, "root"), in which hereditary gemmules, or "germs," collect in cells rather than circulating through the body.

24. Complementary to his crucial insight into the sequestration and continuity of the germ line, in 1887 Weismann recognized the significance of *reduction division* of chromosomes in the production of gametes and its role in mixing up genetic material generation after generation. Weismann's work dovetailed with and built upon an exciting constellation of contemporary discoveries in cell biology in the 1870s and 1880s, including Friedrich Miescher's investigations into the cell nucleus with its phosphorus-rich nucleoproteins (dubbed *nuclein*, first described in 1871), studies of the nature of chromosomes, and research into the processes of mitosis and meiosis by Walther Flemming, Edouard van Beneden, Eduard Strasburger, and others. For a concise overview of this early history, see Ernst Mayr (1982), *The Growth of Biological Thought: Diversity, Evolution, and Inheritance* (Cambridge, MA: Harvard University Press).

25. August Weismann (E. B. Poulton, S. Schönland, and A. E. Shipley, trans.) (1889), *Essays upon Heredity and Kindred Biological Problems* (Oxford: Clarendon Press); A. R. Wallace to E. B. Poulton, 24 October 1888, WCP4354.

26. *Darwinism*, p. 444.

27. *My Life*, 2:22.

28. A. R. Wallace to F. Galton, 3 February 1891, WCP1434; F. Galton to A. R. Wallace, 5 February 1891, WCP1515; F. Galton to A. R. Wallace, 12 February 1891, WCP2442; A. R. Wallace to F. Galton, 7 February 1891, WCP4137.

29. Alfred R. Wallace (1891), "English and American flowers. I," *Fortnightly Review*, n.s. 50, o.s. 56:525–34 (1 October) and "English and American flowers. II. Flowers and forests of the Far West," *Fortnightly Review*, n.s. 50, o.s. 56:796–810 (1 December).

30. A. R. Wallace to E. B. Poulton, 28 May 1889, WCP4366.

31. Poulton recounted his challenge to get Wallace to accept the honorary degree on pp. 31–32 of his obituary notice for Wallace written for the Royal Society of London; see Edward B. Poulton (1924), "Alfred Russel Wallace, 1823–1913," *Proceedings of the Royal Society of London* B 95:1–35.

32. A. R. Wallace to V. I. Wallace, 22 November 1890, WCP215.

33. Anonymous (1892), "The International Horticultural Exhibition, Earl's Court," *London Society*, May 14, 1892.

34. Anonymous (1892), "Notes," *Nature* 46:61 (19 May).

35. A. R. Wallace to V. I. Wallace, 20 May 1892, WCP228.

36. Evidently others in the scientific community felt Wallace should have been elected a Fellow of the Royal Society far sooner too, and the institution was criticized for what was seen as an egregious oversight. But Wallace himself may have stymied earlier efforts to nominate him: George Darwin, rising to the defense of the Society in the matter, wrote to the *Times* of London: "Sir—The election of Mr. Alfred Russel Wallace to the Royal Society last week has been commented on in the public journals as showing the inefficiency of the method by which Fellows are elected. It seems, therefore, only just to the Royal Society to state that it is notorious that Mr. Wallace would have been elected at any time within the last 35 years if he had ever allowed himself to be nominated." G. H. Darwin (1893), *Times* (London), no. 33971, 7 June 1893.

37. W. T. Thiselton-Dyer to A. R. Wallace, 23 October 1892, WCP2457; A. R. Wallace to W. T. Thiselton-Dyer, 25 October 1892, WCP3829; W. T. Thiselton-Dyer to A. R. Wallace, 12 January 1893, WCP2458; M. Foster to A. R. Wallace, 8 June 1893, WCP1498.

38. Edward Bellamy (1887, 1951), *Looking Backward, 2000–1887* (New York: Modern Library), p. 3.

39. A. R. Wallace to E. D. Girdlestone, 11 August 1889, WCP7018; 24 August 1889, WCP3699; see also Wallace's account of his conversion to (and his apologia for) socialism in *My Life*, 2:266–74.

40. Michael Robertson (2018), *The Last Utopians: Four Late Nineteenth-Century Visionaries and Their Legacy* (Princeton, NJ: Princeton University Press), p. 54. See especially chapter 2, "Edward Bellamy's orderly utopia," for an excellent minibiography of Bellamy and an analysis of *Looking Backward* in its cultural context.

41. In a letter to his friend Richard Ely at Johns Hopkins University, subsequently published, Wallace commented that "from boyhood, when I was an ardent admirer of Robert Owen, I have been interested in Socialism, but reluctantly came to the conclusion that it was impracticable, and also, to some extent, repugnant to my ideas of individual liberty and home privacy. But Mr. Bellamy has completely altered my views in this matter." A. R. Wallace to R. T. Ely, c. 1890, WCP5125. The letter was published in the *New York Times* the following February: Anonymous (1891), "An English nationalist," *New York Times*, no. 12303, 1 February, p. 9.

42. *My Life*, 2:399.

43. Wallace described in detail The Grange and its potential in a letter to his son Will: A. R. Wallace to W. G. Wallace, 28 November 1900, WCP31.

44. *My Life*, 2:274.

45. As one especially clear example of Wallace's dedication to women's rights, he wrote the following letter of support that was read at a women's suffrage meeting in Godalming, as printed in the *Times (London)* issue of 11 February 1909, p. 10. See WCP5205:

> Dr. Wallace wrote:—"As long as I have thought or written at all on politics, I have been in favour of woman suffrage. None of the arguments for or against have any weight with me, except the broad one, which may be thus stated:—All the human inhabitants of any one country should have equal rights and liberties before the law; women are human beings; therefore they should have votes as well as men. It matters not to me whether ten millions or only ten claim it—the right and the liberty should exist, even if they do not use it. The term 'Liberal' does not apply to those who refuse this natural and inde-feasible right. *Fiat justitia, ruat coelum.*"

46. Bellamy, *Looking Backward*, p. 218.

47. For excellent in-depth treatments of Wallace's interrelated views on socialism, social reform, women, and eugenics, see chapter 5 in Martin Fichman (2004), *An Elusive Victorian: The Evolution of Alfred Russel Wallace* (Chicago: University of Chicago Press); chapters 6 (Sherrie Lyons) and 7 (Martin Fichman) in Charles H. Smith, James T. Costa, and David Collard, eds. (2019), *An Alfred Russel Wallace Companion* (Chicago: University of Chicago Press); and chapter 14 (Diane Paul) in Smith and Beccaloni, *Natural Selection and Beyond* (Oxford: Oxford University Press).

48. *My Life*, 2:209; Alfred R. Wallace (1890), "Human selection," *Fortnightly Review*, n.s., 48, o.s., 54:325–37; (1892), "Human progress: Past and future," *Arena* 5:145–59.

49. For a succinct introduction to Galton, see Nicholas W. Gillham (2001), "Sir Francis Galton and the birth of eugenics," *Annual Review of Genetics* 35:83–101.

50. In "Hereditary talent and character," Galton imagined a "Utopia . . . in which a system of competitive examination for girls, as well as for youths, had been so developed as to embrace every important quality of mind and body, and where a considerable sum was yearly allotted to the endowment of such marriages as promised to yield children who would grow into eminent

servants of the state." Francis Galton (1865), "Hereditary talent and character," *Macmillan's Magazine* 12:157–66, 165.

51. Darwin, too, held that sympathy, compassion, and care for the weak, infirm, and vulnerable are the highest of human attributes and discusses the evolution of sympathy and other moral attributes in *The Descent of Man*, vol. 1, chaps. 4 and 5. For example: "The aid which we feel impelled to give to the helpless is mainly an incidental result of the instinct of sympathy, which was originally acquired as part of the social instincts, but subsequently rendered, in the manner previously indicated, more tender and more widely diffused. Nor could we check our sympathy, if so urged by hard reason, without deterioration in the noblest part of our nature. . . . If we were intentionally to neglect the weak and helpless, it could only be for a contingent benefit, with a certain and great present evil." *Descent* 1:168–69.

52. Wallace, "Human selection," pp. 325–37, 337. Twenty years later, Wallace was far more vehement in his denunciation of eugenics, stating in one of the last interviews he gave: "Segregation of the unfit, indeed! It is a mere excuse for establishing a medical tyranny. And we have enough of this kind of tyranny already. . . . The world does not want the eugenist to set it straight. Give the people good conditions, improve their environment, and all will tend towards the highest type. Eugenics is simply the meddlesome interference of an arrogant, scientific priestcraft." Frederick Rockell (1912), "The last of the great Victorians: Special interview with Dr. Alfred Russel Wallace," *Millgate Monthly* 7, pt. 2 (83):657–63.

53. In an article in the *Clarion*, Wallace stated that the "great principle" of equality of opportunity "would not necessarily lead to Socialism, but rather to a perfect *individualism* under equal and fair conditions, and this would almost certainly, as I have urged elsewhere, bring about universal *cooperation*, which might or might not lead on to Socialism." Alfred R. Wallace (1905), "If there were a Socialist government—how should it begin?," *Clarion* (London), no. 715:5a–f (18 August).

54. A. R. Wallace to W. Mitten, 13 August 1893, WCP624; A. R. Wallace to V. I. Wallace, 29 October 1894, WCP255.

55. In the late 1890s, Wallace wrote several letters and essays on monetary reform. Arguing for eliminating the gold standard, he devised a kind of value index to stabilize currency. The great American economist, statistician, and monetary theorist Irving Fisher dedicated his 1920 book *Stabilizing the Dollar* (New York: Macmillan) to three persons "who have anticipated me in proposing plans for stabilizing monetary units," naming Wallace foremost. University of Bath economist and historian David Collard discusses Wallace's contributions to the theory of paper money as well as his thought on depressions, trade reciprocity, capital markets, and taxation; see pp. 253–66 in David Collard (2019), "Land and economics," pp. 235–73, in Smith, Costa, and Collard, *An Alfred Russel Wallace Companion*, and (2009), "Alfred Russel Wallace and the political economists," *History of Political Economy* 41(4):605–44. See also Alfred R. Wallace (1898), "Is there scarcity or monopoly of money?" [letter to columnist "Dangle"], *Clarion*, no. 360:348c–e (29 October); "A complete system of paper money" [letter to columnist "Dangle"], *Clarion*, no. 365:389e–f (3 December 1898); "Paper money as a standard of value," *Academy* 55:549–50 (31 December 1898).

56. Croll's pioneering research lay the groundwork for Milutin Milankovitch's work on *orbital forcing* in the 1920s. Building on Croll and others, Milankovitch showed how orbital eccentricity, axial tilt, and axial precession combine to yield cyclical variations in insolation and thus Earth's climate; see James Rodger Fleming (2006), "James Croll in context: The encounter

between climate dynamics and geology in the second half of the nineteenth century," *History of Meteorology* 3:43–54.

57. See chapter 29 in James Campbell Irons, ed. (1896), *Autobiographical Sketch of James Croll LL.D., F.R.S., Etc. With Memoir of His Life and Work* (London: Edward Stanford). See also "Correspondence with James Croll," S531a, at the Alfred Russell Wallace Page, http://people.wku .edu/charles.smith/wallace/S531A.htm.

58. George H. Darwin (1896), "The astronomical theory of the glacial period," *Nature* 53:196–97 (2 January); Alfred R. Wallace (1896), "The cause of an ice age," *Nature* 53:220–21 (9 January).

59. Alfred R. Wallace (1896), "The problem of utility: Are specific characters always or generally useful?" [Read at the Linnean Society of London meeting of 18 June 1896], *Journal of the Linnean Society: Zoology* 25:481–96.

60. George J. Romanes (1889), "Mr. Wallace on Darwinism," *Contemporary Review* 56:244–58.

61. A. R. Wallace to V. I. Wallace, 27 November 1896, WCP278.

62. The statue of Darwin was executed by Henry Hope Pinker and unveiled at the Oxford University Museum by Poulton on 14 June 1899; see https://oumnh.ox.ac.uk/learn-art-0.

63. A. R. Wallace to V. I. Wallace, 1 October 1893, WCP239.

64. A. R. Wallace to W. Mitten, 13 August 1893, WCP624.

65. A. R. Wallace to V. I. Wallace, 14 July 1894, WCP253; A. R. Wallace to V. I. Wallace, 8 July 1894, WCP252.

66. Gérard M. Stampfli, ed. (2001), *Geology of the Western Swiss Alps: A Guide-Book* (Lausanne: Mémoires de Géologie), no. 36, sect. 1.1 and 1.2, pp. 2–9.

67. Lunn's companies, the Polytechnic Touring Association and Sir Henry Lunn Travel, merged in the 1960s to form Lunn Poly, a major British travel agency (now Thomson/TUI).

68. A. R. Wallace to A. Wallace, 9 August 1896, WCP416.

69. A. R. Wallace to W. G. Wallace, 15 July 1898, WCP158.

70. W. G. Wallace to A. R. Wallace, 5 December 1897, WCP1324.

71. A. R. Wallace to L. F. Ward, 12 October 1898, WCP3781.

72. Alfred R. Wallace (1898), "Spiritualism and social duty" [An address delivered at the International Congress of Spiritualists on 23 June 1898, at St. James Hall, London], *Light* 18:334–36.

73. Alfred R. Wallace (1898), *The Wonderful Century: Its Successes and Failures* (London: Swan Sonnenschein).

74. Wallace's spirited antiwar writings were published under such titles as "The causes of war, and the remedies," "Protests against war," "The Transvaal war. Wanted facts," and "Imperial might and human right." Wallace also endorsed a special issue of the *Review of Reviews* titled *Shall We Let Hell Loose in Africa? A South African Catechism*, providing a blurb in an advertisement appearing in the *War against War in South Africa*, 20 October 1899, p. 15. See S567, S569, S571, S572, S574, S576, S579, S581, and S595 in the bibliography of Wallace's writings at Charles Smith's Alfred Russel Wallace Page, http://people.wku.edu/charles.smith/wallace/bibintro.htm.

75. A. R. Wallace to V. I. Wallace, 25 October 1901, WCP328.

76. A. R. Wallace to W. G. Wallace, 2 November 1902, WCP65; 15 November 1902, WCP66.

77. Ernest H. Rann (1909), "Dr. Alfred Russel Wallace at Home" [Interview], *Pall Mall Magazine* 43:274–84 (March 1909). Regarding Tulgey Wood, see pp. 41–42 in George Beccaloni (2008), "Homes sweet homes: A biographical tour of Wallace's many places of residence,"

pp. 7–43, in Smith and Beccaloni, *Natural Selection and Beyond*; G. Beccaloni, personal communication.

78. *My Life*, 2:227–28; pp. 28–29 in Lester, "Homing In," 22–32.

79. With thanks to geologist Dr. Ian West for his extremely informative websites on the geology of southern England; see Geology of the Wessex Coast of Southern England, https://wessexcoastgeology.soton.ac.uk/index.htm.

80. A. R. Wallace to W. G. Wallace, 30 November 1902, WCP67.

81. A. R. Wallace to W. G. Wallace, 19 December 1902, WCP68. Wallace's *Independent* article was published the following February and reprinted the following month in the *Fortnightly Review*: Alfred R. Wallace (1903), "Man's place in the universe," *Independent* (New York) 55:473–83 (26 February); and "Man's place in the universe: As indicated by the new astronomy," *Fortnightly Review*, n.s., 73:395–411 (1 March).

82. See, for example, A. M. Clerke to A. R. Wallace, 17 April 1903, WCP2821; 21 April 1903, WCP2822; 29 April 1903, WCP2824. In 2017 the Royal Astronomical Society established the Agnes Clerke Medal, awarded to individuals who have made outstanding contributions in the history of astronomy or geophysics.

83. Wallace, "Man's place in the universe," pp. 473–83, 483.

84. Alfred R. Wallace (1903), *Man's Place in the Universe: A Study of the Results of Scientific Research in Relation to the Unity of Plurality of Worlds* (London: Chapman and Hall), p. 474.

85. A. R. Wallace to V. I. Wallace, 7 August 1892, WCP1254; 25 November 1894, WCP257.

86. Wallace, *Wonderful Century*, pp. 237–39, 327–28.

87. Suffice it to say that while Lowell did not give up his cherished idea readily, publishing a response, *Mars as the Abode of Life*, in 1908, the issue was soon rendered moot by observations made the very next year with the great sixty-inch telescope at Mount Wilson Observatory in California, a telescope large enough to reveal the putative "canals" as less than regular natural geomorphological features. For detailed and illuminating treatments of Wallace and the question of extraterrestrial life and the Mars episode, see Robert W. Smith (2015), "Alfred Russel Wallace, extraterrestrial life, Mars, and the nature of the universe," *Victorian Review* 41(2):151–75, and (2019), "Wallace and extraterrestrial life," pp. 357–380 in Smith, Costa, and Collard, *An Alfred Russel Wallace Companion*. Wallace would probably have rolled his eyes at the ongoing twentieth-century efforts to ascertain the possibility of (microbial) life on Mars, past or present, though he would surely have found the technology and the exploration effort led by NASA and other space agencies to be absolutely breathtaking.

88. Alfred R. Wallace (1908), [Concerning Wallace's notification of winning the Copley Medal; read at the Royal Society's annual meeting of 30 Nov. 1908], *Times* (London) 38818:9; see also WCP5518.

89. A. R. Wallace to F. Birch, 30 December 1908, WCP1672.

90. In July 2022 Wallace's medals and Order of Merit cross were sold by auction. The winning bid for the Darwin-Wallace Medal of the Linnean Society was £75,000, made by an anonymous buyer. The remaining eight medals and the Order of Merit were bought for some £198,000 by Roan Hackney, British specialist in military antiquities and polar exploration, in hopes that a museum will eventually raise the funds to purchase the set. Indeed, it would be a pity if Wallace's medals and awards were divided up and scattered. See Laura Chesters (2022), "Eight Wallace medals stay together as one bidder buys them all," *Antiques Trade Gazette*, no. 2553 (6 August 2022), and G. Beccaloni's appeal to keep the medals together at

https://wallacefund.myspecies.info/content/please-save-alfred-russel-wallaces-scientific
-medals-or-least-one-them.

91. Alfred R. Wallace (1908/1909), [Acceptance speech on receiving the Darwin-Wallace Medal on 1 July 1908] in *The Darwin-Wallace Celebration Held on Thursday, 1st July 1908*, by the Linnean Society of London (London: Burlington House, Longmans, Green), pp. 5–11.

92. A. R. Wallace to A. Eastwood, 26 February 1913, WCP3971.

93. Indeed, he already had another book project in the works: *Darwin and Wallace* was to be authored by James Marchant with Wallace's assistance—they had gotten as far as developing an outline and had a contract with John Murray when Wallace died. Marchant went on to write Wallace's *Letters and Reminiscences*, published in two volumes, which he dedicated to Annie Wallace. He described the plans for *Darwin and Wallace* in the introduction to *Letters and Reminiscences*, 1:3–4:

> It may be stated here that Wallace had suggested to the present writer that he should undertake a new work, to be called 'Darwin and Wallace,' which was to have been a comparative study of their literary and scientific writings, with an estimate of the present position of the theory of Natural Selection as an adequate explanation of the process of organic evolution. Wallace had promised to give as much assistance as possible in selecting the material without which the task on such a scale would obviously have been impossible. Alas! soon after the agreement with the publishers was signed and in the very month that the plan of the work was to have been shown to Wallace, his hand was unexpectedly stilled in death; and the book remains unwritten.

See also A. R. Wallace to J. Marchant, 27 March 1913, WCP6575.

94. W. G. Wallace to J. Marchant, 5 November 1913, WCP6553. On 1 November 1915 a medallion of Wallace was unveiled near Darwin's final resting place in Westminster Abbey. A photograph of the medallion is inserted between pp. 254 and 255 in Wallace, *Letters and Reminiscences*, vol. 2.

95. Wallace would have been delighted to learn that this species was later found in Switzerland, one of his favorite landscapes, and that this and other discoveries of this species in Europe are helping reconstruct the paleoenvironment of the region; see, for example, Marc Philippe, Jean-Paul Billon-Bruyat, et al. (2010), "New occurrences of the wood *Protocupressinoxylon purbeckensis* Francis: Implications for terrestrial biomes in southwestern Europe at the Jurassic/Cretaceous boundary," *Paleontology* 53(1):201–14. See also Ian West (2016), "The Fossil Forest Exposure," pts. 1 and 2, https://wessexcoastgeology.soton.ac.uk/Fossil-Forest.htm and https://wessexcoastgeology.soton .ac.uk/Fossil-Forest-Purbeck-Trees.htm) and https://wallacefund.myspecies.info/fossil-tree.

Coda

1. H. Lewis McKinney (1966), "Alfred Russel Wallace and the discovery of natural selection," *Journal of the History of Medicine and Allied Sciences* 21:333–57.

2. James T. Costa (2013), *On the Organic Law of Change: A Facsimile Edition and Annotated Transcription of Alfred Russel Wallace's Species Notebook of 1855–1859* (Cambridge, MA: Harvard University Press).

3. James T. Costa (2014), *Wallace, Darwin, and the Origin of Species* (Cambridge, MA: Harvard University Press).

4. Ted Benton (2013), *Alfred Russel Wallace: Explorer, Evolutionist, Public Intellectual—a Thinker for Our Own Times?* (Manchester, UK: Siri Scientific Press).

5. Walt Whitman (1855), *Song of Myself*, canto 52 (self-pub).

FIGURE CREDITS

Figures

Color Plates

Pages

h   (bottom) Courtesy of Smithsonian Institution Libraries/Biodiversity Heritage Library

i   (top) Leslie Costa Collection, with thanks to the Linnean Society of London

i   (bottom) Reproduced with kind permission of Clay Bolt

j   (top left) Courtesy of Smithsonian Institution Libraries/Biodiversity Heritage Library

j   (top right) Courtesy of Ernst Mayr Library, Harvard University/ Biodiversity Heritage Library

j   (bottom) Courtesy of Smithsonian Institution Libraries/Biodiversity Heritage Library

k   (top) Reproduced with kind permission of the Linnean Society of London

k   (bottom) Courtesy of Smithsonian Institution Libraries/Biodiversity Heritage Library

l   (top) Courtesy of Captain Occam/Wikimedia

l   (bottom) Courtesy of the University of Pennsylvania Schoenberg Center for Electronic Text & Image

m   (top) Courtesy of the Peter H. Raven Library, Missouri Botanical Garden/ Biodiversity Heritage Library

m   (bottom) Courtesy of the Natural History Museum, London

n   J. T. Costa Collection/A. R. Wallace, *Geographical Distribution of Animals* (1876)

o   (top left) Copyright Wallace Memorial Fund & G. W. Beccaloni

o   (top right) Courtesy of the Wellcome Library

o   (bottom) Courtesy of the Royal Botanic Gardens, Kew/Wikimedia Commons

p   (top) Courtesy of the BLM-Canyons of the Ancients Visitor Center and Museum

p   (bottom left) Copyright Wallace Memorial Fund & G. W. Beccaloni

p   (bottom right) Copyright Wallace Memorial Fund & G. W. Beccaloni

INDEX

Page numbers in italics refer to figures.

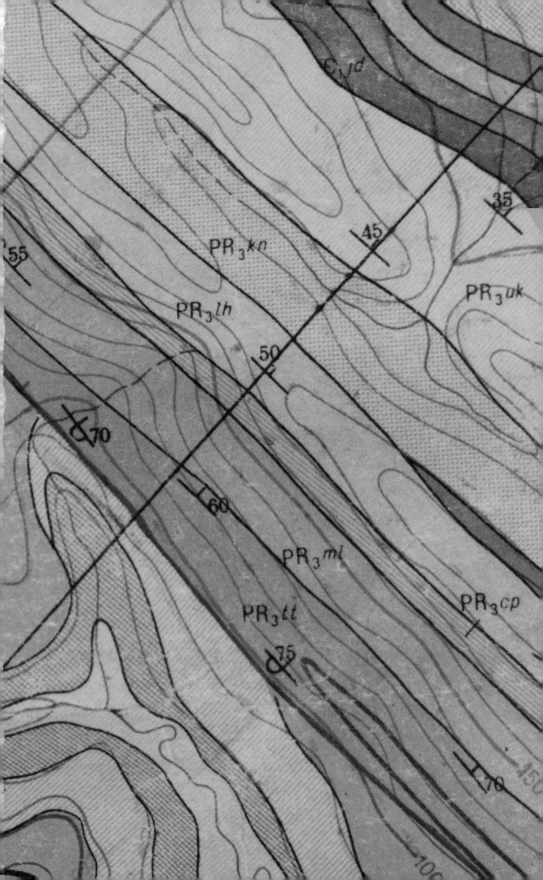

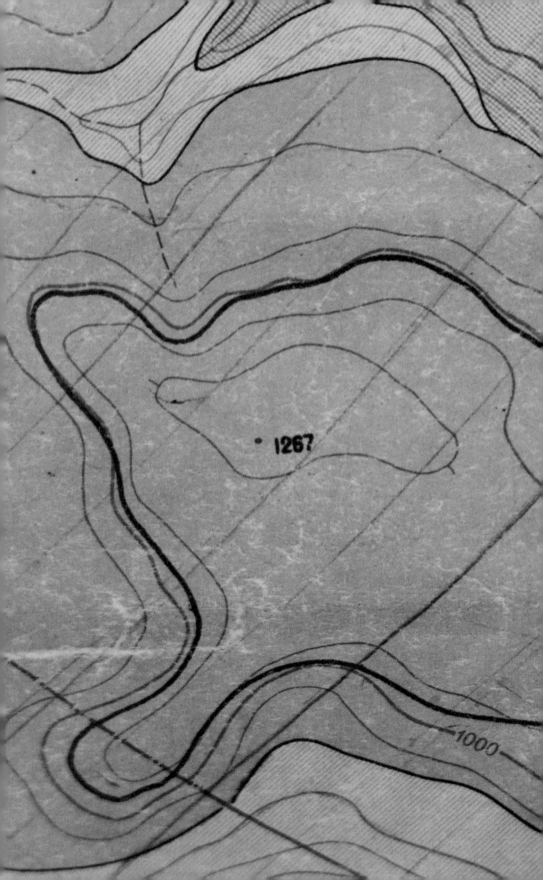